Basic College Mathematics

An Applied Approach

FOURTH EDITION

Richard N. Aufmann
Palomar College, California

Vernon C. Barker
Palomar College, California

HOUGHTON MIFFLIN COMPANY Boston
Dallas Geneva, Illinois
Palo Alto Princeton, New Jersey

Chapter Opener Art designed and rendered by Daniel P. Derdula.

All Interest Grabber Art designed and rendered by Daniel P. Derdula (with airbrushing by Linda Phinney on illustrations for Chapters 6 and 8).

All Interior Math Figures rendered by NETWORKGRAPHICS (135 Fell Court, Hauppauge, New York 11788).

Interior design by George McLean.

Cover photograph by Slide Graphics of New England, Inc.

Printed in the U.S.A.

ISBN Numbers:
Text: 0-395-43194-8
Instructor's Annotated Edition: 0-395-57061-1
Solutions Manual: 0-395-57062-X
Instructor's Resource Manual/Testing Program: 0-395-46298-3
Test Bank for the Computerized Test Generator: 0-395-57063-8
Transparencies: 0-395-57064-6
Videos: 0-395-57065-4

ABCDEFGHIJ-VH-9876543210

Contents

Preface *x*

1 Whole Numbers *1*

SECTION 1.1 **Introduction to Whole Numbers** *3*

Objective A To identify the order relation between two numbers *3*
Objective B To write whole numbers in words and in standard form *4*
Objective C To write whole numbers in expanded form *4*
Objective D To round a whole number to a given place value *5*

SECTION 1.2. **Addition of Whole Numbers** *9*

Objective A To add whole numbers without carrying *9*
Objective B To add whole numbers with carrying *10*
Objective C To solve application problems *12*

SECTION 1.3 **Subtraction of Whole Numbers** *19*

Objective A To subtract whole numbers without borrowing *19*
Objective B To subtract whole numbers with borrowing *20*
Objective C To solve application problems *22*

SECTION 1.4 **Multiplication of Whole Numbers** *27*

Objective A To multiply a number by a single digit *27*
Objective B To multiply larger whole numbers *28*
Objective C To solve application problems *30*

SECTION 1.5 **Division of Whole Numbers** *35*

Objective A To divide by a single digit with no remainder in the quotient *35*
Objective B To divide by a single digit with a remainder in the quotient *37*
Objective C To divide by larger whole numbers *39*
Objective D To solve application problems *41*

SECTION 1.6 **Exponential Notation and the Order of Operations Agreement** *47*

Objective A To simplify expressions containing exponents *47*
Objective B To use the Order of Operations Agreement to simplify expressions *48*

SECTION 1.7 **Prime Numbers and Factoring** *51*

Objective A To factor numbers *51*
Objective B To find the prime factorization of a number *52*

Calculators and Computers 55; Chapter Summary 55; Chapter Review 57; Chapter Test 59

2 Fractions *61*

SECTION 2.1 **The Least Common Multiple and Greatest Common Factor** *63*

Objective A To find the least common multiple (LCM) *63*
Objective B To find the greatest common factor (GCF) *64*

SECTION 2.2 **Introduction to Fractions** *67*

Objective A To write a fraction that represents part of a whole *67*
Objective B To write an improper fraction as a mixed number or a whole number, and a mixed number as an improper fraction *68*

SECTION 2.3 **Writing Equivalent Fractions** *71*

Objective A To find equivalent fractions by raising to higher terms *71*
Objective B To write a fraction in simplest form *72*

SECTION 2.4 **Addition of Fractions and Mixed Numbers** *75*

Objective A To add fractions with the same denominator *75*
Objective B To add fractions with unlike denominators *75*
Objective C To add whole numbers, mixed numbers, and fractions *76*
Objective D To solve application problems *78*

SECTION 2.5 **Subtraction of Fractions and Mixed Numbers** *83*

Objective A To subtract fractions with the same denominator *83*
Objective B To subtract fractions with unlike denominators *83*
Objective C To subtract whole numbers, mixed numbers, and fractions *84*
Objective D To solve application problems *86*

SECTION 2.6 **Multiplication of Fractions and Mixed Numbers** *91*

Objective A To multiply fractions *91*
Objective B To multiply whole numbers, mixed numbers, and fractions *92*
Objective C To solve application problems *94*

SECTION 2.7 **Division of Fractions and Mixed Numbers** *99*

Objective A To divide fractions *99*
Objective B To divide whole numbers, mixed numbers, and fractions *99*
Objective C To solve application problems *101*

SECTION 2.8 **Order, Exponents, and the Order of Operations Agreement** *107*

Objective A To identify the order relation between two fractions *107*
Objective B To simplify expressions containing exponents *107*
Objective C To use the Order of Operations Agreement to simplify expressions *108*

Calculators and Computers 111; Chapter Summary 111; Chapter Review 113; Chapter Test 115; Cumulative Review 117

3 Decimals *119*

SECTION 3.1 **Introduction to Decimals** *121*

Objective A To write decimals in standard form and in words *121*
Objective B To round a decimal to a given place value *122*

SECTION 3.2 **Addition of Decimals** *125*

Objective A To add decimals *125*
Objective B To solve application problems *126*

SECTION 3.3 **Subtraction of Decimals** *129*

Objective A To subtract decimals *129*
Objective B To solve application problems *130*

SECTION 3.4 **Multiplication of Decimals** *133*

Objective A To multiply decimals *133*
Objective B To solve application problems *136*

SECTION 3.5 **Division of Decimals** *141*

Objective A To divide decimals *141*
Objective B To solve application problems *144*

SECTION 3.6 **Comparing and Converting Fractions and Decimals** *149*

Objective A To convert fractions to decimals *149*
Objective B To convert decimals to fractions *149*
Objective C To identify the order relation between two decimals or between a decimal and a fraction *150*

Calculators and Computers 153; Chapter Summary 153; Chapter Review 155; Chapter Test 157; Cumulative Review 159

4 Ratio and Proportion *161*

SECTION 4.1 **Ratio** *163*

Objective A To write the ratio of two quantities in simplest form *163*
Objective B To solve application problems *164*

SECTION 4.2 **Rates** *167*

Objective A To write rates *167*
Objective B To write unit rates *167*
Objective C To solve application problems *168*

SECTION 4.3 **Proportions** *171*

Objective A To determine if a proportion is true *171*
Objective B To solve proportions *172*
Objective C To solve application problems *174*

Calculators and Computers 179; Chapter Summary 180; Chapter Review 181; Chapter Test 183; Cumulative Review 185

5 Percents *187*

SECTION 5.1 **Introduction to Percents** *189*

Objective A To write a percent as a fraction or decimal *189*
Objective B To write a fraction or a decimal as a percent *190*

SECTION 5.2 **Percent Equations: Part I** *193*

Objective A To find the amount when the percent and base are given *193*
Objective B To solve application problems *194*

SECTION 5.3 **Percent Equations: Part II** *197*

Objective A To find the percent when the base and amount are given *197*
Objective B To solve application problems *198*

SECTION 5.4 **Percent Equations: Part III** *201*

Objective A To find the base when the percent and amount are given *201*
Objective B To solve application problems *201*

SECTION 5.5 **Percent Problems: Proportion Method** *205*

Objective A To solve percent problems using proportions *205*
Objective B To solve application problems *206*

Calculators and Computers 209; Chapter Summary 210; Chapter Review 211; Chapter Test 213; Cumulative Review 215

6 Applications for Business and Consumers: A Calculator Approach 217

SECTION 6.1 **Applications to Purchasing** 219
Objective A To find unit cost 219
Objective B To find the most economical purchase 219
Objective C To find total cost 220

SECTION 6.2 **Percent Increase and Percent Decrease** 223
Objective A To find percent increase 223
Objective B To apply percent increase to business—markup 224
Objective C To find percent decrease 226
Objective D To apply percent decrease to business—discount 227

SECTION 6.3 **Interest** 233
Objective A To calculate simple interest 233
Objective B To calculate compound interest 234

SECTION 6.4 **Real Estate Expenses** 237
Objective A To calculate the initial expenses of buying a home 237
Objective B To calculate ongoing expenses of owning a home 238

SECTION 6.5 **Car Expenses** 243
Objective A To calculate the initial expenses of buying a car 243
Objective B To calculate ongoing expenses of owning a car 244

SECTION 6.6 **Wages** 247
Objective A To calculate commisions, total hourly wages, and salaries 247

SECTION 6.7 **Bank Statements** 251
Objective A To calculate checkbook balances 251
Objective B To balance a checkbook 252

Calculators and Computers 261; *Chapter Summary* 261;
Chapter Review 263; *Chapter Test* 265; *Cumulative Review* 267

7 Statistics 269

SECTION 7.1 **Pictographs and Circle Graphs** 271
Objective A To read a pictograph 271
Objective B To read a circle graph 273

SECTION 7.2 **Bar Graphs and Broken-Line Graphs** 279
Objective A To read a bar graph 279
Objective B To read a broken–line graph 280

SECTION 7.3 **Histograms and Frequency Polygons** 283
Objective A To read a histogram 283
Objective B To read a frequency polygon 284

SECTION 7.4 **Means and Medians** 287
Objective A To find the mean of a set of numbers 287
Objective B To find the median of a set of numbers 288

Calculators and Computers 291; *Chapter Summary* 292;
Chapter Review 293; *Chapter Test* 295; *Cumulative Review* 297

8 U.S. Customary Units of Measurement 299

SECTION 8.1 **Length** *301*

Objective A To convert measurements of length in the U.S. Customary System *301*
Objective B To perform arithmetic operations with measurements of length *302*
Objective C To solve application problems *304*

SECTION 8.2 **Weight** *307*

Objective A To convert measurements of weight in the U.S. Customary System *307*
Objective B To perform arithmetic operations with measurements of weight *308*
Objective C To solve application problems *308*

SECTION 8.3 **Capacity** *311*

Objective A To convert measurements of capacity in the U.S. Customary System *311*
Objective B To perform arithmetic operations with measurements of capacity *312*
Objective C To solve application problems *312*

SECTION 8.4 **Energy and Power** *315*

Objective A To use units of energy in the U.S. Customary System *315*
Objective B To use units of power in the U.S. Customary System *316*

Calculators and Computers 319; Chapter Summary 320; Chapter Review 321; Chapter Test 323; Cumulative Review 325

9 The Metric System of Measurement 327

SECTION 9.1 **Length** *329*

Objective A To convert units of length in the metric system of measurement *329*
Objective B To perform arithmetic operations with measurements of length *330*
Objective C To solve application problems *330*

SECTION 9.2 **Mass** *333*

Objective A To convert units of mass in the metric system of measurement *333*
Objective B To perform arithmetic operations with measurements of mass *334*
Objective C To solve application problems *334*

SECTION 9.3 **Capacity** *337*

Objective A To convert units of capacity in the metric system of measurement *337*
Objective B To perform arithmetic operations with measurements of capacity *338*
Objective C To solve application problems *338*

SECTION 9.4 **Energy** *341*

Objective A To use units of energy in the metric system of measurement *341*

SECTION 9.5 **Conversion between the U.S. Customary and the Metric System of Measurement** *345*

Objective A To convert U.S. Customary units of metric units *345*
Objective B To convert metric units to U.S. Customary untis *346*

Calculators and Computers 349; Chapter Summary 350; Chapter Review 351; Chapter Test 353; Cumulative Review 355

10 Geometry 357

SECTION 10.1 **Angles, Lines, and Geometric Figures** *359*

Objective A To define and describe lines and angles *359*
Objective B To define and describe geometric figures *362*
Objective C To solve problems involving angles formed by intersecting lines *365*

SECTION 10.2 **Perimeter** *371*

Objective A To find the perimeter of geometric figures *371*
Objective B To find the perimeter of composite geometric figures *373*
Objective C To solve application problems *374*

SECTION 10.3 **Area** *379*

Objective A To find the area of geometric figures *379*
Objective B To find the area of composite geometric figures *381*
Objective C To solve application problems *382*

SECTION 10.4 **Volume** *387*

Objective A To find the volume of geometric solids *387*
Objective B To find the volume of composite geometric solids *390*
Objective C To solve application problems *392*

SECTION 10.5 **The Pythagorean Theorem** *397*

Objective A To use a table to find the square root of a number *397*
Objective B To find the unknown side of a right triangle using the Pythagorean Theorem *398*
Objective C To solve application problems *400*

SECTION 10.6 **Similar Triangles** *403*

Objective A To solve similar triangles *403*
Objective B To solve application problems *406*

Calculators and Computers 409; Chapter Summary 409; Chapter Review 411; Chapter Test 413; Cumulative Review 415

11 Rational Numbers 417

SECTION 11.1 **Introduction to Integers** *419*

Objective A To identify the order relation between two integers *419*
Objective B To evaluate expressions containing the absolute value symbol *420*

SECTION 11.2 **Addition and Subtraction of Integers** *423*

Objective A To add integers *423*
Objective B To substract integers *425*

SECTION 11.3 **Multiplication and Division of Integers** *429*

Objective A To multiply integers *429*
Objective B To divide integers *430*
Objective C To solve application problems *432*

SECTION 11.4 **Operations with Rational Numbers** *437*

Objective A To add or subtract rational numbers *437*
Objective B To multiply and divide rational numbers *439*

SECTION 11.5 **The Order of Operations Agreement** *445*

Objective A To use the Order of Operations Agreement to simplify expressions *445*

Calculators and Computers 449; *Chapter Summary* 450; *Chapter Review* 451; *Chapter Test* 453; *Cumulative Review* 455

12 Introduction to Algebra 457

SECTION 12.1 **Variable Expressions** 459

Objective A To evaluate variable expressions 459
Objective B To simplify variable expressions containing no parentheses 460
Objective C To simplify variable expressions containing parentheses 463

SECTION 12.2 **Introduction to Equations** 469

Objective A To determine if a given value is a solution of an equation 469
Objective B To solve an equation of the form $x + a = b$ 470
Objective C To solve an equation of the form $ax = b$ 472
Objective D To solve application problems 474

SECTION 12.3 **General Equations: Part I** 479

Objective A To solve an equation of the form $ax + b = c$ 479
Objective B To solve application problems 480

SECTION 12.4 **General Equations: Part II** 485

Objective A To solve an equation of the form $ax + b = cx + d$ 485
Objective B To solve an equation containing parentheses 486

SECTION 12.5 **Translating verbal expressions into mathematical expressions** 491

Objective A To translate a verbal expression into a mathematical expression given the variable 491
Objective B To translate a verbal expression into a mathematical expression by assigning the variable 492

SECTION 12.6 **Translating Sentences into Equations and Solving** 495

Objective A To translate a sentence into an equation and solve 495
Objective B To solve application problems 497

Calculators and Computers 503; *Chapter Summary* 504; *Chapter Review* 505; *Chapter Test* 507; *Cumulative Review* 509; *Final Exam* 511

APPENDIX A1 Addition Table A2
Multiplication Table A2
Table of Square Roots A3
Compound Interest Table A4
Monthly Payment Table A6

Solutions to Student Examples A7

Answers to Odd–Numbered Exercises A47

Index A69

Preface

The fourth edition of Basic College Math: An Applied Approach provides mathematically sound and comprehensive coverage of the topics considered essential in a basic mathematics course. Our strategy in preparing this revision has been to build on the successful features of the third edition, features designed to enhance the student's mastery of math skills. In addition, we have expanded the ancillary package for both the instructor and the student by adding transparencies and videos.

Features

The Interactive Approach

Instructors have long recognized the need for a text that requires the student to use a skill as it is being taught. Basic College Mathematics: An Applied Approach uses an interactive technique that meets this need. Each section is divided into objectives, and every objective contains one or more sets of matched-pair examples. The first example in each set is worked out; the second example is not. By solving this second problem, the student interacts with the text. The complete worked-out solutions to these examples are provided in an appendix at the end of the book, so the student can obtain immediate feedback on and reinforcement of the skill being learned.

Emphasis on Problem-Solving Strategies

Basic College Mathematics: An Applied Approach features a carefully developed approach to problem solving that emphasizes developing strategies to solve problems. For each type of word problem contained in the text, the student is prompted to use a "strategy step" before performing the actual manipulation of numbers. By developing problem-solving strategies, the student will know better how to analyze and solve those word problems encountered in a variety of courses.

Applications

The traditional approach to teaching or reviewing mathematics, which places major emphasis on the manipulation of numbers, is lacking in that it fails to teach students the practical value of mathematics. By contrast, Basic College Mathematics: An Applied Approach emphasizes applications. Wherever appropriate, the last objective in each section presents applications that require the student to use the skills covered in that section to solve practical problems. Also, all of Chapter 6, "Applications for Business and Consumers: A Calculator Approach," and portions of several other chapters are devoted entirely to applications. This carefully-integrated applied approach generates awareness on the student's part of the value of mathematics as a real-life tool.

Complete Integrated Learning System Organized by Objectives

Each chapter begins with a list of the learning objectives included within that chapter. Each of the objectives is then restated in the chapter to remind the student of the current topic of discussion. The same objectives that organize the text organize each ancillary. The Solutions Manual, Computerized Test Generator, Computer Tutor™, Videos, Transparencies, Test Bank, and the Printed Testing Program have all been prepared so that both the student and instructor can easily connect all of the different aids.

Exercises

There are more than 6000 exercises in the text, grouped in the following categories:

- **End-of-section exercise sets,** which are keyed to the corresponding learning objectives, provide ample practice and review of each skill.
- **Chapter review exercises,** which appear at the end of each chapter, help the student integrate all of the skills presented in the chapter.
- **Chapter tests,** which appear at the end of each chapter, are typical one-hour exams that the student can use to prepare for an in-class test.
- **Cumulative review exercises,** which appear at the end of each chapter (beginning with Chapter 2), help the student retain math skills learned in earlier chapters.
- **Final exam,** which follows the last chapter, can be used as a review item or practice final.
- **Calculator Exercises,** which appear in the end-of-section exercises, are included throughout the text. These exercises provide the student with the opportunity to practice using a hand-held calculator and are identified by a special color box printed over the exercise number for each calculator exercise.

Calculator and Computer Enrichment Topics

Each chapter also contains an optional calculator or computer enrichment topic. Calculator topics provide the student with valuable key-stroking instructions and practice in using a hand-held calculator. Computer topics correspond directly to the programs found on the Math ACE (Additional Computer Exercises) Disk. These topics range from adding and subtracting fractions to calculating monthly mortgage payments.

New To This Edition

Topical Coverage

The topic of Estimation has been added to Chapters 1 and 3, along with an abundance of Estimation exercises.

In this edition, Chapters 6 and 7 have been interchanged. This change provides a smoother transition from numerical skills to applications. Chapter 7, Statistics, now includes material on reading pictographs.

The topical coverage in Chapter 10, Geometry, has been expanded to provide for a more thorough treatment of topics. Chapter 11, Rational Numbers, has been completely rewritten and expanded. Chapter 12, Introduction of Algebra, now uses subtraction and division properties of equations, along with the previously included addition and multiplication properties, to solve equations.

In keeping with our commitment to applications, over two hundred word problems have been rewritten to reflect contemporary situations.

Chapter Review and Chapter Test

There is now a Chapter Review and a Chapter Test at the end of each chapter. The Chapter Review is organized by section with the use of section heads; the Chapter Test is organized by objectives. The objective references for the Chapter Test do not appear on the test page, so students are not prompted in any way. Instead, the objective references are given in the Answer Section at the back of the book. Thus the student can refer back to the Answer Section to find out which objective to restudy if necessary.

New Testing Program

Both the Computerized Testing program and the Printed Testing Program have been completely rewritten to provide instructors with the option of creating countless new tests.

Supplements For The Student

Two computerized study aids, the Computer Tutor™ and the Math ACE (Additional Computer Exercises) Disk have been carefully designed for the student. The Computer Tutor™ has been expanded to include nine "you-try-it" examples with specific help screens for each lesson.

The COMPUTER TUTOR™

The Computer Tutor™ is an interactive instructional microcomputer program for student use. Each learning objective in the text is supported by a lesson on the Computer Tutor™. As a reminder of this, a small computer icon appears to the right of each objective title in the text. Lessons on the tutor provide additional instruction and practice and can be used in several ways: (1) to cover material the student missed because of absence from class; (2) to repeat instruction on the skill or concept that the student has not yet mastered; or (3) to review material in preparation for examinations. This tutorial program is available for the IBM PC and compatible computers.

Math ACE (Additional Computer Exercises) Disk

The Math ACE Disk contains a number of computational and drill-and-practice programs that correspond to selected Calculator and Computer Enrichment Topics in the text. These programs are available for the IBM PC and compatible computers.

Supplements For The Instructor

Basic College Mathematics: An Applied Approach has an unusually complete set of teaching aids for the instructor.

Instructor's Annotated Edition

The Instructor's Annotated Edition is an exact replica of the student text except that the answers to all of the exercises are printed in color next to the problems.

Solutions Manual

The Solutions Manual contains worked-out solutions for all end-of-section exercise sets, chapter reviews, chapter tests, cumulative reviews, and the final exam.

Instructor's Resource Manual/Testing Program

The Instructor's Resource Manual/Testing Program contains the printed testing program, which is the first of three sources of testing material available to users of Basic College Mathematics: An Applied Approach. Eight printed tests (in two formats—free response and multiple choice) are provided for each chapter, as are cumulative and final exams. In addition, the Instructor's Manual includes the documentation for all the software ancillaries (Math ACE, the Computer Tutor™, and the Instructor's Computerized Test Generator), as well as suggested course sequences and class assignments.

Instructor's Computerized Test Generator

The Instructor's Computerized Test Generator is the second source of testing material for use with Basic College Mathematics: An Applied Approach. The database contains over 1800 new test items. These questions are unique to the test generator and do not repeat items provided in the Instructor's Resource Manual/Testing Program. Organized according to the keyed objectives in the text, the Test Generator is designed to produce an unlimited number of tests for each chapter of the text, including cumulative tests and final exams. It is available for the IBM PC or compatible computers with editing capabilities for all nongraphic questions.

Printed Test Bank

The Printed Test Bank, the third component of the testing materials, is a printout of all items in the Instructor's Computerized Test Generator. Instructors using the Test Generator can use the test bank to select specific items from the database. Instructors who do not have access to a computer can use the test bank to select items to be included on a test being prepared by hand.

Videotapes

Approximately 20 half-hour videotape lessons accompany Basic College Mathematics: An Applied Approach. These lessons follow the format and style of the text and are closely tied to specific sections of the text.

Transparencies

Approximately 200 transparencies accompany Basic College Mathematics: An Applied Approach. These transparencies contain the complete solutions to every "you-try-it" example in the text.

Acknowledgements

The authors would like to thank the people who have reviewed this manuscript and provided many valuable suggestions:

Geoffrey Akst
Borough of Manhattan Community College, NY

Betty Jo Baker
Lansing Community College, MI

Judith Brower
North Idaho College, ID

Mary Cabral
Middlesex Community College, MA

Carmy Carranga
Indiana University of Pennsylvania, PA

Patricia Confort
Roger Williams College, RI

Michael Contino
California State University, CA

Bob C. Denton
Orange Coast College, CA

Sharon Edgmon
Bakersfield College, CA

Joel Greenstein
New York Technical College, NY

Bonnie-Lou Guertin
Mt. Wachusett Community College, MA

Lynn Hartsell
Dona Ana Branch Community College, NM

Ida E. Hendricks
Shepherd College, WV

Diana L. Hestwood
Minneapolis Community College, MN

Pamela Hunt
Paris Junior College, TX

Arlene Jesky
Rose State College, OK

Sue Korsak
New Mexico State University, NM

Randy Leifson
Pierce College, WA

Virginia M. Licata
Camden County College, NJ

Judy Liles
North Harris County College-South Campus, TX

David Longshore
Victor Valley College, CA

Margaret Luciano
Broome Community College, NY

Carl C. Maneri
Wright State University, OH

Rudy Maglio
DePaul University, IL

Patricia McCann
Franklin University, OH

Dale Miller
Highland Park Community College, MI

Judy Miller
Delta College, MI

Ellen Milosheff
Triton College, IL

Scott L. Mortensen
Dixie College, UT

Michelle Mosman
Des Moines Area Community College, IA

Linda Murphy
Northern Essex Community College, MA

Wendell Neal
Houston Community College, TX

Doris Nice
University of Wisconsin–Parkside, WI

Kent Pearce
Texas Tech University, TX

Sue Porter
Davenport College, MI

Jack Rotman
Lansing Community College, MI

Karen Schwitters
Seminole Community College, FL

Dorothy Smith
Del Mar College, TX

James T. Sullivan
Massachusetts Bay Community College, MA

Lana Taylor
Siena Heights College, MI

William N. Thomas, Jr.
University of Toledo, OH

Dana Mignogna Thompson
Mount Aloysius Junior College, PA

Paul Treuer
University of Minnesota–Duluth, MN

Thomas Wentland
Columbus College, GA

William T. Wheeler
Abraham Baldwin College, GA

Harvey Wilensky
San Diego Miramar College, CA

Professor Warren Wise
Blue Ridge Community College, VA

Wayne Wolfe
Orange Coast College, CA

Kenneth Word
Central Texas College, TX

Justane Valudez-Ortiz
San Jose College, CA

To the Student

Many students feel that they will never understand math while others appear to do very well with little effort. Oftentimes what makes the difference is that successful students take an active role in the learning process.

Learning mathematics requires your *active* participation. Although doing homework is one way you can actively participate, it is not the only way. First, you must attend class regularly and become an active participant in class. Secondly, you must become actively involved with the textbook.

Basic College Mathematics: An Applied Approach was written and designed with you in mind as a participant. Here are some suggestions on how to use the features of this textbook.

There are 12 chapters in this text. Each chapter is divided into sections, and each section is subdivided into learning objectives. Each learning objective is labeled with a letter from A–E.

First, read each objective statement carefully so you will understand the learning goal that is being presented. Next, read the objective material carefully, being sure to note each bold word. These words indicate important concepts that you should familiarize yourself with. Study each in-text example carefully, noting the techniques and strategies used to solve the example.

You will then come to the key learning feature of this text, the *boxed examples.* These examples have been designed to aid you in a very specific way. Notice that in each example box, the example on the left is completely worked out and the example on the right is not. The reason for this is that *you* are expected to work the right-hand example (in the space provided) in order to immediately test your understanding of the material you have just studied.

You should study the worked-out example carefully by working through each step presented. This allows you to focus on each step and reinforces the technique for solving that type of problem. You can then use the worked-out example as a model for solving similar problems.

Next, try to solve the right-hand example using the problem-solving techniques that you have just studied. When you have completed your solution, check your work by turning to the page in the appendix where the complete solution can be found. The page number on which the solution appears is printed at the bottom of the example box in the right-hand corner. By checking your solution, you will know immediately whether or not you fully understand the skill you just studied.

When you have completed studying an objective, do all of the exercises in the exercise set that correspond with that objective. The exercises will be labeled with the same letter as the objective. Math is a subject that needs to be learned in small sections and practiced continually in order to be mastered. Doing all of the exercises in each exercise set will help you master the problem-solving techniques necessary for success.

Once you have completed the exercises to an objective, you should check your answers to the odd-numbered exercises with those found in the back of the book.

After completing a chapter, read the Chapter Summary. This summary highlights the important topics covered in the chapter. Following the Chapter Summary are Chapter Review Exercises, a Chapter Test, and a Cumulative Review (beginning with Chapter 2). Doing the review exercises is an important way of testing your understanding of the chapter. The answer to each review exercise is given in an appendix at back of the book. Each answer is followed by a reference that tells which objective that exercise was taken from. For example, (4.2B) means Section 4.2, Objective B. After checking your answers, restudy any objective that you missed. It may be very helpful to retry some of the exercises for that objective to reinforce your problem-solving techniques.

The Chapter Test should be used to prepare for an exam. We suggest that you try the Chapter Test a few days before your actual exam. Take the test in a quiet place and try to complete the test in the same amount of time you will be allowed for your exam. When taking the Chapter Test, practice the strategies of successful test takers: 1) scan the entire test to get a feel for the questions; 2) read the directions carefully; 3) work the problems that are easiest for you first; and perhaps most importantly, 4) try to stay calm.

When you have completed the Chapter Test, check your answers. If you missed a question, review the material in that objective and rework some of the exercises from that objective. This will strengthen your ability to perform the skills outlined in that objective.

The Cumulative Review allows you to refresh the skills you learned in previous chapters. This is very important in mathematics. By consistently reviewing previous material, you will retain the skills already learned as you build new ones.

Remember, to be successful: attend class regularly; read the textbook carefully; actively participate in class; work with your textbook using the boxed examples for immediate feedback and reinforcement of each skill; do all the homework assignments; review constantly; and work carefully.

1 Whole Numbers

OBJECTIVES

- ► To identify the order relation between two numbers
- ► To write whole numbers in words and in standard form
- ► To write whole numbers in expanded form
- ► To round a whole number to a given place value
- ► To add whole numbers without carrying
- ► To add whole numbers with carrying
- ► To solve application problems
- ► To subtract whole numbers without borrowing
- ► To subtract whole numbers with borrowing
- ► To solve application problems
- ► To multiply a number by a single digit
- ► To multiply larger whole numbers
- ► To solve application problems
- ► To divide by a single digit with no remainder in the quotient
- ► To divide by a single digit with a remainder in the quotient
- ► To divide by larger whole numbers
- ► To solve application problems
- ► To simplify expressions containing exponents
- ► To use the Order of Operations Agreement to simplify expressions
- ► To factor numbers
- ► To find the prime factorization of a number

Family Tree for Numbers

Our number system is called the Hindu-Arabic system because it has its ancestry in India and was refined by the Arabs. But despite the influence of these cultures on our system, there is some evidence that our system may have originated in China around 1400 B.C. That is 34 centuries ago.

The family tree below illustrates the most popular belief of the history of our number system. In the 16th century, symbols for our numbers started to become standardized with the advent, by Gutenberg, of the printing press.

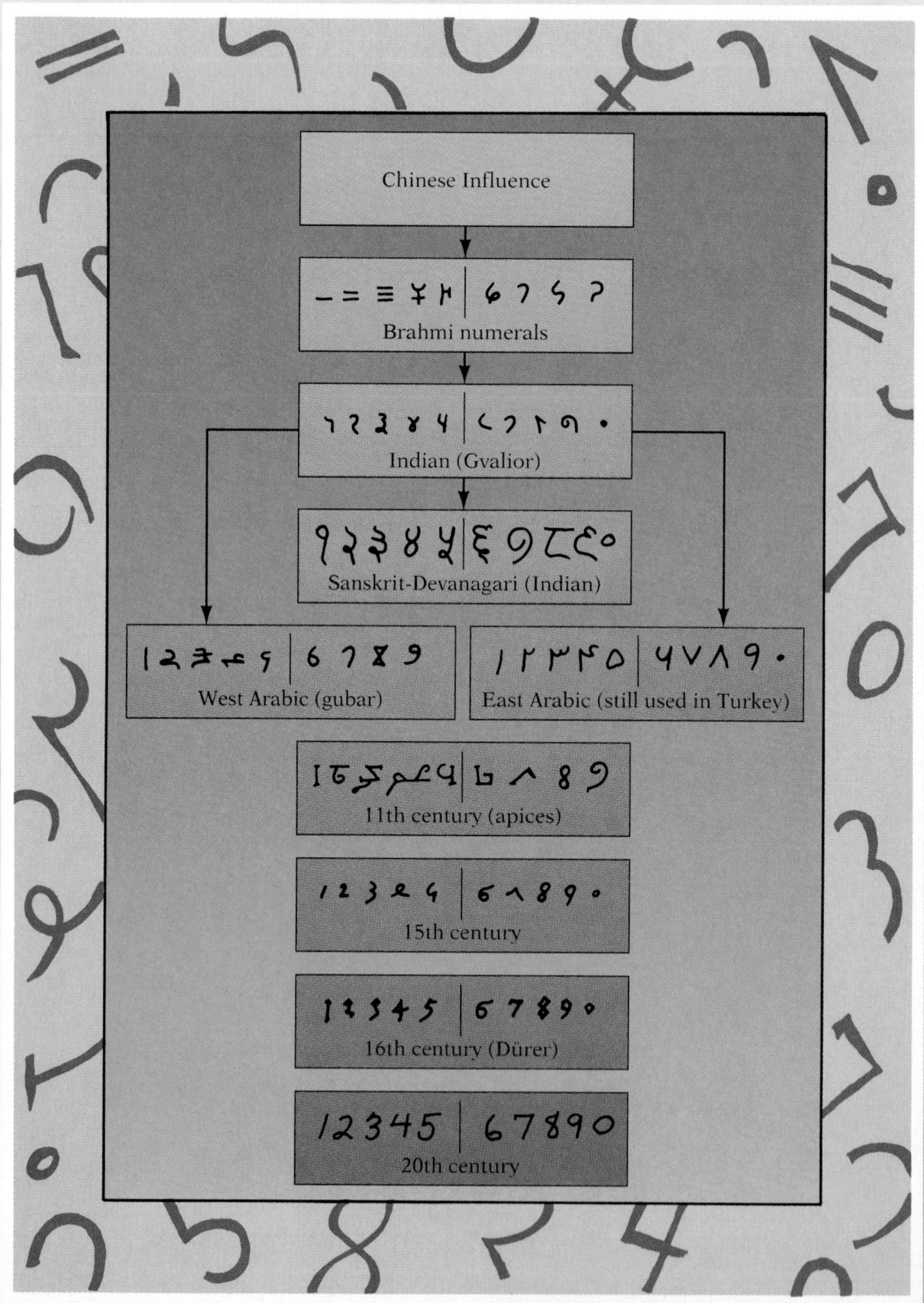

SECTION 1.1 Introduction to Whole Numbers

Objective A To identify the order relation between two numbers

The **whole numbers** are 0, 1, 2, 3, 4, 5, 6, 7, 8, 9, 10, 11, 12, 13, 14, . . .

The three dots mean that the list continues on and on and that there is no largest whole number.

Just as distances are associated with the markings on the edge of a ruler, the whole numbers can be associated with points on a line. This line is called the **number line.**

The Number Line

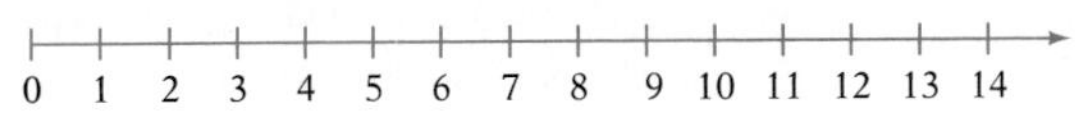

The **graph** of a whole number is shown by placing a heavy dot on the number line directly above the number.

The graph of 7 on the number line.

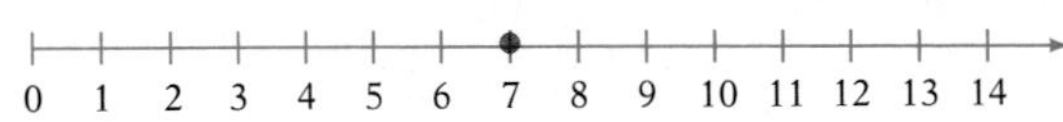

The number line can be used to show the order of whole numbers. A number that appears to the left of a given number is **less than** the given number. The symbol for "is less than" is <. A number that appears to the right of a given number is **greater than** the given number. The symbol for "is greater than" is >.

Four is less than seven.
$4 < 7$

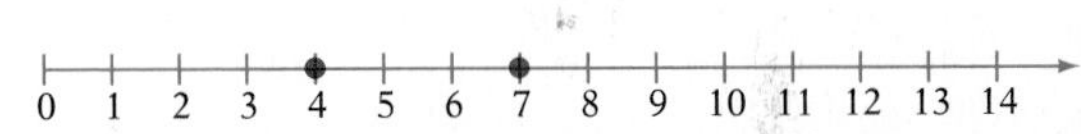

Twelve is greater than seven.
$12 > 7$

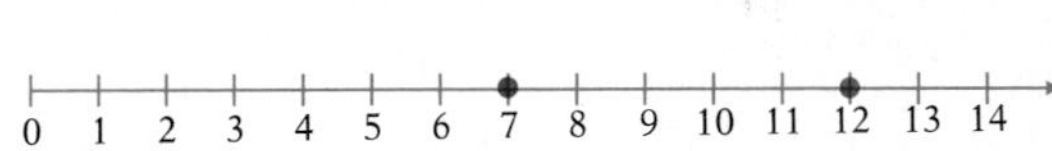

Example 1 Graph 11 on the number line.

Solution

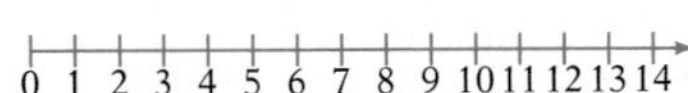

Example 2 Graph 9 on the number line.

Your solution

0 1 2 3 4 5 6 7 8 9 10 11 12 13 14

Example 3 Place the correct symbol, < or >, between the two numbers.

a. 39 24

b. 0 51

Solution **a.** $39 > 24$

b. $0 < 51$

Example 4 Place the correct symbol, < or >, between the two numbers.

a. 45 29

b. 27 0

Your solution **a.**

b.

Solutions on p. A7

Objective B

To write whole numbers in words and in standard form

When a whole number is written using the digits 0, 1, 2, 3, 4, 5, 6, 7, 8, and 9, it is said to be in **standard form.** The position of each digit in the number determines the digit's **place value.** The diagram below shows a **place-value chart** naming the first twelve place values. The number 37,462 is in standard form and has been entered in the chart.

In the number 37,462, the position of the digit 3 determines that its place value is ten-thousands.

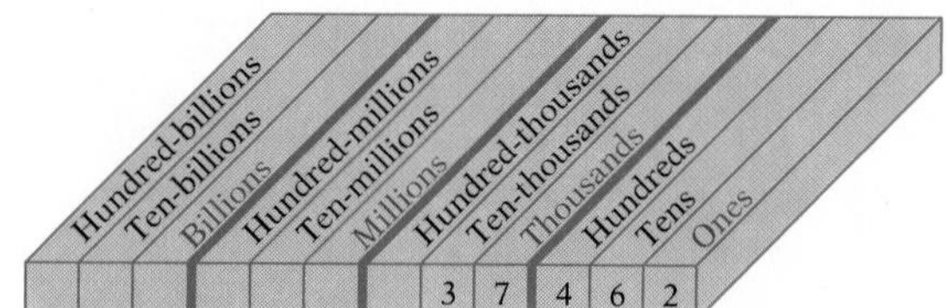

When a number is written in standard form, each group of digits separated by a comma is called a **period.** The number 3,786,451,294 has four periods. The period names are shown in red in the place-value chart above.

To write a number in words, start from the left. Name the number in each period. Then write the period name in place of the comma.

3,786,451,294 is read "three billion seven hundred eighty-six million four hundred fifty-one thousand two hundred ninety-four."

To write a whole number in standard form, write the number named in each period, and replace each period name with a comma.

Four million sixty-two thousand five hundred eighty-four is written 4,062,584. The zero is used as a place holder for the hundred-thousands' place.

Example 5 Write 25,478,083 in words.

Solution Twenty-five million four hundred seventy-eight thousand eighty-three

Example 6 Write 36,462,075 in words.

Your solution

Example 7 Write three hundred three thousand three in standard form.

Solution 303,003

Example 8 Write four hundred fifty-two thousand seven in standard form.

Your solution

Solutions on p. A7

Objective C

To write whole numbers in expanded form

The whole number 26,429 can be written in **expanded form** as

20,000 + 6000 + 400 + 20 + 9

The place-value chart can be used to find the expanded form of a number.

2 Ten-thousands	+	6 Thousands	+	4 Hundreds	+	2 Tens	+	9 Ones
20,000	+	6000	+	400	+	20	+	9

Write the number 420,806 in expanded form.

Notice the effect of having zeros in the number.

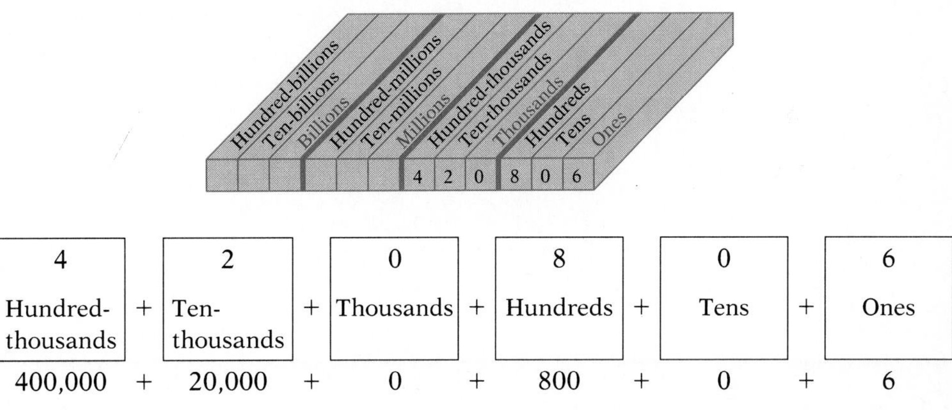

4 Hundred-thousands	+	2 Ten-thousands	+	0 Thousands	+	8 Hundreds	+	0 Tens	+	6 Ones
400,000	+	20,000	+	0	+	800	+	0	+	6

or simply 400,000 + 20,000 + 800 + 6

Example 9 Write 23,859 in expanded form.

Solution 20,000 + 3000 + 800 + 50 + 9

Example 10 Write 68,281 in expanded form.

Your solution

Example 11 Write 709,542 in expanded form.

Solution 700,000 + 9000 + 500 + 40 + 2

Example 12 Write 109,207 in expanded form.

Your solution

Solutions on p. A7

Objective D

To round a whole number to a given place value

When the distance to the moon is given as 240,000 miles, the number represents an approximation to the true distance. Giving an approximate value for an exact number is called **rounding.** A number is always rounded to a given place value.

37 is closer to 40 than it is to 30. 37 rounded to the nearest ten is 40.

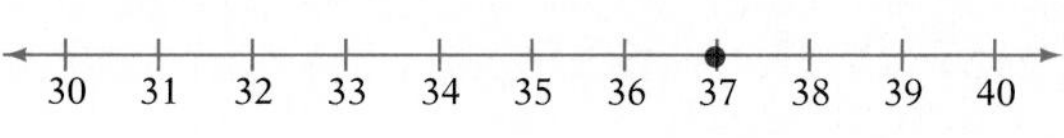

3673 rounded to the nearest ten is 3670. 3673 rounded to the nearest hundred is 3700.

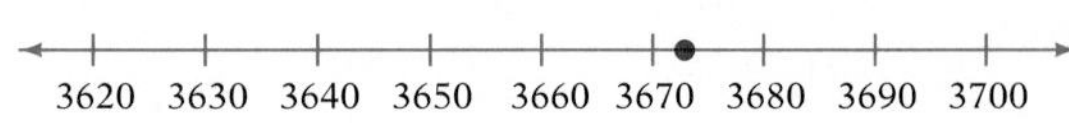

A whole number is rounded to a given place value without using the number line by looking at the first digit to the right of the given place value.

If the digit to the right of the given place value is less than 5, that digit and all digits to the right are replaced by zeros.

Round 13,834 to the nearest hundred.

Given place value

13,834

$3 < 5$

13,834 rounded to the nearest hundred is 13,800.

If the digit to the right of the given place value is greater than or equal to 5, increase the digit in the given place value by 1, and replace all other digits to the right by zeros.

Round 386,217 to the nearest ten-thousand.

Given place value

386,217

$6 > 5$

386,217 rounded to the nearest ten-thousand is 390,000.

Example 13 Round 525,453 to the nearest ten-thousand.

Solution Given place value

525,453

$5 = 5$

525,453 rounded to the nearest ten-thousand is 530,000.

Example 14 Round 368,492 to the nearest ten-thousand.

Your solution

Example 15 Round 1972 to the nearest hundred.

Solution Given place value

1972

$7 > 5$

1972 rounded to the nearest hundred is 2000.

Example 16 Round 3962 to the nearest hundred.

Your solution

Solutions on p. A7

1.1 EXERCISES

Objective A

Graph the number on the number line.

1. 3

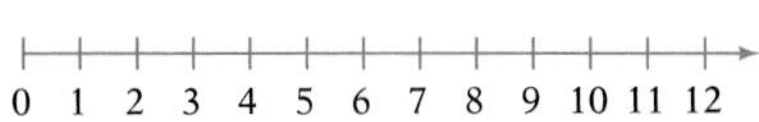

2. 5

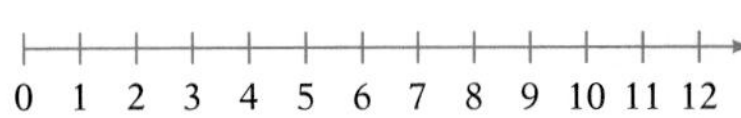

3. 9

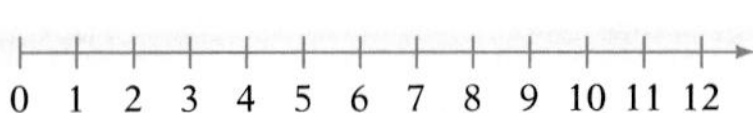

4. 0

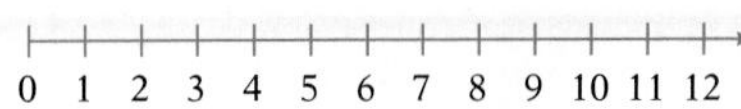

Place the correct symbol, < or >, between the two numbers.

5. 37 49

6. 58 21

7. 101 87

8. 16 5

9. 245 158

10. 2701 2071

11. 0 45

12. 107 0

13. 815 928

14. 2400 24,000

15. 7003 7020

16. 36,010 36,001

Objective B

Write the number in words.

17. 805

18. 609

19. 485

20. 576

21. 2675

22. 3790

23. 42,928

24. 58,473

25. 356,943

26. 498,512

27. 3,697,483

28. 6,842,715

Write the number in standard form.

29. Eighty-five

30. Three hundred fifty-seven

31. Three thousand four hundred fifty-six

32. Sixty-three thousand seven hundred eighty

33. Six hundred nine thousand nine hundred forty-eight

34. Seven million twenty-four thousand seven hundred nine

35. Four million three thousand two

36. Six million five thousand eight

Objective C

Write the number in expanded form.

37. 5287

38. 6295

39. 58,943

40. 453,921

41. 217,586

42. 4053

43. 50,943

44. 80,492

45. 200,583

46. 301,809

47. 403,705

48. 3,000,642

Objective D

Round the number to the given place value.

49. 926 Tens

50. 845 Tens

51. 1439 Hundreds

52. 3973 Hundreds

53. 7238 Thousands

54. 7609 Thousands

55. 43,607 Thousands

56. 52,715 Thousands

57. 647,989 Ten-thousands

58. 253,678 Ten-thousands

SECTION 1.2 Addition of Whole Numbers

Objective A To add whole numbers without carrying

Addition is the process of finding the total of two or more numbers.

By counting, we see that the total of \$3 and \$4 is \$7.

\$3	+	\$4	=	\$7
Addend		**Addend**		**Sum**

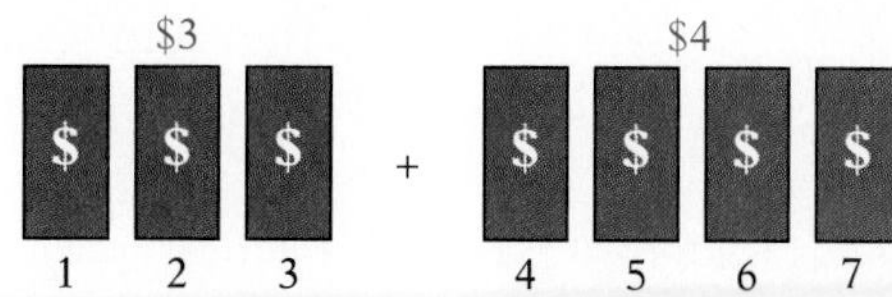

Addition can be illustrated on the number line by using arrows to represent the addends. The size or magnitude of a number can be represented on the number line by an arrow.

The number 3 can be represented anywhere on the number line by an arrow that is 3 units in length.

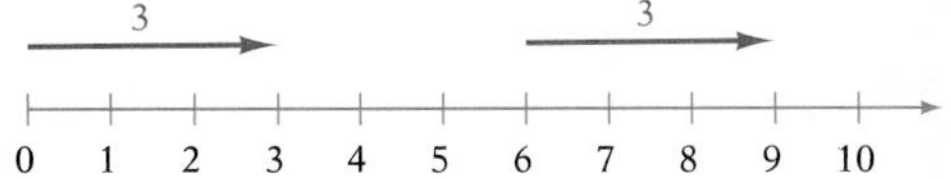

To add on the number line, place the arrows representing the addends head to tail, with the first arrow starting at zero. The sum is represented by an arrow starting at zero and stopping at the tip of the last arrow.

$$3 + 4 = 7$$

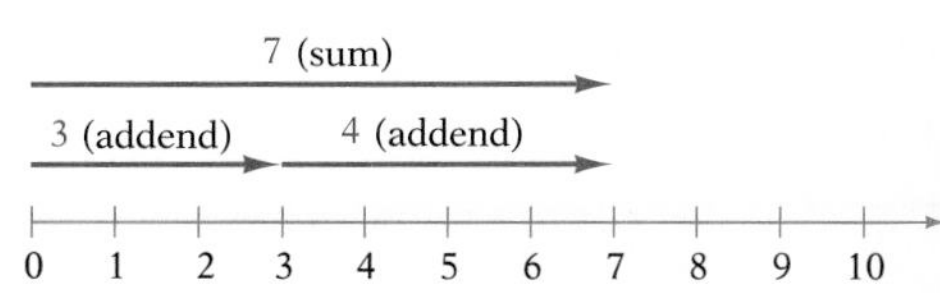

More than two numbers can be added on the number line.

$$3 + 2 + 4 = 9$$

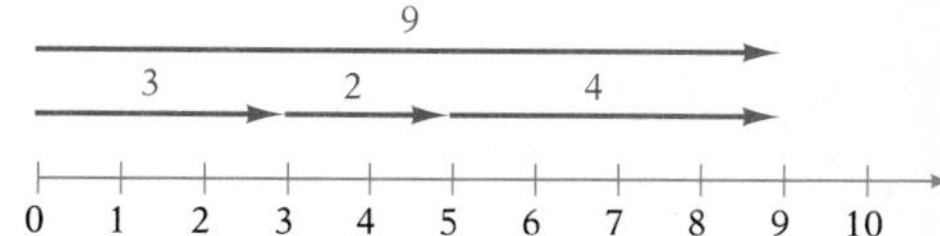

Some special properties of addition that are used frequently are given below.

Addition Property of Zero

Zero added to a number does not change the number.

$$4 + 0 = 4$$
$$0 + 7 = 7$$

Commutative Property of Addition

Two numbers can be added in either order; the sum will be the same.

$$4 + 8 = 8 + 4$$
$$12 = 12$$

Associative Property of Addition

Grouping the addition in any order gives the same result. The parentheses are grouping symbols and have the meaning "do the operations inside the parentheses first."

$$\underbrace{(4 + 2)}_{6} + 3 = 4 + \underbrace{(2 + 3)}_{5}$$
$$6 + 3 = 4 + 5$$
$$9 = 9$$

The number line is not useful for adding large numbers. The basic addition facts for adding one digit to one digit are listed in the Appendix on page A2 for your review. Addition of larger numbers requires the repeated use of the basic addition facts.

To add large numbers, begin by arranging the numbers vertically, keeping the digits of the same place value in the same column.

Add: 321 + 6472

Add the digits in each column.

```
   3 2 1
+6 4 7 2
 6 7 9 3
```

Example 1

Add: 4205 + 72,103

Solution

```
   4 2 0 5
+7 2 1 0 3
 7 6 3 0 8
```

Example 2

Add: 3508 + 92,170

Your solution

Example 3

Add: 131 + 2135 + 3123 + 510

Solution

```
   1 3 1
 2 1 3 5
 3 1 2 3
+  5 1 0
 5 8 9 9
```

Example 4

Add: 102 + 7351 + 1024 + 410

Your solution

Solutions on p. A7

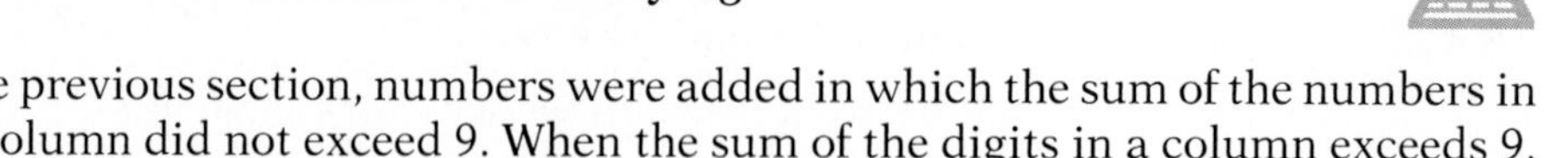

Objective B To add whole numbers with carrying

In the previous section, numbers were added in which the sum of the numbers in any column did not exceed 9. When the sum of the digits in a column exceeds 9, the addition will involve "carrying."

Add: 487 + 369

HUNDREDS TENS ONES

Add the ones' column.
7 + 9 = 16 (1 ten + 6 ones).
Write the 6 in the ones' column and carry the 1 ten to the tens' column.

```
   1
 4 8 7
+3 6 9
     6
```

Add the tens' column.
1 + 8 + 6 = 15 (1 hundred + 5 tens).
Write the 5 in the tens' column and carry the 1 hundred to the hundreds' column.

```
 1 1
 4 8 7
+3 6 9
   5 6
```

Add the hundreds' column.
1 + 4 + 3 = 8 (8 hundreds).
Write the 8 in the hundreds' column.

```
  1 1
  4 8 7
 +3 6 9
 ------
  8 5 6
```

Example 5 Add: 89 + 36 + 98

Solution

```
  2
  89
  36
 +98
 ---
 223
```

Example 6 Add: 95 + 88 + 67

Your solution

Example 7 Add:

```
    41,395
     4,327
   497,625
 +  32,991
```

Solution

```
   1 1 2 2 1
    41,395
     4,327
   497,625
 +  32,991
 ---------
   576,338
```

Example 8 Add:

```
       392
     4,079
    89,035
 +   4,992
```

Your solution

Solutions on p. A7

ESTIMATION

Estimation and Calculators

At some places in the text you will be asked to use your calculator. Effective use of a calculator requires that you estimate the answer to the problem. This helps ensure that you have entered the numbers correctly and pressed the correct keys.

For example, if you use your calculator to find 22,347 + 5896 and the answer in the calculator's display is 131,757,912, you should realize that you have entered some part of the calculation incorrectly. In this case, you pressed [×] instead of [+]. By estimating the answer to a problem, you can help ensure the accuracy of your calculations. The symbol ≈ is used to denote **approximately equal.**

For example, to estimate the answer to 22,347 + 5896, round off each number to the same place value. In this case we will round to the nearest thousands. Then add.

```
   22,347 ≈   22,000
 +  5,896 ≈ +  6,000
              28,000
```

The sum 22,347 + 5896 is approximately 28,000. Knowing this, you would know that 131,757,912 is much too large and therefore incorrect.

To estimate the sum of two numbers, first round each whole number to the same place value and then add. Compare this answer with the calculator's answer.

Objective C To solve application problems

To solve an application problem, first read the problem carefully. The **Strategy** involves identifying the quantity to be found and planning the steps that are necessary to find that quantity. The **Solution** involves performing each operation stated in the Strategy and writing the answer.

Example 9

You had \$1054 in your checking account before you made a deposit of \$870. Find the amount in your checking account after the deposit.

Strategy

To find the amount in the checking account after the deposit, add the amount deposited (\$870) to the amount in the checking account before the deposit (\$1054).

Solution

$$\begin{array}{r} \$1054 \\ +\ \ 870 \\ \hline \$1924 \end{array}$$

The amount in your checking account after the deposit was \$1924.

Example 10

The odometer on your car read 23,972 miles before a 297-mile trip. What was the odometer reading after the trip?

Your strategy

Your solution

Example 11

Your paycheck shows deductions of \$225 for savings, \$98 for taxes, and \$27 for insurance. Find the total of the three deductions.

Strategy

To find the total of the deductions, add the three amounts (\$225, \$98, and \$27).

Solution

$$\begin{array}{r} \$225 \\ 98 \\ +\ 27 \\ \hline \$350 \end{array}$$

The total of the three deductions is \$350.

Example 12

A homemaker has a monthly budget of \$275 for food, \$75 for car expenses, and \$64 for entertainment. Find the total amount budgeted for the three items each month.

Your strategy

Your solution

Solutions on p. A7

1.2 EXERCISES

Objective A

Add.

1. $\begin{array}{r} 6 \\ +5 \\ \hline \end{array}$

2. $\begin{array}{r} 7 \\ +8 \\ \hline \end{array}$

3. $\begin{array}{r} 0 \\ +8 \\ \hline \end{array}$

4. $\begin{array}{r} 3 \\ +0 \\ \hline \end{array}$

5. $\begin{array}{r} 16 \\ +\ 3 \\ \hline \end{array}$

6. $\begin{array}{r} 5 \\ +13 \\ \hline \end{array}$

7. $\begin{array}{r} 40 \\ +\ 9 \\ \hline \end{array}$

8. $\begin{array}{r} 8 \\ +70 \\ \hline \end{array}$

9. $\begin{array}{r} 17 \\ +11 \\ \hline \end{array}$

10. $\begin{array}{r} 25 \\ +63 \\ \hline \end{array}$

11. $\begin{array}{r} 83 \\ +42 \\ \hline \end{array}$

12. $\begin{array}{r} 63 \\ +94 \\ \hline \end{array}$

13. $\begin{array}{r} 231 \\ +462 \\ \hline \end{array}$

14. $\begin{array}{r} 845 \\ +152 \\ \hline \end{array}$

15. $\begin{array}{r} 770 \\ +329 \\ \hline \end{array}$

16. $\begin{array}{r} 542 \\ +920 \\ \hline \end{array}$

17. $\begin{array}{r} 421 \\ +308 \\ \hline \end{array}$

18. $\begin{array}{r} 202 \\ +725 \\ \hline \end{array}$

19. $\begin{array}{r} 658 \\ +831 \\ \hline \end{array}$

20. $\begin{array}{r} 842 \\ +936 \\ \hline \end{array}$

21. $\begin{array}{r} 8092 \\ +6307 \\ \hline \end{array}$

22. $\begin{array}{r} 5024 \\ +7902 \\ \hline \end{array}$

23. $\begin{array}{r} 6721 \\ +9144 \\ \hline \end{array}$

24. $\begin{array}{r} 6503 \\ +9285 \\ \hline \end{array}$

25. $\begin{array}{r} 71{,}092 \\ +85{,}407 \\ \hline \end{array}$

26. $\begin{array}{r} 62{,}304 \\ +94{,}093 \\ \hline \end{array}$

27. $\begin{array}{r} 8004 \\ +6391 \\ \hline \end{array}$

28. $\begin{array}{r} 6002 \\ +7297 \\ \hline \end{array}$

Add.

29. 923,571 + 863,117

30. 580,236 + 516,643

31. 823,453 + 571,346

32. 673,914 + 716,055

33. 5 + 4 + 9

34. 3911 + 4045

35. 9144 + 7632

36. 6648 + 9221

37. 67,453 + 82,546

38. 73,581 + 66,418

39. 88,123 + 80,451

40. 44,765 + 82,033

41. 86,332 + 43,511

42. 97,531 + 91,346

Objective B

Add.

43. 859 + 725

44. 637 + 829

45. 470 + 749

46. 427 + 690

47. 6016 + 5050

48. 4505 + 6009

49. 1897 + 3246

50. 8975 + 2146

51. 36,925 + 65,392

52. 56,772 + 51,239

53. 50,873 + 28,453

54. 34,872 + 46,079

Add.

55. 878 + 737 + 189

$$\begin{array}{r} 878 \\ 737 \\ +189 \\ \hline \end{array}$$

56.

$$\begin{array}{r} 768 \\ 461 \\ +669 \\ \hline \end{array}$$

57.

$$\begin{array}{r} 319 \\ 348 \\ +912 \\ \hline \end{array}$$

58.

$$\begin{array}{r} 292 \\ 579 \\ +315 \\ \hline \end{array}$$

59.

$$\begin{array}{r} 9409 \\ 3253 \\ +7078 \\ \hline \end{array}$$

60.

$$\begin{array}{r} 8188 \\ 8020 \\ +7104 \\ \hline \end{array}$$

61.

$$\begin{array}{r} 2038 \\ 2243 \\ +3139 \\ \hline \end{array}$$

62.

$$\begin{array}{r} 4252 \\ 6882 \\ +5235 \\ \hline \end{array}$$

63.

$$\begin{array}{r} 67{,}428 \\ 32{,}171 \\ +20{,}971 \\ \hline \end{array}$$

64.

$$\begin{array}{r} 52{,}801 \\ 11{,}664 \\ +89{,}638 \\ \hline \end{array}$$

65.

$$\begin{array}{r} 76{,}290 \\ 43{,}761 \\ +87{,}402 \\ \hline \end{array}$$

66.

$$\begin{array}{r} 43{,}901 \\ 98{,}301 \\ +67{,}943 \\ \hline \end{array}$$

67.

$$\begin{array}{r} 45{,}098 \\ 44{,}532 \\ 98{,}123 \\ 44{,}109 \\ +77{,}310 \\ \hline \end{array}$$

68.

$$\begin{array}{r} 32{,}087 \\ 33{,}687 \\ 66{,}301 \\ 63{,}442 \\ +91{,}842 \\ \hline \end{array}$$

69.

$$\begin{array}{r} 56{,}035 \\ 66{,}321 \\ 66{,}281 \\ 99{,}517 \\ +55{,}109 \\ \hline \end{array}$$

70.

$$\begin{array}{r} 55{,}320 \\ 88{,}321 \\ 80{,}332 \\ 87{,}539 \\ +10{,}321 \\ \hline \end{array}$$

Add.

71. $3 + 6 + 9 + 8 + 5$

72. $7 + 6 + 9 + 4 + 6$

73. $27 + 3 + 504$

74. $82 + 5 + 609$

75. $42 + 290 + 302$

76. $67 + 840 + 325$

77. $2709 + 658 + 10{,}935$

78. $8707 + 216 + 90{,}714$

Add.

79. 20,958 + 3218 + 42

80. 80,973 + 5168 + 29

81. 392 + 37 + 10,924 + 621

82. 694 + 62 + 70,129 + 217

83. 294 + 1029 + 7935 + 65

84. 692 + 2107 + 3196 + 92

85. 97 + 7234 + 69,532 + 276

86. 87 + 1698 + 27,317 + 727

87. 62 + 329 + 8954 + 1072

88. 87 + 946 + 6571 + 2103

89. 654 + 7293 + 237 + 33

90. 994 + 91,764 + 872 + 65

Estimate by rounding to the nearest hundreds. Then use your calculator to add.

91. 1234 + 9780 + 6740

92. 919 + 3642 + 8796

93. 241 + 569 + 390 + 1672

94. 107 + 984 + 1035 + 2904

Estimate by rounding to the nearest thousands. Then use your calculator to add.

95.
$$\begin{array}{r} 32{,}461 \\ 9{,}844 \\ +59{,}407 \\ \hline \end{array}$$

96.
$$\begin{array}{r} 29{,}036 \\ 22{,}904 \\ +\ 7{,}903 \\ \hline \end{array}$$

97.
$$\begin{array}{r} 25{,}432 \\ 62{,}941 \\ +70{,}390 \\ \hline \end{array}$$

98.
$$\begin{array}{r} 66{,}541 \\ 29{,}365 \\ +98{,}742 \\ \hline \end{array}$$

Estimate by rounding to the nearest ten-thousands. Then use your calculator to add.

99.
```
  67,421
  82,984
  66,361
  10,792
 +34,037
```

100.
```
  21,896
   4,235
  62,544
  21,892
+  1,334
```

101.
```
  281,421
    9,874
   34,394
  526,398
+  94,631
```

102.
```
  542,698
   97,327
    7,235
   73,667
 +173,201
```

Estimate by rounding to the nearest millions. Then use your calculator to add.

103.
```
  28,627,052
     983,073
+  3,081,496
```

104.
```
   1,792,085
  29,919,301
+  3,406,882
```

105.
```
  12,377,491
   3,409,723
   7,928,026
 +10,705,682
```

106.
```
  46,751,070
   6,095,832
     280,011
+  1,563,897
```

Objective C *Application Problems*

107. The attendance at the Friday night rock concert was 2114, and the attendance at the Saturday night concert was 3678. Find the total attendance at the two concerts.

108. The attendance at the Saturday night professional baseball game was 35,946, and the attendance at the Sunday afternoon game was 27,429. Find the total attendance for the two games.

109. A quarterback threw passes for 42 yards in the first quarter, 117 yards in the second quarter, 66 yards in the third quarter, and 82 yards in the fourth quarter. Find the total number of yards gained by passing.

110. Your basketball team scored 17 points the first quarter, 26 points the second quarter, 14 points the third quarter, and 32 points the fourth quarter. Find the total number of points scored.

111. A computer manufacturer produced 3285 computers in January, 2714 computers in February, and 2182 computers in March. How many computers were manufactured during the 3 months?

112. An account executive received commissions of $3168 during October, $2986 during November, and $819 during December. Find the total commission received for the 3-month period.

113. There were 289,658 paid admissions for the first five games of a world series. There were 63,786 and 67,562 paid admissions for the sixth and seventh games of the series.
a. Find the total paid attendance at the sixth and seventh games.
b. Find the total paid attendance for the entire series.

114. A compact disc retail company sold 1578 compact discs during the first 9 months of the year. The company sold 98 compact discs in October, 167 in November, and 467 in December.
a. How many compact discs were sold during the last 3 months of the year?
b. What was the total number of compact discs sold during the year?

115. A student has $1348 in a checking account to be used for the fall semester. During the summer the student makes deposits of $518, $678, and $463.
a. Find the total amount deposited.
b. Find the new checking account balance.

116. The odometer on a moving van reads 68,692. The driver plans to drive 515 miles the first day, 492 miles the second day, and 278 miles the third day.
a. How many miles will be driven during the 3 days?
b. What will the odometer reading be at the end of the trip?

117. A clothing shop has the following monthly expenses: office supplies—$144, postage—$65, computer rental—$640, salaries—$2338, insurance—$212, rent—$1468, and utilities—$365. Find the total costs for the month.

118. An account executive keeps a record of miles driven for income tax purposes. The miles driven the previous 12 months were: 1544 miles, 2098 miles, 2536 miles, 1984 miles, 1690 miles, 2856 miles, 785 miles, 2348 miles, 1854 miles, 3054 miles, 2213 miles, and 1256 miles. Find the total number of miles driven during the year.

SECTION 1.3 Subtraction of Whole Numbers

Objective A To subtract whole numbers without borrowing

Subtraction is the process of finding the difference between two numbers.

By counting, we see that the difference between \$8 and \$5 is \$3.

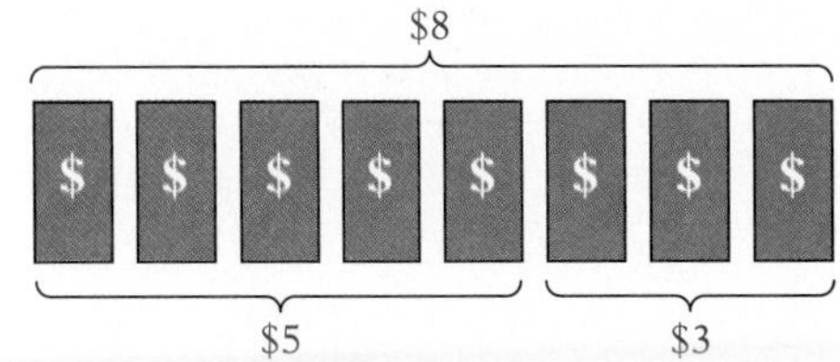

\$8	−	\$5	=	\$3
Minuend		**Subtrahend**		**Difference**

The difference 8 − 5 can be shown on the number line.

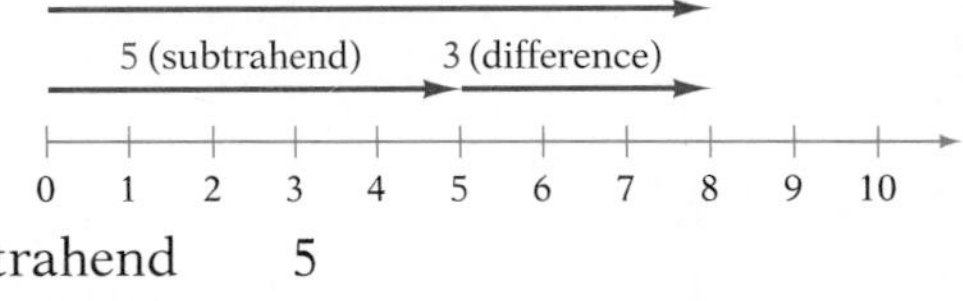

Note from the number line that addition and subtraction are related.

Subtrahend	5
+ Difference	+ 3
= Minuend	8

The fact that the sum of the subtrahend and difference equals the minuend can be used to check subtraction.

To subtract large numbers, begin by arranging the numbers vertically, keeping the digits of the same place value in the same column. Then subtract the digits in each column.

Subtract 8955 − 2432 and check.

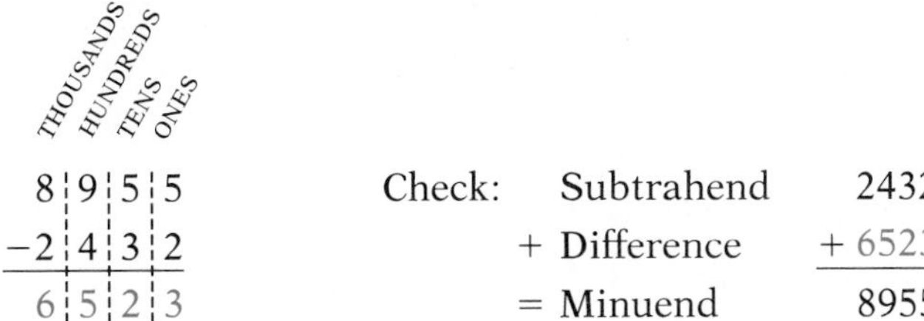

Example 1 Subtract 6594 − 3271 and check.

Solution

```
 6594   Check:  3271
−3271          +3323
 3323           6594
```

Example 2 Subtract 8925 − 6413 and check.

Your solution

Example 3 Subtract 15,762 − 7541 and check.

Solution

```
 15,762   Check:  7,541
− 7,541          +8,221
  8,221          15,762
```

Example 4 Subtract 17,504 − 9302 and check.

Your solution

Solutions on p. A8

Objective B

To subtract whole numbers with borrowing

In all the subtraction problems in the previous objective, for each place value the lower digit was not larger than the upper digit. When the lower digit is larger than the upper digit, subtraction will involve "borrowing."

Subtract: 692 − 378

HUNDREDS TENS ONES	HUNDREDS TENS ONES	HUNDREDS TENS ONES	HUNDREDS TENS ONES
6 9 2 (8+1) −3 7 8	6 9 2 (8+① 10) −3 7 8	6 9 2 (8 12) −3 7 8	6 9 2 (8 12) −3 7 8 3 1 4
Because 8 > 2, borrowing is necessary. 9 tens = 8 tens + 1 ten.	Borrow 1 ten from the tens' column and write 10 in the ones' column.	Add the borrowed 10 to 2.	Subtract the digits in each column.

Subtract 751 − 234 and check.

```
  4 11        4 11
 7 5 1       7 5 1      Check:   1
-2 3 4      -2 3 4              234
            5 1 7              +517
                                751
```

There may be more than one column in which borrowing is necessary.

Subtract 1234 − 485 and check.

```
     2 14          1 12 14       0 11 12 14             11
 1 2 3 4       1 2 3 4       1 2 3 4      Check:   485
-  4 8 5      -  4 8 5      -  4 8 5              +749
       9          4 9          7 4 9              1234
```

Subtraction with a zero in the minuend involves repeated borrowing.

Subtract: 3904 − 1775

```
                    9             9
   8 10          8 10 14       8 10 14
 3 9 0 4       3 9 0 4       3 9 0 4
-1 7 7 5      -1 7 7 5      -1 7 7 5
                             2 1 2 9
```

5 > 4
There is a 0 in the tens' column. Borrow 1 hundred (= 10 tens) from the hundreds' column and write 10 in the tens' column.

Borrow 1 ten from the tens' column and add 10 to the 4 in the ones' column.

Subtract the digits in each column.

Example 5 Subtract 4392 − 678 and check.

Solution

$$\begin{array}{r} \overset{3}{\not4}\ \overset{13}{\not3}\ \overset{8}{\not9}\ \overset{12}{\not2} \\ -\quad 6\ 7\ 8 \\ \hline 3\ 7\ 1\ 4 \end{array}$$

Check:

$$\begin{array}{r} 678 \\ +3714 \\ \hline 4392 \end{array}$$

Example 6 Subtract 3481 − 865 and check.

Your solution

Example 7 Subtract 63,221 − 23,954 and check.

Solution

$$\begin{array}{r} \overset{5}{\not6}\ \overset{12}{\not3},\overset{11}{\not2}\ \overset{11}{\not2}\ \overset{11}{\not1} \\ -2\ 3,9\ 5\ 4 \\ \hline 3\ 9,2\ 6\ 7 \end{array}$$

Check:

$$\begin{array}{r} 23,954 \\ +39,267 \\ \hline 63,221 \end{array}$$

Example 8 Subtract 54,562 − 14,485 and check.

Your solution

Example 9 Subtract 46,005 − 32,167 and check.

Solution

$$\begin{array}{r} 4\ \overset{5}{\not6},\overset{10}{\not0}\ 0\ 5 \\ -3\ 2,1\ 6\ 7 \\ \hline \end{array}$$

There are two zeros in the minuend. Borrow 1 thousand from the thousands' column and write 10 in the hundreds' column.

$$\begin{array}{r} 4\ \overset{5}{\not6},\overset{9}{\not0}\ \overset{10}{\not0}\ 5 \\ -3\ 2,1\ 6\ 7 \\ \hline \end{array}$$

Borrow 1 hundred from the hundreds' column and write 10 in the tens' column.

$$\begin{array}{r} 4\ \overset{5}{\not6},\overset{9}{\not0}\ \overset{9}{\not0}\ \overset{15}{\not5} \\ -3\ 2,1\ 6\ 7 \\ \hline 1\ 3,8\ 3\ 8 \end{array}$$

Borrow 1 ten from the tens' column and add 10 to the 5 in the ones' column.

Check:

$$\begin{array}{r} 32,167 \\ +13,838 \\ \hline 46,005 \end{array}$$

Example 10 Subtract 64,003 − 54,936 and check.

Your solution

Solutions on p. A8

ESTIMATION

Estimating the Difference Between Two Whole Numbers

Estimate and then use your calculator to find 323,502 − 28,912.

To estimate the difference between two numbers, round each number to the same place value. In this case we will round to the nearest ten-thousands. Then subtract. The estimated answer is 290,000.

$$\begin{array}{rcr} 323,502 & \approx & 320,000 \\ -\ 28,912 & \approx & -\ 30,000 \\ \hline & & 290,000 \end{array}$$

Now use your calculator to find the exact result. The exact answer is 294,590.

323502 [−] 28912 [=] 294590

Objective C To solve application problems

Example 11

Your monthly food budget is $215. How much is left in the food budget after $148 is spent on groceries?

Strategy

To find the amount left in the food budget, subtract the amount already spent on groceries ($148) from the monthly food budget ($215).

Solution

```
  $215
-  148
  $ 67
```

$67 is left in the monthly food budget.

Example 12

The down payment on a car costing $3250 is $675. Find the amount that remains to be paid.

Your strategy

Your solution

Example 13

You had a balance of $815 in your checking account. You then wrote checks in the amount of $112 for taxes, $57 for food, and $39 for shoes. What is your new checking account balance?

Strategy

To find your new checking account balance:

- Add to find the total of the three checks ($112 + $57 + $39).
- Subtract the total of the three checks from the old balance ($815).

Solution

```
  $112            $815
    57          -  208
+   39            $607
  $208 total of checks
```

Your new checking account balance is $607.

Example 14

Your total salary is $638. Deductions of $127 for taxes, $18 for insurance, and $35 for savings are taken from your pay. Find your take-home pay.

Your strategy

Your solution

Solutions on p. A8

1.3 EXERCISES

▶ Objective A

Subtract.

1. $\begin{array}{r} 9 \\ -5 \\ \hline \end{array}$ 2. $\begin{array}{r} 8 \\ -7 \\ \hline \end{array}$ 3. $\begin{array}{r} 8 \\ -4 \\ \hline \end{array}$ 4. $\begin{array}{r} 7 \\ -3 \\ \hline \end{array}$ 5. $\begin{array}{r} 10 \\ -0 \\ \hline \end{array}$ 6. $\begin{array}{r} 7 \\ -0 \\ \hline \end{array}$

7. $\begin{array}{r} 11 \\ -4 \\ \hline \end{array}$ 8. $\begin{array}{r} 12 \\ -8 \\ \hline \end{array}$ 9. $\begin{array}{r} 19 \\ -8 \\ \hline \end{array}$ 10. $\begin{array}{r} 15 \\ -6 \\ \hline \end{array}$ 11. $\begin{array}{r} 16 \\ -7 \\ \hline \end{array}$ 12. $\begin{array}{r} 18 \\ -9 \\ \hline \end{array}$

13. $\begin{array}{r} 25 \\ -3 \\ \hline \end{array}$ 14. $\begin{array}{r} 55 \\ -4 \\ \hline \end{array}$ 15. $\begin{array}{r} 68 \\ -8 \\ \hline \end{array}$ 16. $\begin{array}{r} 77 \\ -3 \\ \hline \end{array}$ 17. $\begin{array}{r} 89 \\ -23 \\ \hline \end{array}$ 18. $\begin{array}{r} 68 \\ -43 \\ \hline \end{array}$

19. $\begin{array}{r} 54 \\ -21 \\ \hline \end{array}$ 20. $\begin{array}{r} 88 \\ -57 \\ \hline \end{array}$ 21. $\begin{array}{r} 1202 \\ -701 \\ \hline \end{array}$ 22. $\begin{array}{r} 1305 \\ -404 \\ \hline \end{array}$ 23. $\begin{array}{r} 1763 \\ -801 \\ \hline \end{array}$

24. $\begin{array}{r} 1497 \\ -706 \\ \hline \end{array}$ 25. $\begin{array}{r} 8974 \\ -3972 \\ \hline \end{array}$ 26. $\begin{array}{r} 2836 \\ -1711 \\ \hline \end{array}$ 27. $\begin{array}{r} 8976 \\ -7463 \\ \hline \end{array}$ 28. $\begin{array}{r} 9273 \\ -6142 \\ \hline \end{array}$

29. $17 - 8$ 30. $15 - 7$ 31. $12 - 5$ 32. $93 - 71$ 33. $83 - 52$

34. $77 - 36$ 35. $129 - 82$ 36. $132 - 61$ 37. $969 - 44$ 38. $1347 - 103$

39. $4865 - 304$ 40. $1525 - 702$ 41. $9999 - 6794$ 42. $7806 - 3405$ 43. $8843 - 7621$

44. $8713 - 6512$ 45. $5672 - 4361$ 46. $2900 - 1900$

47. $7727 - 4326$ 48. $5988 - 1713$ 49. $2904 - 1403$

Objective B

Subtract.

50. $\begin{array}{r} 71 \\ -18 \\ \hline \end{array}$ **51.** $\begin{array}{r} 93 \\ -28 \\ \hline \end{array}$ **52.** $\begin{array}{r} 47 \\ -18 \\ \hline \end{array}$ **53.** $\begin{array}{r} 44 \\ -27 \\ \hline \end{array}$ **54.** $\begin{array}{r} 71 \\ -67 \\ \hline \end{array}$

55. $\begin{array}{r} 37 \\ -29 \\ \hline \end{array}$ **56.** $\begin{array}{r} 50 \\ -27 \\ \hline \end{array}$ **57.** $\begin{array}{r} 70 \\ -33 \\ \hline \end{array}$ **58.** $\begin{array}{r} 993 \\ -537 \\ \hline \end{array}$ **59.** $\begin{array}{r} 681 \\ -328 \\ \hline \end{array}$

60. $\begin{array}{r} 250 \\ -192 \\ \hline \end{array}$ **61.** $\begin{array}{r} 840 \\ -783 \\ \hline \end{array}$ **62.** $\begin{array}{r} 768 \\ -194 \\ \hline \end{array}$ **63.** $\begin{array}{r} 679 \\ -519 \\ \hline \end{array}$ **64.** $\begin{array}{r} 770 \\ -395 \\ \hline \end{array}$

65. $\begin{array}{r} 630 \\ -475 \\ \hline \end{array}$ **66.** $\begin{array}{r} 893 \\ -874 \\ \hline \end{array}$ **67.** $\begin{array}{r} 673 \\ -649 \\ \hline \end{array}$ **68.** $\begin{array}{r} 508 \\ -299 \\ \hline \end{array}$ **69.** $\begin{array}{r} 704 \\ -157 \\ \hline \end{array}$

70. 67 − 39 **71.** 82 − 37 **72.** 80 − 32 **73.** 837 − 83

74. 642 − 87 **75.** 470 − 92 **76.** 810 − 380 **77.** 720 − 290

78. 674 − 337 **79.** 3526 − 387 **80.** 1712 − 289 **81.** 4350 − 729

82. 1702 − 948 **83.** 1607 − 869 **84.** 5933 − 3754 **85.** 7293 − 3748

86. 8143 − 2417 **87.** 7236 − 1978 **88.** 5714 − 2367 **89.** 8462 − 3575

90. 9407 − 2918 **91.** 3706 − 2957 **92.** 8605 − 7716 **93.** 8052 − 2709

94. 80,305 − 9176 **95.** 70,702 − 4239 **96.** 10,004 − 9306 **97.** 80,009 − 63,419

98. 70,618 − 41,213 **99.** 80,053 − 27,649 **100.** 70,700 − 21,076 **101.** 80,800 − 42,023

Subtract.

102. 95,432 − 87,857

103. 13,806 − 9439

104. 12,701 − 8624

105. 14,316 − 9427

106. 10,072 − 9514

107. 10,067 − 7428

108. 68,023 − 29,174

109. 3000 − 1785

110. 6000 − 2975

111. 7407 − 2359

112. 9703 − 2347

113. 2600 − 1972

114. 8400 − 3762

115. 9003 − 2471

116. 6004 − 2392

117. 8202 − 3916

118. 7050 − 4137

119. 7015 − 2973

120. 4207 − 1624

121. 7005 − 1796

122. 8003 − 2735

123. 20,005 − 9,627

124. 80,004 − 8,237

Estimate by rounding to the nearest ten-thousands. Then use your calculator to subtract.

125. 80,032 − 19,605

126. 90,765 − 60,928

127. 32,574 − 10,961

128. 96,430 − 59,762

129. 567,423 − 208,444

130. 300,712 − 198,714

Estimate by rounding to the nearest hundred-thousands. Then use your calculator to subtract.

131. 224,198 − 96,441

132. 318,074 − 64,908

133. 556,409 − 99,642

134. 873,925 − 28,744

135. 305,219 − 147,355

136. 608,123 − 238,107

Estimate by rounding to the nearest millions. Then use your calculator to subtract.

137. 3,050,092
−1,967,325

138. 9,000,765
−1,230,768

139. 9,070,605
−3,492,763

140. 9,872,504
−8,947,396

141. 4,682,191
−3,056,822

142. 9,947,812
−9,057,867

Objective C *Application Problems*

143. You have $304 in your checking account. If you write a check for $139, how much is left in your checking account?

144. You have $521 in your checking account. If you write a check for $247, how much is left in your checking account?

145. A tennis coach purchases a video camera that costs $1079 and makes a down payment of $180. Find the amount that remains to be paid.

146. A student purchases a used car that costs $5225 and makes a down payment of $550. Find the amount that remains to be paid.

147. The odometer of a rental car reads 2041 miles after a trip of 396 miles. What was the odometer reading at the start of the trip?

148. At the end of a vacation trip, the odometer of your car read 71,004 miles. If the odometer at the beginning of the trip read 68,238 miles, what was the length of the trip?

149. A steelworker receives a salary of $872 per week. Deductions from the check are $168 for taxes, $32 for union dues, $48 for social security, and $28 for insurance.
a. Find the total amount deducted.
b. Find the steelworker's take home pay.

150. A student had a bank balance of $6531. The student then wrote checks of $3248 for tuition, $116 for fees, and $279 for books.
a. Find the total of the three checks.
b. Find the new bank balance.

151. A family has a monthly budget of $1200. After $329 is spent for food, $128 for clothing, $82 for transportation, and $79 for entertainment, how much is left in the budget?

152. A wholesale distributor has 2405 computers in stock. After sales of 94 computers, 32 computers, and 116 computers, how many computers remain in stock?

153. A basketball team had an income of $13,425,320. Player salaries amounted to $9,125,320, and other expenses amounted to $3,215,860. Find the team's profit for the season.

154. The land area of Canada is 3,851,809 square miles and the land area of the United States is 3,675,633 square miles. How much larger than the United States is Canada?

SECTION 1.4 Multiplication of Whole Numbers

Objective A To multiply a number by a single digit

Six boxes of toasters are ordered. Each box contains eight toasters. How many toasters are ordered?

This problem can be worked by adding 6 eights.

$8 + 8 + 8 + 8 + 8 + 8 = 48$

This problem involves repeated addition of the same number and can be worked by a shorter process called **multiplication.**

Multiplication is the repeated addition of the same number.

8 + 8 + 8 + 8 + 8 + 8 = 48

or

The numbers that are multiplied are called **factors.** The answer is called the **product.**

6	×	8	= 48
Factor		Factor	Product

The product of 6 × 8 can be represented on the number line. The arrow representing the whole number 8 is repeated 6 times. The result is the arrow representing 48.

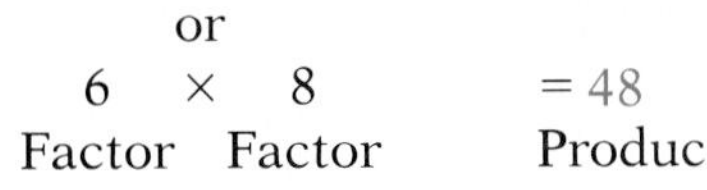

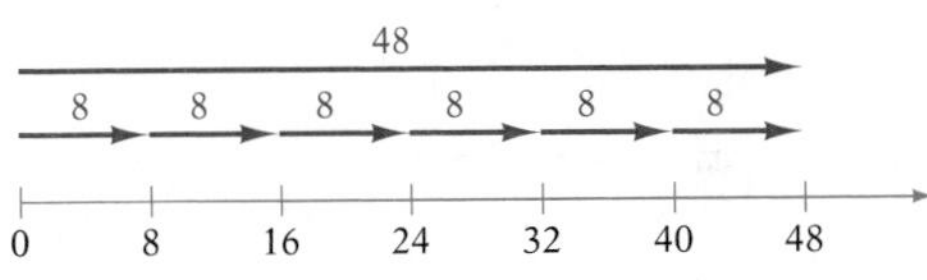

The times sign "×" is one symbol that is used to mean multiplication. Another common symbol used is a dot placed between the numbers.

$$7 \times 8 = 56 \qquad 7 \cdot 8 = 56$$

As with addition, there are some useful properties of multiplication.

Multiplication Property of Zero

The product of a number and zero is zero.

$0 \times 4 = 0$
$7 \times 0 = 0$

Multiplication Property of One

The product of a number and one is the number.

$1 \times 6 = 6$
$8 \times 1 = 8$

Commutative Property of Multiplication

Two numbers can be multiplied in either order. The product will be the same.

$4 \times 3 = 3 \times 4$
$12 = 12$

Associative Property of Multiplication

Grouping the multiplication in any order gives the same result. Do the multiplication inside the parentheses first.

$\underbrace{(4 \times 2)}_{8} \times 3 = 4 \times \underbrace{(2 \times 3)}_{6}$
$24 = 24$

The basic facts for multiplying one digit numbers are listed on page A2 in the Appendix for your review. Multiplication of larger numbers requires the repeated use of the basic multiplication facts.

Multiply: 37 × 4

Multiply 4 × 7. 4 × 7 = 28 (2 tens + 8 ones). Write the 8 in the ones' column and carry the 2 to the tens' column.

$$\begin{array}{r} \scriptstyle 2 \\ 37 \\ \times \ 4 \\ \hline 8 \end{array}$$

The 3 in 37 is 3 tens.
Multiply 4 × 3 tens. 4 × 3 tens = 12 tens
Add the carry digit. + 2 tens
14 tens

$$\begin{array}{r} \scriptstyle 2 \\ 37 \\ \times \ 4 \\ \hline 148 \end{array}$$

Write the 14.

Example 1 Multiply: 64 × 8

Solution

$$\begin{array}{r} \scriptstyle 3 \\ 64 \\ \times \ 8 \\ \hline 512 \end{array}$$

Example 2 Multiply: 78 × 6

Your solution

Example 3 Multiply: 735 × 9

Solution

$$\begin{array}{r} \scriptstyle 3\,4 \\ 735 \\ \times \ \ 9 \\ \hline 6615 \end{array}$$

Example 4 Multiply: 648 × 7

Your solution

Solutions on p. A8

Objective B **To multiply larger whole numbers**

Note the pattern when the following numbers are multiplied.

Multiply the nonzero part of the factors.

Now attach the same number of zeros to the product as the total number of zeros in the factors.

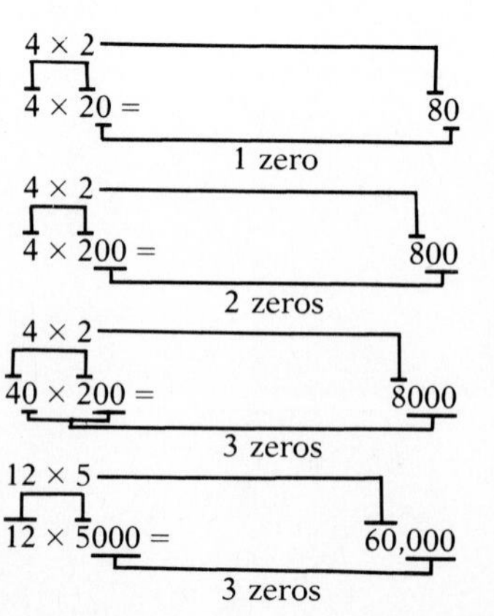

Multiply: 47×23

Multiply by the ones' digit.	Multiply by the tens' digit.	Add.
47	47	47
×23	×23	×23
141 (= 47 × 3)	141	141
	940 (= 47 × 20)	94
	Writing the 0 is optional.	1081

Thousands Hundreds Tens Ones

	Thousands	Hundreds	Tens	Ones	
			4	7	
×			2	3	
		1	4	1	3 × 47
		9	4	0	20 × 47
	1	0	8	1	141 + 940

The place-value chart illustrates the placement of the products.

Note the placement of the products when multiplying by a factor that contains a zero.

Multiply: 439×206

```
  439
 ×206
 2634
 000     0 × 439
878
90,434
```

When working the problem, usually only one zero is written. Writing this zero ensures the proper placement of the products.

```
  439
 ×206
 2634
8780
90,434
```

Example 5 Multiply: 829×603

Solution

```
   829
  ×603
  2487
49740
499,887
```

Example 6 Multiply: 756×305

Your solution

Solution on p. A8

ESTIMATION

Estimating the Product of Two Whole Numbers

Estimate and then use your calculator to find 3267×389.

To estimate a product, round each number so that all the digits are zero except the first digit. Then multiply. The estimated answer is 1,200,000.

$$\begin{array}{rcr} 3267 & \approx & 3000 \\ \times\ 389 & \approx & \times\ 400 \\ \hline & & 1{,}200{,}000 \end{array}$$

Now use your calculator to find the exact answer. The exact answer is 1,270,863.

3267 [×] 389 [=] 1270863

Objective C To solve application problems

Example 7

An auto mechanic receives a salary of $525 each week. How much does the auto mechanic earn in 4 weeks?

Strategy

To find the mechanic's earnings for 4 weeks, multiply the weekly salary ($525) by the number of weeks (4).

Solution

$$\begin{array}{r} \$525 \\ \times \quad 4 \\ \hline \$2100 \end{array}$$

The mechanic earns $2100 in 4 weeks.

Example 8

A new car dealer receives a shipment of 37 cars each month. Find the number of cars the dealer will receive in 12 months.

Your strategy

Your solution

Example 9

A press operator earns $320 for working a 40-hour week. This week the press operator also worked 7 hours of overtime at $13 an hour. Find the press operator's total pay for the week.

Strategy

To find the press operator's income for the week:

- Find the overtime pay by multiplying the hours of overtime (7) by the overtime rate of pay ($13).
- Add the weekly salary ($320) to the overtime pay.

Solution

$$\begin{array}{r} \$13 \\ \times \quad 7 \\ \hline \$91 \end{array} \text{ overtime pay} \qquad \begin{array}{r} \$320 \\ +\quad 91 \\ \hline \$411 \end{array}$$

The press operator earned $411 this week.

Example 10

The manager of a clothing store can buy 80 men's suits for $4800. Each sports jacket will cost the store $23. The manager orders 80 men's suits and 25 sports jackets. What is the total cost of the order?

Your strategy

Your solution

Solutions on p. A9

1.4 EXERCISES

Objective A

Multiply.

1. 3×4	**2.** 2×8	**3.** 5×7	**4.** 6×4	**5.** 5×5
6. 7×7	**7.** 0×7	**8.** 8×0	**9.** 8×9	**10.** 7×6
11. 7×3	**12.** 2×9	**13.** 5×6	**14.** 7×4	**15.** 6×9
16. 8×5	**17.** 82×4	**18.** 79×3	**19.** 55×4	**20.** 72×2
21. 89×7	**22.** 96×5	**23.** 79×9	**24.** 67×7	**25.** 45×9
26. 66×3	**27.** 70×4	**28.** 67×5	**29.** 127×9	**30.** 623×4
31. 802×5	**32.** 607×9	**33.** 300×5	**34.** 600×7	**35.** 906×8
36. 703×9	**37.** 127×5	**38.** 632×3	**39.** 559×4	**40.** 632×8
41. 780×7	**42.** 690×5	**43.** 465×4	**44.** 382×7	**45.** 367×3

Multiply.

46. 524×4

47. 337×5

48. 841×6

49. 6709×7

50. 3608×5

51. 8568×7

52. 5495×4

53. 4780×4

54. 3690×5

55. 9895×2

56. 4697×3

57. $48{,}253 \times 3$

58. $69{,}587 \times 8$

59. $80{,}057 \times 7$

60. $90{,}032 \times 7$

61. $93{,}000 \times 7$

62. $87{,}000 \times 9$

63. $93{,}257 \times 6$

64. $88{,}524 \times 7$

Objective B

Multiply.

65. 16×21

66. 18×24

67. 35×26

68. 27×72

69. 20×32

70. 30×47

71. 40×20

72. 30×70

73. 39×46

74. 37×25

75. 67×23

76. 95×33

77. 693×91

78. 581×72

79. 419×80

80. 727×60

81. 8279×46

82. 9577×35

83. 6938×78

84. 8875×67

85. 7035×57

86. 6702×48

87. 3009×35

88. 6003×57

Multiply.

89. 3300 × 20

90. 5400 × 60

91. 3987 × 29

92. 4765 × 37

93. 809 ×530

94. 607 ×460

95. 800 ×325

96. 700 ×274

97. 987 ×349

98. 688 ×674

99. 312 ×134

100. 423 ×427

101. 379 ×500

102. 684 ×700

103. 985 ×408

104. 758 ×209

105. 3407 × 309

106. 5207 × 902

107. 4258 × 986

108. 6327 × 876

109. 8800 × 140

110. 9900 × 320

111. 2675 × 487

112. 3985 × 364

Estimate and then use your calculator to multiply.

113. 8745 × 63

114. 4732 × 93

115. 39,246 × 29

116. 64,409 × 67

117. 2937 × 206

118. 8941 × 726

119. 3097 ×1025

120. 6379 ×2936

121. 32,508 × 591

122. 62,504 × 923

123. 81,405 × 902

124. 66,735 × 844

Objective C *Application Problems*

125. A plane flying from Los Angeles to Boston uses 865 gallons of jet fuel each hour. How many gallons of jet fuel were used on a 5-hour flight?

126. A compact car averages 43 miles on 1 gallon of gas. How many miles could such a car travel on 13 gallons of gas?

127. A machine at a bottling company can fill and cap 4200 bottles of a soft drink in 1 hour. How many bottles of soft drink can the machine fill and cap in 40 hours?

128. A buyer for a department store purchased 225 suits at $64 each. What is the total cost for the 225 suits?

129. A computer programmer buys a car by making payments of $234 each month for 48 months. Find the total cost of the car.

130. An investor receives a check of $187 each month from a tax-free municipal bond fund. How much will the investor receive over a 12-month period?

131. A manufacturer of video cameras has 4347 video cameras in stock. The company can manufacture 16 video cameras each hour.
a. How many video cameras can be manufactured in 12 hours?
b. How many video cameras will be in stock 12 hours from now, assuming none are sold?

132. A gasoline storage tank contains 75,000 gallons of gasoline. When a valve is opened, a pipeline can deliver 37 gallons of gasoline each minute to the tank.
a. How many gallons of gasoline are delivered to the tank in 28 minutes?
b. How many gallons are in the tank after 28 minutes?

133. The tickets to a Christmas play were $4 for adults and $1 for children. 172 adult tickets and 136 child tickets were sold. Find the total income from the sale of the tickets.

134. A service station attendant earns $286 for working a 40-hour week. Last week the attendant worked an additional 8 hours at $13 an hour. Find the attendant's total pay for last week's work.

135. A computer graphing screen has 640 rows of pixels, and there are 480 pixels in a row. Find the total number of pixels on the screen.

136. A computer graphing screen has 720 rows of pixels, and there are 348 pixels in a row. Find the total number of pixels on the screen.

137. An oil company with 98,460,000 shares is bought by another company. The price per share is $88. How much was paid for the oil company?

138. An earth satellite travels 27,560 miles in making 1 revolution. The satellite makes 16 revolutions in 1 day. How many miles does the satellite travel in 30 days?

SECTION 1.5 Division of Whole Numbers

Objective A To divide by a single digit with no remainder in the quotient

Division is used to separate objects into equal groups.

A store manager wants to distribute 24 new objects equally on 4 shelves. From the diagram, we see that the manager would place 6 objects on each shelf.

The manager's division problem can be written:

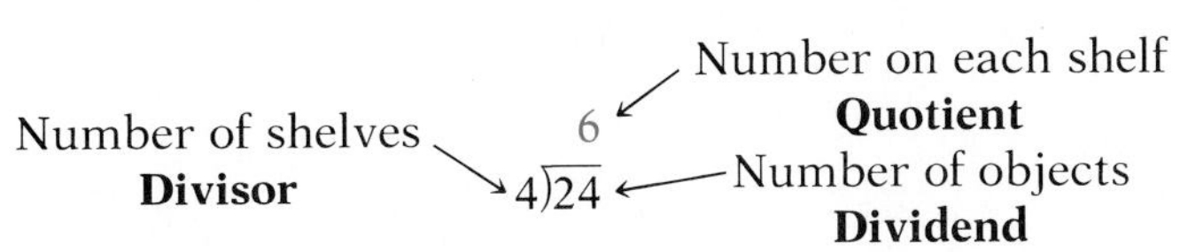

Notice that the quotient multiplied by the divisor equals the dividend.

$4\overline{)24}$ with quotient 6	because	6 Quotient	×	4 Divisor	=	24 Dividend
$9\overline{)54}$ with quotient 6	because	6	×	9	=	54
$8\overline{)40}$ with quotient 5	because	5	×	8	=	40

Here are some important quotients and the properties of 0 in division:

Important Quotients

Any whole number, except zero, divided by itself is 1. $\quad 8\overline{)8} = 1 \quad 14\overline{)14} = 1 \quad 10\overline{)10} = 1$

Any whole number divided by 1 is the whole number. $\quad 1\overline{)9} = 9 \quad 1\overline{)27} = 27 \quad 1\overline{)10} = 10$

Properties of Zero in Division

Zero divided by any other whole number is zero. $\quad 7\overline{)0} = 0 \quad 13\overline{)0} = 0 \quad 10\overline{)0} = 0$

Division by zero is not allowed. $\quad 0\overline{)8} = \boxed{?}$ There is no number whose product with 0 is 8.

When the dividend is a larger whole number, the digits in the quotient are found in steps.

Divide $4\overline{)3192}$ and check.

$$\begin{array}{r} 7 \\ 4\overline{)\,3192} \\ -28 \\ \hline 39 \end{array}$$

Think $4\overline{)31}$.
Subtract 7 × 4.
Bring down the 9.

$$\begin{array}{r} 79 \\ 4\overline{)\,3192} \\ -28 \\ \hline 39 \\ -36 \\ \hline 32 \end{array}$$

Think $4\overline{)39}$.
Subtract 9 × 4.
Bring down the 2.

$$\begin{array}{r} 798 \\ 4\overline{)\,3192} \\ -28 \\ \hline 39 \\ -36 \\ \hline 32 \\ -32 \\ \hline 0 \end{array}$$

Think $4\overline{)32}$.
Subtract 8 × 4.

Check:
$$\begin{array}{r} 798 \\ \times\ \ 4 \\ \hline 3192 \end{array}$$

The place-value chart can be used to show why this method works.

HUNDREDS TENS ONES

$$\begin{array}{rl} 7\ 9\ 8 & \\ 4\overline{)\,3\ 1\ 9\ 2} & \\ -2\ 8\ 0\ 0 & \text{7 hundreds} \times 4 \\ \hline 3\ 9\ 2 & \\ -3\ 6\ 0 & \text{9 tens} \times 4 \\ \hline 3\ 2 & \\ -3\ 2 & \text{8 ones} \times 4 \\ \hline 0 & \end{array}$$

Division is also expressed by using the symbols " ÷ " or "—", which are read "divided by."

54 divided by 9 equals 6.
54 ÷ 9 equals 6.
$\frac{54}{9}$ equals 6.

Example 1 Divide $7\overline{)56}$ and check.

Solution

$$\begin{array}{r} 8 \\ 7\overline{)56} \end{array}$$

Check: $8 \times 7 = 56$

Example 2 Divide $9\overline{)63}$ and check.

Your solution

Example 3 Divide $2808 \div 8$ and check.

Solution

$$\begin{array}{r} 351 \\ 8\overline{)\ 2808} \\ \underline{-24} \\ 40 \\ \underline{-40} \\ 08 \\ \underline{-\ 8} \\ 0 \end{array}$$

Check: $351 \times 8 = 2808$

Example 4 Divide $4077 \div 9$ and check.

Your solution

Example 5 Divide $7\overline{)2856}$ and check.

Solution

$$\begin{array}{r} 408 \\ 7\overline{)\ 2856} \\ \underline{-28} \\ 05 \\ \underline{-\ 0} \\ 56 \\ \underline{-56} \\ 0 \end{array}$$

Think $7\overline{)5}$.
Place 0 in quotient.
Subtract 0×7.
Bring down the 6.

Check: $408 \times 7 = 2856$

Example 6 Divide $9\overline{)6345}$ and check.

Your solution

Solutions on p. A9

Objective B

To divide by a single digit with a remainder in the quotient

Occasionally it is not possible to separate objects into a whole number of equal groups.

A warehouse clerk must place 14 objects into 3 boxes. From the diagram, we see that the clerk would place 4 objects in each box and have 2 objects left over. The 2 is called the **remainder.**

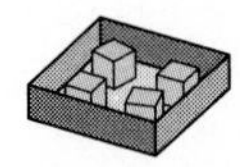
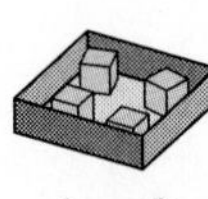
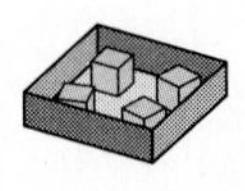

The clerk's division problem could be written:

```
                              ↙ Number in each box
                         4        Quotient
Number of boxes → 3) 14 ← Total number of objects
    Divisor             −12       Dividend
                          2 ← Number left over
                                  Remainder
```

The answer to a division problem with a remainder is frequently written:

$$3\overline{)14}^{\;4\text{ r2}}$$

Notice that $\boxed{\underset{\text{Quotient}}{4} \times \underset{\text{Divisor}}{3}} + \boxed{\underset{\text{Remainder}}{2}} = \boxed{\underset{\text{Dividend}}{14}}$

Example 7 Divide $4\overline{)2522}$ and check.

Solution

```
     630 r2
4) 2522
  −24
    12
   −12
    02     Think 4)2.
   − 0     Place 0 in
     2     quotient.
           Subtract 0 × 4.
```

Check: (630 × 4) + 2 =
2520 + 2 = 2522

Example 8 Divide $6\overline{)5225}$ and check.

Your solution

Example 9 Divide $9\overline{)27,438}$ and check.

Solution

```
      3,048 r6
9) 27,438
  −27
    04       Think 9)4.
   − 0       Subtract
    43       0 × 4.
   −36
     78
    −72
      6
```

Check: (3048 × 9) + 6 =
27,432 + 6 = 27,438

Example 10 Divide $7\overline{)21,409}$ and check.

Your solution

Solutions on p. A9

Objective C To divide by larger whole numbers

When the divisor has more than one digit, estimate at each step by using the first digit of the divisor. If that product is too large, lower the guess by 1 and try again.

Divide $34\overline{)1598}$ and check.

$$\begin{array}{r} 5 \\ 34\overline{)\,1598} \\ -170 \end{array}$$

Think $3\overline{)15}$.
Subtract 5 × 34.

170 is too large. Lower guess by 1 and try again.

$$\begin{array}{r} 4 \\ 34\overline{)\,1598} \\ -136 \\ \hline 238 \end{array}$$

Subtract 4 × 34.

$$\begin{array}{r} 47 \\ 34\overline{)\,1598} \\ -136 \\ \hline 238 \\ -238 \\ \hline 0 \end{array}$$

Think $3\overline{)23}$.
Subtract 7 × 34.

Check:
$$\begin{array}{r} 47 \\ \times 34 \\ \hline 188 \\ 141 \\ \hline 1598 \end{array}$$

Example 11 Divide 7077 ÷ 34 and check.

Solution

$$\begin{array}{r} 208 \text{ r5} \\ 34\overline{)\,7077} \\ -68 \\ \hline 27 \\ -\ 0 \\ \hline 277 \\ -272 \\ \hline 5 \end{array}$$

Think $34\overline{)27}$.
Place 0 in the quotient.
Subtract 0 × 34.

Check: (208 × 34) + 5 =
7072 + 5 = 7077

Example 12 Divide 4578 ÷ 42 and check.

Your solution

Solution on p. A9

Example 13 Divide 21,312 ÷ 56 and check.

Solution

```
     380 r32
56)21,312        Think 5)21.
  -16 8          4 × 56 is too
    4 51         large. Try 3.
   -4 48
       32
     -  0
       32
```

Check: (380 × 56) + 32 =
21,280 + 32 = 21,312

Example 14 Divide 18,359 ÷ 39 and check.

Your solution

Example 15 Divide 427)24,782 and check.

Solution

```
       58 r16
427)24,782
   -21 35
     3 432
    -3 416
        16
```

Check: (58 × 427) + 16 =
24,766 + 16 = 24,782

Example 16 Divide 534)33,219 and check.

Your solution

Example 17 Divide 386)206,149 and check.

Solution

```
        534 r25
386)206,149
   -193 0
     13 14
    -11 58
      1 569
     -1 544
         25
```

Check: (534 × 386) + 25 =
206,124 + 25 = 206,149

Example 18 Divide 515)216,848 and check.

Your solution

Solutions on p. A10

ESTIMATION

Estimating the Quotient of Two Whole Numbers

Estimate and then use your calculator to find 36,936 ÷ 54.

To estimate a quotient, round each number so that all the digits are zero except the first digit. Then divide.

$36{,}936 \div 54 \approx$
$40{,}000 \div 50 = 800$

The estimated answer is 800.

Now use your calculator to find the exact answer.

36 936 [÷] 54 [=] 684

The exact answer is 684.

Objective D To solve application problems

Example 19

A wheat farmer ships 192,600 bushels of wheat in 9 railroad cars. Find the amount of wheat shipped in each car.

Strategy

To find the amount of wheat shipped in each car, divide the number of bushels (192,600) by the number of cars (9).

Solution

$$\begin{array}{r} 21{,}400 \\ 9\overline{)192{,}600} \\ \underline{-18}\phantom{0{,}600} \\ 12\phantom{{,}600} \\ \underline{-\ 9}\phantom{{,}600} \\ 3\ 6 \\ \underline{-3\ 6} \\ 0 \end{array}$$

Each car carried 21,400 bushels of wheat.

Example 20

A retail tire store can stock 270 tires on 15 shelves. How many tires can be stored on each shelf?

Your strategy

Your solution

Solution on p. A10

Example 21

The car you are buying costs $11,216. A down payment of $2000 is required. The remaining balance is paid in 48 equal monthly payments. What is the monthly car payment?

Strategy

To find the monthly payment:

- Find the remaining balance by subtracting the down payment ($2000) from the total cost of the car ($11,216).
- Divide the remaining balance by the number of equal monthly payments (48).

Solution

$$\begin{array}{r} \$11{,}216 \\ -\ \ 2{,}000 \\ \hline \$\ 9{,}216 \end{array} \text{ remaining balance} \qquad \begin{array}{r} \$\ 192 \\ 48\overline{)\$9216} \\ \underline{-48} \\ 441 \\ \underline{-432} \\ 96 \\ \underline{-96} \\ 0 \end{array}$$

The monthly payment is $192.

Example 22

A soft-drink manufacturer produces 12,600 cans of soft drink each hour. Cans are packed 24 to a case. How many cases of soft drink are produced in 8 hours?

Your strategy

Your solution

Solution on p. A10

1.5 EXERCISES

Objective A

Divide.

1. $4\overline{)8}$
2. $3\overline{)9}$
3. $6\overline{)36}$
4. $9\overline{)81}$
5. $5\overline{)25}$
6. $7\overline{)49}$
7. $5\overline{)80}$
8. $6\overline{)96}$
9. $6\overline{)480}$
10. $9\overline{)630}$
11. $4\overline{)840}$
12. $3\overline{)690}$
13. $7\overline{)308}$
14. $7\overline{)203}$
15. $7\overline{)1442}$
16. $9\overline{)6327}$
17. $4\overline{)2120}$
18. $8\overline{)7280}$
19. $9\overline{)8118}$
20. $8\overline{)7264}$
21. $9\overline{)3510}$
22. $7\overline{)6020}$
23. $5\overline{)5280}$
24. $6\overline{)7218}$
25. $6\overline{)15,642}$
26. $6\overline{)19,254}$
27. $4\overline{)39,200}$
28. $3\overline{)16,200}$
29. $3\overline{)64,680}$
30. $4\overline{)50,760}$
31. $6\overline{)21,480}$
32. $5\overline{)18,050}$

Objective B

Divide.

33. $4\overline{)9}$
34. $2\overline{)7}$
35. $5\overline{)27}$
36. $9\overline{)88}$
37. $3\overline{)40}$
38. $6\overline{)97}$
39. $8\overline{)83}$
40. $5\overline{)54}$
41. $7\overline{)632}$
42. $4\overline{)363}$

Divide.

43. $4\overline{)921}$ **44.** $7\overline{)845}$ **45.** $8\overline{)1635}$ **46.** $5\overline{)1548}$ **47.** $7\overline{)9432}$

48. $6\overline{)6352}$ **49.** $9\overline{)7004}$ **50.** $7\overline{)6001}$ **51.** $6\overline{)2350}$ **52.** $8\overline{)6370}$

53. $3\overline{)7252}$ **54.** $4\overline{)6538}$ **55.** $9\overline{)4157}$ **56.** $7\overline{)6172}$

57. $7\overline{)8124}$ **58.** $3\overline{)5162}$ **59.** $5\overline{)3542}$ **60.** $8\overline{)3274}$

61. $4\overline{)15,300}$ **62.** $7\overline{)43,500}$ **63.** $8\overline{)72,354}$ **64.** $5\overline{)43,542}$

Objective C

Divide.

65. $27\overline{)96}$ **66.** $44\overline{)82}$ **67.** $42\overline{)87}$ **68.** $67\overline{)93}$

69. $41\overline{)897}$ **70.** $32\overline{)693}$ **71.** $23\overline{)784}$ **72.** $25\overline{)772}$

73. $74\overline{)600}$ **74.** $92\overline{)500}$ **75.** $70\overline{)329}$ **76.** $50\overline{)467}$

77. $36\overline{)7225}$ **78.** $44\overline{)8821}$ **79.** $19\overline{)3859}$ **80.** $32\overline{)9697}$

81. $23\overline{)4257}$ **82.** $62\overline{)9143}$ **83.** $92\overline{)6234}$ **84.** $87\overline{)4911}$

Divide.

85. $14\overline{)8422}$

86. $42\overline{)4497}$

87. $26\overline{)5209}$

88. $37\overline{)7403}$

89. $88\overline{)3127}$

90. $92\overline{)6177}$

91. $33\overline{)8943}$

92. $27\overline{)4765}$

93. $22\overline{)98{,}654}$

94. $77\overline{)83{,}629}$

95. $64\overline{)38{,}912}$

96. $78\overline{)31{,}434}$

97. $206\overline{)3097}$

98. $504\overline{)6504}$

99. $654\overline{)1217}$

100. $546\overline{)2344}$

101. $169\overline{)8542}$

102. $456\overline{)7723}$

103. $223\overline{)8927}$

104. $467\overline{)9344}$

Estimate and then use your calculator to divide.

105. $76\overline{)389{,}804}$

106. $53\overline{)117{,}925}$

107. $29\overline{)637{,}072}$

108. $67\overline{)738{,}072}$

109. $38\overline{)934{,}648}$

110. $34\overline{)906{,}304}$

111. $309\overline{)876{,}324}$

112. $642\overline{)323{,}568}$

113. $209\overline{)632{,}016}$

114. $614\overline{)332{,}174}$

115. $179\overline{)5{,}734{,}444}$

116. $374\overline{)7{,}712{,}254}$

117. $492\overline{)40{,}836}$

118. $609\overline{)56{,}028}$

119. $396\overline{)26{,}928}$

120. $234\overline{)69{,}030}$

Objective D *Application Problems*

121. Four service organizations collected \$548,000 to promote and provide community services. What amount did each organization receive if the money was divided evenly?

122. An insurance agent used 128 gallons of gas in traveling 3456 miles. Find the number of miles traveled on each gallon of gas.

123. A computer can store 2,211,840 bytes of information on six disks. How many bytes of information can be stored on one disk?

124. A flight engineer has an annual income of \$41,496. Find the engineer's monthly income.

125. A computer analyst doing consulting work received \$5376 for working 168 hours on a project. Find the hourly rate the consultant charged.

126. The total cost of a living room set, including finance charges, is \$2856. This amount is to be repaid in 24 equal payments. Find the amount of each payment.

127. A family makes a down payment of \$1560 on a car costing \$10,536.
   a. What is the remaining balance to be paid?
   b. The balance is to be paid in 48 equal monthly payments. Find the monthly payment.

128. A farmer harvested 39,544 heads of lettuce from one plot of land and 38,456 from a second plot. The lettuce was packed in boxes with 24 heads in each box.
   a. What is the total number of heads of lettuce harvested from both plots?
   b. How many boxes were needed to pack the lettuce from the two plots?

129. An organization was hired to raise funds for five charities. Of the amount collected, \$35,000 was used to pay expenses. The remaining amount of the \$352,765 collected was divided equally among the five charities. What amount did each charity receive?

130. A manufacturer produces and boxes 240 computer keyboards each hour. Four keyboards are put in each box for shipment. How many boxes of computer keyboards can be produced in 12 hours?

131. A large corporation has a life and medical insurance plan that costs the corporation \$158 per month for each employee. The monthly cost of the insurance plan is \$666,128. How many employees participate in the insurance plan?

132. A limited partnership consisting of 2750 investors sold a piece of land for \$5,959,250. Find the amount each investor received.

SECTION 1.6 Exponential Notation and the Order of Operations Agreement

Objective A To simplify expressions containing exponents

Repeated multiplication of the same factor can be written two ways.

$3 \cdot 3 \cdot 3 \cdot 3 \cdot 3$ or $3^5 \leftarrow$ **exponent**

The exponent indicates how many times the factor occurs in the multiplication. The expression 3^5 is in **exponential notation.**

It is important to be able to read numbers written in exponential notation.

$6 = 6^1$ read "six to the first power" or just "six." Usually the exponent 1 is not written.

$6 \cdot 6 = 6^2$ read "six squared" or "six to the second power."

$6 \cdot 6 \cdot 6 = 6^3$ read "six cubed" or "six to the third power."

$6 \cdot 6 \cdot 6 \cdot 6 = 6^4$ read "six to the fourth power."

$6 \cdot 6 \cdot 6 \cdot 6 \cdot 6 = 6^5$ read "six to the fifth power."

Each place value in the place-value chart can be expressed as a power of 10.

Ten = 10 = 10 = 10^1

Hundred = 100 = $10 \cdot 10$ = 10^2

Thousand = 1000 = $10 \cdot 10 \cdot 10$ = 10^3

Ten-Thousand = 10,000 = $10 \cdot 10 \cdot 10 \cdot 10$ = 10^4

Hundred-Thousand = 100,000 = $10 \cdot 10 \cdot 10 \cdot 10 \cdot 10$ = 10^5

Million = 1,000,000 = $10 \cdot 10 \cdot 10 \cdot 10 \cdot 10 \cdot 10$ = 10^6

To simplify a numerical expression containing exponents, write each factor as many times as indicated by the exponent and carry out the indicated multiplication.

$4^3 = 4 \cdot 4 \cdot 4 = 64$

$2^2 \cdot 3^4 = (2 \cdot 2) \cdot (3 \cdot 3 \cdot 3 \cdot 3) = 4 \cdot 81 = 324$

Example 1 Write $3 \cdot 3 \cdot 3 \cdot 5 \cdot 5$ in exponential notation.

Solution $3 \cdot 3 \cdot 3 \cdot 5 \cdot 5 = 3^3 \cdot 5^2$

Example 2 Write $2 \cdot 2 \cdot 2 \cdot 2 \cdot 3 \cdot 3 \cdot 3$ in exponential notation.

Your solution

Example 3 Write $10 \cdot 10 \cdot 10 \cdot 10$ as a power of 10.

Solution $10 \cdot 10 \cdot 10 \cdot 10 = 10^4$

Example 4 Write $10 \cdot 10 \cdot 10 \cdot 10 \cdot 10 \cdot 10 \cdot 10$ as a power of 10.

Your solution

Example 5 Simplify $3^2 \cdot 5^3$

Solution $3^2 \cdot 5^3 = (3 \cdot 3) \cdot (5 \cdot 5 \cdot 5)$
$= 9 \cdot 125 = 1125$

Example 6 Simplify $2^3 \cdot 5^2$.

Your solution

Solutions on p. A11

Objective B

To use the Order of Operations Agreement to simplify expressions

More than one operation may occur in a numerical expression. The answer may be different, depending on the order in which the operations are performed. For example, consider $3 + 4 \times 5$.

Multiply first, then add.

$$3 + \underbrace{4 \times 5}$$
$$\underbrace{3 + 20}$$
$$23$$

Add first, then multiply.

$$\underbrace{3 + 4} \times 5$$
$$\underbrace{7 \times 5}$$
$$35$$

An Order of Operations Agreement is used so that only one answer is possible.

The Order of Operations Agreement:

Step 1 Do all operations inside parentheses.
Step 2 Simplify any number expressions containing exponents.
Step 3 Do multiplication and division as they occur from left to right.
Step 4 Do addition and subtraction as they occur from left to right.

Simplify $3 \times (2 + 1) - 2^2 + 4 \div 2$ by using the Order of Operations Agreement.

$3 \times \underbrace{(2 + 1)} - 2^2 + 4 \div 2$ 1. Perform operations in parentheses.

$3 \times 3 - \underbrace{2^2} + 4 \div 2$ 2. Simplify expressions with exponents.

$\underbrace{3 \times 3} - 4 + 4 \div 2$ 3. Do multiplications and divisions as they occur from left to right.

$9 - 4 + \underbrace{4 \div 2}$

$\underbrace{9 - 4} + 2$ 4. Do additions and subtractions as they occur from left to right.

$\underbrace{5 + 2}$

7

One or more of the above steps may not be needed to simplify an expression. In that case, proceed to the next step in the Order of Operations Agreement.

Simplify $5 + 8 \div 2$. No parentheses or exponents. Proceed to step 3 of the Agreement.

$5 + \underbrace{8 \div 2}$ 3. Do multiplication or division.

$\underbrace{5 + 4}$ 4. Do addition or subtraction.

9

Example 7 Simplify $16 \div (8 - 4) \cdot 9 - 5^2$.

Solution
$16 \div (8 - 4) \cdot 9 - 5^2$
$16 \div 4 \cdot 9 - 5^2$
$16 \div 4 \cdot 9 - 25$
$4 \cdot 9 - 25$
$36 - 25$
11

Example 8 Simplify $5 \cdot (8 - 4) \div 4 - 2$.

Your solution

Solution on p. A11

1.6 EXERCISES

Objective A

Write the number in exponential notation.

1. $2 \cdot 2 \cdot 2$
2. $7 \cdot 7 \cdot 7 \cdot 7 \cdot 7$
3. $6 \cdot 6 \cdot 6 \cdot 7 \cdot 7 \cdot 7 \cdot 7$
4. $3 \cdot 3 \cdot 3 \cdot 5 \cdot 5 \cdot 5 \cdot 5$
5. $6 \cdot 6 \cdot 6 \cdot 11 \cdot 11 \cdot 11 \cdot 11 \cdot 11$
6. $2 \cdot 2 \cdot 2 \cdot 2 \cdot 2 \cdot 2 \cdot 3 \cdot 3 \cdot 3$
7. $3 \cdot 10 \cdot 10 \cdot 10 \cdot 10$
8. $7 \cdot 10 \cdot 10 \cdot 10 \cdot 10 \cdot 10 \cdot 10 \cdot 10$
9. $2 \cdot 2 \cdot 3 \cdot 3 \cdot 3 \cdot 5 \cdot 5 \cdot 5 \cdot 5$
10. $7 \cdot 7 \cdot 11 \cdot 11 \cdot 11 \cdot 13 \cdot 13 \cdot 13 \cdot 13$
11. $2 \cdot 3 \cdot 3 \cdot 7 \cdot 7 \cdot 11$
12. $5 \cdot 7 \cdot 7 \cdot 9 \cdot 9 \cdot 11$
13. $2 \cdot 2 \cdot 7 \cdot 7 \cdot 7 \cdot 7 \cdot 11 \cdot 11 \cdot 11 \cdot 11$
14. $3 \cdot 3 \cdot 7 \cdot 7 \cdot 7 \cdot 7 \cdot 17 \cdot 17 \cdot 17 \cdot 17$

Simplify.

15. 2^3
16. 2^6
17. $2^4 \cdot 5^2$
18. $2^6 \cdot 3^2$
19. $3^2 \cdot 10^2$
20. $2^3 \cdot 10^4$
21. $6^2 \cdot 3^3$
22. $4^3 \cdot 5^2$
23. $5 \cdot 2^3 \cdot 3$
24. $6 \cdot 3^2 \cdot 4$
25. $2^2 \cdot 3^2 \cdot 10$
26. $3^2 \cdot 5^2 \cdot 10$
27. $0^2 \cdot 4^3$
28. $6^2 \cdot 0^3$
29. $3^2 \cdot 10^4$
30. $5^3 \cdot 10^3$
31. $2^2 \cdot 3^3 \cdot 5$
32. $5^2 \cdot 7^3 \cdot 2$
33. $2 \cdot 3^4 \cdot 5^2$
34. $6 \cdot 2^6 \cdot 7^2$
35. $5^2 \cdot 3^2 \cdot 7^2$
36. $4^2 \cdot 9^2 \cdot 6^2$
37. $3^4 \cdot 2^6 \cdot 5$
38. $4^3 \cdot 6^3 \cdot 7$
39. $4^2 \cdot 3^3 \cdot 10^4$
40. $5^2 \cdot 2^3 \cdot 10^3$
41. $6^2 \cdot 4^4 \cdot 10$

Objective B

Simply by using the Order of Operations Agreement.

42. $4 - 2 + 3$

43. $6 - 3 + 2$

44. $6 \div 3 + 2$

45. $8 \div 4 + 8$

46. $6 \cdot 3 + 5$

47. $5 \cdot 9 + 2$

48. $3^2 - 4$

49. $5^2 - 17$

50. $4 \cdot (5 - 3) + 2$

51. $3 + (4 + 2) \div 3$

52. $5 + (8 + 4) \div 6$

53. $8 - 2^2 + 4$

54. $16 \cdot (3 + 2) \div 10$

55. $12 \cdot (1 + 5) \div 12$

56. $10 - 2^3 + 4$

57. $5 \cdot 3^2 + 8$

58. $16 + 4 \cdot 3^2$

59. $12 + 4 \cdot 2^3$

60. $16 + (8 - 3) \cdot 2$

61. $7 + (9 - 5) \cdot 3$

62. $2^2 + 3 \cdot (6 - 2)$

63. $3^3 + 5 \cdot (8 - 6)$

64. $2^2 \cdot 3^2 + 2 \cdot 3$

65. $4 \cdot 6 + 3^2 \cdot 4^2$

66. $16 - 2 \cdot 4$

67. $12 + 3 \cdot 5$

68. $3 \cdot (6 - 2) + 4$

69. $5 \cdot (8 - 4) - 6$

70. $8 - (8 - 2) \div 3$

71. $12 - (12 - 4) \div 4$

72. $8 + 2 - 3 \cdot 2 \div 3$

73. $10 + 1 - 5 \cdot 2 \div 5$

74. $3 \cdot (4 + 2) \div 6$

Some calculators use the Order of Operations Agreement, others do not. Try the following problems with your calculator to see if it uses the Order of Operations Agreement.

75. $2^2 + 8 \div 2$

76. $5^2 + 12 \div 6$

77. $4 \cdot 3 \div 6 + 5$

78. $8 \cdot 2 \div 4 - 3$

79. $3 \cdot (4 - 1) + 3^2$

80. $5 \cdot (7 - 2) - 2^3$

SECTION 1.7 Prime Numbers and Factoring

Objective A

To factor numbers

Whole number factors of a number divide that number evenly (there is no remainder).

1, 2, 3, and 6 are whole number factors of 6 because they divide 6 evenly.

$1\overline{)6}$ quotient 6, $2\overline{)6}$ quotient 3, $3\overline{)6}$ quotient 2, $6\overline{)6}$ quotient 1

Notice that both the divisor and the quotient are factors of the dividend.

To find the factors of a number, try dividing the number by 1, 2, 3, 4, 5, . . . Those numbers that divide the number evenly are its factors. Continue this process until the factors start to repeat.

Find all the factors of 42.

42 ÷ 1 = 42	1 and 42 are factors	
42 ÷ 2 = 21	2 and 21 are factors	
42 ÷ 3 = 14	3 and 14 are factors	
42 ÷ 4	Will not divide evenly	
42 ÷ 5	Will not divide evenly	
42 ÷ 6 = 7	6 and 7 are factors	Factors are repeating; all the factors
42 ÷ 7 = 6	7 and 6 are factors	of 42 have been found.

1, 2, 3, 6, 7, 14, 21, and 42 are factors of 42.

The following rules are helpful in finding the factors of a number.

2 is a factor of a number if the last digit of the number is 0, 2, 4, 6, or 8.

436 ends in 6; therefore, 2 is a factor of 436. (436 ÷ 2 = 218)

3 is a factor of a number if the sum of the digits of the number is divisible by 3.

The sum of the digits of 489 is 4 + 8 + 9 = 21. 21 is divisible by 3. Therefore, 3 is a factor of 489. (489 ÷ 3 = 163)

5 is a factor of a number if the last digit of the number is 0 or 5.

520 ends in 0; therefore, 5 is a factor of 520. (520 ÷ 5 = 104)

Example 1 Find all the factors of 30.

Solution
30 ÷ 1 = 30
30 ÷ 2 = 15
30 ÷ 3 = 10
30 ÷ 4 Will not divide evenly
30 ÷ 5 = 6
30 ÷ 6 = 5

1, 2, 3, 5, 6, 10, 15 and 30 are factors of 30.

Example 2 Find all the factors of 40.

Your solution

Solution on p. A11

Objective B To find the prime factorization of a number

A number is a **prime** number if its only whole number factors are 1 and itself. 7 is prime because its only factors are 1 and 7. If a number is not prime, it is called a **composite** number. Because 6 has factors of 2 and 3, 6 is a composite number. The number 1 is not considered a prime number; therefore it is not included in the following list of prime numbers less than 50.

2, 3, 5, 7, 11, 13, 17, 19, 23, 29, 31, 37, 41, 43, 47

The **prime factorization** of a number is the expression of the number as a product of its prime factors. To find the prime factors of 60, begin with the smallest prime number as a trial divisor and continue with prime numbers as trial divisors until the final quotient is prime. A "T-diagram" is sometimes used as a short version of the procedure.

$$2\overline{)60} = 30 \qquad 2\overline{)30} = 15 \qquad 3\overline{)15} = 5 \text{ Prime}$$

T-diagram Version

60	
2	30
2	15
3	5 Prime

The prime factorization of 60 is $2 \cdot 2 \cdot 3 \cdot 5$.

Finding the prime factorization of larger numbers can be more difficult. Try each prime number as a trial divisor. Stop when the square of the trial divisor is greater than the number being factored.

Find the prime factorization of 106.

$2\overline{)106} = 53$ — 53 cannot be divided evenly by 2, 3, 5, 7, or 11. Prime numbers greater than 11 need not be tested because 11^2 is greater than 53.

The prime factorization of 106 is $2 \cdot 53$.

Example 3 Find the prime factorization of 84.

Solution

$2\overline{)84} = 42$, $2\overline{)42} = 21$, $3\overline{)21} = 7$

$84 = 2 \cdot 2 \cdot 3 \cdot 7$

Example 4 Find the prime factorization of 44.

Your solution

Example 5 Find the prime factorization of 201.

Solution

$3\overline{)201} = 67$ Try only 2, 3, 5, 7, and 11 because $11^2 > 67$.

$201 = 3 \cdot 67$

Example 6 Find the prime factorization of 177.

Your solution

Solutions on p. A11

1.7 EXERCISES

Objective A

Find all the factors of the number.

1. 4 **2.** 6 **3.** 10 **4.** 20 **5.** 5

6. 7 **7.** 12 **8.** 9 **9.** 8 **10.** 16

11. 13 **12.** 17 **13.** 18 **14.** 24 **15.** 25

16. 56 **17.** 36 **18.** 45 **19.** 28 **20.** 32

21. 29 **22.** 33 **23.** 22 **24.** 26 **25.** 44

26. 52 **27.** 49 **28.** 82 **29.** 37 **30.** 79

31. 57 **32.** 69 **33.** 48 **34.** 64 **35.** 87

36. 95 **37.** 46 **38.** 54 **39.** 50 **40.** 75

41. 66 **42.** 77 **43.** 80 **44.** 100 **45.** 84

46. 96 **47.** 85 **48.** 90 **49.** 101 **50.** 123

Find all the factors of the number.

51. 120 **52.** 150 **53.** 125 **54.** 140 **55.** 132

56. 144 **57.** 135 **58.** 180 **59.** 189 **60.** 168

Objective B

Find the prime factorization.

61. 6 **62.** 14 **63.** 17 **64.** 83 **65.** 16

66. 24 **67.** 12 **68.** 27 **69.** 9 **70.** 15

71. 36 **72.** 40 **73.** 19 **74.** 37 **75.** 50

76. 90 **77.** 65 **78.** 115 **79.** 80 **80.** 87

81. 18 **82.** 26 **83.** 28 **84.** 49 **85.** 29

86. 31 **87.** 42 **88.** 62 **89.** 81 **90.** 51

91. 22 **92.** 39 **93.** 101 **94.** 89 **95.** 70

96. 66 **97.** 86 **98.** 74 **99.** 95 **100.** 105

101. 67 **102.** 78 **103.** 55 **104.** 46 **105.** 122

106. 130 **107.** 118 **108.** 102 **109.** 150 **110.** 125

111. 120 **112.** 144 **113.** 160 **114.** 175 **115.** 343

116. 216 **117.** 400 **118.** 625 **119.** 225 **120.** 180

Calculators and Computers

Order of Operations Does your calculator use the Order of Operations Agreement? To find out, try this problem:

$$2 + 4 \cdot 7$$

If your answer is 30, then the calculator uses the Order of Operations Agreement. If your answer is 42, it does not use that agreement.

Even if your calculator does not use the Order of Operations Agreement, you can still correctly evaluate numerical expressions. The parentheses keys, [( ] and [ )], are used for this purpose.

Remember that $2 + 4 \cdot 7$ means $2 + (4 \cdot 7)$ because the multiplication is completed before addition. To evaluate this expression, enter the following:

Enter	*Display*	*Comments*
2	2.	
[+]	2.	
[( ]	(	Not all scientific calculators display the "(" when it is entered.
4	4.	
[×]	4.	
7	7.	
[ )]	28.	The product inside the parentheses.
[=]	30.	The answer to $2 + (4 \times 7)$.

When using your calculator to evaluate numerical expressions, insert parentheses around multiplications or divisions. This has the effect of forcing the calculator to do the operations in the order you want rather than the order the calculator wants.

Evaluate $3 \cdot (15 - 2 \cdot 3) - 36 \div 3$.

Press the keys shown in the box. $3 \times (15 - (2 \times 3)) - (36 \div 3) =$

The answer 15 should be in the display.

Notice that parentheses were used around $15 - 2 \cdot 3$ and around $2 \cdot 3$. The parentheses are important to ensure that multiplication is performed before additions or subtractions from the innermost grouping symbols to the outermost grouping symbols.

Chapter Summary

Key Words The *whole numbers* are 0, 1, 2, 3, 4, 5, 6, 7, 8, 9, 10, . . .

The symbol for *"is less than"* is $<$.

The symbol for *"is greater than"* is $>$.

The position of a digit in a number determines the digit's *place value.*

Giving an approximate value for an exact number is called *"rounding."*

An expression of the form 4^2 is in *exponential notation,* where 4 is the *base* and 2 is the *exponent.*

A number is *prime* if its only whole number factors are 1 and itself.

The *prime factorization* of a number is the expression of the number as a product of its prime factors.

Essential Rules

The Addition Property of Zero — Zero added to a number does not change the number.
5 + 0 = 0 + 5 = 5

The Commutative Property of Addition — Two numbers can be added in either order.
6 + 5 = 5 + 6 = 11

The Associative Property of Addition — Grouping the addition in any order gives the same result.
(2 + 3) + 5 = 2 + (3 + 5) = 10

The Multiplication Property of Zero — The product of a number and zero is zero.
0 × 3 = 3 × 0 = 0

The Multiplication Property of 1 — The product of a number and 1 is the number.
1 × 8 = 8 × 1 = 8

The Commutative Property of Multiplication — Two numbers can be multiplied in any order.
5 × 2 = 2 × 5 = 10

The Associative Property of Multiplication — Grouping multiplication in any order gives the same result.
(3 × 2) × 5 = 3 × (2 × 5) = 30

The Properties of Zero in Division — Zero divided by any other number is zero. Division by zero is not allowed.

Order of Operations Agreement

Step 1 Perform operations inside grouping symbols.
Step 2 Simplify expressions with exponents.
Step 3 Do multiplications and divisions as they occur from left to right.
Step 4 Do additions and subtractions as they occur from left to right.

Chapter Review

SECTION 1.1

1. Place the correct symbol, $<$ or $>$, between the two numbers. 101 87

2. Write 276,057 in words.

3. Write two million eleven thousand forty-four in standard form.

4. Write 10,327 in expanded form.

SECTION 1.2

5. Add: $\begin{array}{r} 298 \\ 461 \\ +322 \\ \hline \end{array}$

6. Add: $\begin{array}{r} 5894 \\ 6301 \\ +\ 298 \\ \hline \end{array}$

7. An insurance account executive receives commissions of $723, $544, $812, and $488 during a 4-week period. Find the total income from commissions for the 4 weeks.

8. You had a balance of $516 in your checking account before making deposits of $88 and $213. Find the total amount deposited, and determine your new checking balance.

SECTION 1.3

9. Subtract: $\begin{array}{r} 4926 \\ -3177 \\ \hline \end{array}$

10. Subtract: $\begin{array}{r} 10{,}134 \\ -\ 4{,}725 \\ \hline \end{array}$

11. After a trip of 972 miles, the odometer of your car read 52,031 miles. What was the odometer reading at the beginning of the trip?

12. Your monthly budget for household expenses is $600. After you spend $137 for food and $67 for utilities, how much is left in your budget?

SECTION 1.4

13. Multiply: $\begin{array}{r} 843 \\ \times\ 27 \\ \hline \end{array}$

14. Multiply: $\begin{array}{r} 2{,}019 \\ \times\ 307 \\ \hline \end{array}$

15. You have a car payment of \$123 per month. What is the total of the car payments over a 12-month period?

16. A sales assistant earns \$240 for working a 40-hour week. Last week the assistant worked an additional 12 hours at \$12 an hour. Find the assistant's total pay for last week's work.

SECTION 1.5

17. Divide: $7\overline{)14{,}945}$

18. Divide: $84\overline{)109{,}763}$

19. A sales executive drove a car 351 miles on 13 gallons of gas. Find the number of miles driven per gallon of gasoline.

20. A car is purchased for \$8940, with a down payment of \$1500. The balance is paid in 48 equal monthly payments. Find the monthly car payment.

SECTION 1.6

21. Write $2 \cdot 2 \cdot 2 \cdot 2 \cdot 5 \cdot 5 \cdot 5$ in exponential notation.

22. Write $5 \cdot 5 \cdot 7 \cdot 7 \cdot 7 \cdot 7 \cdot 7$ in exponential notation.

23. Simplify $3 \cdot 2^3 \cdot 5^2$.

24. Simplify $2^3 - 3 \cdot 2$.

25. Simplify $3^2 + 2^2 \cdot (5 - 3)$

26. Simplify $8 \cdot (6 - 2) \div 4$.

SECTION 1.7

27. Find all the factors of 18.

28. Find all the factors of 30.

29. Find the prime factorization of 42.

30. Find the prime factorization of 72.

Chapter Test

1. Place the correct symbol, < or >, between the two numbers. 21 19

2. Write 207,068 in words.

3. Write one million two hundred four thousand six in standard form.

4. Write 906,378 in expanded form.

5. Round 74,965 to the nearest hundred.

6. Add: 25,492 + 71,306

7. Add: 89,756 + 9,094 + 37,065

8. A family drives 425 miles the first day, 187 miles the second day, and 243 miles the third day of their vacation. The odometer read 47,626 miles at the start of the vacation.
   a. How many miles were driven during the 3 days?
   b. What is the odometer reading at the end of the 3 days?

9. Subtract: 17,495 − 8,162

10. Subtract: 20,736 − 9,854

11. A flight attendant purchases a car costing $16,405 and makes a down payment of $2450. What is the remaining balance to be paid?

12. Multiply: 90,763 × 8

13. Multiply: 9736 × 704

14. An investor receives $237 each month from a corporate bond fund. How much will the investor receive over a 12-month period?

15. Divide: $8\overline{)5624}$

16. Divide: $7\overline{)60,972}$

17. Divide: $97\overline{)108,764}$

18. A farmer harvested 48,290 pounds of lemons from one grove and 23,710 pounds of lemons from another grove. The lemons were packed in boxes with 24 pounds in each box. How many boxes were needed to pack the lemons?

19. Write $3 \cdot 3 \cdot 3 \cdot 7 \cdot 7$ in exponential notation.

20. Simplify $3^3 \cdot 4^2$.

21. Simplify $4^2 \cdot (4 - 2) \div 8 + 5$.

22. Simplify $16 \div 4 \times 2 - (6 - 5)^3$

23. Find all the factors of 20.

24. Find the prime factorization of 84.

25. Find the prime factorization of 102.

2 Fractions

OBJECTIVES

- To find the least common multiple (LCM)
- To find the greatest common factor (GCF)
- To write a fraction that represents part of a whole
- To write an improper fraction as a mixed number or a whole number, and a mixed number as an improper fraction
- To find equivalent fractions by raising to higher terms
- To write a fraction in simplest form
- To add fractions with the same denominator
- To add fractions with unlike denominators
- To add whole numbers, mixed numbers, and fractions
- To solve application problems
- To subtract fractions with the same denominator
- To subtract fractions with unlike denominators
- To subtract whole numbers, mixed numbers, and fractions
- To solve application problems
- To multiply fractions
- To multiply whole numbers, mixed numbers, and fractions
- To solve application problems
- To divide fractions
- To divide whole numbers, mixed numbers, and fractions
- To solve application problems
- To identify the order relation between two fractions
- To simplify expressions containing exponents
- To use the Order of Operations Agreement to simplify expressions

Egyptian Fractions

The Rhind papyrus is one of the earliest written accounts of mathematics.[1] In the papyrus, a scribe named Ahmes gives an early account of the concept of fractions. A portion of the Rhind papyrus is rendered below with its heiroglyphic transcription.

The early Egyptian primarily used unit fractions. This is a fraction in which the numerator is a 1. To write a fraction, a small oval was placed above a series of lines. The number of lines indicated the denominator. Some examples of these fractions are

(oval over four lines) $= \frac{1}{4}$ (special symbol) $= \frac{1}{2}$

In the first example, each line represents a 1. Because there are 4 lines, the fraction is $\frac{1}{4}$.

The second example is the special symbol that was used for the fraction $\frac{1}{2}$.

[1]Papyrus comes from the stem of a plant. The stem was dried and then pounded thin. The resulting material served as a primitive type of paper.

SECTION 2.1 The Least Common Multiple and Greatest Common Factor

Objective A To find the least common multiple (LCM)

The **multiples** of a number are the products of that number and the numbers 1, 2, 3, 4, 5, . . .

$3 \times 1 = 3$
$3 \times 2 = 6$
$3 \times 3 = 9$
$3 \times 4 = 12$ The multiples of 3 are 3, 6, 9, 12, 15, . . .
$3 \times 5 = 15$
⋮

A number that is a multiple of two or more numbers is a **common multiple** of those numbers.

The multiples of 4 are: 4, 8, 12, 16, 20, 24, 28, 32, 36, . . .
The multiples of 6 are: 6, 12, 18, 24, 30, 36, 42, . . .
Some common multiples of 4 and 6 are: 12, 24, and 36.

The **least common multiple** (LCM) is the smallest common multiple of two or more numbers.

The least common multiple of 4 and 6 is 12.

Listing the multiples of each number is one way to find the LCM. Another way to find the LCM uses the prime factorization of each number.

To find the LCM of 450 and 600, find the prime factorization of each number and write the factorization of each number in a table. Circle the largest product in each column. The LCM is the product of the circled numbers.

	2	3	5
450 =	2	(3 · 3)	5 · 5
600 =	(2 · 2 · 2)	3	(5 · 5)

In the column headed by 5, the products are equal. Circle just one product.

The LCM is the product of the circled numbers.
The LCM = $2 \cdot 2 \cdot 2 \cdot 3 \cdot 3 \cdot 5 \cdot 5 = 1800$.

Example 1 Find the LCM of 24, 36, and 50.

Solution

	2	3	5
24 =	(2 · 2 · 2)	3	
36 =	2 · 2	(3 · 3)	
50 =	2		(5 · 5)

The LCM = $2 \cdot 2 \cdot 2 \cdot 3 \cdot 3 \cdot 5 \cdot 5 = 1800$.

Example 2 Find the LCM of 50, 84, and 135.

Your solution

Solution on p. A11

Objective B — To find the greatest common factor (GCF)

Recall that a number that divides another number evenly is a factor of that number. 64 can be evenly divided by 1, 2, 4, 8, 16, 32, and 64. 1, 2, 4, 8, 16, 32, and 64 are factors of 64.

A number that is a factor of two or more numbers is a **common factor** of those numbers.

The factors of 30 are: 1, 2, 3, 5, 6, 10, 15, 30.
The factors of 105 are: 1, 3, 5, 7, 15, 21, 35, 105.
The common factors of 30 and 105 are: 1, 3, 5, and 15.

The **greatest common factor** (GCF) is the largest common factor of two or more numbers.

The greatest common factor of 30 and 105 is 15.

Listing the factors of each number is one way of finding the GCF. Another way to find the GCF uses the prime factorization of each number.

To find the GCF of 126 and 180, find the prime factorization of each number and write the factorization of each number in a table. Circle the smallest product in each column that does not have a blank. The GCF is the product of the circled numbers.

	2	3	5	7
126 =	(2)	(3 · 3)		7
180 =	2 · 2	3 · 3	5	

In the column headed by 3, the products are equal. Circle just one product.
Columns 5 and 7 have a blank, so 5 and 7 are not common factors of 126 and 180. Do not circle any number in these columns.

The GCF is the product of the circled numbers.
The GCF = $2 \cdot 3 \cdot 3 = 18$.

Example 3 Find the GCF of 90, 168, and 420.

Solution

	2	3	5	7
90 =	(2)	3 · 3	5	
168 =	2 · 2 · 2	(3)		7
420 =	2 · 2	3	5	7

The GCF = $2 \cdot 3 = 6$.

Example 4 Find the GCF of 36, 60, and 72.

Your solution

Example 5 Find the GCF of 7, 12, and 20.

Solution

	2	3	5	7
7 =				7
12 =	2 · 2	3		
20 =	2 · 2		5	

Since no numbers are circled, the GCF = 1.

Example 6 Find the GCF of 11, 24, and 30.

Your solution

Solutions on p. A11

2.1 EXERCISES

Objective A

Find the LCM.

1. 5, 8
2. 3, 6
3. 3, 8
4. 2, 5
5. 5, 6
6. 5, 7
7. 4, 6
8. 6, 8
9. 8, 12
10. 12, 16
11. 5, 12
12. 3, 16
13. 8, 14
14. 6, 18
15. 3, 9
16. 4, 10
17. 8, 32
18. 7, 21
19. 9, 36
20. 14, 42
21. 44, 60
22. 120, 160
23. 102, 184
24. 123, 234
25. 4, 8, 12
26. 5, 10, 15
27. 3, 5, 10
28. 2, 5, 8
29. 3, 8, 12
30. 5, 12, 18
31. 9, 36, 64
32. 18, 54, 63
33. 16, 30, 84
34. 9, 12, 15
35. 12, 18, 24
36. 30, 60, 80
37. 6, 9, 15
38. 8, 18, 24
39. 12, 15, 25
40. 13, 26, 39
41. 10, 15, 40
42. 12, 48, 72

Objective B

Find the GCF.

43. 3, 5 **44.** 5, 7 **45.** 6, 9 **46.** 18, 24 **47.** 15, 25

48. 14, 49 **49.** 25, 100 **50.** 16, 80 **51.** 32, 51 **52.** 21, 44

53. 12, 80 **54.** 8, 36 **55.** 16, 140 **56.** 12, 76 **57.** 8, 14

58. 24, 30 **59.** 48, 144 **60.** 44, 96 **61.** 18, 32 **62.** 16, 30

63. 40, 64 **64.** 24, 140 **65.** 60, 82 **66.** 40, 68 **67.** 2, 5, 7

68. 3, 5, 11 **69.** 6, 8, 10 **70.** 7, 14, 49 **71.** 6, 15, 36 **72.** 8, 12, 16

73. 10, 15, 20 **74.** 12, 18, 20 **75.** 24, 40, 72 **76.** 3, 17, 51

77. 17, 31, 81 **78.** 14, 42, 84 **79.** 25, 125, 625 **80.** 12, 68, 92

81. 28, 35, 70 **82.** 1, 49, 153 **83.** 32, 56, 72 **84.** 24, 36, 48

SECTION 2.2 Introduction to Fractions

Objective A To write a fraction that represents part of a whole

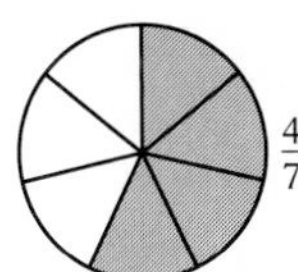

A **fraction** can represent the number of equal parts of a whole.

The shaded portion of the circle is represented by the fraction $\frac{4}{7}$.

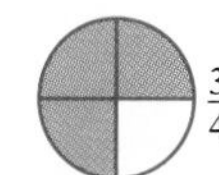

Four sevenths of the circle are shaded.

Each part of a fraction has a name.

$$\textbf{Fraction bar} \rightarrow \frac{4}{7} \begin{array}{l} \leftarrow \textbf{Numerator} \\ \leftarrow \textbf{Denominator} \end{array}$$

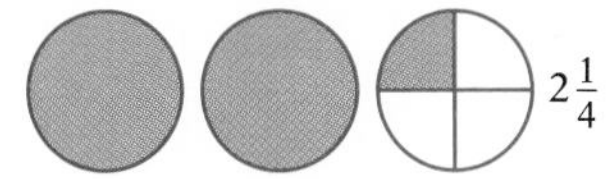

A **proper fraction** is a fraction less than 1. The numerator of a proper fraction is smaller than the denominator. The shaded portion of the circle can be represented by the proper fraction $\frac{3}{4}$.

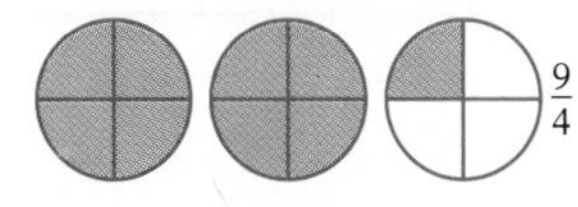

A **mixed number** is a number greater than 1 with a whole number part and a fractional part. The shaded portion of the circles can be represented by the mixed number $2\frac{1}{4}$.

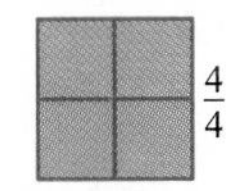

An **improper fraction** is a fraction greater than or equal to 1. The numerator of an improper fraction is greater than or equal to the denominator. The shaded portion of the circles can be represented by the improper fraction $\frac{9}{4}$. The shaded portion of the square can be represented by $\frac{4}{4}$.

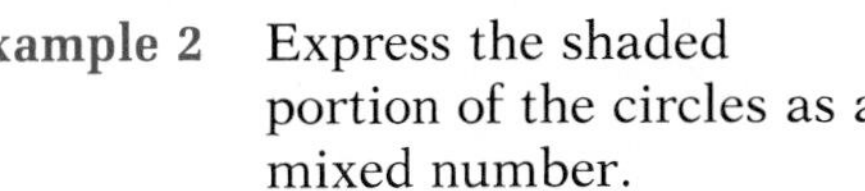

Example 1 Express the shaded portion of the circles as a mixed number.

Solution $3\frac{2}{5}$

Example 2 Express the shaded portion of the circles as a mixed number.

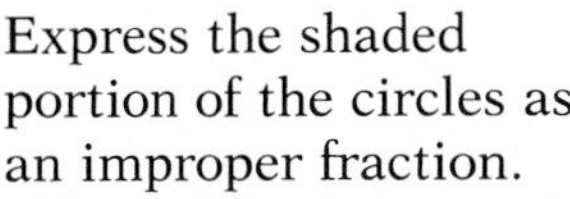

Your solution

Example 3 Express the shaded portion of the circles as an improper fraction.

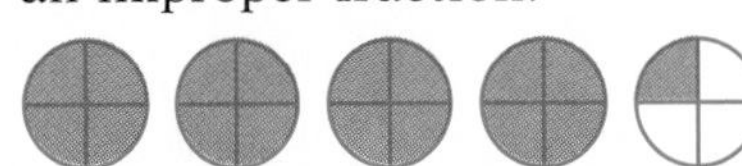

Solution $\frac{17}{5}$

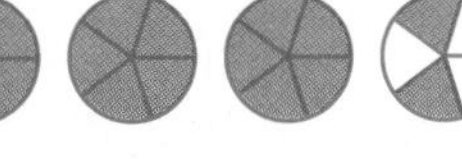

Example 4 Express the shaded portion of the circles as an improper fraction.

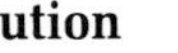

Your solution

Solutions on p. A11

Objective B

To write an improper fraction as a mixed number or a whole number, and a mixed number as an improper fraction

Note from the diagram that the mixed number $2\frac{3}{5}$ and the improper fraction $\frac{13}{5}$ represent the shaded portion of the circles.

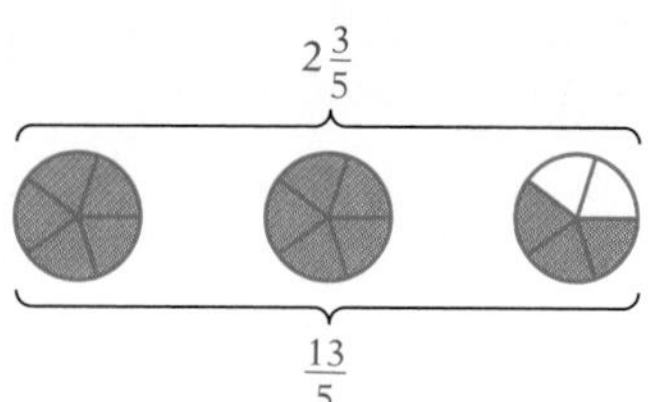

$$2\frac{3}{5} = \frac{13}{5}$$

An improper fraction can be written as a mixed number.

Write $\frac{13}{5}$ as a mixed number.

Divide the numerator by the denominator.

$$5\overline{)\,13}\quad \text{quotient } 2; \quad -10; \quad 3$$

To write the fractional part of the mixed number, write the remainder over the divisor.

$$5\overline{)\,13}\quad \text{quotient } 2\frac{3}{5}; \quad -10; \quad 3$$

Write the answer.

$$\frac{13}{5} = 2\frac{3}{5}$$

To write a mixed number as an improper fraction, multiply the denominator of the fractional part by the whole number part. The sum of this product and the numerator of the fractional part is the numerator of the improper fraction. The denominator remains the same.

Write $7\frac{3}{8}$ as an improper fraction. $\quad 7\frac{3}{8} = \frac{(8 \times 7) + 3}{8} = \frac{56 + 3}{8} = \frac{59}{8} \qquad 7\frac{3}{8} = \frac{59}{8}$

Example 5 Write $\frac{21}{4}$ as a mixed number.

Solution

$$4\overline{)\,21}\quad \text{quotient } 5; \quad -20; \quad 1 \qquad \frac{21}{4} = 5\frac{1}{4}$$

Example 6 Write $\frac{22}{5}$ as a mixed number.

Your solution

Example 7 Write $\frac{18}{6}$ as a whole number.

Solution

$$6\overline{)\,18}\quad \text{quotient } 3; \quad -18; \quad 0$$

Note: the remainder is zero. $\quad \frac{18}{6} = 3$

Example 8 Write $\frac{28}{7}$ as a whole number.

Your solution

Example 9 Write $21\frac{3}{4}$ as an improper fraction.

Solution $21\frac{3}{4} = \frac{84 + 3}{4} = \frac{87}{4}$

Example 10 Write $14\frac{5}{8}$ as an improper fraction.

Your solution

Solutions on p. A11

2.2 EXERCISES

Objective A

Express the shaded portion of the circle as a fraction.

1. 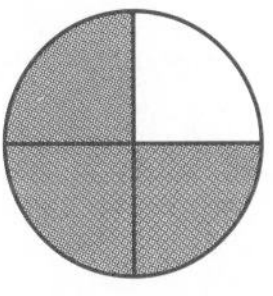

2. 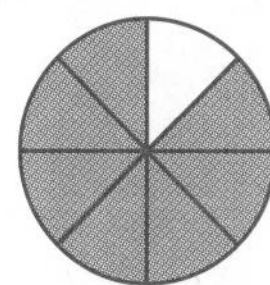

3. 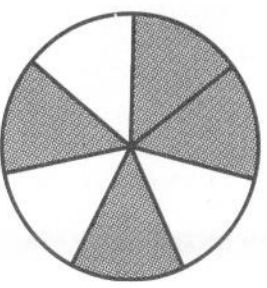

4. 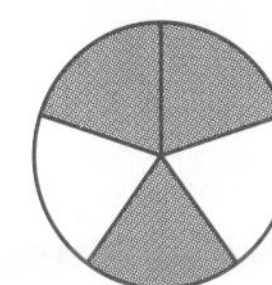

Express the shaded portion of the circles as a mixed number.

5. 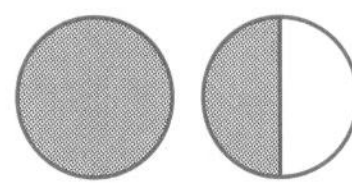

6. 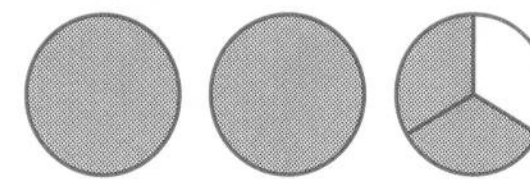

7. 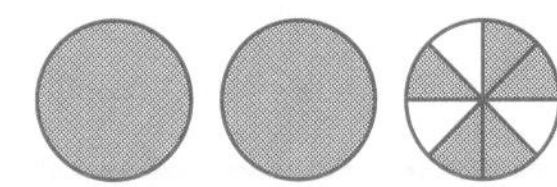

8. 

9. 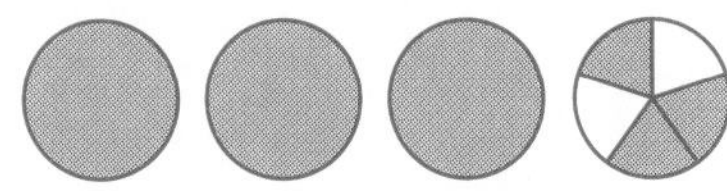

10. 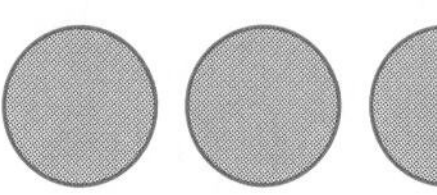

Express the shaded portion of the circles as an improper fraction.

11. 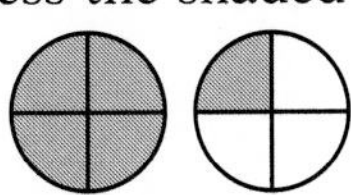

12. 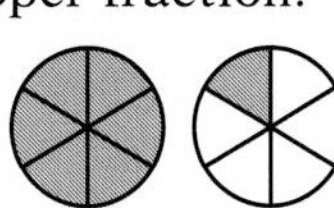

13. 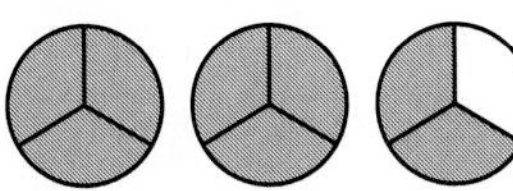

14. 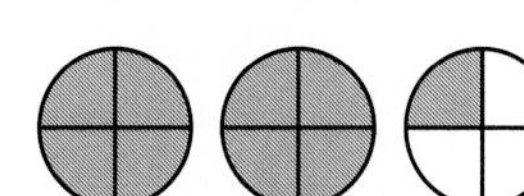

15. 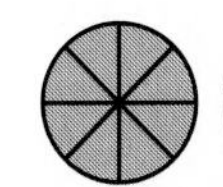  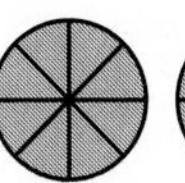 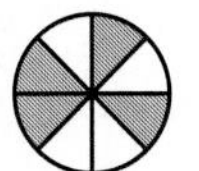

16. 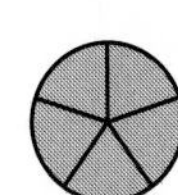   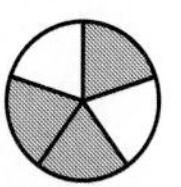

Objective B

Write the improper fraction as a mixed number or whole number.

17. $\frac{11}{4}$ **18.** $\frac{16}{3}$ **19.** $\frac{20}{4}$ **20.** $\frac{18}{9}$ **21.** $\frac{9}{8}$ **22.** $\frac{13}{4}$

23. $\frac{23}{10}$ **24.** $\frac{29}{2}$ **25.** $\frac{48}{16}$ **26.** $\frac{51}{3}$ **27.** $\frac{8}{7}$ **28.** $\frac{16}{9}$

29. $\frac{7}{3}$ **30.** $\frac{9}{5}$ **31.** $\frac{16}{1}$ **32.** $\frac{23}{1}$ **33.** $\frac{17}{8}$ **34.** $\frac{31}{16}$

35. $\frac{12}{5}$ **36.** $\frac{19}{3}$ **37.** $\frac{9}{9}$ **38.** $\frac{40}{8}$ **39.** $\frac{72}{8}$ **40.** $\frac{3}{3}$

Write the mixed number as an improper fraction.

41. $2\frac{1}{3}$ **42.** $4\frac{2}{3}$ **43.** $6\frac{1}{2}$ **44.** $8\frac{2}{3}$ **45.** $6\frac{5}{6}$ **46.** $7\frac{3}{8}$

47. $9\frac{1}{4}$ **48.** $6\frac{1}{4}$ **49.** $10\frac{1}{2}$ **50.** $15\frac{1}{8}$ **51.** $8\frac{1}{9}$ **52.** $3\frac{5}{12}$

53. $5\frac{3}{11}$ **54.** $3\frac{7}{9}$ **55.** $2\frac{5}{8}$ **56.** $12\frac{2}{3}$ **57.** $1\frac{5}{8}$ **58.** $5\frac{3}{7}$

59. $11\frac{1}{9}$ **60.** $12\frac{3}{5}$ **61.** $3\frac{3}{8}$ **62.** $4\frac{5}{9}$ **63.** $6\frac{7}{13}$ **64.** $8\frac{5}{14}$

SECTION 2.3 Writing Equivalent Fractions

Objective A To find equivalent fractions by raising to higher terms

Equal fractions with different denominators are called **equivalent fractions.**

$\frac{4}{6}$ is equivalent to $\frac{2}{3}$.

$\frac{2}{3}$ was rewritten as the equivalent fraction $\frac{4}{6}$.

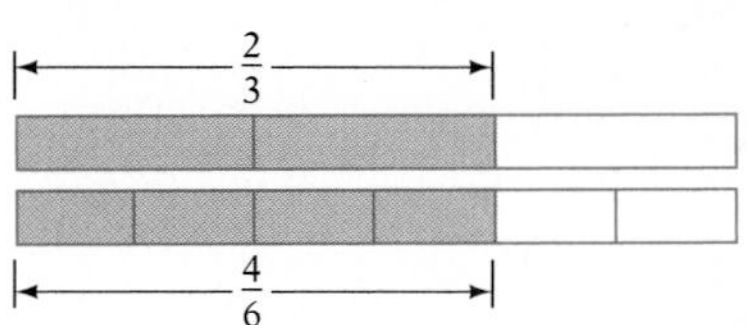

Remember that the Multiplication Property of 1 stated that the product of a number and 1 is the number. This is true for fractions as well as whole numbers. This property can be used to write equivalent fractions.

$$\frac{2}{3} \times 1 = \frac{2}{3} \times \frac{1}{1} = \frac{2 \cdot 1}{3 \cdot 1} = \frac{2}{3}$$

$\frac{2}{3} \times 1 = \frac{2}{3} \times \boxed{\frac{2}{2}} = \frac{2 \cdot 2}{3 \cdot 2} = \frac{4}{6}$ $\quad \frac{4}{6}$ is equivalent to $\frac{2}{3}$.

$\frac{2}{3} \times 1 = \frac{2}{3} \times \boxed{\frac{4}{4}} = \frac{2 \cdot 4}{3 \cdot 4} = \frac{8}{12}$ $\quad \frac{8}{12}$ is equivalent to $\frac{2}{3}$.

$\frac{2}{3}$ was rewritten as the equivalent fractions $\frac{4}{6}$ and $\frac{8}{12}$.

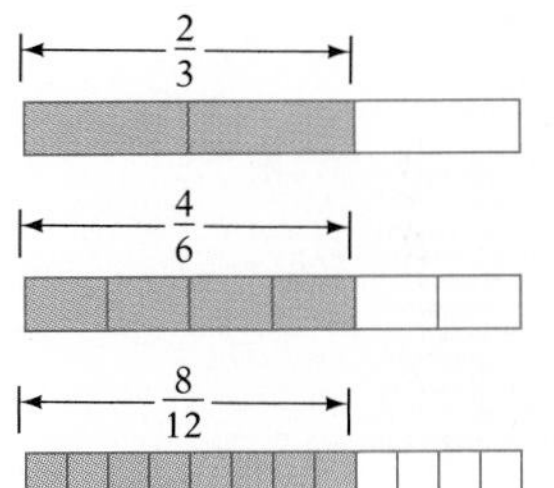

Write a fraction that is equivalent to $\frac{5}{8}$ and has a denominator of 32.

Divide the larger denominator by the smaller.

$$8\overline{)32} = 4$$

Multiply the numerator and denominator of the given fraction by the quotient (4).

$$\frac{5}{8} = \frac{5 \cdot 4}{8 \cdot 4} = \frac{20}{32}$$

$\frac{20}{32}$ is equivalent to $\frac{5}{8}$.

Example 1 Write $\frac{2}{3}$ as an equivalent fraction that has a denominator of 42.

Solution $3\overline{)42} = 14$ $\quad \frac{2}{3} = \frac{2 \cdot 14}{3 \cdot 14} = \frac{28}{42}$

$\frac{28}{42}$ is equivalent to $\frac{2}{3}$.

Example 2 Write $\frac{3}{5}$ as an equivalent fraction that has a denominator of 45.

Your solution

Example 3 Write 4 as a fraction that has a denominator of 12.

Solution Write 4 as $\frac{4}{1}$.

$1\overline{)12} = 12$ $\quad 4 = \frac{4 \cdot 12}{1 \cdot 12} = \frac{48}{12}$

$\frac{48}{12}$ is equivalent to 4.

Example 4 Write 6 as a fraction that has a denominator of 18.

Your solution

Solutions on p. A12

Objective B To write a fraction in simplest form

2

A fraction is in **simplest form** when there are no common factors in the numerator and the denominator.

The fractions $\frac{4}{6}$ and $\frac{2}{3}$ are equivalent fractions.

$\frac{4}{6}$ has been written in simplest form as $\frac{2}{3}$.

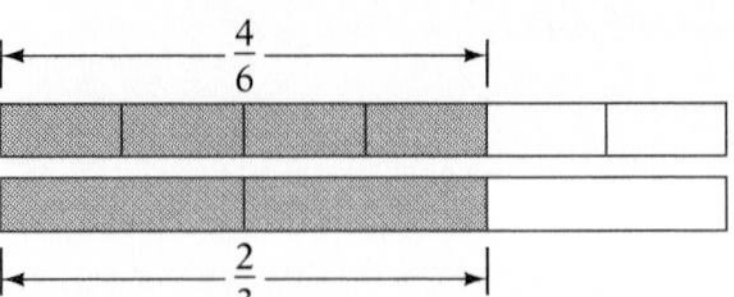

The Multiplication Property of 1 can be used to write fractions in simplest form. Write the numerator and denominator of the given fraction as a product of factors. Write factors common to both the numerator and denominator as an improper fraction equivalent to 1.

$$\frac{4}{6}=\frac{2\cdot 2}{2\cdot 3}=\frac{2}{2}\cdot\frac{2}{3}=\frac{2}{2}\cdot\frac{2}{3}=1\cdot\frac{2}{3}=\frac{2}{3}$$

The process of eliminating common factors is usually written as shown at the right.

$$\frac{4}{6}=\frac{\overset{1}{\cancel{2}}\cdot 2}{\underset{1}{\cancel{2}}\cdot 3}=\frac{2}{3}$$

To write a fraction in simplest form, eliminate the common factors.

$$\frac{18}{30}=\frac{\overset{1}{\cancel{2}}\cdot\overset{1}{\cancel{3}}\cdot 3}{\underset{1}{\cancel{2}}\cdot\underset{1}{\cancel{3}}\cdot 5}=\frac{3}{5}$$

An improper fraction should be changed to a mixed number.

$$\frac{22}{6}=\frac{\overset{1}{\cancel{2}}\cdot 11}{\underset{1}{\cancel{2}}\cdot 3}=\frac{11}{3}=3\frac{2}{3}$$

Example 5 Write $\frac{15}{40}$ in simplest form.

Solution $\frac{15}{40}=\frac{3\cdot\overset{1}{\cancel{5}}}{2\cdot 2\cdot 2\cdot\underset{1}{\cancel{5}}}=\frac{3}{8}$

Example 6 Write $\frac{16}{24}$ in simplest form.

Your solution

Example 7 Write $\frac{6}{42}$ in simplest form.

Solution $\frac{6}{42}=\frac{\overset{1}{\cancel{2}}\cdot\overset{1}{\cancel{3}}}{\underset{1}{\cancel{2}}\cdot\underset{1}{\cancel{3}}\cdot 7}=\frac{1}{7}$

Example 8 Write $\frac{8}{56}$ in simplest form.

Your solution

Example 9 Write $\frac{8}{9}$ in simplest form.

Solution $\frac{8}{9}=\frac{2\cdot 2\cdot 2}{3\cdot 3}=\frac{8}{9}$

$\frac{8}{9}$ is in simplest form because there are no common factors in the numerator and denominator.

Example 10 Write $\frac{15}{32}$ in simplest form.

Your solution

Example 11 Write $\frac{30}{12}$ in simplest form.

Solution $\frac{30}{12}=\frac{\overset{1}{\cancel{2}}\cdot\overset{1}{\cancel{3}}\cdot 5}{\underset{1}{\cancel{2}}\cdot 2\cdot\underset{1}{\cancel{3}}}=\frac{5}{2}=2\frac{1}{2}$

Example 12 Write $\frac{48}{36}$ in simplest form.

Your solution

Solutions on p. A12

2.3 EXERCISES

Objective A

Write an equivalent fraction with the given denominator.

1. $\frac{1}{2} = \frac{}{10}$
2. $\frac{1}{4} = \frac{}{16}$
3. $\frac{3}{16} = \frac{}{48}$
4. $\frac{5}{9} = \frac{}{81}$
5. $\frac{3}{8} = \frac{}{32}$
6. $\frac{7}{11} = \frac{}{33}$
7. $\frac{3}{17} = \frac{}{51}$
8. $\frac{7}{10} = \frac{}{90}$
9. $\frac{3}{4} = \frac{}{16}$
10. $\frac{5}{8} = \frac{}{32}$
11. $3 = \frac{}{9}$
12. $5 = \frac{}{25}$
13. $\frac{1}{3} = \frac{}{60}$
14. $\frac{1}{16} = \frac{}{48}$
15. $\frac{11}{15} = \frac{}{60}$
16. $\frac{3}{50} = \frac{}{300}$
17. $\frac{2}{3} = \frac{}{18}$
18. $\frac{5}{9} = \frac{}{36}$
19. $\frac{5}{7} = \frac{}{49}$
20. $\frac{7}{8} = \frac{}{32}$
21. $\frac{5}{9} = \frac{}{18}$
22. $\frac{11}{12} = \frac{}{36}$
23. $7 = \frac{}{3}$
24. $9 = \frac{}{4}$
25. $\frac{7}{9} = \frac{}{45}$
26. $\frac{5}{6} = \frac{}{42}$
27. $\frac{15}{16} = \frac{}{64}$
28. $\frac{11}{18} = \frac{}{54}$
29. $\frac{3}{14} = \frac{}{98}$
30. $\frac{5}{6} = \frac{}{144}$
31. $\frac{5}{8} = \frac{}{48}$
32. $\frac{7}{12} = \frac{}{96}$
33. $\frac{5}{14} = \frac{}{42}$
34. $\frac{2}{3} = \frac{}{42}$
35. $\frac{17}{24} = \frac{}{144}$
36. $\frac{5}{13} = \frac{}{169}$
37. $\frac{3}{8} = \frac{}{408}$
38. $\frac{9}{16} = \frac{}{272}$
39. $\frac{17}{40} = \frac{}{800}$
40. $\frac{9}{25} = \frac{}{1000}$
41. $\frac{5}{12} = \frac{}{72}$
42. $\frac{9}{17} = \frac{}{119}$
43. $\frac{17}{30} = \frac{}{210}$
44. $\frac{11}{21} = \frac{}{189}$

Objective B

Write the fraction in simplest form.

45. $\frac{4}{12}$ **46.** $\frac{8}{22}$ **47.** $\frac{22}{44}$ **48.** $\frac{2}{14}$ **49.** $\frac{2}{12}$ **50.** $\frac{6}{25}$

51. $\frac{50}{75}$ **52.** $\frac{40}{36}$ **53.** $\frac{12}{8}$ **54.** $\frac{36}{9}$ **55.** $\frac{0}{30}$ **56.** $\frac{10}{10}$

57. $\frac{9}{22}$ **58.** $\frac{14}{35}$ **59.** $\frac{75}{25}$ **60.** $\frac{8}{60}$ **61.** $\frac{16}{84}$ **62.** $\frac{14}{45}$

63. $\frac{20}{44}$ **64.** $\frac{12}{35}$ **65.** $\frac{8}{36}$ **66.** $\frac{28}{44}$ **67.** $\frac{12}{16}$ **68.** $\frac{48}{35}$

69. $\frac{16}{12}$ **70.** $\frac{24}{18}$ **71.** $\frac{24}{40}$ **72.** $\frac{44}{60}$ **73.** $\frac{8}{88}$

74. $\frac{9}{90}$ **75.** $\frac{144}{36}$ **76.** $\frac{140}{297}$ **77.** $\frac{48}{144}$ **78.** $\frac{32}{120}$

79. $\frac{60}{100}$ **80.** $\frac{33}{110}$ **81.** $\frac{36}{16}$ **82.** $\frac{80}{45}$ **83.** $\frac{32}{160}$

84. $\frac{192}{640}$ **85.** $\frac{19}{51}$ **86.** $\frac{23}{146}$ **87.** $\frac{81}{168}$ **88.** $\frac{66}{220}$

SECTION 2.4 Addition of Fractions and Mixed Numbers

Objective A

To add fractions with the same denominator

Fractions with the same denominator are added by adding the numerators and placing the sum over the common denominator. After adding, write the sum in simplest form.

Add: $\frac{2}{7} + \frac{4}{7}$

$$\begin{array}{r} \frac{2}{7} \\ +\frac{4}{7} \\ \hline \frac{6}{7} \end{array}$$

Add the numerators and place the sum over the common denominator.

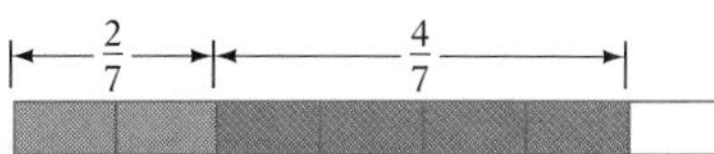

$$\frac{2}{7} + \frac{4}{7} = \frac{2+4}{7} = \frac{6}{7}$$

Example 1 Add: $\frac{5}{12} + \frac{11}{12}$

Solution

$$\begin{array}{r} \frac{5}{12} \\ +\frac{11}{12} \\ \hline \frac{16}{12} = \frac{4}{3} = 1\frac{1}{3} \end{array}$$

Example 2 Add: $\frac{3}{8} + \frac{7}{8}$

Your solution

Solution on p. A12

Objective B

To add fractions with unlike denominators

To add fractions with unlike denominators, first rewrite the fractions as equivalent fractions with a common denominator. The common denominator is the LCM of the denominators of the fractions.

Add: $\frac{1}{2} + \frac{1}{3}$

The common denominator is the LCM of 2 and 3. LCM = 6. The LCM of denominators is sometimes called the **least common denominator** (LCD).

Write equivalent fractions using the LCM.

$$\begin{array}{r} \frac{1}{2} = \frac{3}{6} \\ +\frac{1}{3} = \frac{2}{6} \\ \hline \end{array}$$

Add the fractions.

$$\begin{array}{r} \frac{1}{2} = \frac{3}{6} \\ +\frac{1}{3} = \frac{2}{6} \\ \hline \frac{5}{6} \end{array}$$

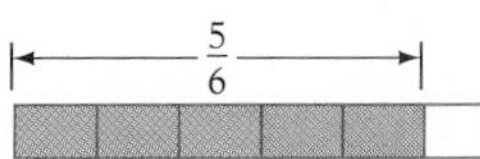

Example 3 Add: $\frac{3}{8} + \frac{7}{12}$

Solution

$$\frac{3}{8} = \frac{9}{24}$$
$$+\frac{7}{12} = \frac{14}{24}$$
$$\frac{23}{24}$$

The LCM of 8 and 12 is 24.

Example 4 Add: $\frac{5}{12} + \frac{9}{16}$

Your solution

Example 5 Add: $\frac{5}{8} + \frac{7}{9}$

Solution

$$\frac{5}{8} = \frac{45}{72}$$
$$+\frac{7}{9} = \frac{56}{72}$$
$$\frac{101}{72} = 1\frac{29}{72}$$

The LCM of 8 and 9 is 72.

Example 6 Add: $\frac{7}{8} + \frac{11}{15}$

Your solution

Example 7 Add: $\frac{2}{3} + \frac{3}{5} + \frac{5}{6}$

Solution

$$\frac{2}{3} = \frac{20}{30}$$
$$\frac{3}{5} = \frac{18}{30}$$
$$+\frac{5}{6} = \frac{25}{30}$$
$$\frac{63}{30} = 2\frac{3}{30} = 2\frac{1}{10}$$

The LCM of 3, 5, and 6 is 30.

Example 8 Add: $\frac{3}{4} + \frac{4}{5} + \frac{5}{8}$

Your solution

Solutions on p. A12

Objective C — To add whole numbers, mixed numbers, and fractions

The sum of a whole number and a fraction is a mixed number.

Add: $2 + \frac{2}{3}$

$$\boxed{2} + \frac{2}{3} = \boxed{\frac{6}{3}} + \frac{2}{3} = \frac{8}{3} = 2\frac{2}{3}$$

To add a whole number and a mixed number, write the fraction, then add the whole numbers.

Add: $7\frac{2}{5} + 4$

Write the fraction.

$$7\frac{2}{5}$$
$$+4$$
$$\frac{2}{5}$$

Add the whole numbers.

$$7\frac{2}{5}$$
$$+4$$
$$11\frac{2}{5}$$

To add two mixed numbers, add the fractional parts and then add the whole numbers. Remember to reduce the sum to simplest form.

Add: $5\frac{4}{9} + 6\frac{14}{15}$

Add the fractional parts.

$$\begin{array}{r} 5\frac{4}{9} = 5\frac{20}{45} \\ +6\frac{14}{15} = 6\frac{42}{45} \\ \hline \frac{62}{45} \end{array}$$

The LCM of 9 and 15 is 45.

Add the whole numbers.

$$\begin{array}{r} 5\frac{4}{9} = 5\frac{20}{45} \\ +6\frac{14}{15} = 6\frac{42}{45} \\ \hline 11\frac{62}{45} \end{array} = 11 + 1\frac{17}{45} = 12\frac{17}{45}$$

Example 9 Add: $5 + \frac{3}{8}$

Solution $5 + \frac{3}{8} = 5\frac{3}{8}$

Example 10 Add: $7 + \frac{6}{11}$

Your solution

Example 11 Add: $17 + 3\frac{3}{8}$

Solution

$$\begin{array}{r} 17 \\ +\ 3\frac{3}{8} \\ \hline 20\frac{3}{8} \end{array}$$

Example 12 Add: $29 + 17\frac{5}{12}$

Your solution

Example 13 Add: $5\frac{2}{3} + 11\frac{5}{6} + 12\frac{7}{9}$

Solution

$$\begin{array}{r} 5\frac{2}{3} = 5\frac{12}{18} \\ 11\frac{5}{6} = 11\frac{15}{18} \\ +12\frac{7}{9} = 12\frac{14}{18} \\ \hline 28\frac{41}{18} \end{array} = 30\frac{5}{18}$$

LCM = 18.

Example 14 Add: $7\frac{4}{5} + 6\frac{7}{10} + 13\frac{11}{15}$

Your solution

Example 15 Add: $11\frac{5}{8} + 7\frac{5}{9} + 8\frac{7}{15}$

Solution

$$\begin{array}{r} 11\frac{5}{8} = 11\frac{225}{360} \\ 7\frac{5}{9} = 7\frac{200}{360} \\ +\ 8\frac{7}{15} = 8\frac{168}{360} \\ \hline 26\frac{593}{360} \end{array} = 27\frac{233}{360}$$

LCM = 360.

Example 16 Add: $9\frac{3}{8} + 17\frac{7}{12} + 10\frac{14}{15}$

Your solution

Solutions on p. A12

Objective D To solve application problems

Example 17

A rain gauge collected $2\frac{1}{3}$ inches of rain in October, $5\frac{1}{2}$ inches in November, and $3\frac{3}{8}$ inches in December. Find the total rainfall for the 3 months.

Strategy

To find the total rainfall for the 3 months, add the 3 amounts of rainfall ($2\frac{1}{3}$, $5\frac{1}{2}$, and $3\frac{3}{8}$).

Solution

$$\begin{aligned} 2\frac{1}{3} &= 2\frac{8}{24} \\ 5\frac{1}{2} &= 5\frac{12}{24} \\ +3\frac{3}{8} &= 3\frac{9}{24} \\ \hline & 10\frac{29}{24} = 11\frac{5}{24} \end{aligned}$$

The total rainfall for the three months was $11\frac{5}{24}$ inches.

Example 18

You spend $4\frac{1}{2}$ hours in class, $3\frac{3}{4}$ hours studying, and $1\frac{1}{3}$ hours eating each day. Find the total time spent each day on these three activities.

Your strategy

Your solution

Example 19

A student worked 4 hours, $2\frac{1}{3}$ hours, and $5\frac{2}{3}$ hours this week on a part-time job. The student is paid \$6 an hour. How much did the student earn this week?

Strategy

To find how much the student earned:

- Find the total number of hours worked.
- Multiply the total number of hours worked by the hourly wage (\$6).

Solution

$$\begin{aligned} &4 \\ &2\frac{1}{3} \\ +&5\frac{2}{3} \\ \hline &11\frac{3}{3} = 12 \text{ hours worked} \end{aligned} \qquad \begin{aligned} &12 \\ \times&\$6 \\ \hline &\$72 \end{aligned}$$

The student earned \$72 this week.

Example 20

A carpenter worked $1\frac{2}{3}$ hours of overtime on Monday, $3\frac{1}{3}$ hours of overtime on Tuesday, and 2 hours of overtime on Wednesday. At \$24 an hour, find the carpenter's overtime pay for the 3 days.

Your strategy

Your solution

Solutions on p. A13

2.4 EXERCISES

Objective A

Add.

1. $\frac{2}{7} + \frac{1}{7}$
2. $\frac{3}{11} + \frac{5}{11}$
3. $\frac{1}{2} + \frac{1}{2}$
4. $\frac{1}{3} + \frac{2}{3}$
5. $\frac{8}{11} + \frac{7}{11}$
6. $\frac{9}{13} + \frac{7}{13}$
7. $\frac{8}{5} + \frac{9}{5}$
8. $\frac{5}{3} + \frac{7}{3}$
9. $\frac{3}{5} + \frac{8}{5} + \frac{3}{5}$
10. $\frac{3}{8} + \frac{5}{8} + \frac{7}{8}$
11. $\frac{3}{4} + \frac{1}{4} + \frac{5}{4}$
12. $\frac{2}{7} + \frac{4}{7} + \frac{5}{7}$
13. $\frac{3}{8} + \frac{7}{8} + \frac{1}{8}$
14. $\frac{5}{12} + \frac{7}{12} + \frac{1}{12}$
15. $\frac{4}{15} + \frac{7}{15} + \frac{11}{15}$
16. $\frac{3}{4} + \frac{3}{4} + \frac{1}{4}$
17. $\frac{3}{16} + \frac{5}{16} + \frac{7}{16}$
18. $\frac{5}{18} + \frac{11}{18} + \frac{17}{18}$
19. $\frac{3}{11} + \frac{5}{11} + \frac{7}{11}$
20. $\frac{5}{7} + \frac{4}{7} + \frac{5}{7}$

Objective B

Add.

21. $\frac{1}{2} + \frac{2}{3}$
22. $\frac{2}{3} + \frac{1}{4}$
23. $\frac{3}{14} + \frac{5}{7}$
24. $\frac{3}{5} + \frac{7}{10}$
25. $\frac{5}{9} + \frac{7}{15}$
26. $\frac{8}{15} + \frac{7}{20}$
27. $\frac{1}{6} + \frac{7}{9}$
28. $\frac{3}{8} + \frac{9}{14}$
29. $\frac{5}{12} + \frac{5}{16}$
30. $\frac{13}{40} + \frac{12}{25}$
31. $\frac{3}{20} + \frac{7}{30}$
32. $\frac{5}{12} + \frac{7}{30}$
33. $\frac{2}{3} + \frac{6}{19}$
34. $\frac{1}{2} + \frac{3}{29}$
35. $\frac{3}{7} + \frac{4}{21}$
36. $\frac{5}{14} + \frac{35}{84}$
37. $\frac{3}{14} + \frac{6}{49}$
38. $\frac{3}{10} + \frac{7}{45}$
39. $\frac{2}{7} + \frac{3}{8}$
40. $\frac{5}{8} + \frac{4}{9}$

Add.

41. $\frac{1}{3}+\frac{5}{6}+\frac{7}{9}$

42. $\frac{2}{3}+\frac{5}{6}+\frac{7}{12}$

43. $\frac{5}{6}+\frac{1}{12}+\frac{5}{16}$

44. $\frac{2}{9}+\frac{7}{15}+\frac{4}{21}$

45. $\frac{2}{3}+\frac{1}{5}+\frac{7}{12}$

46. $\frac{3}{4}+\frac{4}{5}+\frac{7}{12}$

47. $\frac{1}{4}+\frac{4}{5}+\frac{5}{9}$

48. $\frac{2}{3}+\frac{3}{5}+\frac{7}{8}$

49. $\frac{5}{16}+\frac{11}{18}+\frac{17}{24}$

50. $\frac{3}{10}+\frac{14}{15}+\frac{9}{25}$

51. $\frac{2}{3}+\frac{5}{8}+\frac{7}{9}$

52. $\frac{1}{3}+\frac{2}{9}+\frac{7}{8}$

53. $\frac{5}{6}+\frac{1}{2}+\frac{3}{11}$

54. $\frac{9}{11}+\frac{1}{2}+\frac{1}{6}$

55. $\frac{3}{4}+\frac{5}{6}+\frac{7}{8}$

Objective C

Add.

56. $1\frac{1}{2} + 2\frac{1}{6}$

57. $2\frac{2}{5} + 3\frac{3}{10}$

58. $4\frac{1}{2} + 5\frac{7}{12}$

59. $3\frac{3}{8} + 2\frac{5}{16}$

60. $2\frac{7}{9} + 3\frac{5}{12}$

61. $4\frac{7}{15} + 3\frac{11}{12}$

62. $4 + 5\frac{2}{7}$

63. $6\frac{8}{9} + 12$

64. $3\frac{5}{8} + 2\frac{11}{20}$

65. $4\frac{5}{12} + 6\frac{11}{18}$

66. $10\frac{2}{7} + 7\frac{21}{35}$

67. $16\frac{2}{3} + 8\frac{1}{4}$

68. $16\frac{5}{8}+3\frac{7}{20}$

69. $2\frac{3}{8}+4\frac{9}{20}$

70. $2\frac{7}{10}+7\frac{11}{15}$

71. $3\frac{3}{8}+8\frac{5}{12}$

72. $7\frac{5}{12}+2\frac{9}{16}$

73. $9\frac{1}{2}+3\frac{3}{11}$

74. $6\frac{1}{3}+2\frac{3}{13}$

75. $8\frac{21}{40}+6\frac{21}{32}$

76. $8\frac{29}{30}+7\frac{11}{40}$

77. $17\frac{5}{16}+3\frac{11}{24}$

78. $17\frac{3}{8}+7\frac{7}{20}$

79. $14\frac{7}{12}+29\frac{13}{21}$

Add.

80. $5\frac{7}{8} + 27\frac{5}{12}$

81. $7\frac{5}{6} + 3\frac{5}{9}$

82. $7\frac{5}{9} + 2\frac{7}{12}$

83. $3\frac{1}{2} + 2\frac{3}{4} + 1\frac{5}{6}$

84. $2\frac{1}{2} + 3\frac{2}{3} + 4\frac{1}{4}$

85. $3\frac{1}{3} + 7\frac{1}{5} + 2\frac{1}{7}$

86. $3\frac{1}{2} + 3\frac{1}{5} + 8\frac{1}{9}$

87. $6\frac{5}{9} + 6\frac{5}{12} + 2\frac{5}{18}$

88. $2\frac{3}{8} + 4\frac{7}{12} + 3\frac{5}{16}$

89. $2\frac{1}{8} + 4\frac{2}{9} + 5\frac{17}{18}$

90. $6\frac{5}{6} + 17\frac{2}{9} + 18\frac{5}{27}$

91. $4\frac{7}{20} + \frac{17}{80} + 25\frac{23}{60}$

92. $7\frac{5}{12} + 1\frac{11}{24} + 9\frac{21}{36}$

93. $3\frac{5}{6} + 7\frac{7}{8} + 3\frac{1}{12}$

94. $7\frac{2}{5} + 3\frac{7}{10} + 5\frac{11}{15}$

Objective D *Application Problems*

95. A family of four find that $\frac{1}{3}$ of their income is spent on housing, $\frac{1}{8}$ is spent on transportation, and $\frac{1}{4}$ is spent on food. Find the total fractional amount of their income that is spent on these three items.

96. A table top that is $1\frac{1}{8}$ inch thick is covered by a $\frac{3}{16}$-inch veneer. Find the total thickness of the table top after the veneer is applied.

97. At the beginning of the year, the stock in a cellular phone company was selling for $\$26\frac{5}{8}$ per share. The price of the stock gained $\$13\frac{3}{4}$ per share during a 6-month period. Find the price of the stock at the end of 6 months.

98. At the beginning of the week, an airlines stock was selling at $\$158\frac{3}{8}$ per share. During the week, the stock gained $\$28\frac{3}{4}$ per share. Find the price of the stock at the end of the week.

99. Find the length of the shaft.

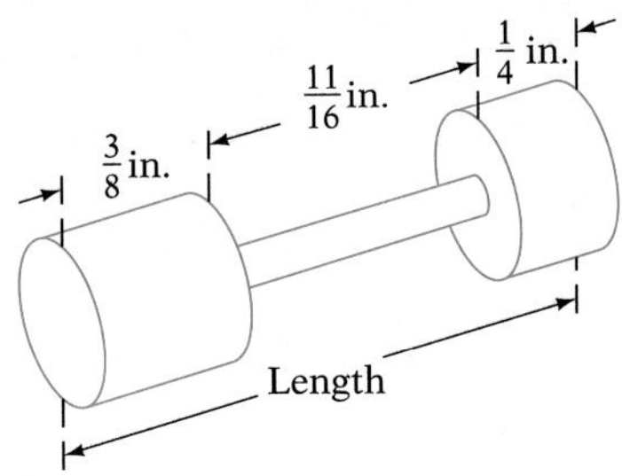

100. Find the length of the shaft.

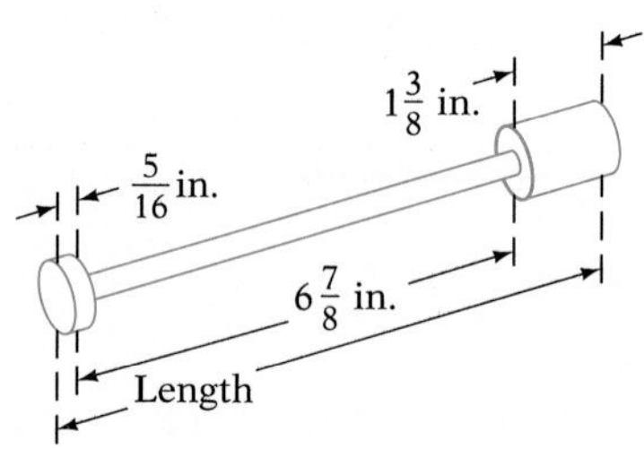

101. You are working a part-time job that pays \$7 an hour. You worked 5, $3\frac{3}{4}$, $2\frac{1}{3}$, $1\frac{1}{4}$, and $7\frac{2}{3}$ hours during the last 5 days.

a. Find the total number of hours worked during the last 5 days.
b. Find your total wages for the 5 days.

102. A nurse worked $2\frac{2}{3}$ hours of overtime on Monday, $1\frac{1}{4}$ hours on Wednesday, $1\frac{1}{3}$ hours on Friday, and $6\frac{3}{4}$ hours on Saturday.

a. Find the total number of overtime hours worked during the week.
b. At an overtime hourly wage of \$22 an hour, how much overtime pay does the nurse receive?

103. A Southern California mountain had $5\frac{3}{4}$ inches of snow in December, $15\frac{1}{2}$ inches in January, and $9\frac{5}{8}$ inches in February. Find the total snowfall for the 3 months.

104. A utility stock was sold for \$$26\frac{3}{16}$. Monthly gains for 3 months were \$$1\frac{1}{2}$, \$$\frac{5}{16}$, and \$$2\frac{5}{8}$ per share. Find the value of the stock at the end of the 3 months.

SECTION 2.5 Subtraction of Fractions and Mixed Numbers

Objective A To subtract fractions with the same denominator

Fractions with the same denominator are subtracted by subtracting the numerators and placing the difference over the common denominator. After subtracting, write the fraction in simplest form.

Subtract: $\frac{5}{7} - \frac{3}{7}$

$$\begin{array}{r} \frac{5}{7} \\ -\frac{3}{7} \\ \hline \frac{2}{7} \end{array}$$

Subtract the numerators and place the difference over the common denominator.

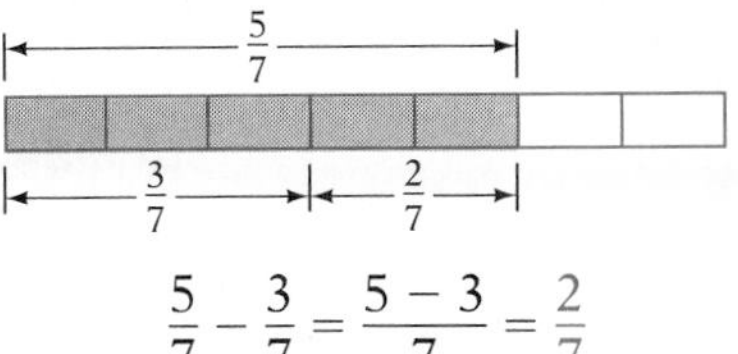

$$\frac{5}{7} - \frac{3}{7} = \frac{5-3}{7} = \frac{2}{7}$$

Example 1 Subtract: $\frac{17}{30} - \frac{11}{30}$

Solution

$$\begin{array}{r} \frac{17}{30} \\ -\frac{11}{30} \\ \hline \frac{6}{30} = \frac{1}{5} \end{array}$$

Example 2 Subtract: $\frac{16}{27} - \frac{7}{27}$

Your solution

Solution on p. A13

Objective B To subtract fractions with unlike denominators

To subtract fractions with unlike denominators, first rewrite the fractions as equivalent fractions with a common denominator. As with adding fractions, the common denominator is the LCM of the denominators of the fractions.

Subtract: $\frac{5}{6} - \frac{1}{4}$

The common denominator is the LCM of 6 and 4. LCM = 12.

Build equivalent fractions using the LCM.

$$\begin{array}{r} \frac{5}{6} = \frac{10}{12} \\ -\frac{1}{4} = \frac{3}{12} \\ \hline \end{array}$$

Subtract the fractions.

$$\begin{array}{r} \frac{5}{6} = \frac{10}{12} \\ -\frac{1}{4} = \frac{3}{12} \\ \hline \frac{7}{12} \end{array}$$

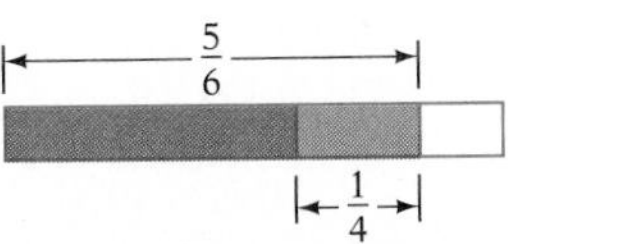

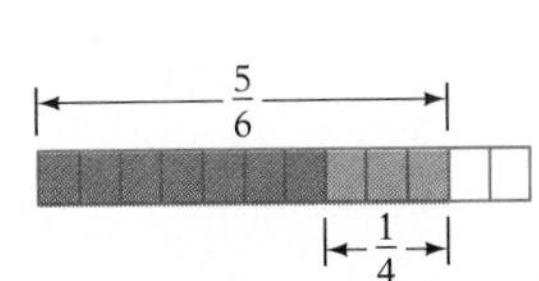

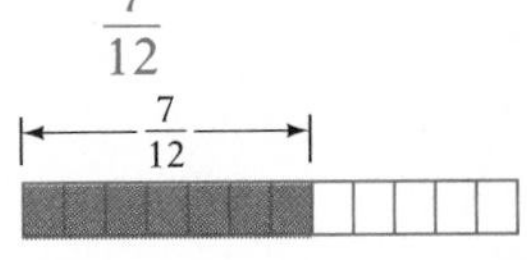

Example 3 Subtract: $\frac{2}{3} - \frac{1}{6}$

Solution

$$\begin{array}{r} \frac{2}{3} = \frac{4}{6} \\ -\frac{1}{6} = \frac{1}{6} \\ \hline \frac{3}{6} = \frac{1}{2} \end{array} \quad \text{LCM} = 6$$

Example 4 Subtract: $\frac{5}{6} - \frac{3}{8}$

Your solution

Example 5 Subtract: $\frac{11}{16} - \frac{5}{12}$

Solution

$$\begin{array}{r} \frac{11}{16} = \frac{33}{48} \\ -\frac{5}{12} = \frac{20}{48} \\ \hline \frac{13}{48} \end{array} \quad \text{LCM} = 48$$

Example 6 Subtract: $\frac{13}{18} - \frac{7}{24}$

Your solution

Solutions on p. A13

Objective C To subtract whole numbers, mixed numbers, and fractions

To subtract mixed numbers without borrowing, subtract the fractional parts and then subtract the whole numbers.

Subtract: $5\frac{5}{6} - 2\frac{3}{4}$

Subtract the fractional parts.

$$\begin{array}{r} 5\frac{5}{6} = 5\frac{10}{12} \\ -2\frac{3}{4} = 2\frac{9}{12} \\ \hline \frac{1}{12} \end{array}$$

The LCM of 6 and 4 is 12.

Subtract the whole numbers.

$$\begin{array}{r} 5\frac{5}{6} = 5\frac{10}{12} \\ -2\frac{3}{4} = 2\frac{9}{12} \\ \hline 3\frac{1}{12} \end{array}$$

As with whole numbers, subtraction of mixed numbers sometimes involves borrowing.

Subtract: $5 - 2\frac{5}{8}$

Borrow 1 from 5.

$$\begin{array}{r} 5 = \overset{4}{\cancel{5}}1 \\ -2\frac{5}{8} = 2\frac{5}{8} \\ \hline \end{array}$$

Write 1 as a fraction so that the fractions have the same denominators.

$$\begin{array}{r} 5 = 4\frac{8}{8} \\ -2\frac{5}{8} = 2\frac{5}{8} \\ \hline \end{array}$$

Subtract the mixed numbers.

$$\begin{array}{r} 5 = 4\frac{8}{8} \\ -2\frac{5}{8} = 2\frac{5}{8} \\ \hline 2\frac{3}{8} \end{array}$$

Subtract: $7\frac{1}{6} - 2\frac{5}{8}$

Write equivalent fractions using the LCM.

$$7\frac{1}{6} = 7\frac{4}{24}$$
$$-2\frac{5}{8} = 2\frac{15}{24}$$

Borrow 1 from 7. Add the 1 to $\frac{4}{24}$. Write $1\frac{4}{24}$ as $\frac{28}{24}$.

$$7\frac{1}{6} = \overset{6}{\not 7}1\frac{4}{24} = 6\frac{28}{24}$$
$$-2\frac{5}{8} = 2\frac{15}{24} = 2\frac{15}{24}$$

Subtract the mixed numbers.

$$7\frac{1}{6} = 6\frac{28}{24}$$
$$-2\frac{5}{8} = 2\frac{15}{24}$$
$$4\frac{13}{24}$$

Example 7 Subtract: $15\frac{7}{8} - 12\frac{2}{3}$

Solution LCM = 24.

$$15\frac{7}{8} = 15\frac{21}{24}$$
$$-12\frac{2}{3} = 12\frac{16}{24}$$
$$3\frac{5}{24}$$

Example 8 Subtract: $17\frac{5}{9} - 11\frac{5}{12}$

Example 9 Subtract: $9 - 4\frac{3}{11}$

Solution LCM = 11.

$$9 = 8\frac{11}{11}$$
$$-4\frac{3}{11} = 4\frac{3}{11}$$
$$4\frac{8}{11}$$

Example 10 Subtract: $8 - 2\frac{4}{13}$

Your solution

Example 11 Subtract: $11\frac{5}{12} - 2\frac{11}{16}$

Solution LCM = 48.

$$11\frac{5}{12} = 11\frac{20}{48} = 10\frac{68}{48}$$
$$-\ 2\frac{11}{16} = 2\frac{33}{48} = 2\frac{33}{48}$$
$$8\frac{35}{48}$$

Example 12 Subtract: $21\frac{7}{9} - 7\frac{11}{12}$

Your solution

Example 13 Subtract: $72\frac{5}{42} - 49\frac{37}{63}$

Solution LCM = 126.

$$72\frac{5}{42} = 72\frac{15}{126} = 71\frac{141}{126}$$
$$-49\frac{37}{63} = 49\frac{74}{126} = 49\frac{74}{126}$$
$$22\frac{67}{126}$$

Example 14 Subtract: $81\frac{17}{80} - 17\frac{29}{80}$

Your solution

Solutions on p. A13

Objective D To solve application problems

Example 15

A $2\frac{2}{3}$-inch piece is cut from a $6\frac{5}{8}$-inch board. How much of the board is left?

Strategy

To find the length remaining, subtract the length of the piece cut from the total length of the board.

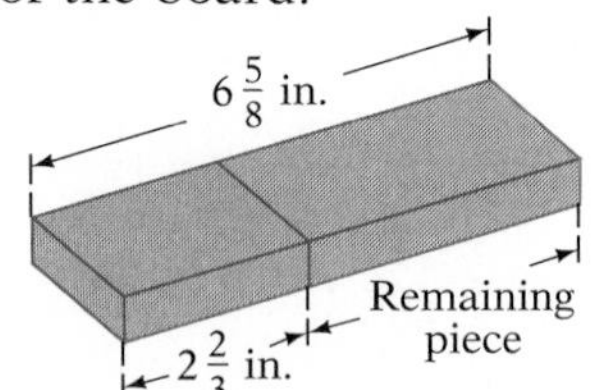

Solution

$$\begin{array}{r} 6\frac{5}{8} = 6\frac{15}{24} = 5\frac{39}{24} \\ -2\frac{2}{3} = 2\frac{16}{24} = 2\frac{16}{24} \\ \hline 3\frac{23}{24} \end{array}$$

$3\frac{23}{24}$ inches of the board are left.

Example 16

A flight from New York to Los Angeles takes $5\frac{1}{2}$ hours. After the plane is in the air for $2\frac{3}{4}$ hours, how much flight time remains?

Your strategy

Your solution

Example 17

Two painters are staining a house. In 1 day one painter stains $\frac{1}{3}$ of the house, and the other stains $\frac{1}{4}$ of the house. How much of the job remains to be done?

Strategy

To find how much of the job remains:

- Find the total amount of the house already stained $\left(\frac{1}{3} + \frac{1}{4}\right)$.
- Subtract the amount already stained from 1, which represents the complete job.

Solution

$$\begin{array}{r} \frac{1}{3} = \frac{4}{12} \\ +\frac{1}{4} = \frac{3}{12} \\ \hline \frac{7}{12} \end{array} \qquad \begin{array}{r} 1 = \frac{12}{12} \\ -\frac{7}{12} = \frac{7}{12} \\ \hline \frac{5}{12} \end{array}$$

$\frac{5}{12}$ of the house remains to be stained.

Example 18

A patient is put on a diet to lose 24 pounds in 3 months. The patient loses $7\frac{1}{2}$ pounds the first month and $5\frac{3}{4}$ pounds the second month. How much weight must be lost the third month to achieve the goal?

Your strategy

Your solution

Solutions on pp. A13-A14

2.5 EXERCISES

Objective A

Subtract.

1. $\frac{9}{17} - \frac{7}{17}$

2. $\frac{11}{15} - \frac{3}{15}$

3. $\frac{11}{12} - \frac{7}{12}$

4. $\frac{13}{15} - \frac{4}{15}$

5. $\frac{9}{20} - \frac{7}{20}$

6. $\frac{27}{40} - \frac{13}{40}$

7. $\frac{48}{55} - \frac{13}{55}$

8. $\frac{42}{65} - \frac{17}{65}$

9. $\frac{11}{24} - \frac{5}{24}$

10. $\frac{23}{30} - \frac{13}{30}$

11. $\frac{17}{42} - \frac{5}{42}$

12. $\frac{29}{48} - \frac{13}{48}$

Objective B

Subtract.

13. $\frac{2}{3} - \frac{1}{6}$

14. $\frac{7}{8} - \frac{5}{16}$

15. $\frac{5}{8} - \frac{2}{7}$

16. $\frac{5}{6} - \frac{3}{7}$

17. $\frac{5}{7} - \frac{3}{14}$

18. $\frac{7}{10} - \frac{3}{5}$

19. $\frac{5}{9} - \frac{7}{15}$

20. $\frac{8}{15} - \frac{7}{20}$

21. $\frac{7}{9} - \frac{1}{6}$

22. $\frac{9}{14} - \frac{3}{8}$

23. $\frac{5}{12} - \frac{5}{16}$

24. $\frac{12}{25} - \frac{13}{40}$

25. $\frac{7}{30} - \frac{3}{20}$

26. $\frac{5}{12} - \frac{7}{30}$

27. $\frac{2}{3} - \frac{6}{19}$

28. $\frac{1}{2} - \frac{3}{29}$

29. $\frac{5}{9} - \frac{1}{12}$

30. $\frac{11}{16} - \frac{5}{12}$

Subtract.

31. $\frac{46}{51} - \frac{3}{17}$

32. $\frac{9}{16} - \frac{17}{32}$

33. $\frac{21}{35} - \frac{5}{14}$

34. $\frac{19}{40} - \frac{3}{16}$

35. $\frac{29}{60} - \frac{3}{40}$

36. $\frac{53}{70} - \frac{13}{35}$

37. $\frac{2}{9} - \frac{1}{42}$

38. $\frac{11}{15} - \frac{5}{27}$

39. $\frac{7}{16} - \frac{5}{24}$

40. $\frac{11}{18} - \frac{7}{24}$

41. $\frac{7}{8} - \frac{5}{9}$

42. $\frac{5}{6} - \frac{3}{5}$

43. $\frac{21}{29} - \frac{3}{8}$

44. $\frac{17}{31} - \frac{4}{9}$

Objective C

Subtract.

45. $5\frac{7}{12} - 2\frac{5}{12}$

46. $16\frac{11}{15} - 11\frac{8}{15}$

47. $72\frac{21}{23} - 16\frac{17}{23}$

48. $19\frac{16}{17} - 9\frac{7}{17}$

49. $6\frac{1}{3} - 2$

50. $5\frac{7}{8} - 1$

51. $10 - 6\frac{1}{3}$

52. $3 - 2\frac{5}{21}$

53. $6\frac{2}{5} - 4\frac{4}{5}$

54. $16\frac{3}{8} - 10\frac{7}{8}$

55. $25\frac{4}{9} - 16\frac{7}{9}$

56. $8\frac{3}{7} - 2\frac{6}{7}$

57. $16\frac{2}{5} - 8\frac{4}{9}$

58. $23\frac{7}{8} - 16\frac{2}{3}$

59. $6 - 4\frac{3}{5}$

Subtract.

60. $12 - 2\frac{7}{9}$

61. $16\frac{3}{20} - 1\frac{7}{20}$

62. $16\frac{3}{10} - 7\frac{9}{10}$

63. $25\frac{5}{12} - 4\frac{17}{24}$

64. $16\frac{5}{8} - 13\frac{11}{12}$

65. $6\frac{1}{8} - 3$

66. $32\frac{4}{15} - 17$

67. $32 - 6\frac{7}{15}$

68. $47 - 9\frac{7}{20}$

69. $14\frac{3}{5} - 7\frac{8}{9}$

70. $17\frac{3}{4} - 8\frac{10}{11}$

71. $3\frac{9}{32} - 1\frac{11}{40}$

72. $7\frac{11}{30} - 5\frac{13}{36}$

73. $9\frac{13}{40} - 7\frac{12}{25}$

74. $13\frac{27}{50} - 5\frac{39}{40}$

75. $65\frac{8}{35} - 16\frac{11}{14}$

76. $82\frac{4}{33} - 16\frac{5}{22}$

77. $101\frac{2}{9} - 16$

78. $77\frac{5}{18} - 61$

79. $17 - 7\frac{8}{13}$

80. $7 - 3\frac{22}{45}$

81. $25\frac{7}{15} - 14\frac{19}{36}$

82. $14\frac{17}{24} - 10\frac{31}{36}$

83. $61\frac{5}{21} - 28\frac{5}{9}$

84. $26\frac{7}{20} - 13\frac{11}{14}$

85. $137\frac{3}{5} - 69\frac{7}{12}$

86. $267\frac{4}{7} - 129\frac{2}{9}$

87. $13\frac{7}{25} - 6\frac{9}{20}$

88. $68\frac{5}{18} - 13\frac{17}{24}$

Objective D *Application Problems*

89. Molding $2\frac{2}{3}$ feet long is cut from an 8-foot piece of molding. What is the length of the remaining piece of molding?

90. The inseam of a pair of jeans is 30 inches long. After being washed the inseam of the jeans shrinks $\frac{5}{8}$ inch. Find the length of the pants after they are washed.

91. A plane trip from San Francisco to Washington takes $5\frac{1}{4}$ hours. After the plane is in the air $2\frac{1}{2}$ hours, how much time remains before landing?

92. A clerk can ride a bicycle to work in $\frac{3}{8}$ of an hour. The clerk can walk the distance in $1\frac{1}{4}$ hours. How much longer does it take to walk rather than ride to work?

93. A metal platform in a bridge is $280\frac{1}{2}$ feet during the summer. On a cold winter day the metal platform shrinks to $280\frac{3}{8}$ feet. How much did the bridge shrink due to the cold weather?

94. A housepainter painted $\frac{2}{5}$ of a house during 1 day. How much of the paint job remains to be done?

95. Find the missing dimension.

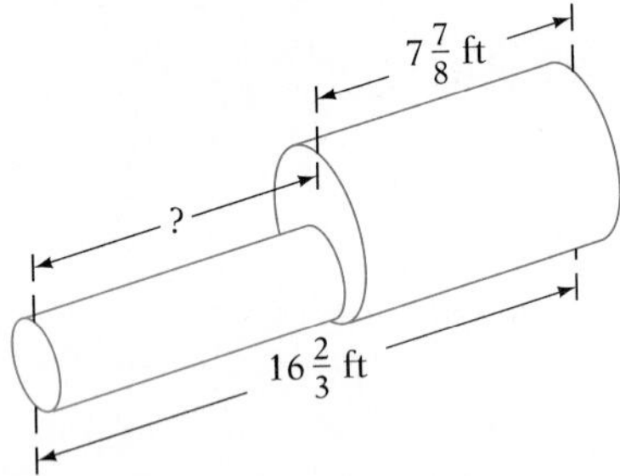

96. Find the missing dimension.

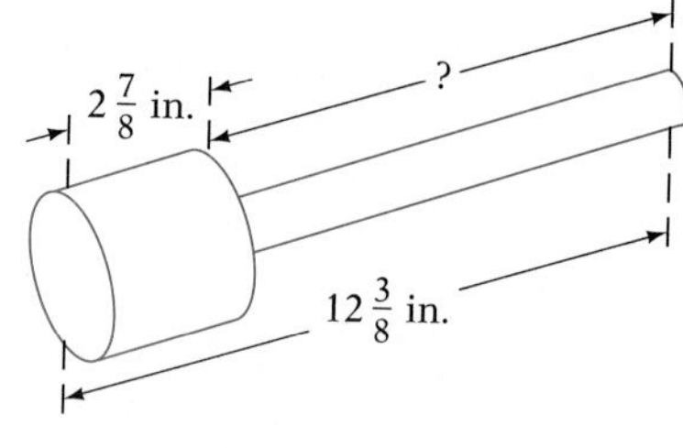

97. A 3-day $27\frac{1}{2}$-mile backpack hike is planned. The hikers plan to travel $7\frac{3}{8}$ miles the first day and $10\frac{1}{3}$ miles the second day.
a. How many miles do the hikers plan to travel the first 2 days?
b. How many miles are left to travel on the third day?

98. A 12-mile walkathon for charity has three checkpoints. The first checkpoint is $3\frac{3}{8}$ miles from the starting point. The second checkpoint is $4\frac{1}{3}$ miles from the first checkpoint.
a. How many miles is it from the starting point to the second checkpoint?
b. How many miles is it from the second checkpoint to the finish line?

99. A patient with high blood pressure is put on a diet to lose 38 pounds in 3 months. The patient loses $12\frac{1}{2}$ pounds the first month and $16\frac{3}{8}$ pounds the second month. How much weight must be lost the third month for the goal to be achieved?

100. To reach a desired weight of 172 pounds, a wrestler needs to lose $12\frac{3}{4}$ pounds. If $5\frac{1}{4}$ are lost the first week and $4\frac{1}{4}$ pounds the second week, how many more pounds must be lost for the desired weight to be reached?

SECTION 2.6 Multiplication of Fractions and Mixed Numbers

Objective A To multiply fractions

The product of two fractions is the product of the numerators over the product of the denominators.

Multiply: $\frac{2}{3} \times \frac{4}{5}$

Multiply the numerators.
Multiply the denominators.

$$\frac{2}{3} \times \frac{4}{5} = \frac{2 \cdot 4}{3 \cdot 5} = \frac{8}{15}$$

The product $\frac{2}{3} \times \frac{4}{5}$ can be read "$\frac{2}{3}$ times $\frac{4}{5}$" or "$\frac{2}{3}$ of $\frac{4}{5}$."

Reading the times sign as "of" is useful in application problems involving fractions and in diagramming the product of fractions.

$\frac{4}{5}$ of the bar is shaded.

Shade $\frac{2}{3}$ of the $\frac{4}{5}$ already shaded.

$\frac{8}{15}$ of the bar is then shaded.

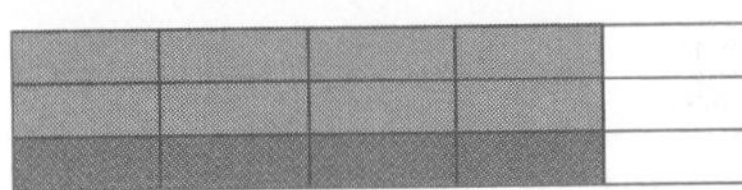

$\frac{2}{3}$ of $\frac{4}{5} = \frac{2}{3} \times \frac{4}{5} = \frac{8}{15}$

After multiplying two fractions, write the product in simplest form.

Multiply: $\frac{3}{4} \times \frac{14}{15}$

Multiply the fractions.	Write the prime factorization of each number.	Eliminate the common factors.	Multiply the numbers remaining in the numerator. Multiply the numbers remaining in the denominator.
$\frac{3}{4} \times \frac{14}{15} = \frac{3 \cdot 14}{4 \cdot 15}$	$= \frac{3 \cdot 2 \cdot 7}{2 \cdot 2 \cdot 3 \cdot 5}$	$= \frac{\overset{1}{\cancel{3}} \cdot \overset{1}{\cancel{2}} \cdot 7}{\underset{1}{\cancel{2}} \cdot 2 \cdot \underset{1}{\cancel{3}} \cdot 5}$	$\frac{7}{10}$

Example 1

Multiply: $\frac{9}{10} \times \frac{2}{3}$

Solution

$$\frac{9}{10} \times \frac{2}{3} = \frac{9 \cdot 2}{10 \cdot 3} = \frac{\overset{1}{\cancel{3}} \cdot 3 \cdot \overset{1}{\cancel{2}}}{\underset{1}{\cancel{2}} \cdot 5 \cdot \underset{1}{\cancel{3}}} = \frac{3}{5}$$

Example 2

Multiply: $\frac{12}{25} \times \frac{5}{16}$

Your solution

Solution on p. A14

Example 3

Multiply: $\frac{4}{15} \times \frac{5}{28}$

Solution

$$\frac{4}{15} \times \frac{5}{28} = \frac{4 \cdot 5}{15 \cdot 28} = \frac{\overset{1}{\cancel{2}} \cdot \overset{1}{\cancel{2}} \cdot \overset{1}{\cancel{5}}}{3 \cdot \underset{1}{\cancel{5}} \cdot \underset{1}{\cancel{2}} \cdot \underset{1}{\cancel{2}} \cdot 7} = \frac{1}{21}$$

Example 4

Multiply: $\frac{4}{21} \times \frac{7}{44}$

Your solution

Example 5

Multiply: $\frac{9}{20} \times \frac{33}{35}$

Solution

$$\frac{9}{20} \times \frac{33}{35} = \frac{9 \cdot 33}{20 \cdot 35} = \frac{3 \cdot 3 \cdot 3 \cdot 11}{2 \cdot 2 \cdot 5 \cdot 5 \cdot 7} = \frac{297}{700}$$

Example 6

Multiply: $\frac{2}{21} \times \frac{10}{33}$

Your solution

Example 7

Multiply: $\frac{14}{9} \times \frac{12}{7}$

Solution

$$\frac{14}{9} \times \frac{12}{7} = \frac{14 \cdot 12}{9 \cdot 7} = \frac{2 \cdot \overset{1}{\cancel{7}} \cdot 2 \cdot 2 \cdot \overset{1}{\cancel{3}}}{3 \cdot \underset{1}{\cancel{3}} \cdot \underset{1}{\cancel{7}}} = \frac{8}{3} = 2\frac{2}{3}$$

Example 8

Multiply: $\frac{16}{5} \times \frac{15}{24}$

Your solution

Solution on p. A14

Objective B To multiply whole numbers, mixed numbers, and fractions

To multiply a whole number by a fraction or mixed number, first write the whole number as a fraction with a denominator of 1.

Multiply: $4 \times \frac{3}{7}$

Write 4 with a denominator of 1; then multiply the fractions.

$$4 \times \frac{3}{7} = \frac{4}{1} \times \frac{3}{7} = \frac{4 \cdot 3}{1 \cdot 7} = \frac{2 \cdot 2 \cdot 3}{7} = \frac{12}{7} = 1\frac{5}{7}$$

When one or more of the factors in a product is a mixed number, write the mixed number as an improper fraction before multiplying.

Multiply: $2\frac{1}{3} \times \frac{3}{14}$

Write $2\frac{1}{3}$ as an improper fraction; then multiply the fractions.

$$2\frac{1}{3} \times \frac{3}{14} = \frac{7}{3} \times \frac{3}{14} = \frac{7 \cdot 3}{3 \cdot 14} = \frac{\overset{1}{\cancel{7}} \cdot \overset{1}{\cancel{3}}}{\underset{1}{\cancel{3}} \cdot 2 \cdot \underset{1}{\cancel{7}}} = \frac{1}{2}$$

Multiply: $5\frac{3}{5} \times 4\frac{2}{3}$

Write the mixed numbers as improper fractions; then multiply the fractions.

$$5\frac{3}{5} \times 4\frac{2}{3} = \frac{28}{5} \times \frac{14}{3} = \frac{28 \cdot 14}{5 \cdot 3} = \frac{2 \cdot 2 \cdot 7 \cdot 2 \cdot 7}{5 \cdot 3} = \frac{392}{15} = 26\frac{2}{15}$$

Example 9

Multiply: $4\frac{5}{6} \times \frac{12}{13}$

Solution

$$4\frac{5}{6} \times \frac{12}{13} = \frac{29}{6} \times \frac{12}{13} = \frac{29 \cdot 12}{6 \cdot 13}$$

$$= \frac{29 \cdot \overset{1}{\cancel{2}} \cdot 2 \cdot \overset{1}{\cancel{3}}}{\underset{1}{\cancel{2}} \cdot \underset{1}{\cancel{3}} \cdot 13} = \frac{58}{13} = 4\frac{6}{13}$$

Example 10

Multiply: $5\frac{2}{5} \times \frac{5}{9}$

Your solution

Example 11

Multiply: $\frac{3}{8} \times 6$

Solution

$$\frac{3}{8} \times 6 = \frac{3}{8} \times \frac{6}{1} = \frac{3 \cdot 6}{8 \cdot 1} = \frac{3 \cdot \overset{1}{\cancel{2}} \cdot 3}{2 \cdot \underset{1}{\cancel{2}} \cdot 2} = \frac{9}{4} = 2\frac{1}{4}$$

Example 12

Multiply: $9 \times \frac{7}{12}$

Your solution

Example 13

Multiply: $5\frac{2}{3} \times 4\frac{1}{2}$

Solution

$$5\frac{2}{3} \times 4\frac{1}{2} = \frac{17}{3} \times \frac{9}{2} = \frac{17 \cdot 9}{3 \cdot 2}$$

$$= \frac{17 \cdot \overset{1}{\cancel{3}} \cdot 3}{\underset{1}{\cancel{3}} \cdot 2} = \frac{51}{2} = 25\frac{1}{2}$$

Example 14

Multiply: $3\frac{2}{5} \times 6\frac{1}{4}$

Your solution

Example 15

Multiply: $4\frac{2}{5} \times 7$

Solution

$$4\frac{2}{5} \times 7 = \frac{22}{5} \times \frac{7}{1} = \frac{22 \cdot 7}{5 \cdot 1}$$

$$= \frac{2 \cdot 11 \cdot 7}{5} = \frac{154}{5} = 30\frac{4}{5}$$

Example 16

Multiply: $3\frac{2}{7} \times 6$

Your solution

Solutions on p. A14

Objective C To solve application problems

Example 17

An electrician earns $150 for each day worked. What are the electrician's earnings for working $4\frac{1}{2}$ days?

Strategy

To find the electrician's total earnings, multiply the daily earnings ($150) by the number of days worked $\left(4\frac{1}{2}\right)$.

Solution

$$150 \times 4\frac{1}{2} = \frac{150}{1} \times \frac{9}{2} = \frac{150 \cdot 9}{1 \cdot 2} = 675$$

The electrician's earnings are $675.

Example 18

A house increased in value $3\frac{1}{2}$ times during the last 10 years. The price of the house 10 years ago was $30,000. What is the value of the house today?

Your strategy

Your solution

Example 19

The value of a small office building and the land on which it is built is $90,000. The value of the land is $\frac{1}{4}$ the total value. What is the value of the building (in dollars)?

Strategy

To find the value of the building:

- Find the value of the land $\left(\frac{1}{4} \times \$90,000\right)$.
- Subtract the value of the land from the total value.

Solution

$$\$90,000 \times \frac{1}{4} = \frac{\$90,000}{4} = \$22,500 \quad \text{value of the land}$$

$$\begin{array}{r} \$90,000 \\ -22,500 \\ \hline \$67,500 \end{array}$$

The value of the building is $67,500.

Example 20

A paint company buys a drying chamber and an air compressor for spray painting. The total cost of the two items is $60,000. The drying chamber's cost is $\frac{4}{5}$ of the total cost. What is the cost of the air compressor?

Your strategy

Your solution

Solutions on p. A15

2.6 EXERCISES

Objective A

Multiply.

1. $\frac{2}{3} \times \frac{7}{8}$
2. $\frac{1}{2} \times \frac{2}{3}$
3. $\frac{5}{16} \times \frac{7}{15}$
4. $\frac{3}{8} \times \frac{6}{7}$
5. $\frac{1}{2} \times \frac{5}{6}$
6. $\frac{1}{6} \times \frac{1}{8}$
7. $\frac{2}{5} \times \frac{5}{6}$
8. $\frac{11}{12} \times \frac{6}{7}$
9. $\frac{11}{12} \times \frac{3}{5}$
10. $\frac{2}{5} \times \frac{4}{9}$
11. $\frac{1}{6} \times \frac{6}{7}$
12. $\frac{3}{5} \times \frac{10}{11}$
13. $\frac{1}{5} \times \frac{5}{8}$
14. $\frac{6}{7} \times \frac{14}{15}$
15. $\frac{4}{9} \times \frac{15}{16}$
16. $\frac{8}{9} \times \frac{27}{4}$
17. $\frac{3}{5} \times \frac{3}{10}$
18. $\frac{5}{6} \times \frac{1}{2}$
19. $\frac{3}{8} \times \frac{5}{12}$
20. $\frac{3}{10} \times \frac{6}{7}$
21. $\frac{16}{9} \times \frac{27}{8}$
22. $\frac{5}{8} \times \frac{16}{15}$
23. $\frac{3}{2} \times \frac{4}{9}$
24. $\frac{5}{3} \times \frac{3}{7}$
25. $\frac{5}{16} \times \frac{12}{11}$
26. $\frac{7}{8} \times \frac{3}{14}$
27. $\frac{2}{9} \times \frac{1}{5}$
28. $\frac{1}{10} \times \frac{3}{8}$
29. $\frac{5}{12} \times \frac{6}{7}$
30. $\frac{1}{3} \times \frac{9}{2}$
31. $\frac{15}{8} \times \frac{16}{3}$
32. $\frac{5}{6} \times \frac{4}{15}$
33. $\frac{1}{2} \times \frac{2}{15}$
34. $\frac{3}{8} \times \frac{5}{16}$
35. $\frac{2}{9} \times \frac{15}{17}$
36. $\frac{5}{7} \times \frac{14}{15}$
37. $\frac{3}{8} \times \frac{15}{41}$
38. $\frac{5}{12} \times \frac{42}{65}$
39. $\frac{16}{33} \times \frac{55}{72}$
40. $\frac{8}{3} \times \frac{21}{32}$
41. $\frac{12}{5} \times \frac{5}{3}$
42. $\frac{17}{9} \times \frac{81}{17}$
43. $\frac{16}{85} \times \frac{125}{84}$
44. $\frac{19}{64} \times \frac{48}{95}$

Objective B

Multiply.

45. $4 \times \frac{3}{8}$

46. $14 \times \frac{5}{7}$

47. $\frac{2}{3} \times 6$

48. $\frac{5}{12} \times 40$

49. $\frac{1}{3} \times 1\frac{1}{3}$

50. $\frac{2}{5} \times 2\frac{1}{2}$

51. $1\frac{7}{8} \times \frac{4}{15}$

52. $2\frac{1}{5} \times \frac{5}{22}$

53. $55 \times \frac{3}{10}$

54. $\frac{5}{14} \times 49$

55. $4 \times 2\frac{1}{2}$

56. $9 \times 3\frac{1}{3}$

57. $2\frac{1}{7} \times 3$

58. $5\frac{1}{4} \times 8$

59. $3\frac{2}{3} \times 5$

60. $4\frac{2}{9} \times 3$

61. $\frac{1}{2} \times 3\frac{3}{7}$

62. $\frac{3}{8} \times 4\frac{4}{5}$

63. $6\frac{1}{8} \times \frac{4}{7}$

64. $5\frac{1}{3} \times \frac{5}{16}$

65. $5\frac{1}{8} \times 5$

66. $6\frac{1}{9} \times 2$

67. $\frac{3}{8} \times 4\frac{1}{2}$

68. $\frac{5}{7} \times 2\frac{1}{3}$

69. $6 \times 2\frac{2}{3}$

70. $6\frac{1}{8} \times 0$

71. $1\frac{1}{3} \times 2\frac{1}{4}$

72. $2\frac{5}{8} \times \frac{3}{23}$

73. $1\frac{1}{2} \times 5\frac{1}{2}$

74. $3\frac{1}{8} \times 2\frac{1}{4}$

75. $6\frac{1}{2} \times 4\frac{3}{13}$

76. $7\frac{3}{5} \times 4\frac{1}{6}$

77. $5\frac{1}{2} \times 1\frac{3}{11}$

78. $2\frac{2}{5} \times 1\frac{7}{12}$

79. $3\frac{1}{2} \times 7$

80. $5\frac{2}{3} \times 12$

81. $3\frac{1}{3} \times 6\frac{3}{5}$

82. $5\frac{3}{7} \times 3\frac{1}{2}$

83. $4\frac{3}{13} \times 7\frac{4}{5}$

84. $7\frac{5}{12} \times 3\frac{3}{7}$

Multiply.

85. $2\frac{5}{8} \times 3\frac{2}{5}$

86. $5\frac{3}{16} \times 5\frac{1}{3}$

87. $3\frac{1}{7} \times 2\frac{1}{8}$

88. $16\frac{5}{8} \times 1\frac{1}{16}$

89. $9\frac{5}{8} \times 7\frac{6}{7}$

90. $2\frac{1}{3} \times 3\frac{1}{3}$

91. $2 \times 4\frac{1}{8}$

92. $3\frac{2}{3} \times 5\frac{2}{11}$

93. $5 \times \frac{2}{3}$

94. $0 \times 5\frac{2}{3}$

95. $2\frac{2}{7} \times 3\frac{1}{2}$

96. $4\frac{1}{2} \times \frac{1}{3}$

97. $2\frac{2}{3} \times 3\frac{1}{4}$

98. $\frac{3}{4} \times 5\frac{1}{3}$

99. $5\frac{1}{5} \times 3\frac{3}{4}$

100. $4\frac{1}{7} \times 2\frac{1}{3}$

101. $\frac{5}{8} \times 5\frac{3}{5}$

102. $6\frac{1}{3} \times 1\frac{2}{19}$

103. $1\frac{1}{2} \times 1\frac{1}{3}$

104. $3\frac{5}{8} \times 0$

105. $2\frac{2}{5} \times 3\frac{1}{12}$

106. $2\frac{2}{3} \times \frac{3}{20}$

107. $5\frac{1}{5} \times 3\frac{1}{13}$

108. $3\frac{3}{4} \times 2\frac{3}{20}$

109. $10\frac{1}{4} \times 3\frac{1}{5}$

110. $12\frac{3}{5} \times 1\frac{3}{7}$

111. $5\frac{3}{7} \times 5\frac{1}{4}$

112. $6\frac{1}{2} \times 1\frac{3}{13}$

113. $8\frac{2}{3} \times 2\frac{1}{13}$

114. $10\frac{2}{5} \times 3\frac{5}{26}$

115. $3\frac{1}{5} \times 2\frac{2}{3}$

116. $15\frac{1}{3} \times 6\frac{1}{2}$

117. $5\frac{3}{8} \times 6\frac{2}{5}$

118. $3\frac{1}{2} \times 2\frac{5}{14}$

119. $21\frac{5}{8} \times 24$

120. $16 \times 3\frac{3}{4}$

121. $84 \times 3\frac{3}{14}$

122. $5\frac{1}{8} \times 41$

123. $11\frac{1}{3} \times 7\frac{3}{34}$

124. $8\frac{7}{9} \times 6\frac{6}{11}$

Objective C *Application Problems*

125. A car can travel $23\frac{1}{2}$ miles on 1 gallon of gas. How far can the car travel on $\frac{1}{2}$ gallon of gas?

126. A person can walk $3\frac{1}{2}$ miles in 1 hour. How far can the person walk in $\frac{1}{3}$ hour?

127. A compact car travels 38 miles on each gallon of gasoline. How many miles can the car travel on $9\frac{1}{2}$ gallons of gasoline?

128. A board is $5\frac{3}{4}$ feet long. One third of the board is cut off. What is the length of the piece cut off?

129. A cook is using a recipe that calls for $3\frac{1}{2}$ cups of flour. The cook wants to double the recipe. How much flour should the cook use?

130. The F-1 engine in the first stage of the Saturn 5 rocket burns 214,000 gallons of propellant in 1 minute. The first stage burns $2\frac{1}{2}$ minutes before burnout. How much propellant is used before burnout?

131. The stock of a high technology company is selling for $\$68\frac{3}{8}$. Find the cost of 40 shares of the stock.

132. A soft drink company budgets $\frac{1}{10}$ of its income each month for advertising. In July the company had an income of $24,500. What is the amount budgeted for advertising in July?

133. A family budgets $\frac{2}{5}$ of its monthly income of $3200 per month for housing and utilities.
a. What amount is budgeted for housing and utilities?
b. What amount remains for purposes other than housing and utilities?

134. A student read $\frac{2}{3}$ of a book containing 432 pages.
a. How many pages did the student read?
b. How many pages remain to be read?

135. The parents of the school choir members are making robes for the choir. Each robe requires $2\frac{5}{8}$ yards of material at a cost of $8 per yard. Find the total cost of 24 choir robes.

136. A college spends $\frac{5}{8}$ of its monthly income on employee salaries. During one month the college had an income of $712,000. How much of the monthly income remained after the employee's salaries were paid?

SECTION 2.7 Division of Fractions and Mixed Numbers

Objective A To divide fractions

The **reciprocal** of a fraction is the fraction with the numerator and denominator interchanged. The process of interchanging the numerator and denominator of a fraction is called **inverting** the fraction.

The reciprocal of $\frac{2}{3}$ is $\frac{3}{2}$.

To find the reciprocal of a whole number, first write the whole number as a fraction with a denominator of 1; then find the reciprocal of that fraction.

The reciprocal of 5 is $\frac{1}{5}$. $\left(\text{Think } 5 = \frac{5}{1}\right)$

Reciprocals are used to rewrite division problems as related multiplication problems. Look at the following two problems:

$8 \div 2 = 4$ $8 \times \frac{1}{2} = 4$

8 divided by 2 is 4. 8 times the reciprocal of 2 is 4.

"Divided by" means the same as "times the reciprocal of." Thus "÷ 2" can be replaced with "× $\frac{1}{2}$," and the answer will be the same. Fractions are divided by making this replacement.

Divide: $\frac{2}{3} \div \frac{3}{4}$ $\frac{2}{3} \div \frac{3}{4} = \frac{2}{3} \times \frac{4}{3} = \frac{2 \cdot 4}{3 \cdot 3} = \frac{2 \cdot 2 \cdot 2}{3 \cdot 3} = \frac{8}{9}$

Example 1 Divide: $\frac{5}{8} \div \frac{4}{9}$

Solution $\frac{5}{8} \div \frac{4}{9} = \frac{5}{8} \times \frac{9}{4} = \frac{5 \cdot 9}{8 \cdot 4}$

$= \frac{5 \cdot 3 \cdot 3}{2 \cdot 2 \cdot 2 \cdot 2 \cdot 2} = \frac{45}{32} = 1\frac{13}{32}$

Example 2 Divide: $\frac{3}{7} \div \frac{2}{3}$

Your solution

Example 3 Divide: $\frac{3}{5} \div \frac{12}{25}$

Solution $\frac{3}{5} \div \frac{12}{25} = \frac{3}{5} \times \frac{25}{12} = \frac{3 \cdot 25}{5 \cdot 12}$

$= \frac{\overset{1}{\cancel{3}} \cdot \overset{1}{\cancel{5}} \cdot 5}{\underset{1}{\cancel{5}} \cdot 2 \cdot 2 \cdot \underset{1}{\cancel{3}}} = \frac{5}{4} = 1\frac{1}{4}$

Example 4 Divide: $\frac{3}{4} \div \frac{9}{10}$

Your solution

Solutions on p. A15

Objective B To divide whole numbers, mixed numbers, and fractions

To divide a fraction and a whole number, first write the whole number as a fraction with a denominator of 1.

Divide: $\frac{3}{7} \div 5$

Write 5 with a denominator of 1; then divide the fractions.

$\frac{3}{7} \div \boxed{5} = \frac{3}{7} \div \boxed{\frac{5}{1}} = \frac{3}{7} \times \frac{1}{5} = \frac{3 \cdot 1}{7 \cdot 5} = \frac{3}{35}$

When one of the numbers in a quotient is a mixed number, write the mixed number as an improper fraction before dividing.

Divide: $4\frac{2}{3} \div \frac{8}{15}$

Write $4\frac{2}{3}$ as an improper fraction; then divide the fractions.

$$4\frac{2}{3} \div \frac{8}{15} = \frac{14}{3} \div \frac{8}{15} = \frac{14}{3} \times \frac{15}{8} = \frac{14 \cdot 15}{3 \cdot 8} = \frac{\overset{1}{\cancel{2}} \cdot 7 \cdot \overset{1}{\cancel{3}} \cdot 5}{\underset{1}{\cancel{3}} \cdot 2 \cdot \underset{1}{\cancel{2}} \cdot 2} = \frac{35}{4} = 8\frac{3}{4}$$

Divide: $1\frac{13}{15} \div 4\frac{4}{5}$

Write the mixed numbers as improper fractions. Then divide the fractions.

$$1\frac{13}{15} \div 4\frac{4}{5} = \frac{28}{15} \div \frac{24}{5} = \frac{28}{15} \times \frac{5}{24} = \frac{28 \cdot 5}{15 \cdot 24} = \frac{\overset{1}{\cancel{2}} \cdot \overset{1}{\cancel{2}} \cdot 7 \cdot \overset{1}{\cancel{5}}}{3 \cdot \underset{1}{\cancel{5}} \cdot \underset{1}{\cancel{2}} \cdot \underset{1}{\cancel{2}} \cdot 2 \cdot 3} = \frac{7}{18}$$

Example 5 Divide: $6 \div \frac{3}{4}$

Solution

$$6 \div \frac{3}{4} = \frac{6}{1} \div \frac{3}{4} = \frac{6}{1} \times \frac{4}{3}$$
$$= \frac{2 \cdot \overset{1}{\cancel{3}} \cdot 2 \cdot 2}{\underset{1}{\cancel{3}}} = \frac{8}{1} = 8$$

Example 6 Divide: $8 \div \frac{2}{3}$

Your solution

Example 7 Divide: $\frac{4}{9} \div 5$

Solution

$$\frac{4}{9} \div 5 = \frac{4}{9} \div \frac{5}{1} = \frac{4}{9} \times \frac{1}{5}$$
$$= \frac{4 \cdot 1}{9 \cdot 5} = \frac{2 \cdot 2}{3 \cdot 3 \cdot 5} = \frac{4}{45}$$

Example 8 Divide: $\frac{5}{7} \div 6$

Your solution

Example 9 Divide: $\frac{3}{8} \div 2\frac{1}{10}$

Solution

$$\frac{3}{8} \div 2\frac{1}{10} = \frac{3}{8} \div \frac{21}{10} = \frac{3}{8} \times \frac{10}{21}$$
$$= \frac{3 \cdot 10}{8 \cdot 21} = \frac{\overset{1}{\cancel{3}} \cdot \overset{1}{\cancel{2}} \cdot 5}{\underset{1}{\cancel{2}} \cdot 2 \cdot 2 \cdot \underset{1}{\cancel{3}} \cdot 7} = \frac{5}{28}$$

Example 10 Divide: $12\frac{3}{5} \div 7$

Your solution

Example 11 Divide: $2\frac{3}{4} \div 1\frac{5}{7}$

Solution

$$2\frac{3}{4} \div 1\frac{5}{7} = \frac{11}{4} \div \frac{12}{7} = \frac{11}{4} \times \frac{7}{12} = \frac{11 \cdot 7}{4 \cdot 12}$$
$$= \frac{11 \cdot 7}{2 \cdot 2 \cdot 2 \cdot 2 \cdot 3} = \frac{77}{48} = 1\frac{29}{48}$$

Example 12 Divide: $3\frac{2}{3} \div 2\frac{2}{5}$

Your solution

Solutions on p. A15

Example 13 Divide: $1\frac{13}{15} \div 4\frac{1}{5}$

Solution

$$1\frac{13}{15} \div 4\frac{1}{5} = \frac{28}{15} \div \frac{21}{5} = \frac{28}{15} \times \frac{5}{21} = \frac{28 \cdot 5}{15 \cdot 21}$$

$$= \frac{2 \cdot 2 \cdot \overset{1}{\cancel{7}} \cdot \overset{1}{\cancel{5}}}{3 \cdot \underset{1}{\cancel{5}} \cdot 3 \cdot \underset{1}{\cancel{7}}} = \frac{4}{9}$$

Example 14 Divide: $2\frac{5}{6} \div 8\frac{1}{2}$

Your solution

Example 15 Divide: $4\frac{3}{8} \div 7$

Solution

$$4\frac{3}{8} \div 7 = \frac{35}{8} \div \frac{7}{1} = \frac{35}{8} \times \frac{1}{7}$$

$$= \frac{35 \cdot 1}{8 \cdot 7} = \frac{5 \cdot \overset{1}{\cancel{7}}}{2 \cdot 2 \cdot 2 \cdot \underset{1}{\cancel{7}}} = \frac{5}{8}$$

Example 16 Divide: $6\frac{2}{5} \div 4$

Your solution

Solutions on p. A15

Objective C To solve application problems

Example 17

A car used $15\frac{1}{2}$ gallons of gasoline on a 310-mile trip. How many miles can this car travel on 1 gallon of gasoline?

Strategy

To find the number of miles, divide the number of miles traveled by the number of gallons of gasoline used.

Solution

$$310 \div 15\frac{1}{2} = \frac{310}{1} \div \frac{31}{2}$$

$$= \frac{310}{1} \times \frac{2}{31}$$

$$= \frac{310 \cdot 2}{1 \cdot 31}$$

$$= 20$$

The car travels 20 miles on 1 gallon of gasoline.

Example 18

An investor purchased a $2\frac{1}{2}$-ounce gold coin for \$975. What was the price for 1 ounce?

Your strategy

Your solution

Solution on p. A15

Example 19

A 12-foot board is cut into pieces $2\frac{1}{4}$ feet long for use as bookshelves. What is the length of the remaining piece after as many shelves as possible are cut?

Strategy

To find the length of the remaining piece:

- Divide the total length by the length of each shelf $\left(12 \div 2\frac{1}{4}\right)$. This will give you the number of shelves cut, with a certain fraction of a shelf left over.
- Multiply the fraction left over by the length of a shelf to determine the actual length of the piece remaining.

Solution

$$12 \div 2\frac{1}{4} = \frac{12}{1} \div \frac{9}{4} = \frac{12}{1} \times \frac{4}{9}$$
$$= \frac{12 \cdot 4}{1 \cdot 9} = \frac{16}{3} = 5\frac{1}{3}$$

5 pieces $2\frac{1}{4}$ feet long

1 piece $\frac{1}{3}$ of $2\frac{1}{4}$ feet long

$$\frac{1}{3} \times 2\frac{1}{4} = \frac{1}{3} \times \frac{9}{4} = \frac{1 \cdot 9}{3 \cdot 4} = \frac{3}{4}$$

The length of the piece remaining is $\frac{3}{4}$ foot.

Example 20

A 16-foot board is cut into pieces $3\frac{1}{3}$ feet long for shelves for a bookcase. What is the length of the remaining piece after as many shelves as possible are cut?

Your strategy

Your solution

Solution on p. A15

2.7 EXERCISES

Objective A

Divide.

1. $\frac{1}{3} \div \frac{2}{5}$
2. $\frac{3}{7} \div \frac{3}{2}$
3. $\frac{31}{40} \div \frac{11}{15}$
4. $\frac{3}{7} \div \frac{3}{7}$
5. $0 \div \frac{1}{2}$
6. $0 \div \frac{3}{4}$
7. $\frac{16}{33} \div \frac{4}{11}$
8. $\frac{5}{24} \div \frac{15}{36}$
9. $\frac{11}{15} \div \frac{1}{12}$
10. $\frac{2}{9} \div \frac{16}{19}$
11. $\frac{15}{16} \div \frac{16}{39}$
12. $\frac{2}{15} \div \frac{3}{5}$
13. $\frac{8}{9} \div \frac{4}{5}$
14. $\frac{11}{15} \div \frac{5}{22}$
15. $\frac{12}{13} \div \frac{4}{9}$
16. $\frac{1}{9} \div \frac{2}{3}$
17. $\frac{10}{21} \div \frac{5}{7}$
18. $\frac{2}{5} \div \frac{4}{7}$
19. $\frac{3}{8} \div \frac{5}{12}$
20. $\frac{5}{9} \div \frac{15}{32}$
21. $\frac{1}{2} \div \frac{1}{4}$
22. $\frac{1}{3} \div \frac{1}{9}$
23. $\frac{1}{5} \div \frac{1}{10}$
24. $\frac{4}{15} \div \frac{2}{5}$
25. $\frac{7}{15} \div \frac{14}{5}$
26. $\frac{5}{8} \div \frac{15}{2}$
27. $\frac{14}{3} \div \frac{7}{9}$
28. $\frac{7}{4} \div \frac{9}{2}$
29. $\frac{5}{9} \div \frac{25}{3}$
30. $\frac{5}{16} \div \frac{3}{8}$
31. $\frac{2}{3} \div \frac{1}{3}$
32. $\frac{4}{9} \div \frac{1}{9}$
33. $\frac{5}{7} \div \frac{2}{7}$
34. $\frac{5}{6} \div \frac{1}{9}$
35. $\frac{2}{3} \div \frac{2}{9}$
36. $\frac{5}{12} \div \frac{5}{6}$
37. $4 \div \frac{2}{3}$
38. $\frac{2}{3} \div 4$
39. $\frac{3}{2} \div 3$
40. $3 \div \frac{3}{2}$

Divide.

41. $\frac{15}{8} \div \frac{5}{16}$

42. $\frac{15}{8} \div \frac{5}{32}$

43. $16 \div \frac{24}{31}$

44. $\frac{8}{19} \div \frac{7}{38}$

Objective B

Divide.

45. $\frac{5}{6} \div 25$

46. $22 \div \frac{3}{11}$

47. $6 \div 3\frac{1}{3}$

48. $5\frac{1}{2} \div 11$

49. $3\frac{1}{3} \div \frac{3}{8}$

50. $6\frac{1}{2} \div \frac{1}{2}$

51. $\frac{3}{8} \div 2\frac{1}{4}$

52. $\frac{5}{12} \div 4\frac{4}{5}$

53. $1\frac{1}{2} \div 1\frac{3}{8}$

54. $2\frac{1}{4} \div 1\frac{3}{8}$

55. $1\frac{3}{5} \div 2\frac{1}{10}$

56. $2\frac{5}{6} \div 1\frac{1}{9}$

57. $2\frac{1}{3} \div 3\frac{2}{3}$

58. $4\frac{1}{2} \div 2\frac{1}{6}$

59. $7\frac{1}{2} \div 2\frac{2}{3}$

60. $8\frac{1}{4} \div 2\frac{3}{4}$

61. $3\frac{5}{9} \div 32$

62. $4\frac{1}{5} \div 21$

63. $6\frac{8}{9} \div \frac{31}{36}$

64. $5\frac{3}{5} \div \frac{7}{10}$

65. $\frac{11}{12} \div 2\frac{1}{3}$

66. $\frac{7}{8} \div 3\frac{1}{4}$

67. $\frac{5}{16} \div 5\frac{3}{8}$

68. $\frac{9}{14} \div 3\frac{1}{7}$

69. $27 \div \frac{3}{8}$

70. $35 \div \frac{7}{24}$

71. $\frac{3}{8} \div 2\frac{3}{4}$

72. $\frac{11}{18} \div 2\frac{2}{9}$

73. $\frac{21}{40} \div 3\frac{3}{10}$

74. $\frac{6}{25} \div 4\frac{1}{5}$

75. $2\frac{1}{16} \div 2\frac{1}{2}$

76. $7\frac{3}{5} \div 1\frac{7}{12}$

77. $1\frac{2}{3} \div \frac{3}{8}$

78. $16 \div \frac{2}{3}$

79. $2\frac{1}{2} \div 3\frac{5}{14}$

80. $1\frac{5}{8} \div 4$

81. $13\frac{3}{8} \div \frac{1}{4}$

82. $16 \div 1\frac{1}{2}$

83. $9 \div \frac{7}{8}$

84. $15\frac{1}{3} \div 2\frac{2}{9}$

Divide.

85. $16\frac{5}{8} \div 1\frac{2}{3}$

86. $24\frac{4}{5} \div 2\frac{3}{5}$

87. $1\frac{1}{3} \div 5\frac{8}{9}$

88. $13\frac{2}{3} \div 0$

89. $15\frac{1}{3} \div 7\frac{1}{6}$

90. $2\frac{3}{49} \div 1\frac{11}{14}$

91. $0 \div 1\frac{1}{2}$

92. $3\frac{5}{12} \div 2\frac{1}{6}$

93. $4\frac{5}{16} \div 2\frac{3}{8}$

94. $17\frac{4}{5} \div 9\frac{3}{10}$

95. $6\frac{1}{3} \div 5$

96. $15 \div \frac{1}{8}$

97. $\frac{1}{12} \div 12$

98. $14\frac{3}{8} \div 6\frac{7}{12}$

99. $3\frac{5}{11} \div 3\frac{4}{5}$

100. $1\frac{25}{36} \div 2\frac{11}{24}$

101. $19\frac{1}{2} \div \frac{1}{2}$

102. $\frac{1}{4} \div 2\frac{5}{6}$

103. $5\frac{3}{8} \div 2\frac{7}{16}$

104. $7\frac{4}{13} \div 2\frac{3}{26}$

105. $82\frac{3}{4} \div 6\frac{1}{8}$

106. $21\frac{5}{14} \div 7\frac{3}{28}$

107. $19\frac{3}{7} \div 4\frac{5}{14}$

108. $16\frac{5}{21} \div 3\frac{9}{42}$

109. $82\frac{3}{5} \div 19\frac{1}{10}$

110. $45\frac{3}{5} \div 15$

111. $102 \div 1\frac{1}{2}$

112. $44\frac{5}{12} \div 32\frac{5}{6}$

113. $0 \div 3\frac{1}{2}$

114. $8\frac{2}{7} \div 1$

115. $6\frac{9}{16} \div 1\frac{3}{32}$

116. $5\frac{5}{14} \div 3\frac{4}{7}$

117. $8\frac{8}{9} \div 2\frac{13}{18}$

118. $10\frac{1}{5} \div 1\frac{7}{10}$

119. $7\frac{3}{8} \div 1\frac{27}{32}$

120. $4\frac{5}{11} \div 2\frac{1}{3}$

121. $15\frac{3}{8} \div 7\frac{3}{4}$

122. $7\frac{2}{9} \div 1\frac{17}{18}$

123. $3\frac{7}{8} \div 7\frac{3}{4}$

124. $9\frac{7}{12} \div \frac{5}{12}$

Objective C *Application Problems*

125. Individual cereal boxes contain $\frac{3}{4}$ ounce of cereal. How many boxes can be filled with 600 ounces of cereal?

126. A box contains 20 ounces of cereal. How many $1\frac{1}{4}$-ounce portions can be served from this box?

127. A sales executive used $18\frac{2}{5}$ gallons of gas on a 460-mile trip. How many miles can the sales executive travel on 1 gallon of gas?

128. A commuter plane used $63\frac{3}{4}$ gallons of fuel on a 5-hour flight. How many gallons were used each hour?

129. A $\frac{5}{8}$-carat diamond was purchased for $1200. What would a similar diamond weighing 1 carat cost?

130. An investor bought $8\frac{1}{3}$ acres of land for $200,000. What was the cost of each acre?

131. A utility stock is offered for $$4\frac{3}{4}$ per share. How many shares can you buy for $285?

132. A new play opened, and 1200 people attended the opening at the concert hall. The hall was $\frac{2}{3}$ full. What is the capacity of the concert hall?

133. A contractor purchased $9\frac{3}{4}$ acres of land for a building project. One and one-half acres were set aside for a park.

a. How many acres are available for building?

b. How many $\frac{1}{4}$-acre parcels of land can be sold after the land for the park is removed?

134. A chef purchased a roast that weighed $10\frac{3}{4}$ pounds. After the fat was trimmed and the bone removed, the roast weighed $9\frac{1}{3}$ pounds.

a. What was the total weight of the fat and bone?

b. How many $\frac{1}{3}$-pound meat servings can the chef cut from the roast?

135. One-tenth of a shipment of $18\frac{1}{3}$ pounds of grapes were spoiled. How many $\frac{1}{2}$-pound packages of unspoiled grapes can be packaged from this shipment?

136. A 15-foot board is cut into pieces $3\frac{1}{2}$ feet long for a bookcase. What is the length of the piece remaining after as many shelves as possible have been cut?

2.8 EXERCISES

Objective A

Place the correct symbol, < or >, between the two numbers.

1. $\frac{11}{40}\quad\frac{19}{40}$
2. $\frac{92}{103}\quad\frac{19}{103}$
3. $\frac{2}{3}\quad\frac{5}{7}$
4. $\frac{2}{5}\quad\frac{3}{8}$
5. $\frac{5}{8}\quad\frac{7}{12}$
6. $\frac{11}{16}\quad\frac{17}{24}$
7. $\frac{7}{9}\quad\frac{11}{12}$
8. $\frac{5}{12}\quad\frac{7}{15}$
9. $\frac{13}{14}\quad\frac{19}{21}$
10. $\frac{13}{18}\quad\frac{7}{12}$
11. $\frac{7}{24}\quad\frac{11}{30}$
12. $\frac{13}{36}\quad\frac{19}{48}$

Objective B

Simplify.

13. $\left(\frac{3}{8}\right)^2$
14. $\left(\frac{5}{12}\right)^2$
15. $\left(\frac{2}{9}\right)^3$
16. $\left(\frac{5}{7}\right)^3$
17. $\left(\frac{1}{2}\right)\cdot\left(\frac{2}{3}\right)^2$
18. $\left(\frac{2}{3}\right)\cdot\left(\frac{1}{2}\right)^4$
19. $\left(\frac{1}{3}\right)^2\cdot\left(\frac{3}{5}\right)^3$
20. $\left(\frac{2}{5}\right)^2\cdot\left(\frac{5}{8}\right)^3$
21. $\left(\frac{2}{5}\right)^3\cdot\left(\frac{5}{7}\right)^2$
22. $\left(\frac{5}{9}\right)^3\cdot\left(\frac{18}{25}\right)^2$
23. $\left(\frac{1}{3}\right)^4\cdot\left(\frac{9}{11}\right)^2$
24. $\left(\frac{8}{15}\right)^3\cdot\left(\frac{5}{8}\right)^4$
25. $\left(\frac{1}{2}\right)^6\cdot\left(\frac{32}{35}\right)^2$
26. $\left(\frac{2}{3}\right)^4\cdot\left(\frac{81}{100}\right)^2$
27. $\left(\frac{1}{6}\right)\cdot\left(\frac{6}{7}\right)^2\cdot\left(\frac{2}{3}\right)$
28. $\left(\frac{2}{7}\right)\cdot\left(\frac{7}{8}\right)^2\cdot\left(\frac{8}{9}\right)$
29. $3\cdot\left(\frac{3}{5}\right)^3\cdot\left(\frac{1}{3}\right)^2$
30. $4\cdot\left(\frac{3}{4}\right)^3\cdot\left(\frac{4}{7}\right)^2$
31. $11\cdot\left(\frac{3}{8}\right)^3\cdot\left(\frac{8}{11}\right)^2$
32. $5\cdot\left(\frac{3}{5}\right)^3\cdot\left(\frac{2}{3}\right)^4$
33. $\left(\frac{2}{7}\right)^2\cdot\left(\frac{7}{9}\right)^2\cdot\left(\frac{9}{11}\right)^2$

Objective C

Simplify using the Order of Operations Agreement.

34. $\frac{1}{2} - \frac{1}{3} + \frac{2}{3}$

35. $\frac{2}{5} + \frac{3}{10} - \frac{2}{3}$

36. $\frac{1}{3} \div \frac{1}{2} + \frac{3}{4}$

37. $\frac{3}{5} \div \frac{6}{7} + \frac{4}{5}$

38. $\frac{4}{5} + \frac{3}{7} \cdot \frac{14}{15}$

39. $\frac{2}{3} + \frac{5}{8} \cdot \frac{16}{35}$

40. $\left(\frac{3}{4}\right)^2 - \frac{5}{12}$

41. $\left(\frac{3}{5}\right)^3 - \frac{3}{25}$

42. $\frac{5}{6} \cdot \left(\frac{2}{3} - \frac{1}{6}\right) + \frac{7}{18}$

43. $\frac{3}{4} \cdot \left(\frac{11}{12} - \frac{7}{8}\right) + \frac{5}{16}$

44. $\frac{7}{12} - \left(\frac{2}{3}\right)^2 + \frac{5}{8}$

45. $\frac{11}{16} - \left(\frac{3}{4}\right)^2 + \frac{7}{12}$

46. $\frac{3}{4} \cdot \left(\frac{4}{9}\right)^2 + \frac{1}{2}$

47. $\frac{9}{10} \cdot \left(\frac{2}{3}\right)^3 + \frac{2}{3}$

48. $\left(\frac{1}{2} + \frac{3}{4}\right) \div \frac{5}{8}$

49. $\left(\frac{2}{3} + \frac{5}{6}\right) \div \frac{5}{9}$

50. $\frac{3}{8} \div \left(\frac{5}{12} + \frac{3}{8}\right)$

51. $\frac{7}{12} \div \left(\frac{2}{3} + \frac{5}{9}\right)$

52. $\left(\frac{3}{8}\right)^2 \div \left(\frac{3}{7} + \frac{3}{14}\right)$

53. $\left(\frac{5}{6}\right)^2 \div \left(\frac{5}{12} + \frac{2}{3}\right)$

54. $\frac{2}{5} \div \frac{3}{8} \cdot \frac{4}{5}$

55. $\frac{7}{8} - \left(\frac{1}{6} + \frac{5}{12}\right) \div \frac{7}{8}$

56. $\left(\frac{3}{4}\right)^2 \cdot \left(\frac{1}{4} + \frac{3}{8}\right) \div \frac{5}{12}$

57. $\left(\frac{2}{3}\right)^2 + \left(\frac{1}{2} - \frac{1}{3}\right) \div \frac{3}{8}$

58. $\left(1\frac{1}{3} - \frac{5}{6}\right) + \frac{7}{8} \div \left(\frac{1}{2}\right)^3$

Calculators and Computers

Fractions The program TO ADD OR SUBTRACT FRACTIONS can be found on the Math ACE disk. This program will allow you to practice adding and subtracting fractions.

The program is designed to provide you with a problem. Then, using a pencil and paper, you are to solve the problem. After you have solved the problem, press the RETURN key; the solution will be displayed. Compare this solution with your solution.

Chapter Summary

Key Words The *least common multiple* (LCM) is the smallest common multiple of two or more numbers.

The *greatest common factor* (GCF) is the largest common factor of two or more numbers.

A *fraction* can represent the number of equal parts of a whole.

A *proper fraction* is a fraction less than 1.

A *mixed number* is a number greater than 1 with a whole number part and a fractional part.

An *improper fraction* is a fraction greater than or equal to 1.

Equal fractions with different denominators are called *equivalent fractions.*

A fraction is in *simplest form* when there are no common factors in the numerator and the denominator.

The *reciprocal* of a fraction is the fraction with the numerator and denominator interchanged.

Inverting is the process of finding the reciprocal of a fraction.

Essential Rules ***Addition of Fractions with Like Denominators***

To add fractions with like denominators, add the numerators and place the sum over the common denominator.

Addition of Fractions with Unlike Denominators

To add fractions with unlike denominators, first rewrite the fractions as equivalent fractions with the same denominator. Then add the numerators and place the sum over the common denominator.

Subtraction of Fractions with Like Denominators

To subtract fractions with like denominators, subtract the numerators and place the difference over the common denominator.

Subtraction of Fractions with Unlike Denominators

To subtract fractions with unlike denominators, rewrite the fractions as equivalent fractions with the same denominator. Then subtract the numerators and place the difference over the common denominator.

Multiplication of Fractions

To multiply two fractions, multiply the numerators and place the product over the product of the denominators.

Division of Fractions

To divide two fractions, multiply by the reciprocal of the divisor.

Chapter Review

SECTION 2.1

1. Find the LCM of 18 and 12.

2. Find the LCM of 18 and 27.

3. Find the GCF of 20 and 48.

4. Find the GCF of 15 and 25.

SECTION 2.2

5. Express the shaded portion of the circles as an improper fraction.

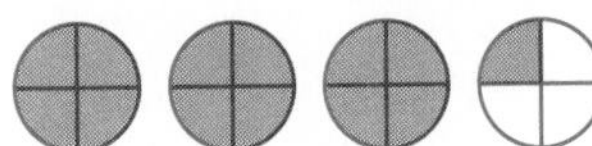

6. Express the shaded portion of the circles as a mixed number.

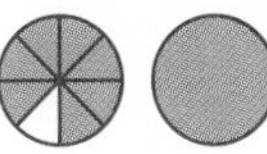

7. Write $\frac{17}{5}$ as a mixed number.

8. Write $2\frac{5}{7}$ as an improper fraction.

SECTION 2.3

9. Write an equivalent fraction with the given denominator.
$\frac{2}{3} = \frac{\quad}{36}$

10. Write an equivalent fraction with the given denominator.
$\frac{8}{11} = \frac{\quad}{44}$

11. Write $\frac{30}{45}$ in simplest form.

12. Write $\frac{16}{44}$ in simplest form.

SECTION 2.4

13. Add: $\frac{3}{8} + \frac{5}{8} + \frac{1}{8}$

14. Add: $\frac{2}{3} + \frac{5}{6} + \frac{2}{9}$

15. Add: $\frac{3}{8} + 1\frac{2}{3} + 3\frac{5}{6}$

16. Add: $4\frac{4}{9} + 2\frac{1}{6} + 11\frac{17}{27}$

17. During 3 months of the rainy season, $5\frac{7}{8}$, $6\frac{2}{3}$, and $8\frac{3}{4}$ inches of rain fell. Find the total rainfall for the 3 months.

SECTION 2.5

18. Subtract: $\frac{11}{18} - \frac{5}{18}$

19. Subtract: $\frac{17}{24} - \frac{3}{16}$

20. Subtract: $\begin{array}{r} 18\frac{1}{6} \\ -\ 3\frac{5}{7} \\ \hline \end{array}$

21. Subtract: $\begin{array}{r} 16 \\ -\ 5\frac{7}{8} \\ \hline \end{array}$

22. A 15-mile race has three checkpoints. The first checkpoint is $4\frac{1}{2}$ miles from the starting point. The second checkpoint is $5\frac{3}{4}$ miles from the first checkpoint. How many miles is the second checkpoint from the finish line?

SECTION 2.6

23. Multiply: $\frac{5}{12} \times \frac{4}{25}$

24. Multiply: $\frac{11}{50} \times \frac{25}{44}$

25. Multiply: $2\frac{1}{3} \times 3\frac{7}{8}$

26. Multiply: $2\frac{1}{4} \times 7\frac{1}{3}$

27. A compact car gets 36 miles on each gallon of gasoline. How many miles can the car travel on $6\frac{3}{4}$ gallons of gasoline?

SECTION 2.7

28. Divide: $\frac{5}{6} \div \frac{5}{12}$

29. Divide: $\frac{15}{28} \div \frac{5}{7}$

30. Divide: $1\frac{1}{3} \div \frac{2}{3}$

31. Divide: $8\frac{2}{3} \div 2\frac{3}{5}$

32. A home building contractor bought $4\frac{2}{3}$ acres for \$168,000. What was the cost of each acre?

SECTION 2.8

33. Place the correct symbol, < or >, between the two numbers.
$\frac{11}{18} \quad \frac{17}{24}$

34. Simplify: $\left(\frac{3}{4}\right)^3 \cdot \frac{20}{27}$

35. Simplify: $\frac{2}{7}\left(\frac{5}{8} - \frac{1}{3}\right) \div \frac{3}{5}$

36. Simplify: $\left(\frac{4}{5} - \frac{2}{3}\right)^2 \div \frac{4}{15}$

Chapter Test

1. Find the LCM of 24 and 40.

2. Find the GCF of 24 and 80.

3. Express the shaded portion of the circles as an improper fraction.

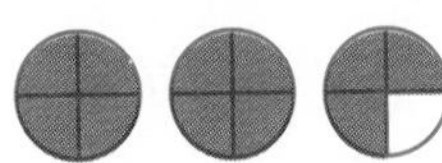

4. Write $\frac{18}{5}$ as a mixed number.

5. Write $9\frac{4}{5}$ as an improper fraction.

6. Write an equivalent fraction with the given denominator.

$\frac{5}{8} = \frac{}{72}$

7. Write $\frac{40}{64}$ in simplest form.

8. Add: $\frac{7}{12} + \frac{11}{12} + \frac{5}{12}$

9. Add:

$$\begin{array}{r} \frac{5}{6} \\ \frac{7}{9} \\ +\frac{1}{15} \\ \hline \end{array}$$

10. Add:

$$\begin{array}{r} 12\frac{5}{12} \\ +\ 9\frac{17}{20} \\ \hline \end{array}$$

11. The rainfall for a 3-month period was $11\frac{1}{2}$ inches, $7\frac{5}{8}$ inches, and $2\frac{1}{3}$ inches. Find the total rainfall for the 3 months.

12. Subtract: $\frac{17}{24} - \frac{11}{24}$

13. Subtract:

$$\begin{array}{r} \frac{9}{16} \\ -\frac{5}{12} \\ \hline \end{array}$$

14. Subtract:

$$\begin{array}{r} 23\frac{1}{8} \\ -\ 9\frac{9}{44} \\ \hline \end{array}$$

15. An investor bought 100 shares of a utility stock at $\$24\frac{1}{2}$. The stock gained $\$5\frac{5}{8}$ during the first month of ownership and lost $\$2\frac{1}{4}$ during the second month. Find the value of 1 share of the utility stock at the end of the second month.

16. Multiply: $\frac{9}{11} \times \frac{44}{81}$

17. Multiply: $5\frac{2}{3} \times 1\frac{7}{17}$

18. An electrician earns $120 for each day worked. What is the total of the electrician's earnings for working $3\frac{1}{2}$ days?

19. Divide: $\frac{5}{9} \div \frac{7}{18}$

20. Divide: $6\frac{2}{3} \div 3\frac{1}{6}$

21. An investor bought $7\frac{1}{4}$ acres of land for a housing project. One and three-fourths acres were set aside for a park, and the remaining land was developed into $\frac{1}{2}$-acre lots. How many lots were available for sale?

22. Place the correct symbol, < or >, between the two numbers.
$\frac{3}{8} \quad \frac{5}{12}$

23. Simplify: $\left(\frac{2}{3}\right)^4 \cdot \frac{27}{32}$

24. Simplify: $\left(\frac{3}{4}\right)^2 \div \left(\frac{2}{3} + \frac{5}{6}\right) - \frac{1}{12}$

25. Simplify: $\left(\frac{1}{4}\right)^3 \div \left(\frac{1}{8}\right)^2 - \frac{1}{6}$

Cumulative Review

1. Round 290,496 to the nearest thousand.

2. Subtract: $\begin{array}{r} 390{,}047 \\ -\ 98{,}769 \\ \hline \end{array}$

3. Multiply: $\begin{array}{r} 926 \\ \times\ 79 \\ \hline \end{array}$

4. Divide: $57\overline{)30{,}792}$

5. Simplify: $4 \cdot (6 - 3) \div 6 - 1$

6. Find the prime factorization of 44.

7. Find the LCM of 30 and 42.

8. Find the GCF of 60 and 80.

9. Write $7\frac{2}{3}$ as an improper fraction.

10. Write $\frac{25}{4}$ as a mixed number.

11. Write an equivalent fraction with the given denominator.
$\frac{5}{16} = \frac{}{48}$

12. Write $\frac{24}{60}$ in simplest form.

13. Add: $\frac{7}{12} + \frac{9}{16}$

14. Add: $\begin{array}{r} 3\frac{7}{8} \\ 7\frac{5}{12} \\ +2\frac{15}{16} \\ \hline \end{array}$

15. Subtract: $\frac{11}{12} - \frac{3}{8}$

16. Subtract: $\begin{array}{r} 5\frac{1}{6} \\ -3\frac{7}{18} \\ \hline \end{array}$

17. Multiply: $\frac{3}{8} \times \frac{14}{15}$

18. Multiply: $3\frac{1}{8} \times 2\frac{2}{5}$

19. Divide: $\frac{7}{16} \div \frac{5}{12}$

20. Divide: $6\frac{1}{8} \div 2\frac{1}{3}$

21. Simplify: $\left(\frac{1}{2}\right)^3 \cdot \frac{8}{9}$

22. Simplify: $\left(\frac{1}{2} + \frac{1}{3}\right) \div \left(\frac{2}{5}\right)^2$

23. A student had \$1359 in a checking account. During the week, the student wrote checks of \$128, \$54, and \$315. Find the amount in the checking account at the end of the week.

24. The tickets for a movie were \$5 for an adult and \$2 for a student. Find the total income from the sale of 87 adult tickets and 135 student tickets.

25. Find the total weight of three packages that weigh $1\frac{1}{2}$ pounds, $7\frac{7}{8}$ pounds, and $2\frac{2}{3}$ pounds.

26. A board $2\frac{5}{8}$ feet long is cut from a board $7\frac{1}{3}$ feet long. What is the length of the remaining piece?

27. A car travels 27 miles on each gallon of gasoline. How many miles can the car travel on $8\frac{1}{3}$ gallons of gasoline?

28. A contractor purchased $10\frac{1}{3}$ acres of land to build a housing development. The contractor donated 2 acres for a park. How many $\frac{1}{3}$-acre parcels can be sold from the remaining land?

3 Decimals

OBJECTIVES

- To write decimals in standard form and in words
- To round a decimal to a given place value
- To add decimals
- To solve application problems
- To subtract decimals
- To solve application problems
- To multiply decimals
- To solve application problems
- To divide decimals
- To solve application problems
- To convert fractions to decimals
- To convert decimals to fractions
- To identify the order relation between two decimals or between a decimal and a fraction

Decimal Fractions

How would you like to add $\frac{37,544}{23,465} + \frac{5184}{3456}$? These two fractions are very cumbersome, and it would take even a mathematician some time to get the answer.

Well, around 1550, help was on the way with the publication of a book called *La Disme* ("The Tenth"), which urged the use of decimal fractions. A decimal fraction is one in which the denominator is 10, 100, 1000, 10,000, and so on.

This book suggested that all whole numbers were "units" and when written would end with the symbol ⓪. For example, the number 294⓪ would be the number two hundred ninety-four. This is very much like the way numbers are currently written (except for the ⓪).

For a fraction between 0 and 1, a unit was broken down into parts called "primes." The fraction three-tenths would be written

$\frac{3}{10}$ = 3①

The ① was used to mean the end of the primes, or what are now called tenths.

Each prime was further broken down to "seconds," and each second was broken down to "thirds," and so on. The primes ended with ①, the seconds ended with ②, and the thirds ended with ③.

Examples of these numbers in the old notation and our modern fraction notation are shown below.

$\frac{37}{100}$ = 3①7② $\frac{257}{1000}$ = 2①5②7③

After completing this chapter, you might come back to the problem in the first line. Use decimals instead of fractions to find the answer. The answer is 3.1.

SECTION 3.1 Introduction to Decimals

Objective A To write decimals in standard form and in words

The smallest human bone is found in the middle ear and measures 0.135 inch in length. The number 0.135 is in **decimal notation.**

Note the relationship between fractions and numbers written in decimal notation.

Three-tenths	Three-hundredths	Three-thousandths
$\frac{3}{10} = 0.\underline{3}$	$\frac{3}{100} = 0.\underline{03}$	$\frac{3}{1000} = 0.\underline{003}$
1 zero 1 decimal place	2 zeros 2 decimal places	3 zeros 3 decimal places

A number written in decimal notation has three parts.

351	.	7089
Whole number part	**Decimal point**	**Decimal part**

A number written in decimal notation is often called simply a **decimal.** The position of a digit in a decimal determines the digit's place value.

In the decimal 351.7089, the position of the digit 9 determines that its place value is ten-thousandths.

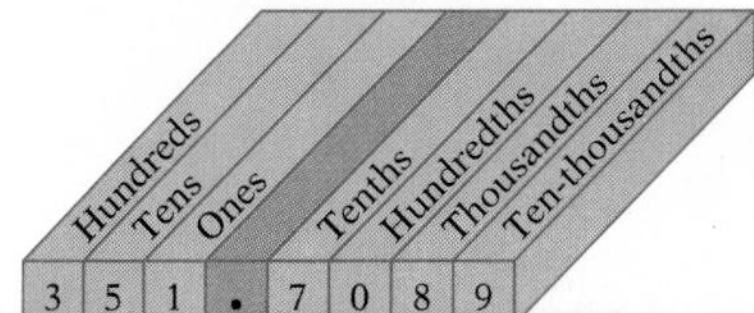

When writing a decimal in words, write the decimal part as if it were a whole number; then name the place value of the last digit.

0.6481	Six thousand four hundred eighty-one ten-thousandths
549.238	Five hundred forty-nine and two hundred thirty-eight thousandths (The decimal point is read as "and.")

To write a decimal in standard form, zeros may have to be inserted after the decimal point so that the last digit is in the given place-value position.

Five and thirty-eight <u>hundredths</u>

8 is in the hundredths' place. 5.3<u>8</u>

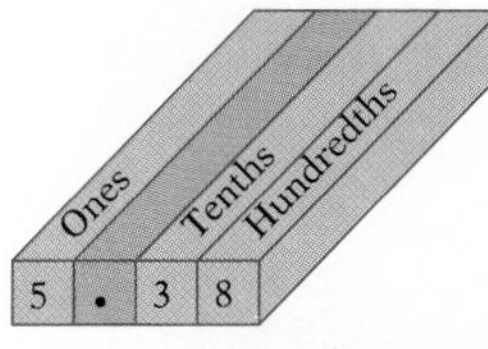

Nineteen and four <u>thousandths</u>

4 is in the thousandths' place. Insert 2 zeros so that 4 is in the thousandths place. 19.00<u>4</u>

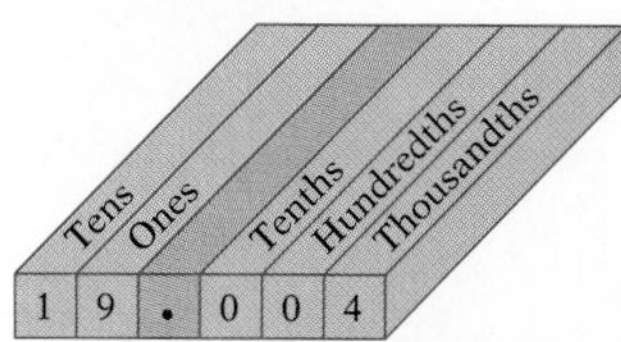

Seventy-one <u>ten-thousandths</u>

1 is in the ten-thousandths' place. Insert 2 zeros so that 1 is in the ten-thousandths place. 0.007<u>1</u>

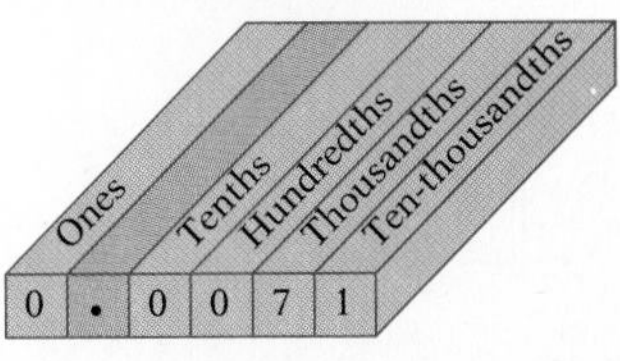

Example 1 Write 307.4027 in words.

Solution Three hundred seven and four thousand twenty-seven ten-thousandths

Example 2 Write 209.05838 in words.

Your solution

Example 3 Write six hundred seven and seven hundred eight hundred-thousandths in standard form.

Solution 607.00708

Example 4 Write forty-two thousand and two hundred seven millionths in standard form.

Your solution

Solutions on p. A16

Objective B To round a decimal to a given place value

Rounding decimals is similar to rounding whole numbers except that the digits to the right of the given place value are dropped instead of being replaced by zeros.

If the digit to the right of the given place value is less than 5, that digit and all digits to the right are dropped. If the digit to the right of the given place value is greater than or equal to 5, increase the given place value by 1 and drop all digits to its right.

Round 26.3799 to the nearest hundredth.

26.3799 (Given place value: 7)

$9 > 5$ Increase 7 by one and drop all digits to the right of 7.

26.3799 rounded to the nearest hundredth is 26.38.

Example 5 Round 0.39275 to the nearest ten-thousandth.

Solution

0.39275 (Given place value: 7)

$5 = 5$

0.3928

Example 6 Round 4.349254 to the nearest hundredth.

Your solution

Example 7 Round 42.0237412 to the nearest hundred-thousandth.

Solution

42.0237412 (Given place value: 4)

$1 < 5$

42.02374

Example 8 Round 3.290532 to the nearest hundred-thousandth.

Your solution

Solutions on p. A16

3.1 EXERCISES

Objective A

Write each decimal in words.

1. 0.27
2. 0.92
3. 1.005
4. 3.067
5. 36.4
6. 59.7
7. 0.00035
8. 0.00092
9. 32.107
10. 65.206
11. 12.1294
12. 26.3297
13. 10.007
14. 20.009
15. 52.00095
16. 64.00037
17. 0.0293
18. 0.0717
19. 6.324
20. 8.916
21. 276.3297
22. 418.3115
23. 216.0729
24. 976.0317
25. 4625.0379
26. 2986.0925
27. 1.00001
28. 3.00003

Write each decimal in standard form.

29. Seven hundred sixty-two thousandths
30. Two hundred ninety-five thousandths
31. Sixty-two millionths
32. Forty-one millionths
33. Eight and three hundred four ten-thousandths
34. Four and nine hundred seven ten-thousandths

Write each decimal in standard form.

35. Three hundred four and seven hundredths

36. Eight hundred ninety-six and four hundred seven thousandths

37. Three hundred sixty-two and forty-eight thousandths

38. Seven hundred eighty-four and eighty-four thousandths

39. Three thousand forty-eight and two thousand two ten-thousandths

40. Seven thousand sixty-one and nine thousand one ten-thousandths

41. Twenty-eight and four thousand three hundred seventy-five hundred-thousandths

42. Sixty-seven and eight thousand seven hundred twenty-three hundred-thousandths

43. One thousand three hundred eight and forty-five ten-thousandths

44. Twenty-seven thousand eight and one thousand five hundred-thousandths

Objective B

Round each decimal to the given place value.

45. 7.359 Tenths

46. 6.405 Tenths

47. 23.009 Tenths

48. 89.19204 Tenths

49. 22.68259 Hundredths

50. 16.30963 Hundredths

51. 480.325 Hundredths

52. 670.974 Hundredths

53. 1.03925 Thousandths

54. 7.072854 Thousandths

55. 1946.3745 Thousandths

56. 62.009435 Thousandths

57. 0.029876 Ten-thousandths

58. 0.012346 Ten-thousandths

59. 1.702596 Nearest whole number

60. 2.079239 Hundred-thousandths

61. 0.0102903 Millionths

62. 0.1009754 Millionths

SECTION 3.2 Addition of Decimals

Objective A To add decimals

To add decimals, write the numbers so that the decimal points are on a vertical line. Add as for whole numbers, and write the decimal point in the sum directly below the decimal points in the addends.

Add: 0.237 + 4.9 + 27.32

Note that by placing the decimal points on a vertical line, digits of the same place value are added.

	Tens	Ones	.	Tenths	Hundredths	Thousandths
	1	1				
		0	.	2	3	7
		4	.	9		
+	2	7	.	3	2	
	3	2	.	4	5	7

Example 1 Add:
42.3 + 162.903 + 65.0729

Solution

```
 111
  42.3
 162.903
+ 65.0729
 270.2759
```

Example 2 Add:
4.62 + 27.9 + 0.62054

Your solution

Example 3 Add: 0.83 + 7.942 + 15

Solution

```
 11
  0.83
  7.942
+15.
 23.772
```

Example 4 Add: 6.05 + 12 + 0.374

Your solution

Solutions on p. A16

ESTIMATION

Estimating the Sum of Two or More Decimals

Estimate and then use your calculator to find 23.037 + 16.7892.

To estimate the sum of two or more numbers, round each number to the same place value. In this case we will round to the nearest whole number. Then add. The estimated answer is 40.

$$\begin{array}{rcr} 23.037 & \approx & 23 \\ +16.7892 & \approx & +17 \\ \hline & & 40 \end{array}$$

Now use your calculator to find the exact result. The exact answer is 39.8262.

23.037 [+] 16.7892 [=] 39.8262

Objective B To solve application problems

Example 5

A student bought 10.5, 8.7, and 9.2 gallons of gasoline in 1 month. How much gas did the student buy during the month?

Strategy

To find the total amount of gasoline purchased, add the three amounts (10.5, 8.7, and 9.2)

$$\begin{array}{r} 10.5 \\ 8.7 \\ +\ 9.2 \\ \hline 28.4 \end{array}$$

The student bought 28.4 gallons of gasoline during the month.

Example 6

A student spent $850.65 for tuition, $282.97 for books, $650 for a room, and $835 for a meal ticket. Find the total cost of the four items.

Your strategy

Your solution

Example 7

A food server earned a salary of $138.50 for working 5 days this week. The food server also received $22.92, $15.80, $19.65, $39.20, and $27.70 in tips during the 5 days. Find the total income for the week.

Strategy

To find the total income, add the tips ($22.92, $15.80, $19.65, $39.20, and $27.70) to the salary ($138.50).

Solution

$$\begin{array}{r} \$138.50 \\ 22.92 \\ 15.80 \\ 19.65 \\ 39.20 \\ +\ 27.70 \\ \hline \$263.77 \end{array}$$

The food server's total income for the week was $263.77.

Example 8

An insurance executive earns a salary of $425 every 4 weeks. During the past 4-week period, the executive received commissions of $485.60, $599.46, $326.75, and $725.42. Find the executive's total income for the past 4-week period.

Your strategy

Your solution

Solutions on p. A16

3.2 EXERCISES

Objective A

Add.

1. 16.008 + 2.0385 + 132.06
2. 17.32 + 1.0579 + 16.5
3. 1.792 + 67 + 27.0526
4. 8.772 + 1.09 + 26.5027
5. 3.02 + 62.7 + 3.924
6. 9.06 + 4.976 + 59.6
7. 82.006 + 9.95 + 0.927
8. 0.826 + 8.76 + 79.005
9. 4.307 + 99.82 + 9.078
10. 2.543 + 7.906 + 2.07 + 1.4092
11. 82.907 + 0.0039 + 234.78308
12. 0.0072 + 96 + 235.09873

13. 0.3 + 0.07
14. 0.29 + 0.4
15. 1.007 + 2.1
16. 7.3 + 9.005

17. 4.9257, 27.05, + 9.0063
18. 8.72, 99.073, + 2.9736
19. 62.4, 9.827, + 692.44
20. 8, 89.43, + 7.0659

Estimate by rounding to the nearest whole number. Then use your calculator to add.

21. 342.42, 89.625, + 176.2
22. 219.9, 0.872, + 13.42
23. 823.9, 82.65, + 46.923
24. 678.92, 97.6, + 5.423

Estimate by rounding to the nearest tenths. Then use your calculator to add.

25. .23, .39, + .85
26. .42, .355, + .84
27. .8543, .8772, + .8591
28. .8972, .1324, + .3096

Objective B *Application Problems*

29. A sales executive received a monthly salary of $655.40 and sales commissions during the month of $1436.78 and $617.32. Find the the sales executive's monthly income.

30. A family receives an electric bill of $79.42, a gas bill of $44.78, and a garbage collection bill of $17.53. Find the total bill for the three services.

31. Find the length of the shaft.

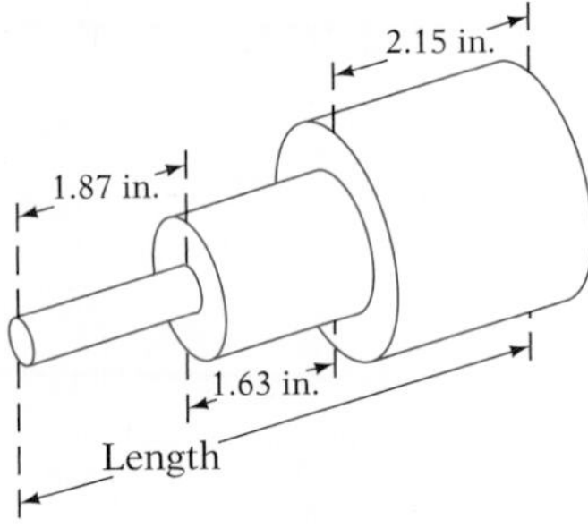

32. Find the length of the shaft.

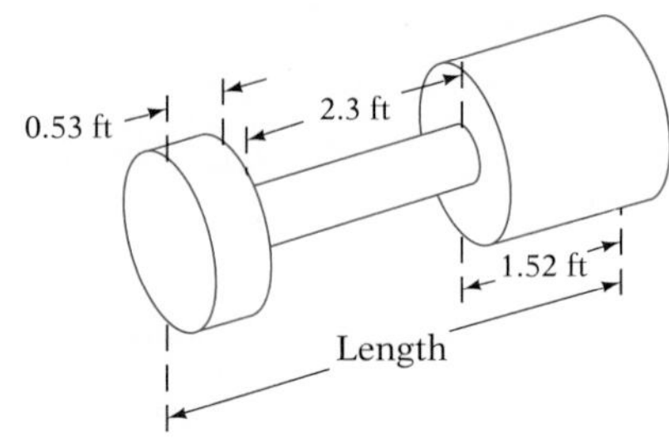

33. A city's rainfall for the three summer months was 7.33 inches, 12.93 inches, and 10.15 inches. Find the total rainfall for the 3 months.

34. A commuter used 12.4 gallons of gas the first week, 9.8 gallons the second week, 13.5 gallons the third week, and 11.2 gallons the fourth week. Find the total amount of gas used for the 4 weeks.

35. The odometer on a family's car reads 1817.4 miles. The car was driven 8.2 miles on Friday, 79.3 miles on Saturday, and 114.8 miles on Sunday.
a. How many miles was the car driven during the 3 days?
b. Find the odometer reading at the end of the 3 days.

36. On a short minivacation, you spent $7.46, $9.93, and $6.54 for gasoline. The motel bills were $32.60 and $45.18.
a. Find the total amount spent for gasoline.
b. Find the total amount spent for gasoline and the motel.

37. A food server earns a salary of $178.50 for working a 5-day week. In 1 week, the food server received $27.12, $18.50, $34.10, $22.75, and $29.80 in tips. Find the total income for the week.

38. You have $2142.07 in your checking account. You make deposits of $214.96, $31.42, $1092.43, and $99.27. Find the amount in your checking account after you make the deposits.

39. The average number of points scored per game by each of 10 members of a professional basketball team over an 82-game season is as follows: 31.3, 19.7, 15.4, 11.3, 9.2, 8.7, 6.5, 4.4, 1.2 and 0.3. Find the average number of points the basketball team scored per game.

40. An account executive's commission checks for 6 months are $2542.61, $2509.92, $1977.05, $1656.78, $2946.44, and $2153.88. Find the total commission income for the 6 months.

SECTION 3.3 Subtraction of Decimals

Objective A To subtract decimals

To subtract decimals, write the numbers so that the decimal points are on a vertical line. Subtract as for whole numbers and write the decimal point in the difference directly below the decimal point in the subtrahend.

Subtract 21.532 − 9.875 and check.

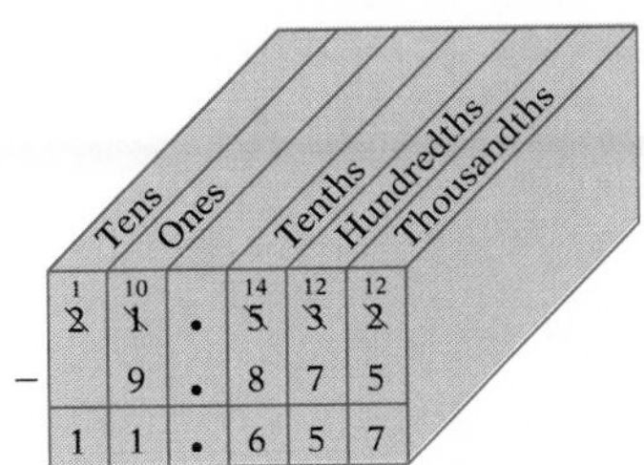

Note that by placing the decimal points on a vertical line, digits of the same place value are subtracted.

Check: Subtrahend 9.875
+ Difference +11.657
= Minuend 21.532

Subtract 4.3 − 1.7942 and check.

If necessary, insert zeros in the minuend before subtracting.

```
 3  12 9 9 10
 4 . 3 0 0 0     Check:   1.7942
−1 . 7 9 4 2             +2.5058
 2 . 5 0 5 8              4.3000
```

Example 1 Subtract 39.047 − 7.96 and check.

Solution

```
  8  9 14
 3 9 . 0 4 7   Check:   7.96
−   7 . 9 6           +31.087
 3 1 . 0 8 7           39.047
```

Example 2 Subtract 72.039 − 8.47 and check.

Your solution

Example 3 Subtract 29 − 9.23 and check.

Solution

```
 1 18  9 10
 2 9 . 0 0   Check:   9.23
−   9 . 2 3         +19.77
 1 9 . 7 7           29.00
```

Example 4 Subtract 35 − 9.67 and check.

Your solution

Example 5 Subtract 1.2 − 0.8235 and check.

Solution

```
 0  11 9 9 10
 1 . 2 0 0 0   Check:   0.8235
−0 . 8 2 3 5           +0.3765
 0 . 3 7 6 5            1.2000
```

Example 6 Subtract 3.7 − 1.9715 and check.

Your solution

Solutions on p. A17

ESTIMATION

Estimating the Difference Between Two Decimals

Estimate and then use your calculator to find 820.2306 − 475.74815.

To estimate the difference between two numbers, round each number to the same place value. In this case we will round to the nearest tens. Then subtract. The estimated answer is 340.

$$\begin{array}{rcr} 820.2306 & \approx & 820 \\ -475.74815 & \approx & -480 \\ \hline & & 340 \end{array}$$

Now use your calculator to find the exact result. The exact answer is 344.48245.

820.2306 [−] 475.74815 [=] 344.48245

Objective B To solve application problems

Example 7

You buy a book for $15.87. How much change do you receive from a $20.00 bill?

Strategy

To find the amount of change, subtract the cost of the book ($15.87) from $20.00

Solution

$$\begin{array}{r} \$20.00 \\ -15.87 \\ \hline \$\ 4.13 \end{array}$$

You receive $4.13 in change.

Example 8

Your breakfast costs $3.85. How much change do you receive from a $5.00 bill?

Your strategy

Your solution

Example 9

You had a balance of $62.41 in your checking account. You then bought a cassette for $8.95, film for $3.17, and a skateboard for $39.77. After paying for these items with a check, how much do you have left in your checking account?

Strategy

To find the new balance:

- Find the total cost of the three items ($8.95 + $3.17 + $39.77).
- Subtract the total cost from the old balance ($62.41).

Solution

$$\begin{array}{r} \$\ 8.95 \\ 3.17 \\ +39.77 \\ \hline \$\ 51.89 \end{array} \text{ total cost} \qquad \begin{array}{r} \$\ 62.41 \\ -51.89 \\ \hline \$\ 10.52 \end{array}$$

The new balance is $10.52.

Example 10

You had a balance of $2472.69 in your checking account. You then wrote checks for $1025.60, $79.85, and $162.47. Find the new balance in your checking account.

Your strategy

Your solution

Solutions on p. A17

3.3 EXERCISES

Objective A

Subtract and check.

1. 24.037 − 18.41
2. 26.029 − 19.31
3. 123.07 − 9.4273
4. 214 − 7.143
5. 16.5 − 9.7902
6. 13.2 − 8.6205
7. 235.79 − 20.093
8. 463.27 − 40.095
9. 63.005 − 9.1274
10. 23.004 − 7.2175
11. 92 − 19.2909
12. 41.2405 − 25.2709
13. 7.01 − 2.325
14. 8.07 − 5.392
15. 19.0035 − 8.967
16. 0.675 − 0.32
17. 0.924 − 0.91
18. 3 − 1.296
19. 7.507 − 3.419
20. 0.32 − 0.0058
21. 0.78 − 0.0073
22. 3.005 − 1.982
23. 6.007 − 2.734
24. 352.16 − 90.994
25. 872 − 80.753
26. 724.32 − 69
27. 625.46 − 77.509
28. 362.394 − 19.4672
29. 421.385 − 17.5293
30. 19 − 10.372
31. 23.4 − 0.921

Estimate by rounding to the nearest tens. Then use your calculator to subtract.

32. 620.59 − 132.79
33. 835.07 − 244.82
34. 67.3 − 19.793
35. 84.1 − 48.906

Estimate by rounding to the nearest whole number (nearest ones). Then use your calculator to subtract.

36. $\begin{array}{r} 93.079256 \\ -66.09249 \\ \hline \end{array}$

37. $\begin{array}{r} 3.7529 \\ -1.00784 \\ \hline \end{array}$

38. $\begin{array}{r} 76.53902 \\ -45.73005 \\ \hline \end{array}$

39. $\begin{array}{r} 9.07325 \\ -1.924 \\ \hline \end{array}$

Objective B *Application Problems*

40. A patient has a fever of 102.3° F. Normal temperature is 98.6° F. How many degrees above normal is the patient's temperature?

41. The manager at a fast-food restaurant takes a reading of the cash register tape each hour. At 1:00 P.M. the tape reads $967.54; at 2:00 P.M. the tape reads $1437.15. Find the amount of sales between 1:00 P.M. and 2:00 P.M.

42. A carpenter's plane removes 0.017 inch of wood from a board that is 1.7 inches thick. Find the resulting thickness of the board.

43. A student assistant received a wage increase from $4.75 per hour to $5.24 an hour. Find the per hour raise the assistant received.

44. Find the missing dimension.

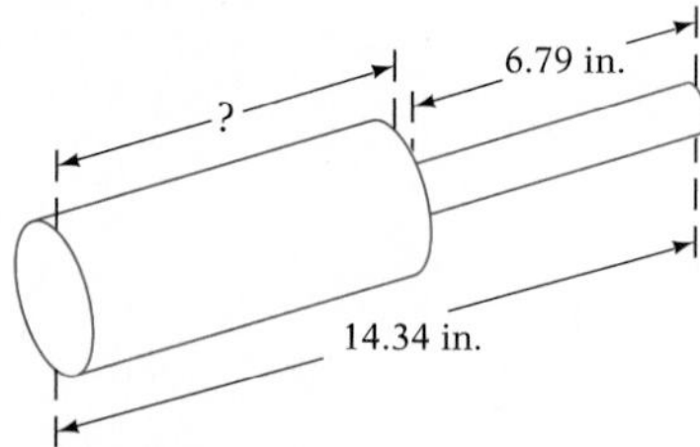

45. Find the missing dimension.

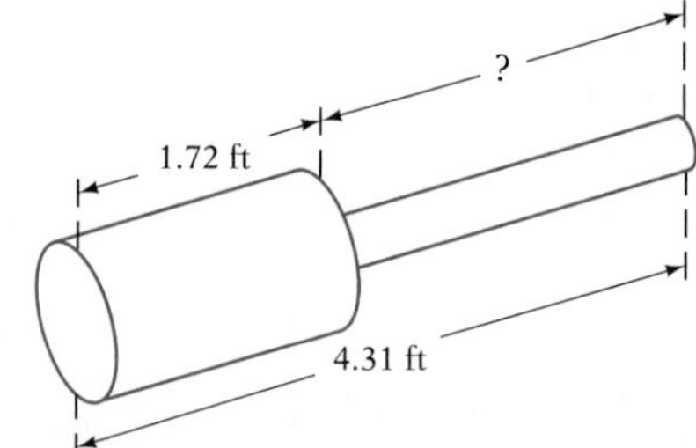

46. You earn a salary of $381.24 per week. You have deductions of $30.50 for social security, $18.60 for medical insurance, and $7.18 for union dues.
a. Find the total amount deducted from your pay.
b. Find your take-home pay.

47. You had a balance of $1029.74 in your checking account. You then wrote checks for $67.92, $43.10, and $496.34.
a. Find the total amount of the checks written.
b. Find the new balance in your checking account.

48. The price of gasoline is $1.22 per gallon after the price rose $.07 one month and $.12 the second month. Find the price of gasoline before the increase in price.

49. Rainfall for the last 3 months of the year was 1.42 inches, 5.39 inches, and 3.55 inches. The normal rainfall for the last 3 months of the year is 11.22 inches. How many inches below normal was the rainfall?

50. A department store had a monthly income of $43,817.62. Salaries amounted to $4214.18, overhead $7124.87, and cost of goods $19,426.44. Find the amount remaining after the three expenses were paid.

51. A telephone company has an income of $4,392,618.58 and pays $692,974.44 in taxes. Find the after-tax income.

SECTION 3.4 Multiplication of Decimals

Objective A To multiply decimals

Decimals are multiplied as if they were whole numbers; then the decimal point is placed in the product. Writing the decimals as fractions shows where to write the decimal point in the product.

$$0.\underline{3} \times 5 = \frac{3}{10} \times \frac{5}{1} = \frac{15}{10} = 1.\underline{5}$$

1 decimal place 1 decimal place

$$0.\underline{3} \times 0.\underline{5} = \frac{3}{10} \times \frac{5}{10} = \frac{15}{100} = 0.\underline{15}$$

1 decimal place 1 decimal place 2 decimal places

$$0.\underline{3} \times 0.\underline{05} = \frac{3}{10} \times \frac{5}{100} = \frac{15}{1000} = 0.\underline{015}$$

1 decimal place 2 decimal places 3 decimal places

To multiply decimals, multiply the numbers as in whole numbers. Write the decimal point in the product so that the number of decimal places in the product is the sum of the decimal places in the factors.

Multiply: 21.4 × 0.36

21.4	1 decimal place
×0.36	2 decimal places
1284	
642	
7.704	3 decimal places

Multiply: 357 × 0.29

357	no decimal places
×0.29	2 decimal places
3213	
714	
103.53	2 decimal places

Multiply: 0.037 × 0.08

0.037	3 decimal places
× 0.08	2 decimal places
0.00296	5 decimal places

Two zeros must be inserted between the 2 and the decimal point so that there are 5 decimal places in the product.

To multiply a decimal by a power of 10 (10, 100, 1000, . . .), move the decimal point to the right the same number of places as there are zeros in the power of 10.

$3.8925 \times 1\underline{0} = 38.925$

1 zero 1 decimal place

$3.8925 \times 100 = 389.25$
2 zeros — 2 decimal places

$3.8925 \times 1000 = 3892.5$
3 zeros — 3 decimal places

$3.8925 \times 10{,}000 = 38{,}925.$
4 zeros — 4 decimal places

$3.8925 \times 100{,}000 = 389{,}250.$
5 zeros — 5 decimal places

Note that a zero must be inserted before the decimal point.

Note that if the power of 10 is written in exponential notation, the exponent indicates how many places to move the decimal point.

$3.8925 \times 10^1 = 38.925$
1 decimal place

$3.8925 \times 10^2 = 389.25$
2 decimal places

$3.8925 \times 10^3 = 3892.5$
3 decimal places

$3.8925 \times 10^4 = 38{,}925.$
4 decimal places

$3.8925 \times 10^5 = 389{,}250.$
5 decimal places

Example 1 Multiply: 920×3.7

Solution

```
  920
× 3.7
 644 0
2760
3404.0
```

Example 2 Multiply: 870×4.6

Your solution

Example 3 Multiply: 0.37×0.9

Solution

```
 0.37
× 0.9
0.333
```

Example 4 Multiply: 0.28×0.7

Your solution

Solutions on p. A17

Example 5 Multiply: 0.00079 × 0.025

Solution

```
   0.00079
 ×   0.025
       395
      158
0.00001975
```

Example 6 Multiply: 0.000086 × 0.057

Your solution

Example 7 Multiply: 49.093 × 0.079

Solution

```
  49.093
 × 0.079
  441837
 343651
3.878347
```

Example 8 Multiply: 79.063 × 0.47

Your solution

Example 9 Multiply: 3.69 × 2.07

Solution

```
  3.69
 ×2.07
  2583
 7380
7.6383
```

Example 10 Multiply: 4.68 × 6.03

Your solution

Example 11 Multiply: 42.07 × 10,000

Solution 42.07 × 10,000 = 420,700

Example 12 Multiply: 6.9 × 1000

Your solution

Example 13 Multiply: 3.01×10^3

Solution $3.01 \times 10^3 = 3010$

Example 14 Multiply: 4.0273×10^2

Your solution

Solutions on p. A17

ESTIMATION

Estimating the Product of Two Decimals

Estimate and then use your calculator to find 28.259 × 0.029.

To estimate a product, round each number so that there is one non-zero digit. Then multiply.

```
  28.259 ≈    30
× 0.029  ≈ ×0.03
            0.90
```

The estimated answer is 0.90.

Now use you calculator to find the exact answer.

28.259 [×] 0.029 [=] 0.819511

The exact answer is 0.819511.

Objective B To solve application problems

Example 15

It costs \$.036 an hour to operate an electric motor. How much does it cost to operate the motor 120 hours?

Strategy

To find the cost of running the motor 120 hours, multiply the hourly cost (\$.036) by the number of hours the motor is run (120).

Solution

```
  $.036
 × 120
   720
   36
 $4.320
```

The cost of running the motor 120 hours is \$4.32.

Example 16

The cost of electricity to run a freezer 1 hour is \$.035. This month the freezer has run 210 hours. Find the total cost of running the freezer this month.

Your strategy

Your solution

Example 17

A clerk earns a salary of \$280 for a 40-hour work week. This week the clerk worked 12 hours of overtime, at a rate of \$10.50 for each hour of overtime worked. Find the clerk's total income for the week.

Strategy

To find the clerk's total income for the week:

- Find the overtime pay by multiplying the hourly overtime rate(\$10.50) by the number of hours of overtime worked (12).
- Add the overtime pay to the weekly salary (\$280).

Solution

```
  $10.50                 $280.00
 ×    12               + 126.00
   21 00                 $406.00
  105 0
 $126.00 overtime pay
```

The clerk's total income for this week is \$406.00.

Example 18

You make a down payment of \$175 on a stereo and agree to make payments of \$37.18 a month for the next 18 months. Find the total cost of the stereo.

Your strategy

Your solution

Solutions on p. A18

3.4 Exercises

Objective A

Multiply.

1. 0.9 $\times$0.4	**2.** 0.7 $\times$0.9	**3.** 0.5 $\times$0.6	**4.** 0.3 $\times$0.7	**5.** 0.5 $\times$0.5
6. 0.7 $\times$0.7	**7.** 0.9 $\times$0.5	**8.** 0.2 $\times$0.6	**9.** 7.7 $\times$0.9	**10.** 3.4 $\times$0.4
11. 9.2 $\times$0.2	**12.** 2.6 $\times$0.7	**13.** 7.2 $\times$0.6	**14.** 6.8 $\times$0.4	**15.** 7.4 $\times$0.1
16. 3.8 $\times$0.1	**17.** 7.9 $\times$ 5	**18.** 9.3 $\times$ 7	**19.** 0.68 $\times$ 4	**20.** 0.83 $\times$ 9
21. 0.67 $\times$ 0.9	**22.** 0.84 $\times$ 0.3	**23.** 0.16 $\times$ 0.6	**24.** 0.47 $\times$ 0.8	**25.** 2.5 $\times$5.4
26. 3.9 $\times$1.9	**27.** 8.4 $\times$9.5	**28.** 7.6 $\times$5.8	**29.** 0.83 $\times$ 5.2	**30.** 0.24 $\times$ 2.7
31. 0.46 $\times$ 3.9	**32.** 0.78 $\times$ 6.8	**33.** 0.2 $\times$0.3	**34.** 0.3 $\times$0.3	**35.** 0.24 $\times$ 0.3
36. 0.17 $\times$ 0.5	**37.** 1.47 $\times$ .09	**38.** 6.37 $\times$0.05	**39.** 8.92 $\times$0.004	**40.** 6.75 $\times$0.007

Multiply.

41. 0.49×0.16

42. 0.38×0.21

43. 7.6×0.01

44. 5.1×0.01

45. 8.62×4

46. 5.83×7

47. 64.5×9

48. 37.8×8

49. 2.19×9.2

50. 1.25×5.6

51. 1.85×0.023

52. 37.8×0.052

53. 0.478×0.37

54. 0.526×0.22

55. 48.3×0.0041

56. 67.2×0.0086

57. 2.437×6.1

58. 4.237×0.54

59. 0.413×0.0016

60. 0.517×0.0029

61. 94.73×0.57

62. 89.23×0.62

63. 8.005×0.067

64. 9.032×0.019

65. 4.29×0.1

66. 6.78×0.1

67. 5.29×0.4

68. 6.78×0.5

69. 0.68×0.7

70. 0.56×0.9

71. 1.4×0.73

72. 6.3×0.37

73. 5.2×7.3

74. 7.4×2.9

75. 3.8×0.61

76. 7.2×0.72

77. 0.32×10

78. 6.93×10

79. 0.065×100

80. 0.039×100

81. 6.2856×1000

Multiply.

82. 3.2954×1000 **83.** 3.2×1000 **84.** $0.006 \times 10{,}000$ **85.** $3.57 \times 10{,}000$

86. 8.52×10^1 **87.** 0.63×10^1 **88.** 82.9×10^2

89. 0.039×10^2 **90.** 6.8×10^3 **91.** 4.9×10^4

92. 6.83×10^4 **93.** 0.067×10^2 **94.** 0.052×10^2

Estimate and then use your calculator to multiply.

95. 28.5×3.2 **96.** 86.3×4.4 **97.** 2.38×0.44 **98.** 9.82×0.77

99. 0.866×4.5 **100.** 0.239×8.2 **101.** 4.34×2.59 **102.** 6.87×9.98

103. 8.434×0.044 **104.** 7.037×0.094 **105.** 28.44×1.12 **106.** 86.57×7.33

107. 49.6854×39.0672 **108.** 2.00547×9.672 **109.** 0.00456×0.009542 **110.** $7.00637 \times .0128$

Objective B *Application Problems*

111. It costs \$.18 per mile to rent a car. Find the cost to rent a car that is driven 114 miles.

112. The cost of operating an electric motor for 1 hour is \$.027. How much does it cost to operate the motor 56 hours? (Round to the nearest cent.)

113. Four hundred empty soft drink cans weigh 18.75 pounds. A recycling center pays \$.75 per pound for the cans. Find the amount received for the 400 cans (round to the nearest cent).

114. A recycling center pays \$.035 per pound for newspapers. Find the amount received from recycling 420 pounds of newspapers.

115. A broker's fee for buying stock is 0.045 times the price of the stock. An investor buys 100 shares of stock at \$38.50 per share. Find the broker's fee.

116. A broker's fee for buying stock is 0.052 times the price of the stock. You buy 100 shares of stock at \$62.75 per share. Find the broker's fee (round to the nearest cent).

117. You buy a car for \$2000 down and payments of \$127.50 each month for 36 months.
a. Find the amount of the payments over the 36 months.
b. Find the total cost of the car.

118. A nurse earns a salary of \$344 for a 40-hour work week. This week the nurse worked 15 hours of overtime at a rate of \$12.90 for each hour of overtime worked.
a. Find the amount of overtime pay.
b. Find the nurse's total income for the week.

119. A rental car agency charges \$12 a day and \$.12 per mile for renting a car. You rented a car for 3 days and drove 235 miles. Find the total cost of renting the car.

120. A taxi cost \$1.50 and \$0.20 for each $\frac{1}{8}$ mile driven. Find the cost of hiring a taxi to get from the airport to the hotel—a distance of 5.5 miles.

121. The escape velocity of a rocket from the earth is 7.4 miles per second. At this speed, how far will the rocket travel in 1 day? (*Hint:* multiply the speed by the number of seconds in 1 day.)

122. The cash price of a house is \$104,200. The house can be bought with a 30-year loan with payments of \$822.45 a month for 30 years. Find the total amount of the payments.

SECTION 3.5 Division of Decimals

Objective A To divide decimals

To divide decimals, move the decimal point in the divisor to the right to make it a whole number. Move the decimal in the dividend the same number of places to the right. Place the decimal point in the quotient directly over the decimal point in the dividend, and then divide as in whole numbers.

Divide: $3.25\overline{)15.275}$

$3.25.\overline{)15.27.5}$

Move the decimal point 2 places to the right in the divisor and then in the dividend. Place the decimal point in the quotient.

```
          4.7
325.) 1527.5
     -1300
       227 5
      -227 5
           0
```

Moving the decimal point the same number of decimal places in the divisor and dividend does not change the value of the quotient because this process is the same as multiplying the numerator and denominator of a fraction by the same number. In the example above,

$$3.25\overline{)15.275} = \frac{15.275}{3.25} = \frac{15.275 \times 100}{3.25 \times 100} = \frac{1527.5}{325} = 325\overline{)1527.5}$$

When dividing decimals, the quotient is usually rounded off to a specified place value, rather than writing the quotient with a remainder.

Divide: $0.3\overline{)0.56}$
Round to the nearest hundredth.

```
       1.866 ≈ 1.87
0.3.) 0.5.600
     - 3
       2 6
      -2 4
         20
        -18
         20
        -18
```

The division must be carried to the thousandths' place to round the quotient to the nearest hundredth. Therefore, zeros must be inserted in the dividend so that the quotient has a digit in the thousandths' place.

Divide: 57.93 ÷ 3.24 Round to the nearest thousandth.

```
          17.8796 ≈ 17.880
3.24.) 57.93.0000   ← Zeros must be inserted in the dividend so
      -32 4           that the quotient has a digit in the ten-
       25 53          thousandths' place.
      -22 68
        2 85 0
       -2 59 2
          25 80
         -22 68
           3 120
          -2 916
             2040
            -1944
```

To divide a decimal by a power of 10 (10, 100, 1000, . . .), move the decimal point to the left the same number of places as there are zeros in the power of 10.

34.65 ÷ 10 = 3.465
1 zero — 1 decimal place

34.65 ÷ 100 = 0.3465
2 zeros — 2 decimal places

34.65 ÷ 1000 = 0.03465
3 zeros — 3 decimal places

Note that a zero must be inserted between the 3 and the decimal point.

34.65 ÷ 10,000 = 0.003465
4 zeros — 4 decimal places

Note that two zeros must be inserted between the 3 and the decimal point.

If the power of 10 is written in exponential notation, the exponent indicates how many places to move the decimal point.

$34.65 \div 10^1 = 3.465$ 1 decimal place

$34.65 \div 10^2 = 0.3465$ 2 decimal places

$34.65 \div 10^3 = 0.03465$ 3 decimal places

$34.65 \div 10^4 = 0.003465$ 4 decimal places

Example 1 Divide: 0.1344 ÷ 0.032

Solution

```
              4.2
0.032.)0.134.4
        -128
           6 4
          -6 4
             0
```

Example 2 Divide: 0.1404 ÷ 0.052

Your solution

Solution on p. A18

Example 3 Divide: 58.092 ÷ 82
Round to the nearest thousandth.

Solution

$$\begin{array}{r} 0.7084 \approx 0.708 \\ 82\overline{)\ 58.0920} \\ \underline{-57\ 4} \\ 69 \\ \underline{-\ 0} \\ 692 \\ \underline{-656} \\ 360 \\ \underline{-328} \end{array}$$

Example 4 Divide: 37.042 ÷ 76
Round to the nearest thousandth.

Your solution

Example 5 Divide: 420.9 ÷ 7.06
Round to the nearest tenth.

Solution

$$\begin{array}{r} 59.61 \approx 59.6 \\ 7.06.\overline{)\ 420.90.00} \\ \underline{-353\ 0} \\ 67\ 90 \\ \underline{-63\ 54} \\ 4\ 36\ 0 \\ \underline{-4\ 23\ 6} \\ 12\ 40 \\ \underline{-\ 7\ 06} \end{array}$$

Example 6 Divide: 370.2 ÷ 5.09
Round to the nearest tenth.

Your solution

Example 7 Divide: 402.75 ÷ 1000

Solution 402.75 ÷ 1000 = 0.40275

Example 8 Divide: 309.21 ÷ 10,000

Your solution

Example 9 Divide: $0.625 \div 10^2$

Solution $0.625 \div 10^2 = 0.00625$

Example 10 Divide: $42.93 \div 10^4$

Your solution

Solutions on p. A18

ESTIMATION

Estimating the Quotient of Two Decimals

Estimate and then use your calculator to find 282.18 ÷ 0.485.

To estimate a quotient, round each number so that there is one non-zero digit. Then divide.

282.18 ÷ 0.485 ≈ 300 ÷ 0.5 = 600.

The estimated answer is 600.

Now use your calculator to find the exact answer.

282.18 [÷] 0.485 [=] 581.814433

The exact answer is 581.814433.

Objective B To solve application problems

Example 11

A retail tire store pays $3.76 excise tax on each tire sold. During 1 month the store paid $300.80 in excise tax. How many tires were sold during the month?

Strategy

To find the number of tires sold, divide the total excise tax paid ($300.80) by the excise tax paid for each tire ($3.76).

Solution

```
                80.
$3.76.)  $300.80.
          -300 8
              00
             - 0
               0
```

The store sold 80 tires during the month.

Example 12

A trucker receives an annual salary of $23,741.40 in 12 equal monthly payments. Find the trucker's monthly income.

Your strategy

Your solution

Example 13

A deluxe microwave oven is on sale for $638.96. You make a down payment of $200 and agree to make 6 equal monthly payments. Find the amount of each monthly payment.

Strategy

To find the amount of each monthly payment:

- Find the balance owed after making the down payment ($638.96 − $200.00).
- Find the monthly payment by dividing the balance by the number of equal monthly payments (6).

Solution

```
 $638.96                  $73.16
-$200.00              6) $438.96
 $438.96 balance          -42
         owed              18
                          -18
                            0 9
                          -  6
                              36
                             -36
                               0
```

The monthly payment is $73.16.

Example 14

A tax of $11.55 is paid on each lamp a store buys. During 1 month the store paid $138.60 in taxes. Each lamp was sold for $95.60. Find the store's total income from the sale of the lamps.

Your strategy

Your solution

Solutions on pp. A18-A19

3.5 EXERCISES

Objective A

Divide.

1. $3\overline{)2.46}$
2. $7\overline{)3.71}$
3. $0.8\overline{)3.84}$
4. $0.9\overline{)6.93}$
5. $0.7\overline{)62.3}$
6. $0.4\overline{)52.8}$
7. $0.4\overline{)24}$
8. $0.5\overline{)65}$
9. $7\overline{)0.42}$
10. $3\overline{)0.12}$
11. $0.4\overline{)27.88}$
12. $0.8\overline{)3.312}$
13. $0.7\overline{)59.01}$
14. $0.9\overline{)8.721}$
15. $0.5\overline{)16.15}$
16. $0.8\overline{)77.6}$
17. $0.7\overline{)3.542}$
18. $0.6\overline{)2.436}$
19. $6.3\overline{)8.19}$
20. $3.2\overline{)7.04}$
21. $3.6\overline{)0.396}$
22. $2.7\overline{)0.648}$
23. $6.9\overline{)26.22}$
24. $1.7\overline{)84.66}$
25. $6.1\overline{)64.05}$
26. $3.9\overline{)8.034}$
27. $0.41\overline{)2.501}$
28. $0.59\overline{)1.003}$
29. $0.04\overline{)27.2}$
30. $0.06\overline{)44.4}$
31. $0.08\overline{)23.2}$
32. $0.018\overline{)0.1314}$
33. $0.064\overline{)0.2304}$

Divide. Round to the nearest tenth.

34. 55.62 ÷ 8.8
35. 25.43 ÷ 5.4
36. 5.427 ÷ 9.5
37. 1.837 ÷ 1.4
38. 18.4 ÷ 7.3
39. 52.9 ÷ 8.1
40. 0.183 ÷ 0.17
41. 0.381 ÷ 0.47
42. 6.924 ÷ 0.053

Divide. Round to the nearest hundredth.

43. 4.817 ÷ 16
44. 6.467 ÷ 8
45. 0.0418 ÷ 0.53
46. 19.08 ÷ 0.45
47. 21.792 ÷ 0.96
48. 38.665 ÷ 0.95
49. 13.97 ÷ 25.4
50. 27.738 ÷ 60.3
51. 3.171 ÷ 45.3

Divide. Round to the nearest thousandth.

52. 1.028 ÷ 54
53. 6.729 ÷ 27
54. 0.0437 ÷ 0.5
55. 75.469 ÷ 77.8
56. 34.31 ÷ 95.3
57. 0.2695 ÷ 2.67
58. 0.4871 ÷ 4.72
59. 0.1142 ÷ 17.2
60. 0.2307 ÷ 26.7

Divide. Round to the nearest whole number.

61. 16.5 ÷ 4
62. 89.76 ÷ 90
63. 1.94 ÷ 0.3
64. 1.0478 ÷ 0.413
65. 2.148 ÷ 0.519
66. 0.79 ÷ 0.778
67. 3.092 ÷ 0.075
68. 392 ÷ 6.9
69. 8.729 ÷ 0.075

Divide.

70. $4.07 \div 10$

71. $0.039 \div 10$

72. $42.67 \div 10$

73. $389.7 \div 100$

74. $1.037 \div 100$

75. $237.835 \div 100$

76. $8.295 \div 1000$

77. $82{,}547 \div 1000$

78. $825.37 \div 1000$

79. $8.35 \div 10^1$

80. $0.32 \div 10^1$

81. $87.65 \div 10^1$

82. $23.627 \div 10^2$

83. $2.954 \div 10^2$

84. $0.0053 \div 10^2$

85. $289.32 \div 10^3$

86. $1.8932 \div 10^3$

87. $0.139 \div 10^3$

Estimate and then use your calculator to divide. Round your calculated answer to the nearest ten-thousandth.

88. $42.42 \div 3.8$

89. $69.8 \div 7.2$

90. $389 \div 0.44$

91. $642 \div 0.83$

92. $6.394 \div 3.5$

93. $8.429 \div 4.2$

94. $1.235 \div 0.021$

95. $7.456 \div 0.072$

96. $95.443 \div 1.32$

97. $423.0925 \div 4.0927$

98. $1.000523 \div 429.07$

99. $0.03629 \div 0.00054$

Objective B *Application Problems*

100. A professional baseball player got 109 hits out of 423 at bats. Find the player's batting average. (Round to three decimal places.)

101. A high school football player gained 162 yards on 26 carries. Find the numbers of yards gained per carry. (Round to two decimal places.)

102. A park ranger earns $39,440.64 for 12 month's work. How much does the park ranger earn in 1 month?

103. Gasoline tax is $.17 per gallon. Find the number of gallons of gasoline used during a month in which $17.34 was paid in taxes.

104. A family pays $1567.70 a year in car insurance. The insurance is paid in 5 equal payments. Find the amount of each payment.

105. A carpenter is building bookcases that are 3.4 feet long. How many complete shelves can be cut from a 12-foot board?

106. A photography supplies store pays a state tax of $7.26 on each SLR camera it sells for $132.40. This month the total tax paid on SLR cameras was $784.08.

a. How many cameras were sold during the month?
b. Find the store's total income from the sale of cameras during the month.

107. You buy a home entertainment center for $1242.58. The down payment is $400, and the balance is to be paid in 18 equal monthly payments.

a. Find the amount to be paid in monthly payments.
b. Find the amount of each monthly payment.

108. A tire store pays $353.92 in excise taxes during the month on the sale of 75R-14 steel-belted tires. A tax of $3.16 is paid for each 75R-14 tire sold. Each tire sells for $37.45. Find the tire store's total income from the sale of 75R-14 tires during the month.

109. A tax of $6.92 is paid on each box of electric shavers a store purchases. One month the store paid $256.04 in taxes on electric shavers. Each box of shavers sold for $192.60. Find the store's total income from the sale of the electric shavers.

110. An oil company has 3,541,221,500 shares of stock. The company paid $6,090,990,120 in dividends. Find the dividend for each share of stock. (Round to two decimal places)

111. The total budget for the United States is $900 billion. If each person in the United States were to pay the same amount of taxes, how much would each person pay to raise that amount of money? Assume that there are 270 million people in the United States. (Round to the nearest dollar.)

SECTION 3.6 Comparing and Converting Fractions and Decimals

Objective A To convert fractions to decimals

Every fraction can be written as a decimal. To write a fraction as a decimal, divide the numerator of the fraction by the denominator. The quotient can be rounded to the desired place value.

Convert $\frac{3}{7}$ to a decimal.

$7\overline{)3.00000}$ = 0.42857 $\quad \frac{3}{7}$ rounded to the nearest hundredth is 0.43.

$\frac{3}{7}$ rounded to the nearest thousandth is 0.429.

$\frac{3}{7}$ rounded to the nearest ten-thousandth is 0.4286

Convert $3\frac{2}{9}$ to a decimal. Round to the nearest thousandth.

$3\frac{2}{9} = \frac{29}{9}$ $\quad 9\overline{)29.0000}$ = 3.2222 $\quad 3\frac{2}{9}$ rounded to the nearest thousandth is 3.222.

Example 1 Convert $\frac{3}{8}$ to a decimal. Round to the nearest hundredth.

Solution $8\overline{)3.000}$ = $0.375 \approx 0.38$

Example 2 Convert $\frac{9}{16}$ to a decimal. Round to the nearest tenth.

Your solution

Example 3 Convert $2\frac{3}{4}$ to a decimal. Round to the nearest tenth.

Solution $2\frac{3}{4} = \frac{11}{4}$ $\quad 4\overline{)11.00}$ = $2.75 \approx 2.8$

Example 4 Convert $4\frac{1}{6}$ to a decimal. Round to the nearest hundredth.

Your solution

Solutions on p. A19

Objective B To convert decimals to fractions

To convert a decimal to a fraction, remove the decimal point and place the decimal part over a denominator equal to the place value of the last digit in the decimal.

$0.47 = \frac{47}{100}$ (hundredths)

$7.45 = 7\frac{45}{100} = 7\frac{9}{20}$ (hundredths)

$0.275 = \frac{275}{1000} = \frac{11}{40}$ (thousandths)

$0.16\frac{2}{3} = \frac{16\frac{2}{3}}{100} = 16\frac{2}{3} \div 100 = \frac{50}{3} \times \frac{1}{100} = \frac{1}{6}$ (hundredths)

Example 5 Convert 0.82 and 4.75 to fractions.

Solution $0.82 = \frac{82}{100} = \frac{41}{50}$

$4.75 = 4\frac{75}{100} = 4\frac{3}{4}$

Example 6 Convert 0.56 and 5.35 to fractions.

Your solution

Example 7 Convert $0.15\frac{2}{3}$ to a fraction.

Solution $0.15\frac{2}{3} = \frac{15\frac{2}{3}}{100} = 15\frac{2}{3} \div 100$

$= \frac{47}{3} \times \frac{1}{100} = \frac{47}{300}$

Example 8 Convert $0.12\frac{7}{8}$ to a fraction.

Your solution

Solutions on p. A19

Objective C

To identify the order relation between two decimals or between a decimal and a fraction

Decimals, like whole numbers and fractions, can be graphed as points on the number line. The number line can be used to show the order of decimals. A decimal that appears to the right of a given number is greater than the given number. A decimal that appears to the left of a given number is less than the given number.

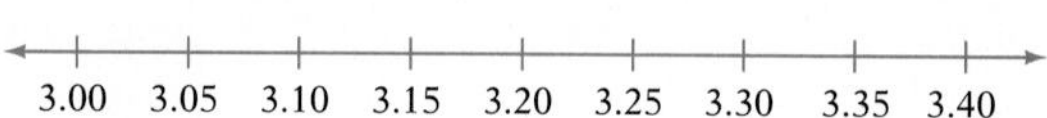

Note that 3, 3.0, and 3.00 represent the same number.

Find the order relation between $\frac{3}{8}$ and 0.38.

$\frac{3}{8} = 0.375$

$0.375 < 0.38$

$\frac{3}{8} < 0.38$

Example 9 Place the correct symbol, $<$ or $>$, between the numbers.
17.005 17.050

Solution $17.005 < 17.050$

Example 10 Place the correct symbol, $<$ or $>$, between the numbers.
64.009 64.99

Your solution

Example 11 Place the correct symbol, $<$ or $>$, between the numbers.
$\frac{5}{16}$ 0.32

Solution $\frac{5}{16} \approx 0.313$

$0.313 < 0.32$

$\frac{5}{16} < 0.32$

Example 12 Place the correct symbol, $<$ or $>$, between the numbers.
0.63 $\frac{5}{8}$

Your solution

Solutions on p. A19

3.6 EXERCISES

Objective A

Convert the fraction to a decimal. Round to the nearest thousandth.

1. $\frac{5}{8}$
2. $\frac{7}{12}$
3. $\frac{2}{3}$
4. $\frac{5}{6}$
5. $\frac{1}{6}$
6. $\frac{7}{8}$
7. $\frac{5}{12}$
8. $\frac{9}{16}$
9. $\frac{7}{4}$
10. $\frac{5}{3}$
11. $1\frac{1}{2}$
12. $2\frac{1}{3}$
13. $\frac{16}{4}$
14. $\frac{36}{9}$
15. $\frac{3}{1000}$
16. $\frac{5}{10}$
17. $7\frac{2}{25}$
18. $16\frac{7}{9}$
19. $37\frac{1}{2}$
20. $87\frac{1}{2}$
21. $\frac{3}{8}$
22. $\frac{11}{16}$
23. $\frac{5}{24}$
24. $\frac{4}{25}$
25. $3\frac{1}{3}$
26. $8\frac{2}{5}$
27. $5\frac{4}{9}$
28. $3\frac{1}{12}$
29. $\frac{5}{16}$
30. $\frac{11}{12}$
31. $2\frac{3}{14}$
32. $1\frac{7}{16}$

Objective B

Convert the decimal to a fraction.

33. 0.8
34. 0.4
35. 0.32
36. 0.48
37. 0.125
38. 0.485
39. 1.25
40. 3.75
41. 16.9
42. 17.5
43. 8.4
44. 10.7
45. 8.437
46. 9.279
47. 2.25

Convert the decimal to a fraction.

48. 7.75 **49.** $0.15\frac{1}{3}$ **50.** $0.17\frac{2}{3}$ **51.** $0.87\frac{7}{8}$ **52.** $0.12\frac{5}{9}$

53. 1.68 **54.** 7.38 **55.** 0.045 **56.** 0.085 **57.** 16.72

58. 82.32 **59.** 0.33 **60.** 0.57 **61.** $0.33\frac{1}{3}$ **62.** $0.66\frac{2}{3}$

63. $0.27\frac{3}{7}$ **64.** $0.29\frac{4}{7}$

Objective C

Place the correct symbol, < or >, between the numbers.

65. 0.15 0.5 **66.** 0.6 0.45 **67.** 6.65 6.56 **68.** 3.89 3.98

69. 2.504 2.054 **70.** 0.025 0.105 **71.** $\frac{3}{8}$ 0.365 **72.** $\frac{4}{5}$ 0.802

73. $\frac{2}{3}$ 0.65 **74.** 0.85 $\frac{7}{8}$ **75.** $\frac{5}{9}$ 0.55 **76.** $\frac{7}{12}$ 0.58

77. 0.62 $\frac{7}{15}$ **78.** $\frac{11}{12}$ 0.92 **79.** 0.161 $\frac{1}{7}$ **80.** 0.623 0.6023

81. 0.86 0.855 **82.** 0.87 0.087 **83.** 1.005 0.5 **84.** 0.033 0.3

85. $\frac{5}{24}$ 0.202 **86.** $\frac{7}{36}$ 0.198 **87.** 0.429 $\frac{4}{9}$ **88.** 0.835 $\frac{11}{16}$

Calculators and Computers

Using a Calculator to Add or Subtract Fractions

You may recall from your study of this chapter that every fraction can be represented as a decimal. By using a calculator, it is easy to find the decimal equivalent of a fraction. Simply divide the numerator by the denominator.

To find the decimal equivalent of $\frac{7}{30}$, follow these steps:

Enter	*Display*	*Comments*
7	7.	
[÷]	7.	
30	30.	
[=]	0.2333333	The answer to $\frac{7}{30}$.

Note: The result displayed in the calculator will only be a decimal *approximation* to many fractions. Using the decimal equivalent of a fraction, it is possible to add or subtract fractions using your calculator. The parentheses keys, [(] and [)], will be used for this calculation.

To add $\frac{23}{37} + \frac{5}{19}$ using your calculator, follow these steps:

Enter	*Display*	*Comments*
(23 ÷ 37)	0.621621	The first fraction.
[+]	0.621621	
(5 ÷ 19)	0.2631578	The second fraction.
[=]	0.8847795	
[×]	0.8847795	
(37 × 19)	703	The product of the two denominators. (Write this number down.)
[=]	621.99991	Round to the nearest whole number (622). The answer is $\frac{622}{703}$.

Chapter Summary

Key Words

A number written in *decimal notation* has three parts: a whole number part, a decimal point, and the decimal part.

The position of a digit in a *decimal* determines the digit's *place value.*

Essential Rules

To Write a Decimal in Words

To write a decimal in words, write the decimal part as if it were a whole number. Then name the place value of the last digit.

To Add Decimals

To add decimals, write the numbers so that the decimal points are on a vertical line. Add as in whole numbers, and place the decimal point in the sum directly below the decimal point in the addends.

To Subtract Decimals

To subtract decimals, place the numbers so that the decimal points are on a vertical line. Subtract as for whole numbers, and write the decimal point in the difference directly below the decimal point in the subtrahend.

To Multiply Decimals

To multiply decimals, multiply the numbers as in whole numbers. Place the decimal point in the product so that the number of decimal places in the product is the sum of the decimal places in the factors.

To Divide Decimals

To divide decimals, move the decimal point in the divisor to make it a whole number. Move the decimal in the dividend the same number of places to the right. Place the decimal point in the quotient directly over the decimal point in the dividend. Then divide as in whole numbers.

To Write a Fraction as a Decimal

To write a fraction as a decimal, divide the numerator of the fraction by the denominator. Round the quotient to the desired number of places.

To Convert a Decimal to a Fraction

To convert a decimal to a fraction, remove the decimal point and place the decimal part over the denominator equal to the place value of the last digit in the decimal.

Chapter Review

SECTION 3.1

1. Write 22.0092 in words.

2. Write 342.37 in words.

3. Write twenty-two and ninety-two ten-thousandths in standard form.

4. Write three and six thousand seven hundred fifty-three hundred-thousandths in standard form.

5. Round 7.93704 to the nearest hundredth.

6. Round 0.05678235 to the nearest hundred-thousandths.

SECTION 3.2

7. Add: 3.42 + 0.794 + 32.5

8. Add:
$$\begin{array}{r} 369.41 \\ 88.3 \\ 9.774 \\ +366.99 \\ \hline \end{array}$$

9. You have $237.44 in your checking account. You make deposits of $56.88, $127.40, and $56.30. Find the amount in your checking account after you make the deposits.

SECTION 3.3

10. Subtract: 27.31 − 4.4465

11. Subtract:
$$\begin{array}{r} 7.796 \\ -2.9175 \\ \hline \end{array}$$

12. You had a balance of $895.68 in your checking account. You then wrote checks of $145.72 and $88.45. Find the new balance in your checking account.

SECTION 3.4

13. Multiply: 3.08×2.9

14. Multiply: $\begin{array}{r} 34.79 \\ \times\ 0.74 \\ \hline \end{array}$

15. The state income tax on the business you own is \$560 plus 0.08 times your profit. You made a profit of \$63,000 last year. Find the amount of income tax you paid last year.

SECTION 3.5

16. Divide: $3.6515 \div 0.067$

17. Divide: $0.053\overline{)0.349482}$

18. A car costing \$5944.20 is bought with a down payment of \$1500 and 36 equal monthly payments. Find the amount of each monthly payment.

SECTION 3.6

19. Convert $\frac{7}{9}$ to a decimal. Round to the nearest thousandths.

20. Convert $2\frac{1}{3}$ to a decimal. Round to the nearest hundredths.

21. Convert 0.375 to a fraction.

22. Convert $0.66\frac{2}{3}$ to a fraction.

23. Place the correct symbol, $<$ or $>$, between the two numbers.
0.055 0.1

24. Place the correct symbol, $<$ or $>$, between the two numbers.
$\frac{5}{8}$ 0.62

Chapter Test

1. Write 45.0302 in words.

2. Write two hundred nine and seven thousand eighty-six hundred-thousandths in standard form.

3. Round 7.0954625 to the nearest thousandth.

4. Round 0.07395 to the nearest ten-thousandth.

5. Add:
$$\begin{array}{r} 62.3 \\ 4.007 \\ +189.65 \\ \hline \end{array}$$

6. Add:
$$\begin{array}{r} 270.93 \\ 97. \\ 1.976 \\ +\ 88.675 \\ \hline \end{array}$$

7. You received a salary of $363.75, a commission of $954.82, and a bonus of $225. Find your total income.

8. Subtract:
$$\begin{array}{r} 13.027 \\ -8.94 \\ \hline \end{array}$$

9. Subtract:
$$\begin{array}{r} 37.003 \\ -\ 9.23674 \\ \hline \end{array}$$

10. Find the missing dimension.

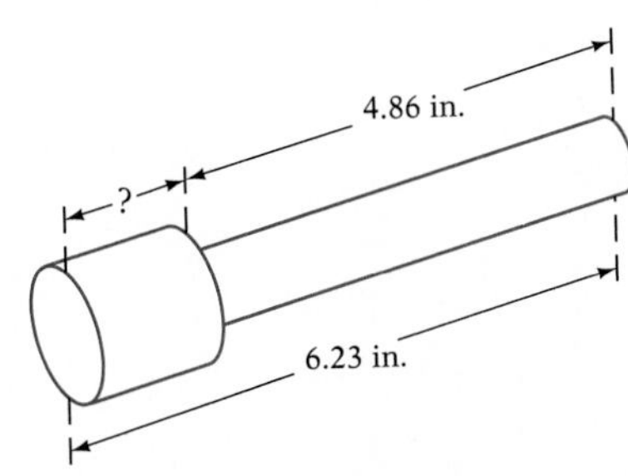

11. Multiply:
$$\begin{array}{r} 1.37 \\ 0.004 \\ \hline \end{array}$$

12. Multiply: $\begin{array}{r} 0.369 \\ \times \ \ 6.7 \\ \hline \end{array}$

13. A long distance telephone call cost $.85 for the first 3 minutes and $.42 for each additional minute. Find the cost of a 12-minute long distance telephone call.

14. Divide: $0.006\overline{)1.392}$

15. Divide: $0.037\overline{)0.0569}$
Round to the nearest thousandth.

16. A car is bought for $6392.60, with a down payment of $1250. The balance is paid in 36 monthly payments. Find the amount of each monthly payment.

17. Convert $\frac{9}{13}$ to a decimal. Round to the nearest thousandth.

18. Convert 0.825 to a fraction.

19. Place the correct symbol, < or >, between the two numbers.
0.66 0.666

20. Place the correct symbol, < or >, between the two numbers.
$\frac{3}{8}$ 0.35

Cumulative Review

1. Divide: $89\overline{)20{,}932}$

2. Simplify: $2^3 \cdot 4^2$

3. Simplify: $2^2 - (7 - 3) \div 2 + 1$

4. Find the LCM of 9, 12, and 24.

5. Write $\frac{22}{5}$ as a mixed number.

6. Write $4\frac{5}{8}$ as an improper fraction.

7. Build an equivalent fraction with the given denominator.
$\frac{5}{12} = \frac{}{60}$

8. Add: $\frac{3}{8} + \frac{5}{12} + \frac{9}{16}$

9. Add: $5\frac{7}{12} + 3\frac{7}{18}$

10. Subtract: $9\frac{5}{9} - 3\frac{11}{12}$

11. Multiply: $\frac{9}{16} \times \frac{4}{27}$

12. Multiply: $2\frac{1}{8} \times 4\frac{5}{17}$

13. Divide: $\frac{11}{12} \div \frac{3}{4}$

14. Divide: $2\frac{3}{8} \div 2\frac{1}{2}$

15. Simplify: $\left(\frac{2}{3}\right)^2 \cdot \left(\frac{3}{4}\right)^3$

16. Simplify: $\left(\frac{2}{3}\right)^2 - \left(\frac{2}{3} - \frac{1}{2}\right) + 2$

17. Write 65.0309 in words.

18. Add:
$$\begin{array}{r} 379.006 \\ 27.523 \\ 9.8707 \\ +\ 88.2994 \\ \hline \end{array}$$

19. Subtract:
$$\begin{array}{r} 29.005 \\ -\ 7.9286 \\ \hline \end{array}$$

20. Multiply:
$$\begin{array}{r} 9.074 \\ \times\ 6.09 \\ \hline \end{array}$$

21. Divide: $8.09\overline{)17.42963}$ Round to the nearest thousandth.

22. Convert $\frac{11}{15}$ to a decimal. Round to the nearest thousandth.

23. Convert $0.16\frac{2}{3}$ to a fraction.

24. Place the correct symbol, < or >, between the two numbers.
$\frac{8}{9}$ 0.98

25. An airplane had 204 passengers aboard. During a stop, 97 passengers got off the plane and 127 passengers got on the plane. How many passengers were on the continuing flight?

26. An investor purchased stock at $\$32\frac{1}{8}$ per share. During the first 2 months of ownership, the stock lost $\$\frac{3}{4}$ and then gained $\$1\frac{1}{2}$. Find the value of each share of stock at the end of the 2 months.

27. You have a checking balance of $814.35. You then write checks for $42.98, $16.43, and $137.56. Find your checking account balance after you write the checks.

28. A machine lathe takes 0.017 inch from a brass bushing that is 1.412 inches thick. Find the resulting thickness of the bushing.

29. The state income tax on your business is $820 plus 0.08 times your profit. You made a profit of $64,860 last year. Find the amount of income tax you paid last year.

30. You bought a camera costing $210.96. The down payment is $20, and the balance is to be paid in 8 equal monthly payments. Find the monthly payment.

4 Ratio and Proportion

OBJECTIVES

- To write the ratio of two quantities in simplest form
- To solve application problems
- To write rates
- To write unit rates
- To solve application problems
- To determine if a proportion is true
- To solve proportions
- To solve application problems

Musical Scales

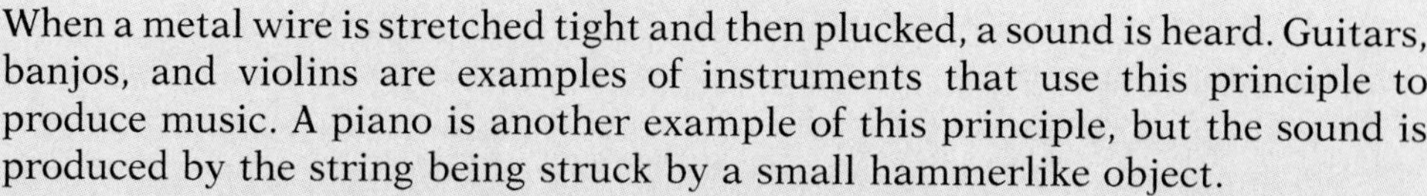

When a metal wire is stretched tight and then plucked, a sound is heard. Guitars, banjos, and violins are examples of instruments that use this principle to produce music. A piano is another example of this principle, but the sound is produced by the string being struck by a small hammerlike object.

After the string is plucked or struck, the string begins to vibrate. The number of times the string vibrates in 1 second is called the frequency of the vibration, or the pitch. Normally humans can hear vibrations as low as 16 cps (cycles per second) and as high as 20,000 cps. The longer the string, the lower the pitch of the sound; the shorter the string, the higher the pitch of the sound. In fact, a string half as long as another string will vibrate twice as fast.

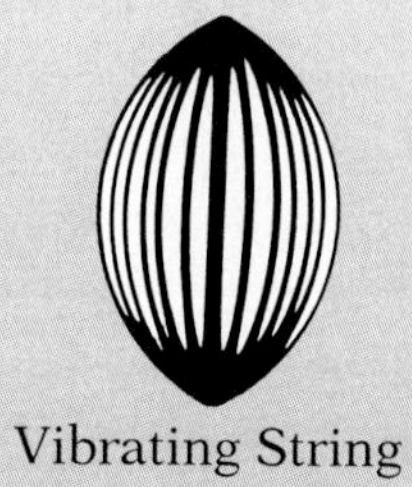

Vibrating String

Most music that is heard today is based on what is called the chromatic or twelve-tone scale. For this scale, a vibration of 261 cps is called middle C. A string half as long as the string for middle C vibrates twice as fast and produces a musical note one octave higher.

To produce the notes between the two C's, strings are placed that produce the desired pitch. Recall that as the string gets smaller, the pitch increases. A top view of a grand piano illustrates how the strings vary in length.

A well-tempered chromatic scale is one in which the string lengths are chosen so that the ratios of the frequencies of adjacent notes are the same.

$$\frac{C}{C^{\#}} = \frac{C^{\#}}{D} = \frac{D}{D^{\#}} = \frac{D^{\#}}{E} = \frac{E}{F} = \frac{F}{F^{\#}} = \frac{F^{\#}}{G} = \frac{G}{G^{\#}} = \frac{G^{\#}}{A} = \frac{A}{A^{\#}} = \frac{A^{\#}}{B} = \frac{B}{C}$$

The common ratio for the chromatic scale is approximately $\frac{1}{1.0595}$.

SECTION 4.1 Ratio

Objective A To write the ratio of two quantities in simplest form

Quantities such as 3 feet, 12 cents, or 9 cars are number quantities written with **units.**

3 feet
12 cents
9 cars
units

These are only some examples of units. Shirts, dollars, trees, miles, and gallons are further examples.

A **ratio** is the comparison of two quantities that have the *same* units. This comparison can be written three different ways:

1. As a fraction
2. As two numbers separated by a colon (:)
3. As two numbers separated by the word TO

The ratio of the lengths of two boards, one 8 feet long and the other 10 feet long, can be written as

1. $\frac{8 \text{ feet}}{10 \text{ feet}} = \frac{8}{10} = \frac{4}{5}$
2. 8 feet:10 feet = 8:10 = 4:5
3. 8 feet TO 10 feet = 8 TO 10 = 4 TO 5

A ratio is in **simplest form** when the two numbers do not have a common factor. Notice that the units are not written.

This ratio means that the smaller board is $\frac{4}{5}$ the length of the longer board.

Example 1

Write the comparison \$6 to \$8 as a ratio in simplest form using a fraction, a colon, and the word TO.

Solution $\frac{\$6}{\$8} = \frac{6}{8} = \frac{3}{4}$

\$6:\$8 = 6:8 = 3:4

\$6 TO \$8 = 6 TO 8 = 3 TO 4

Example 2

Write the comparison 20 pounds to 24 pounds as a ratio in simplest form using a fraction, a colon, and the word TO.

Your solution

Example 3

Write the comparison 18 quarts to 6 quarts as a ratio in simplest form using a fraction, a colon, and the word TO.

Solution $\frac{18 \text{ quarts}}{6 \text{ quarts}} = \frac{18}{6} = \frac{3}{1}$

18 quarts:6 quarts
18:6 = 3:1

18 quarts TO 6 quarts
18 TO 6 = 3 TO 1

Example 4

Write the comparison 64 miles to 8 miles as a ratio in simplest form using a fraction, a colon, and the word TO.

Your solution

Solutions on p. A19

Objective B To solve application problems

Use the table below for Examples 5 and 6.

Board Feet of Wood at a Lumber Store			
Pine	*Ash*	*Oak*	*Cedar*
20,000	18,000	10,000	12,000

Example 5

Find the ratio, as a fraction in simplest form, of the number of board feet of pine to board feet of oak.

Strategy

To find the ratio, write the ratio of board feet of pine (20,000) to board feet of oak (10,000) in simplest form.

Solution

$\frac{20{,}000}{10{,}000} = \frac{2}{1}$

The ratio is $\frac{2}{1}$.

Example 6

Find the ratio, as a fraction in simplest form, of the number of board feet of cedar to board feet of ash.

Your strategy

Your solution

Example 7

The cost of building a patio cover was $250 for labor and $350 for materials. What is the ratio, as a fraction in simplest form, of the cost of materials to the total cost for labor and materials?

Strategy

To find the ratio, write the ratio of the cost of materials ($350) to the total cost ($250 + $350) in simplest form.

Solution

$\frac{\$350}{\$250 + \$350} = \frac{350}{600} = \frac{7}{12}$

The ratio is $\frac{7}{12}$.

Example 8

A company spends $20,000 a month for television advertising and $15,000 a month for radio advertising. What is the ratio, as a fraction in simplest form, of the cost of radio advertising to the total cost of radio and television advertising?

Your strategy

Your Solution

Solutions on p. A19

4.1 EXERCISES

Objective A

Write the comparison as a ratio in simplest form using a fraction, a colon (:), and the word TO.

1. 3 pints to 15 pints

2. 6 pounds to 8 pounds

3. $40 to $20

4. 10 feet to 2 feet

5. 3 miles to 8 miles

6. 2 hours to 3 hours

7. 37 hours to 24 hours

8. 29 inches to 12 inches

9. 6 minutes to 6 minutes

10. 8 days to 12 days

11. 35 cents to 50 cents

12. 28 inches to 36 inches

13. 30 minutes to 60 minutes

14. 25 cents to 100 cents

15. 32 ounces to 16 ounces

16. 12 quarts to 4 quarts

17. 3 cups to 4 cups

18. 6 years to 7 years

19. $5 to $3

20. 30 yards to 12 yards

21. 12 quarts to 18 quarts

22. $20 to $28

23. 14 days to 7 days

24. 9 feet to 3 feet

25. 12 months to 12 months

26. 75 ounces to 75 ounces

27. 8 tons to 16 tons

28. 8 cents to 40 cents

29. 27 feet to 18 feet

30. 28 days to 20 days

Objective B *Application Problems*

Solve. Write ratios as fractions in simplest form.

Family Budget						
Housing $800	*Food* $400	*Transportation* $300	*Taxes* $350	*Utilities* $150	*Miscellaneous* $400	*Total* $2400

31. Use the table to find the ratio of housing cost to total income.

32. Use the table to find the ratio of food to total income.

33. Use the table to find the ratio of utilities cost to food cost.

34. Use the table to find the ratio of transportation cost to housing cost.

35. A school had 400 attend the Friday night performance of a play, and 325 attended the performance on Saturday afternoon. Write the ratio of the number of people attending on Friday to the number of people attending the play Saturday afternoon.

36. One computer prints 80 letters on one line, and a second computer prints 96 letters on one line. Write the ratio of the number of letters per line of the second computer to the number of letters per line of the first computer.

37. A transformer has 40 turns in the primary coil and 480 turns in the secondary coil. State the ratio of the turns in the primary to the number of turns in the secondary coil.

38. A student bought a computer system for $2400; 5 years later the student sold the computer for $900. Find the ratio of the amount the student received for the computer to the cost of the computer.

39. A house had an original value of $90,000 but increased in value to $110,000 in 5 years.
  a. Find the amount of the increase.
  b. What is the ratio of the increase to the original value?

40. A decorator bought one box of ceramic floor tile for $21 and a box of wood tile for $33.
  a. What was the total cost of the box of ceramic tile and the box of wood tile?
  b. What is the ratio of the cost of the wood tile to the total cost?

41. The price of gasoline jumped from $0.98 to $1.26 in 1 year. What is the ratio of the increase in price to the original price?

42. An investor paid $32,000 for an antique car; 5 years later the investor sold the car for $44,000. What is the ratio of the increase in value to the original value the investor paid.

SECTION 4.2 Rates

Objective A To write rates

A **rate** is a comparison of two quantities that have *different* units. A rate is written as a fraction.

A distance runner ran 26 miles in 4 hours. The distance-to-time rate is written

$\frac{26 \text{ miles}}{4 \text{ hours}} = \frac{13 \text{ miles}}{2 \text{ hours}}$

A rate is in **simplest form** when the numbers that form the rate have no common factors. Notice that the units are written as part of the rate.

Example 1 Write "6 roof supports for every 9 feet" as a rate in simplest form.

Solution $\frac{6 \text{ supports}}{9 \text{ feet}} = \frac{2 \text{ supports}}{3 \text{ feet}}$

Example 2 Write "15 pounds of fertilizer for 12 trees" as a rate in simplest form.

Your solution

Solution on p. A20

Objective B To write unit rates

A **unit rate** is a rate in which the number in the denominator is 1.

$\frac{\$3.25}{1 \text{ pound}}$ or \$3.25/pound is read "\$3.25 per pound."

To find unit rates, divide the number in the numerator of the rate by the number in the denominator of the rate.

A car traveled 344 miles on 16 gallons of gasoline. To find the miles per gallon (unit rate), divide the numerator of the rate by the denominator of the rate.

$\frac{344 \text{ miles}}{16 \text{ gallons}}$ is the rate.

$16\overline{)344.0}$ = 21.5   21.5 miles/gallon is the unit rate.

Example 3 Write "300 feet in 8 seconds" as a unit rate.

Solution $\frac{300 \text{ feet}}{8 \text{ seconds}}$ $8\overline{)300.0}$ = 37.5

37.5 feet /second

Example 4 Write "260 miles in 8 hours" as a unit rate.

Your solution

Solution on p. A20

Objective C To solve application problems

Example 5

A grocery store sells 3 grapefruit for 98 cents. What is the cost per grapefruit to the nearest tenth of a cent?

Strategy

To find the cost per grapefruit, divide the cost for 3 grapefruit (98¢) by the number of grapefruit (3).

Solution

$$3\overline{)98.00} = 32.66 \approx 32.7$$

The cost is 32.7¢ per grapefruit.

Example 6

A cyclist rode 47 miles in 3 hours. What is the miles-per-hour rate to the nearest tenth?

Your strategy

Your solution

Example 7

An investor purchased 100 shares of stock for $1500. One year later, the investor sold the 100 shares for $1800. What was the investor's profit per share?

Strategy

To find the investor's profit per share:

- Find the total profit by subtracting the original cost ($1500) from the selling price ($1800).
- Find the profit per share (unit rate) by dividing the total profit by the number of shares of stock (100).

Solution

$1800
−1500
$ 300 total profit

$$100\overline{)\$300} = \$3$$

The investor's profit per share was $3.

Example 8

A jeweler purchased 5 ounces of gold for $1980. Later, the jeweler sold the 5 ounces for $2075. What was the jeweler's profit per ounce?

Your strategy

Your solution

Solutions on p. A20

4.2 EXERCISES

Objective A

Write as a rate in simplest form.

1. 3 pounds of meat for 4 people
2. 30 ounces in 24 glasses
3. $80 for 12 boards
4. 84 cents for 6 bars of soap
5. 300 miles on 15 gallons
6. 88 feet in 8 seconds
7. 20 children in 8 families
8. 48 leaves on 9 plants
9. 16 gallons in 2 hours
10. 25 ounces in 5 minutes

Objective B

Write as a unit rate.

11. 10 feet in 4 seconds
12. 816 miles in 6 days
13. $1300 earned in 4 weeks
14. $27,000 earned in 12 months
15. 1100 trees planted on 10 acres
16. 3750 words on 15 pages
17. $32.97 earned in 7 hours
18. $315.70 earned in 22 hours
19. 628.8 miles in 12 hours
20. 388.8 miles in 8 hours
21. 344.4 miles on 12.3 gallons of gasoline
22. 409.4 miles on 11.5 gallons of gasoline
23. $349.80 for 212 pounds
24. $11.05 for 3.4 pounds
25. $2231.25 for 3.75 tons
26. $14,259 for 5.25 tons

Objective C *Application Problems*

27. A student drives 326.6 miles on 11.5 gallons of gas. Find the number of miles driven per gallon of gas.

28. You drive 246.6 miles in 4.5 hours. Find the number of miles driven per hour.

29. The Saturn-5 rocket uses 534,000 gallons of fuel in 2.5 minutes. How much fuel does the rocket use in 1 minute?

30. A ski instructor works 6 months at a ski resort and earns $13,500. What is the ski instructor's wage per month?

31. An investor owning 400 shares of an automobile corporation receives $1160 in dividends. Find the dividend per share.

32. An investor purchased 350 shares of stock for $12,600. What is the cost per share?

33. The total cost of making 5000 compact discs was $26,536.32. Of the discs made, 122 did not meet company standards.
    a. How many compact discs did meet company standards?
    b. What was the cost per disc for those discs that met company standards?

34. A family purchased a 250-pound side of beef for $365.75 and had it packaged. During the packaging, 75 pounds of beef were discarded as waste.
    a. How many pounds of beef were packaged?
    b. What was the cost per pound for the packaged beef?

35. A fruit stand purchased 250 boxes of strawberries for $162.50. All the strawberries were sold for $312.50. What was the profit per box of strawberries?

36. A company's cost to produce 1000 ties was $3250. The company sold the ties for $9490. What was the company's profit per tie?

37. It takes 500 seconds for the light from the sun to reach the earth. The distance from the sun to the earth is 93 million miles. Find the number of miles that light travels per second.

38. A space satellite travels 418,216 miles in 24 hours. How far does the satellite travel per hour? (Round to the nearest tenth.)

SECTION 4.3 Proportions

Objective A To determine if a proportion is true

A **proportion** is the equality of two ratios or rates.

$\frac{50 \text{ miles}}{4 \text{ gallons}} = \frac{25 \text{ miles}}{2 \text{ gallons}}$ Note that the units of the numerators are the same and the units of the denominators are the same.

$\frac{3}{6} = \frac{1}{2}$

A proportion is **true** if the fractions are equal when written in lowest terms.

In any true proportion, the "cross products" are equal.

Is $\frac{2}{3} = \frac{8}{12}$ a true proportion?

$\frac{2}{3} \quad \frac{8}{12}$ → $3 \times 8 = 24$, $2 \times 12 = 24$

Cross products *are* equal. $\frac{2}{3} = \frac{8}{12}$ is a true proportion.

A proportion is **not true** if the fractions are not equal when reduced to lowest terms.

If the cross products are not equal, then the proportion is not true.

Is $\frac{4}{5} = \frac{8}{9}$ a true proportion?

$\frac{4}{5} \quad \frac{8}{9}$ → $5 \times 8 = 40$, $4 \times 9 = 36$

Cross products *are not* equal. $\frac{4}{5} = \frac{8}{9}$ is not a true proportion.

Example 1 Use cross products to determine if $\frac{5}{8} = \frac{10}{16}$ is a true proportion.

Solution $\frac{5}{8} \quad \frac{10}{16}$ → $8 \times 10 = 80$, $5 \times 16 = 80$

The proportion is true.

Example 2 Use cross products to determine if $\frac{6}{10} = \frac{9}{15}$ is a true proportion.

Your solution

Example 3 Use cross products to determine if $\frac{62 \text{ miles}}{4 \text{ gallons}} = \frac{33 \text{ miles}}{2 \text{ gallons}}$ is a true proportion.

Solution $\frac{62}{4} \quad \frac{33}{2}$ → $4 \times 33 = 132$, $62 \times 2 = 124$

The proportion is not true.

Example 4 Use cross products to determine if $\frac{\$32}{6 \text{ hours}} = \frac{\$90}{8 \text{ hours}}$ is a true proportion.

Your solution

Solutions on p. A20

Objective B To solve proportions

Sometimes one of the numbers in a proportion is unknown. In this case, it is necessary to *solve* the proportion.

To **solve** a proportion, find a number to replace the unknown so that the proportion is true. Do this by following two steps.

Solve $\frac{n}{3} = \frac{6}{9}$ and check.

Find the cross products.
$n \times 9 = 3 \times 6$
$n \times 9 = 18$

To find n, recall the relationship between multiplication and division.

$n \times 9 = 18$ can be written $9\overline{)18}^{\,n}$
Therefore
$n = 18 \div 9 = 2$

Check:

$\frac{2}{3} \bowtie \frac{6}{9}$ → $3 \times 6 = 18$
→ $2 \times 9 = 18$

Example 5

Solve $\frac{n}{12} = \frac{25}{60}$ and check.

Solution

$n \times 60 = 12 \times 25$
$n \times 60 = 300$
$n = 300 \div 60$
$n = 5$

Check:

$\frac{5}{12} \bowtie \frac{25}{60}$ → $12 \times 25 = 300$
→ $5 \times 60 = 300$

Example 6

Solve $\frac{n}{14} = \frac{3}{7}$ and check.

Your solution

Example 7

Solve $\frac{4}{9} = \frac{n}{16}$. Write the answer to the nearest tenth.

Solution

$4 \times 16 = 9 \times n$
$64 = 9 \times n$
$64 \div 9 = n$
$7.1 \approx n$

Note: A rounded answer is an approximation. Therefore the answer to a check will not be exact.

Example 8

Solve $\frac{5}{8} = \frac{n}{20}$. Write the answer to the nearest tenth.

Your solution

Solutions on p. A20

Example 9

Solve $\frac{28}{52} = \frac{7}{n}$ and check.

Solution

$28 \times n = 52 \times 7$
$28 \times n = 364$
$n = 364 \div 28$
$n = 13$

Check:

$\frac{28}{52} \;\; \frac{7}{13}$ → $52 \times 7 = 364$
→ $28 \times 13 = 364$

Example 10

Solve $\frac{15}{20} = \frac{12}{n}$ and check.

Your solution

Example 11

Solve $\frac{15}{n} = \frac{8}{3}$. Write the answer to the nearest hundredth.

Solution

$15 \times 3 = n \times 8$
$45 = n \times 8$
$45 \div 8 = n$
$5.63 \approx n$

Example 12

Solve $\frac{12}{n} = \frac{7}{4}$. Write the answer to the nearest hundredth.

Your solution

Example 13

Solve $\frac{n}{9} = \frac{3}{1}$ and check.

Solution

$n \times 1 = 9 \times 3$
$n \times 1 = 27$
$n = 27 \div 1$
$n = 27$

Check:

$\frac{27}{9} \;\; \frac{3}{1}$ → $9 \times 3 = 27$
→ $27 \times 1 = 27$

Example 14

Solve $\frac{n}{12} = \frac{4}{1}$ and check.

Your solution

Example 15

Solve $\frac{5}{9} = \frac{15}{n}$ and check.

Solution

$5 \times n = 9 \times 15$
$5 \times n = 135$
$n = 135 \div 5$
$n = 27$

Check:

$\frac{5}{9} \;\; \frac{15}{27}$ → $9 \times 15 = 135$
→ $5 \times 27 = 135$

Example 16

Solve $\frac{3}{8} = \frac{12}{n}$ and check.

Your solution

Solutions on p. A20

Objective C To solve application problems

Example 17

A mason determines that 9 cement blocks are required for a retaining wall 2 feet long. At this rate, how many cement blocks are required for a retaining wall that is 24 feet long?

Strategy

To find the number of cement blocks for a retaining wall 24 feet long, write and solve a proportion using n to represent the number of blocks required.

Solution

$$\frac{9 \text{ cement blocks}}{2 \text{ feet}} = \frac{n \text{ cement blocks}}{24 \text{ feet}}$$

$$9 \times 24 = 2 \times n$$
$$216 = 2 \times n$$
$$216 \div 2 = n$$
$$108 = n$$

108 cement blocks are required for a 24-foot retaining wall.

Example 18

Twenty-four jars can be packed in 6 identical boxes. At this rate, how many jars can be packed in 15 boxes?

Your strategy

Your solution

Example 19

The dosage of a certain medication is 2 ounces for every 50 pounds of body weight. How many ounces of this medication are required for a person who weighs 175 pounds?

Strategy

To find the number of ounces of medication for a person weighing 175 pounds, write and solve a proportion using n to represent the number of ounces of medication for a 175-pound person.

Solution $\frac{2 \text{ ounces}}{50 \text{ pounds}} = \frac{n \text{ ounces}}{175 \text{ pounds}}$

$$2 \times 175 = 50 \times n$$
$$350 = 50 \times n$$
$$350 \div 50 = n$$
$$7 = n$$

7 ounces of medication are required for a 175-pound person.

Example 20

Three tablespoons of a liquid plant fertilizer are to be added to every 4 gallons of water. How many tablespoons of fertilizer are required for 10 gallons of water?

Your strategy

Your solution

Solutions on pp. A20–A21

4.3 EXERCISES

Objective A

Determine if the proportion is true or not true.

1. $\frac{4}{8} = \frac{10}{20}$

2. $\frac{39}{48} = \frac{13}{16}$

3. $\frac{7}{8} = \frac{11}{12}$

4. $\frac{15}{7} = \frac{17}{8}$

5. $\frac{27}{8} = \frac{9}{4}$

6. $\frac{3}{18} = \frac{4}{19}$

7. $\frac{45}{135} = \frac{3}{9}$

8. $\frac{3}{4} = \frac{54}{72}$

9. $\frac{16}{3} = \frac{48}{9}$

10. $\frac{15}{5} = \frac{3}{1}$

11. $\frac{7}{40} = \frac{7}{8}$

12. $\frac{9}{7} = \frac{6}{5}$

13. $\frac{\text{50 miles}}{\text{2 gallons}} = \frac{\text{25 miles}}{\text{1 gallon}}$

14. $\frac{\text{16 feet}}{\text{10 seconds}} = \frac{\text{24 feet}}{\text{15 seconds}}$

15. $\frac{\text{6 minutes}}{\text{5 cents}} = \frac{\text{30 minutes}}{\text{25 cents}}$

16. $\frac{\text{16 pounds}}{\text{12 days}} = \frac{\text{20 pounds}}{\text{14 days}}$

17. $\frac{\$15}{\text{4 pounds}} = \frac{\$45}{\text{12 pounds}}$

18. $\frac{\text{270 trees}}{\text{6 acres}} = \frac{\text{90 trees}}{\text{2 acres}}$

19. $\frac{\text{300 feet}}{\text{4 rolls}} = \frac{\text{450 feet}}{\text{7 rolls}}$

20. $\frac{\text{1 gallon}}{\text{4 quarts}} = \frac{\text{7 gallons}}{\text{28 quarts}}$

21. $\frac{\$65}{\text{5 days}} = \frac{\$26}{\text{2 days}}$

22. $\frac{\text{80 miles}}{\text{2 hours}} = \frac{\text{110 miles}}{\text{3 hours}}$

23. $\frac{\text{7 tiles}}{\text{4 feet}} = \frac{\text{42 tiles}}{\text{20 feet}}$

24. $\frac{\text{15 feet}}{\text{3 yards}} = \frac{\text{90 feet}}{\text{18 yards}}$

Determine if the proportion is true or not true.

25. $\frac{60 \text{ gallons}}{4 \text{ hours}} = \frac{160 \text{ gallons}}{12 \text{ hours}}$

26. $\frac{2500 \text{ words}}{15 \text{ pages}} = \frac{5800 \text{ words}}{35 \text{ pages}}$

27. $\frac{\$24{,}000}{12 \text{ months}} = \frac{\$36{,}000}{18 \text{ months}}$

28. $\frac{250 \text{ trees}}{2 \text{ acres}} = \frac{1500 \text{ trees}}{12 \text{ acres}}$

Objective B

Solve. Round to the nearest hundredth.

29. $\frac{n}{4} = \frac{6}{8}$ **30.** $\frac{n}{7} = \frac{9}{21}$ **31.** $\frac{12}{18} = \frac{n}{9}$ **32.** $\frac{7}{21} = \frac{35}{n}$ **33.** $\frac{10}{12} = \frac{n}{24}$

34. $\frac{6}{n} = \frac{24}{36}$ **35.** $\frac{3}{n} = \frac{15}{10}$ **36.** $\frac{n}{45} = \frac{17}{135}$ **37.** $\frac{9}{4} = \frac{18}{n}$ **38.** $\frac{7}{15} = \frac{21}{n}$

39. $\frac{n}{6} = \frac{2}{3}$ **40.** $\frac{5}{12} = \frac{n}{144}$ **41.** $\frac{n}{5} = \frac{7}{8}$ **42.** $\frac{4}{n} = \frac{9}{5}$ **43.** $\frac{8}{5} = \frac{n}{6}$

44. $\frac{n}{11} = \frac{32}{4}$ **45.** $\frac{3}{4} = \frac{8}{n}$ **46.** $\frac{5}{12} = \frac{n}{8}$ **47.** $\frac{36}{20} = \frac{12}{n}$ **48.** $\frac{15}{n} = \frac{65}{100}$

49. $\frac{n}{15} = \frac{21}{12}$ **50.** $\frac{40}{n} = \frac{15}{8}$ **51.** $\frac{32}{n} = \frac{1}{3}$ **52.** $\frac{5}{8} = \frac{42}{n}$

53. $\frac{18}{11} = \frac{16}{n}$ **54.** $\frac{25}{4} = \frac{n}{12}$ **55.** $\frac{28}{8} = \frac{12}{n}$ **56.** $\frac{n}{30} = \frac{65}{120}$

57. $\frac{0.3}{5.6} = \frac{n}{25}$ **58.** $\frac{1.3}{16} = \frac{n}{30}$ **59.** $\frac{0.7}{9.8} = \frac{3.6}{n}$ **60.** $\frac{1.9}{7} = \frac{13}{n}$

61. $\frac{18.6}{23.59} = \frac{88.5}{n}$ **62.** $\frac{4065}{699} = \frac{n}{3260}$ **63.** $\frac{8.23}{n} = \frac{1.025}{7.093}$ **64.** $\frac{n}{4.72} = \frac{3827}{69.4}$

Objective C *Application Problems*

Solve. Round to the nearest hundredth.

65. A 20-ounce box of raisin bran contains 2400 calories. How many calories are in a 1.4-ounce serving of the cereal?

66. A 10-ounce container of egg noodles contains 700 milligrams of cholesterol. How many milligrams of cholesterol are in a 2-ounce serving of egg noodles?

67. A homeowner uses 2 pounds of fertilizer for every 100 square feet of lawn. At this rate, how many pounds of fertilizer are used on a lawn that measures 2500 square feet?

68. A nursery provides a liquid plant food by adding 1 gallon of water for each 2 ounces of plant food. At this rate, how many ounces of plant food are required for 25 gallons of water?

69. A savings and loan company receives $10.84 each month for every $1000 it loans. At this rate, what is the monthly payment for a $15,000 loan?

70. A life insurance policy cost $6.87 for every $1000 of insurance. At this rate, what is the cost of $25,000 of insurance?

71. The property tax on a $90,000 home is $1800. At this rate, what is the property tax on a home worth $125,000?

72. The sales tax on a $90 purchase is $4.50. At this rate, what is the sales tax on a $300 purchase?

73. A brick wall 20 feet in length contains 1040 bricks. At the same rate, how many bricks would it take to build a wall 48 feet in length?

74. A farmer has 50 acres of wheat that yield 2900 bushels of wheat. At the same rate, find the number of bushels of wheat the farmer can expect from another 240 acres of wheat.

75. The scale on a map is 1 inch equals 12 miles. Find the distance between two cities that are 9.5 inches apart on the map.

76. The scale on the plans for a new house is 1 inch equals 3 feet. Find the length and width of a room that measures 5 inches by 8 inches on the drawing.

77. The dosage for a medication is $\frac{1}{3}$ ounce for every 40 pounds of body weight. At this rate, how many ounces of medication are required for a person who weighs 180 pounds?

Solve. Round to the nearest hundredth.

78. A lawyer on a reducing program has lost 3 pounds in 5 weeks. At the same rate, how long will it take the lawyer to lose 27 pounds?

79. A pre-election survey showed that 2 out of every 3 eligible voters would cast ballots in the county election. At this rate, how many people in a county of 240,000 eligible voters would vote in the election?

80. A paint manufacturer suggests using 1 gallon of paint for every 400 square feet of a wall. At this rate, how many gallons of paint would be required for a room that has 1400 square feet of wall?

81. An automobile recall was based on tests that showed 24 braking defects in 1000 cars. At this rate, how many defects would be found in 40,000 cars?

82. From previous experience, a manufacturer knows that in an average production run of 2000 circuit boards, 60 will be defective. What number of defective circuit boards can be expected from a run of 25,000 circuit boards?

83. You own 240 shares of a computer stock. The company declares a stock split of 5 shares for every 3 owned. How many shares of stock do you own after the stock split?

84. An investor own 50 shares of a utility stock that pays dividends of $125. At this rate, what dividend would the investor receive after buying 300 additional shares of stock in the utility company?

85. A community college estimates that the ratio of student time to administrative time on a computer is 3:2. During the month in which the computer was used 200 hours for administration, how many hours were used by students?

86. A management consulting firm recommends that the ratio of midmanagement salaries to junior management salaries be 5:4. Using this recommendation, what is the yearly midmanagement salary when the junior management yearly salary is $32,000?

87. Seawater contains 3.5 parts of salt for every 100 parts of seawater. At this rate, what is the amount of salt that can be recovered from 50,000 pounds of seawater?

88. A survey indicates that 19 out of 30 eligible voters will vote in the next election. How many votes will be cast out of 528,000 eligible voters?

Calculators and Computers

The Memory Key The memory key on a calculator is useful when calculating with unit rates. The unit rate is stored in the memory of the calculator and then recalled when it is needed for a computation.

The cost of a car loan is an example of a unit rate. A bank charges a certain dollar amount each month for each dollar that is borrowed. For example, to repay a 5-year loan at an annual interest rate of 9.5%, the borrower must pay $0.0210019 each month for each $1 borrowed.

1. Find the monthly payment for an $8500 loan.
2. Find the monthly payment for a $6200 loan.
3. Find the monthly payment for a $7800 loan.

To answer question 1, enter the following on your calculator:

0.0210019 [M+] 8500 [×] [MR] [=]

The numbers 178.5161 will be displayed on your calculator. The monthly payment is $178.52 when rounded to the nearest cent.

The [M+] key means "Add to Memory" and the [MR] key means "Recall from Memory". Another key on your calculator is the [MC] key, which means "Clear the Memory".

To answer the second question, it is not necessary to enter the payment rate because it is already stored in the calculator's memory. Just enter the following:

6200 [×] [MR] [=]

The number 130.2117 will be displayed on your calculator. The monthly payment is $130.21 rounded to the nearest cent.

Enter the following to answer the third question:

7800 [×] [MR] [=]

The number 163.8148 will be displayed on your calculator. The monthly payment is $163.81 rounded to the nearest cent.

The advantage of using the memory key is that various monthly payments can be calculated for different loan amounts without the payment rate having to be re-entered each time.

Chapter Summary

Key Words Quantities such as 8 feet and 60 miles are number quantities written with *units.*

A *ratio* is the comparison of two quantities that have the same units.

A ratio is in *simplest form* when the two numbers that form the ratio have no common factors.

A *rate* is the comparison of two quantities that have different units.

A rate is in *simplest form* when the numbers that form the rate have no common factors.

A *unit rate* is a rate in which the number in the denominator is 1.

A *proportion* is the equality of two ratios or rates.

Essential Rules

To Find Unit Rates

To find unit rates, divide the number in the numerator of the rate by the number in the denominator of the rate.

To Solve a Proportion

One of the numbers in a proportion may be unknown. To solve a proportion, find a number to replace the unknown so that the proportion is true.

Ways to Express a Ratio

A ratio can be written three different ways:

1. As a fraction
2. As two numbers separated by a colon (:)
3. As two numbers separated by the word TO

Chapter Review

SECTION 4.1

1. Write the comparison 8 feet to 28 feet as a ratio in simplest form using a fraction, a colon (:), and the word TO.

2. Write the comparison 6 inches to 15 inches as a ratio in simplest form using a fraction, a colon (:), and the word TO.

3. Write the comparison 12 days to 12 days as a ratio in simplest form using a fraction, a colon (:), and the word TO.

4. Write the comparison 32 dollars to 80 dollars as a ratio in simplest form using a fraction, a colon (:), and the word TO.

5. The high temperature during a 24-hour period was 84 degrees and the low temperature was 42 degrees. Write the ratio of the high temperature to the low temperature for the 24-hour period.

6. A house had an original value of $80,000 but increased in value to $120,000 in 2 years. Find the ratio of the increase to the original value.

7. In 5 years the price of a calculator went from $40 to $24. What is the ratio of the decrease in price to the original price?

8. A retail computer store spends $30,000 a year on TV advertising and $12,000 on newspaper advertising. Find the ratio of TV advertising to newspaper advertising.

SECTION 4.2

9. Write "100 miles in 3 hours" as a rate in simplest form.

10. Write "$15 in 4 hours" as a rate in simplest form.

11. Write "250 miles in 4 hours" as a unit rate.

12. Write "$300 earned in 40 hours" as a unit rate.

13. Write $326.4 miles on 12 gallons" as a unit rate.

14. Write "$8.75 for 5 pounds" as a unit rate.

15. An investor purchased 80 shares of stock for \$3580. What is the cost per share?

16. You drove 198.8 miles in 3.5 hours. Find the average number of miles driven per hour.

17. A 15-pound turkey costs \$10.20. What is the cost per pound?

18. The total cost of manufacturing 1000 radios was \$36,600. Of the radios made, 24 did not pass inspection. Find the cost per radio of the radios that did pass inspection.

SECTION 4.3

19. Determine if the proportion is true or not true.
$\frac{5}{7} = \frac{25}{35}$

20. Determine if the proportion is true or not true.
$\frac{2}{9} = \frac{10}{45}$

21. Determine if the proportion is true or not true.
$\frac{8}{15} = \frac{32}{60}$

22. Determine if the proportion is true or not true.
$\frac{3}{8} = \frac{10}{24}$

23. Solve the proportion.
$\frac{n}{8} = \frac{9}{2}$

24. Solve the proportion.
$\frac{16}{n} = \frac{4}{17}$

25. Solve the proportion.
Round to hundredths.
$\frac{24}{11} = \frac{n}{30}$

26. Solve the proportion.
Round to hundredths.
$\frac{18}{35} = \frac{10}{n}$

27. A homeowner used 1.5 pounds of fertilizer for every 200 square feet of lawn. How many pounds of fertilizer are used on a lawn that measures 3000 square feet?

28. An insurance policy costs \$3.87 for every \$1000 of insurance. At this rate, what is the cost of \$50,000 of insurance?

29. The property tax on a \$45,000 home is \$900. At the same rate, what is the property tax on a home valued at \$120,000?

30. A brick wall 40 feet in length contains 448 concrete blocks. At the same rate, how many blocks would it take to build a wall that is 120 feet in length?

Chapter Test

1. Write the comparison 12 days to 8 days as a ratio in simplest form using a fraction, a colon (:), and the word TO.

2. Write the comparison 18 feet to 30 feet as a ratio in simplest form using a fraction, a colon (:), and the word TO.

3. Write the comparison 27 dollars to 81 dollars as a ratio in simplest form using a fraction, a colon (:), and the word TO.

4. Write the comparison 40 miles to 240 miles as a ratio in simplest form using a fraction, a colon (:), and the word TO.

5. The average summer temperature in a California desert is 112 degrees. In a city 100 miles away, the average summer temperature is 86 degrees. Write the ratio of the city temperature to the desert temperature.

6. An automobile sales company spends $25,000 each month for television advertising and $40,000 each month for radio advertising. Find the ratio, as a fraction in simplest form, of the cost of radio advertising to the total cost of advertising.

7. Write "18 supports for every 8 feet" as a rate in simplest form.

8. Write "$81 for 12 boards" as a rate in simplest form.

9. Write "$22,036.80 earned in 12 months" as a unit rate.

10. Write "256.2 miles on 8.4 gallons of gas" as a unit rate.

11. 40 feet of lumber costs $69.20. What is the per foot cost of the lumber?

12. A plane travels 2421 miles in 4.5 hours. Find the plane's speed in miles per hour.

13. Determine if the proportion is true or not true.
$\frac{5}{14} = \frac{25}{70}$

14. Determine if the proportion is true or not true.
$\frac{40}{125} = \frac{5}{25}$

15. Solve the proportion.
$\frac{n}{18} = \frac{9}{4}$

16. Solve the proportion.
$\frac{5}{12} = \frac{60}{n}$

17. A research scientist estimates that the human body contains 88 pounds of water for every 100 pounds of body weight. At this rate, estimate the number of pounds of water in a college student weighing 150 pounds.

18. The dosage of a medicine is $\frac{1}{4}$ ounce for every 50 pounds of body weight. How many ounces of this medication are required for a person who weighs 175 pounds?

19. The property tax on a house valued at $75,000 is $1500. At the same rate, find the property tax on a house valued at $140,000.

20. Fifty shares of a utility stock has a dividend of $62.50. At the same rate, find the dividend of 500 shares of the utility stock.

Cumulative Review

1. Subtract: $\begin{array}{r} 20{,}095 \\ -10{,}937 \\ \hline \end{array}$

2. Write $2 \cdot 2 \cdot 2 \cdot 2 \cdot 3 \cdot 3 \cdot 3$ in exponential notation.

3. Simplify $4 - (5 - 2)^2 \div 3 + 2$.

4. Find the prime factorization of 160.

5. Find the LCM of 9, 12, and 18.

6. Find the GCF of 28 and 42.

7. Reduce $\frac{40}{64}$ to simplest form.

8. Add: $3\frac{5}{6} + 4\frac{7}{15}$

9. Subtract: $10\frac{1}{6} - 4\frac{5}{9}$

10. Multiply: $\frac{11}{12} \times 3\frac{1}{11}$

11. Divide: $3\frac{1}{3} \div \frac{5}{7}$

12. Simplify: $\left(\frac{2}{5} + \frac{3}{4}\right) \div \frac{3}{2}$

13. Write 4.0709 in words.

14. Round 2.09762 to the nearest hundredth.

15. Divide: $8.09\overline{)16.0976}$
Round to the nearest thousandth.

16. Convert $0.06\frac{2}{3}$ to a fraction.

Cumulative Review

17. Write the comparison 25 miles to 200 miles as a ratio in simplest form.

18. Write "87 cents for 6 bars of soap" as a rate in simplest form.

19. Write "250.5 miles on 7.5 gallons of gas" as a unit rate.

20. Solve $\frac{40}{n} = \frac{160}{17}$. Round to the nearest hundredth.

21. A car traveled 457.6 miles in 8 hours. Find the car's speed in miles per hour.

22. Solve the proportion.
$\frac{12}{5} = \frac{n}{15}$

23. You had $1024 in your checking account. You then wrote checks for $192 and $88. What is your new checking account balance?

24. A farmer buys a tractor for $22,760. A down payment of $5000 is required. The balance remaining is paid in 48 equal monthly installments. What is the monthly payment?

25. An assignment for a liberal arts student is to read a book containing 175 pages. The student reads $\frac{2}{5}$ of the book during Thanksgiving vacation. How many pages of the assignment remain to be read?

26. A building contractor bought $2\frac{1}{3}$ acres of land for $84,000. What was the cost of each acre?

27. A shopper bought a shirt for $21.79 and a tie for $8.59. A $50 bill was used to pay for the purchases. Find the amount of change.

28. A college baseball player had 42 hits in 155-at-bats. Find the baseball player's batting average. (Round to three decimal places.)

29. A soil conservationist estimates that a river bank is eroding at the rate of 3 inches every 6 months. At this rate, how many inches will be eroded in 50 months?

30. The dosage of a medicine is $\frac{1}{2}$ ounce for every 50 pounds of body weight. How many ounces of this medication are required for a person who weighs 160 pounds?

5 Percents

OBJECTIVES

- To write a percent as a fraction or decimal
- To write a fraction or a decimal as a percent
- To find the amount when the percent and base are given
- To solve application problems
- To find the percent when the base and amount are given
- To solve application problems
- To find the base when the percent and amount are given
- To solve application problems
- To solve percent problems using proportions
- To solve application problems

The Percent Symbol

The idea of using percent dates back many hundreds of years. Percents are used in business for all types of consumer and business loans; in chemistry to measure the percent concentration of an acid; in economics to measure the increases or decreases of the consumer price index (CPI); and in many other areas that affect our daily lives.

The word percent comes from the Latin phrase *per centum,* which means "by the hundred." The symbol that is used today for percent is %, but this was not always the symbol.

The present symbol apparently is a result of abbreviations for the word "percent."

One abbreviation was p. cent; later, p. 100 and p. $\overset{o}{c}$ were used.

From p. $\overset{o}{c}$, the abbreviation changed to p $\frac{o}{o}$ around the 17th century. This probably was a result of continual writing of p. $\overset{o}{c}$ and the eventual closing of the "c" to make an "o." By the 19th century, the "p" in front of the symbol p$\frac{o}{o}$ was no longer written. The bar that separated the o's became a slash, and the modern symbol % became widely used.

SECTION 5.1 Introduction to Percents

Objective A To write a percent as a fraction or decimal

Percent means "parts of 100." In the figure at the right, there are 100 parts. Because 13 of the 100 parts are shaded, 13% of the figure is shaded.

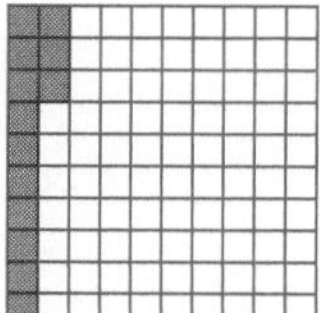

In most applied problems involving percents, it is necessary to rewrite either a percent as a fraction or a decimal or a fraction or a decimal as a percent.

To write a percent as a fraction, remove the percent sign and multiply by $\frac{1}{100}$.

$$13\% = 13 \times \frac{1}{100} = \frac{13}{100}$$

To write a percent as a decimal, remove the percent sign and multiply by 0.01.

$$13\% \quad = \quad 13 \times 0.01 \quad = \quad 0.13$$

Move the decimal point two places to the left. Then remove the percent sign.

Example 1 Write 120% as a fraction and as a decimal.

Solution $120\% = 120 \times \frac{1}{100} = \frac{120}{100}$

$= 1\frac{1}{5}$

$120\% = 120 \times 0.01 = 1.2$

Note that percents larger than 100 are greater than 1.

Example 2 Write 125% as a fraction and as a decimal.

Your solution

Example 3 Write $16\frac{2}{3}\%$ as a fraction.

Solution $16\frac{2}{3}\% = 16\frac{2}{3} \times \frac{1}{100}$

$= \frac{50}{3} \times \frac{1}{100} = \frac{50}{300} = \frac{1}{6}$

Example 4 Write $33\frac{1}{3}\%$ as a fraction.

Your solution

Example 5 Write 0.5% as a decimal.

Solution $0.5\% = 0.5 \times 0.01 = 0.005$

Example 6 Write 0.25% as a decimal.

Your solution

Solutions on p. A21

Objective B To write a fraction or a decimal as a percent

A decimal or a fraction can be written as a percent by multiplying by 100%.

Write $\frac{3}{8}$ as a percent.

$$\frac{3}{8} = \frac{3}{8} \times 100\% = \frac{3}{8} \times \frac{100}{1}\% = \frac{300}{8}\% = 37\frac{1}{2}\% \text{ or } 37.5\%$$

Write 0.37 as a percent.

$$0.37 = 0.37 \times 100\% = 37\%$$

Move the decimal point two places to the right. Then write the percent sign.

Example 7 Write 0.015 as a percent.

Solution $0.015 = 0.015 \times 100\%$
1.5%

Example 8 Write 0.048 as a percent.

Your solution

Example 9 Write 2.15 as a percent.

Solution $2.15 = 2.15 \times 100\% = 215\%$

Example 10 Write 3.67 as a percent.

Your solution

Example 11 Write $0.33\frac{1}{3}$ as a percent.

Solution $0.33\frac{1}{3} = 0.33\frac{1}{3} \times 100\%$
$= 33\frac{1}{3}\%$

Example 12 Write $0.62\frac{1}{2}$ as a percent.

Your solution

Example 13 Write $\frac{2}{3}$ as a percent. Write the remainder in fractional form.

Solution $\frac{2}{3} = \frac{2}{3} \times 100\% = \frac{200}{3}\%$
$= 66\frac{2}{3}\%$

Example 14 Write $\frac{5}{6}$ as a percent. Write the remainder in fractional form.

Your solution

Example 15 Write $2\frac{2}{7}$ as a percent. Round to the nearest tenth.

Solution $2\frac{2}{7} = \frac{16}{7} = \frac{16}{7} \times 100\%$
$= \frac{1600}{7}\% \approx 228.6\%$

Example 16 Write $1\frac{4}{9}$ as a percent. Round to the nearest tenth.

Your solution

Solutions on p. A21

5.1 EXERCISES

Objective A

Write as a fraction and as a decimal.

1. 25%
2. 40%
3. 130%
4. 150%
5. 100%
6. 87%
7. 73%
8. 45%
9. 383%
10. 425%
11. 70%
12. 55%
13. 88%
14. 64%
15. 32%
16. 18%

Write as a fraction.

17. $66\frac{2}{3}\%$
18. $12\frac{1}{2}\%$
19. $83\frac{1}{3}\%$
20. $3\frac{1}{8}\%$
21. $11\frac{1}{9}\%$
22. $\frac{3}{8}\%$
23. $45\frac{5}{11}\%$
24. $15\frac{3}{8}\%$
25. $4\frac{2}{7}\%$
26. $5\frac{3}{4}\%$
27. $6\frac{2}{3}\%$
28. $8\frac{2}{3}\%$

Write as a decimal.

29. 6.5%
30. 12.3%
31. 0.55%
32. 2%
33. 8.25%
34. 5.05%
35. 6.75%
36. 3.08%
37. 0.45%
38. 6.4%
39. 80.4%
40. 16.7%

Objective B

Write as a percent.

41. 0.16 **42.** 0.73 **43.** 0.05 **44.** 0.13 **45.** 0.01 **46.** 0.95

47. 0.70 **48.** 1.07 **49.** 1.24 **50.** 2.07 **51.** 0.004 **52.** 0.37

53. 0.006 **54.** 1.012 **55.** 3.106 **56.** 0.12

Write as a percent. Round to the nearest tenth.

57. $\frac{27}{50}$ **58.** $\frac{37}{100}$ **59.** $\frac{1}{3}$ **60.** $\frac{2}{5}$

61. $\frac{5}{8}$ **62.** $\frac{1}{8}$ **63.** $\frac{1}{6}$ **64.** $1\frac{1}{2}$

65. $\frac{7}{40}$ **66.** $1\frac{2}{3}$ **67.** $1\frac{7}{9}$ **68.** $\frac{7}{8}$

Write as a percent. Write the remainder in fractional form.

69. $\frac{15}{50}$ **70.** $\frac{12}{25}$ **71.** $\frac{7}{30}$ **72.** $\frac{1}{3}$

73. $\frac{4}{7}$ **74.** $\frac{1}{12}$ **75.** $\frac{3}{11}$ **76.** $\frac{2}{9}$

77. $2\frac{3}{8}$ **78.** $1\frac{2}{3}$ **79.** $2\frac{1}{6}$ **80.** $\frac{7}{8}$

SECTION 5.2 Percent Equations: Part I

Objective A To find the amount when the percent and base are given

A real estate broker receives a payment that is 4% of an \$85,000 sale. To find the amount the broker receives requires answering the question "4% of \$85,000 is what?"

This sentence can be written using mathematical symbols and then solved for the unknown number.

4%	of	\$85,000	is	what?
↓	↓	↓	↓	↓
percent 4%	×	base \$85,000	=	amount n

of is written as × (times)
is is written as = (equals)
what is written as n (the unknown number)

$$0.04 \times \$85{,}000 = n$$
$$\$3400 = n$$

Notice that the percent is written as a decimal.

The broker receives a payment of \$3400.

The solution was found by solving the basic percent equation for amount.

The Basic Percent Equation

Percent	×	base	=	amount

In most cases, the percent is written as a decimal before the basic percent equation is solved. However, some percents are more easily written as a fraction than as a decimal. For example,

$33\frac{1}{3}\% = \frac{1}{3}$ $66\frac{2}{3}\% = \frac{2}{3}$ $16\frac{2}{3}\% = \frac{1}{6}$ $83\frac{1}{3}\% = \frac{5}{6}$

Example 1 Find 5.7% of 160.

Solution $n = 0.057 \times 160$
$n = 9.12$

Note the words "what is" are missing from the problem but are implied by the word "Find."

Example 2 Find 6.3% of 150.

Your solution

Example 3 What is $33\frac{1}{3}\%$ of 90?

Solution $n = \frac{1}{3} \times 90$
$n = 30$

Example 4 What is $16\frac{2}{3}\%$ of 66?

Your solution

Solutions on p. A21

Objective B

The solution to percent problems requires identifying the three elements of the basic percent equation. Recall that these three parts are percent, base, and amount. Usually the base will follow the phrase "percent of."

Example 5

The harvest from a small orange grove was 560 crates of oranges. 5% of the 560 crates were damaged during shipment. How many crates were damaged?

Strategy

To find how many crates were damaged, write and solve a basic percent equation using n to represent the number of crates damaged (amount). The percent is 5% and the base is 560.

Solution

$$5\% \times 560 = n$$
$$0.05 \times 560 = n$$
$$28 = n$$

28 crates were damaged during shipment.

Example 6

A person buys a television that costs $450 and pays a sales tax that is 6% of the cost. What is the sales tax on the television?

Your strategy

Your solution

Example 7

A quality control inspector found that 1.2% of 2500 telephones inspected were defective. How many telephones were not defective?

Strategy

To find the number of nondefective phones:

- Find the number of defective phones. Write and solve a basic percent equation using n to represent the number of defective phones (amount). The percent is 1.2% and the base is 2500.
- Subtract the number of defective phones from the number of phones inspected (2500).

Solution

$$1.2\% \times 2500 = n$$
$$0.012 \times 2500 = n$$
$$30 = n \text{ defective phones}$$

$$2500 - 30 = 2470$$

2470 telephones were not defective.

Example 8

An electrician's hourly wage is $13.50 before an 8% raise. What is the new hourly wage?

Your strategy

Your solution

Solutions on p. A21

5.2 EXERCISES

Objective A

Solve.

1. 8% of 100 is what?
2. 16% of 50 is what?
3. 27% of 40 is what?
4. 52% of 95 is what?
5. 0.05% of 150 is what?
6. 0.075% of 625 is what?
7. 125% of 64 is what?
8. 210% of 12 is what?
9. Find 10.7% of 485.
10. Find 12.8% of 625.
11. What is 0.25% of 3000?
12. What is 0.06% of 250?
13. 80% of 16.25 is what?
14. 26% of 19.5 is what?
15. What is $1\frac{1}{2}$% of 250?
16. What is $5\frac{3}{4}$% of 65?
17. $16\frac{2}{3}$% of 120 is what?
18. $83\frac{1}{3}$% of 246 is what?
19. What is $33\frac{1}{3}$% of 630?
20. What is $66\frac{2}{3}$% of 891?
21. Which is larger: 5% of 95, or 75% of 6?
22. Which is larger: 82% of 16, or 20% of 65?
23. Which is larger: 22% of 120, or 84% of 32?
24. Which is larger: 112% of 5, or 0.45% of 800?
25. Which is smaller: 15% of 80, or 95% of 15?
26. Which is smaller: 2% of 1500, or 72% of 40?
27. Find 31.294% of 82,460.
28. Find 123.94% of 275,976.

Objective B *Application Problems*

29. A salesclerk receives a salary of \$2240 per month, and 18% of this amount is deducted for income tax. Find the amount deducted for income tax.

30. In a city election for mayor, the successful candidate received 53% of the 385,000 votes cast. How many votes did the successful candidate receive?

31. An antique shop owner expects to receive $16\frac{2}{3}$% of the shop's sales as profit. What is the profit in a month when the total sales are \$24,000?

32. A student survey in a community college found that $33\frac{1}{3}$% of its 7500 beginning students did not complete their first year of college. How many of the beginning students did not complete the first year of college?

33. A recreational vehicle dealer offers a 7% rebate on some RV models. What rebate would a buyer receive on an RV that cost \$24,000?

34. A farmer is given an income tax credit of 10% of the cost of some farm machinery. What tax credit would the farmer receive on farm equipment that cost \$85,000?

35. You purchase a car for \$9500 and must pay a sales tax of 6% of the cost.
    a. What is the sales tax?
    b. What is the total cost of the car including sales tax?

36. During the packaging process for oranges, spoiled oranges are discarded by an inspector. In 1 day an inspector found that 4.8% of the 20,000 pounds of oranges inspected were spoiled.
    a. How many pounds of oranges were spoiled?
    b. How many pounds of oranges were not spoiled?

37. An amusement park has 550 employees and must hire an additional 22% for the vacation season. What is the total number of employees needed for the vacation season?

38. An entertainment system was purchased for \$1350. A down payment of 15% was required. Find the amount remaining to be paid after the down payment had been paid.

39. An office building has an appraised value of \$2,862,400. The real estate taxes are 1.75% of the appraised value of the building. Find the real estate taxes.

40. A company pays out 35% of its profits in dividends. Find the dividends paid out to investors when the company had a profit of \$22,540,000.

SECTION 5.3 Percent Equations: Part II

Objective A To find the percent when the base and amount are given

A recent promotional game at a grocery store listed the probability of winning a prize as "1 chance in 2." A percent can be used to describe the chance of winning. This requires answering the question, "What percent of 2 is 1?"

The chance of winning can be found by solving the basic percent equation for *percent.*

What percent of 2 is 1?

percent		base		amount
n	$\times$	2	$=$	1

$$n \times 2 = 1$$
$$n = 1 \div 2$$
$$n = 0.5$$
$$n = 50\%$$

The solution must be written as a percent to answer the question.

There is a 50% chance of winning a prize.

Example 1 What percent of 40 is 30?

Solution
$$n \times 40 = 30$$
$$n = 30 \div 40$$
$$n = 0.75$$
$$n = 75\%$$

Example 2 What percent of 32 is 16?

Your solution

Example 3 What percent of 12 is 27?

Solution
$$n \times 12 = 27$$
$$n = 27 \div 12$$
$$n = 2.25$$
$$n = 225\%$$

Example 4 What percent of 15 is 48?

Your solution

Example 5 25 is what percent of 75?

Solution
$$25 = n \times 75$$
$$25 \div 75 = n$$
$$0.33\tfrac{1}{3} = n$$
$$33\tfrac{1}{3}\% = n$$

Example 6 30 is what percent of 45?

Your solution

Solutions on p. A22

Objective B To solve application problems

To solve percent problems, remember that it is necessary to identify the percent, base, and amount. Usually the base will follow the phrase "percent of."

Example 7

A family has an income of \$1500 each month and makes a car payment of \$180 a month. What percent of the income is the car payment?

Strategy

To find what percent of the income the car payment is, write and solve the basic percent equation using n to represent the unknown percent. The base is \$1500 and the amount is \$180.

Solution

$$n \times \$1500 = \$180$$
$$n = \$180 \div \$1500$$
$$n = 0.12 = 12\%$$

The car payment is 12% of the monthly income.

Example 8

A housepainter had an income of \$20,000 and paid \$3000 in income tax. What percent of the income is the income tax?

Your strategy

Your solution

Example 9

A nurse missed 25 questions out of 200 on a state license exam. What percent of the questions did the nurse answer correctly?

Strategy

To find what percent of the questions the nurse answered correctly:

- Find the number of questions the nurse answered correctly (200 − 25).
- Write and solve a basic percent equation using n to represent the unknown percent. The base is 200 and the amount is the number of questions answered correctly.

Solution 200 − 25 = 175 questions answered correctly

$$n \times 200 = 175$$
$$n = 175 \div 200$$
$$n = 0.875 = 87.5\%$$

The nurse answered 87.5% of the questions correctly.

Example 10

A survey of 1000 people showed that 667 people favored a candidate for governor of the state. What percent of the people surveyed did not favor the candidate?

Your strategy

Your solution

Solutions on p. A22

5.3 EXERCISES

Objective A

Solve.

1. What percent of 75 is 24?
2. What percent of 80 is 20?
3. 15 is what percent of 90?
4. 24 is what percent of 60?
5. What percent of 12 is 24?
6. What percent of 6 is 9?
7. What percent of 16 is 6?
8. What percent of 24 is 18?
9. 18 is what percent of 100?
10. 54 is what percent of 100?
11. 5 is what percent of 2000?
12. 8 is what percent of 2500?
13. What percent of 6 is 1.2?
14. What percent of 2.4 is 0.6?
15. 16.4 is what percent of 4.1?
16. 5.3 is what percent of 50?
17. 1 is what percent of 40?
18. 0.3 is what percent of 20?
19. What percent of 48 is 18?
20. What percent of 11 is 88?
21. What percent of 2800 is 7?
22. What percent of 400 is 12?
23. 4.2 is what percent of 175?
24. 41.79 is what percent of 99.5?
25. What percent of 86.5 is 8.304?
26. What percent of 1282.5 is 2.565?
27. 632,000 is what percent of 1,469,300?
28. 16.235 is what percent of 19.6375?

Objective B *Application Problems*

29. A company spends $4500 of its $90,000 budget for advertising. What percent of the budget is spent for advertising?

30. Last month a thrift store had an income of $2812.50. The thrift store pays $900 a month for rent. What percent of last month's income was spent for rent?

31. A car is sold for $8900. The buyer of the car pays a down payment of $1780. What percent of the selling price is the down payment?

32. A utility company's stock is trading at $\$24\frac{1}{2}$ and pays a dividend of $1.96. What percent of the stock price is the dividend?

33. The value of your home in 1984 was $78,500. The value of your home in 1990 was $141,300. What percent of the 1984 value is the 1990 value?

34. A television survey of 4000 families found that 2500 liked a new TV show. What percent of the families surveyed liked the new TV show?

35. A house in a fast-growing city in the southwest was bought for $124,000. A year later the house was sold for $155,000.
    a. Find the increase in price.
    b. What percent of the original price is the increase?

36. A 3.5-pound beef roast contains 0.7 pounds of fat.
    a. What percent of the beef roast is fat?
    b. At the same rate, how many pounds of fat would be in a side of beef weighing 550 pounds?

37. To receive a license to sell insurance, an insurance account executive must answer 70% of the 250 questions on a test correctly. An account executive answered 177 questions correctly. Was this amount enough to pass the test?

38. A test of the breaking strength of concrete slabs for freeway construction found that 3 of the 200 tested did not meet safety requirements. What percent of the slabs tested did meet safety requirements?

39. The surface area of the earth, including land and water, is approximately 197,470,000 square miles. The land area of the earth is estimated as 57,266,300 square miles. What percent of the total area of the earth is the land area?

40. A 50.5-acre piece of raw land is appraised at a value of $5,900,000. The land is bought for $4,950,000. What percent of the appraised value is the purchase price? (Round to the nearest tenth of a percent.)

SECTION 5.4 Percent Equations: Part III

Objective A To find the base when the percent and amount are given

Each year an investor receives a payment that equals 12% of the value of an investment. This year that payment amounted to \$480. To find the value of the investment this year, the investor must find "12% of what value is \$480?"

The value of the investment can be found by solving the basic percent equation for the base.

12% of what is \$480

percent 12%	×	base n	=	amount \$480

$$0.12 \times n = \$480$$
$$n = \$480 \div 0.12$$
$$n = \$4000$$

This year the investment is worth \$4000.

Example 1 18% of what is 900?

Solution $0.18 \times n = 900$

$n = 900 \div 0.18$

$n = 5000$

Example 2 86% of what is 215?

Your solution

Example 3 30 is 1.5% of what?

Solution $30 = 0.015 \times n$

$30 \div 0.015 = n$

$2000 = n$

Example 4 15 is 2.5% of what?

Your solution

Example 5 $33\frac{1}{3}\%$ of what is 7?

Solution $\frac{1}{3} \times n = 7$

$n = 7 \div \frac{1}{3}$

$n = 21$

Note the percent is written as a fraction.

Example 6 $16\frac{2}{3}\%$ of what is 5?

Your solution

Solutions on p. A22

Objective B To solve application problems

To solve percent problems, it is necessary to identify the percent, base, and amount. Usually the base will follow the phrase "percent of."

Example 7

A business office bought a used copy machine for $450, which was 75% of the original cost. What was the original cost of the copier?

Strategy

To find the original cost of the copier, write and solve the basic percent equation using n to represent the original cost (base). The percent is 75% and the amount is $450.

Solution

$75\% \times n = \$450$
$0.75 \times n = \$450$
$n = \$450 \div 0.75$
$n = \$600$

The original cost of the copier was $600.

Example 8

A used car has a value of $3876, which is 51% of the car's original value. What was the car's original value?

Your strategy

Your solution

Example 9

A carpenter's wage this year is $19.80 per hour, which is 110% of last year's wage. What was the increase in the hourly wage over last year?

Strategy

To find the increase in the hourly wage over last year:

- Find last year's wage. Write and solve the basic percent equation using n to represent last year's wage (base). The percent is 110% and the amount is $19.80.
- Subtract last year's wage from this year's wage ($19.80).

Solution

$110\% \times n = \$19.80$
$1.10 \times n = \$19.80$
$n = \$19.80 \div 1.10$
$n = \$18.00$ last year's wage

$\$19.80 - \$18.00 = \$1.80$

The increase in the hourly wage is $1.80.

Example 10

A sporting goods store has a tennis racket on sale for $44.80, which is 80% of the original price. What is the difference between the original price and the sale price?

Your strategy

Your solution

Solutions on pp. A22–A23

5.4 EXERCISES

Objective A

Solve.

1. 12% of what is 9?
2. 38% of what is 171?
3. 8 is 16% of what?
4. 54 is 90% of what?
5. 10 is 10% of what?
6. 37 is 37% of what?
7. 30% of what is 25.5?
8. 25% of what is 21.5?
9. 2.5% of what is 30?
10. 10.4% of what is 52?
11. 125% of what is 24?
12. 180% of what is 21.6?
13. 18 is 240% of what?
14. 24 is 320% of what?
15. 4.8 is 15% of what?
16. 87.5 is 50% of what?
17. 25.6 is 12.8% of what?
18. 45.014 is 63.4% of what?
19. 0.7% of what is 0.56?
20. 0.25% of what is 1?
21. 30% of what is 2.7?
22. 78% of what is 3.9?
23. 84 is $16\frac{2}{3}\%$ of what?
24. 120 is $33\frac{1}{3}\%$ of what?
25. $66\frac{2}{3}\%$ of what is 72?
26. $83\frac{1}{3}\%$ of what is 13.5?
27. 6.59% of what is 469.35?
28. 182.3% of what is 46,253?

Objective B *Application Problems*

29. A mechanic estimates that the brakes of an RV still have 6000 miles of wear. This amount is 12% of the estimated safe-life use of the brakes. What is the estimated life of the brakes?

30. A used mobile home was purchased for $18,000. This amount was 64% of the new mobile home cost. What was the cost of a new mobile home?

31. A city has a population of 42,000. This amount is 75% of what the population was 5 years ago. What was the city's population 5 years ago?

32. A ski resort had a snowfall of 198 inches during a year. This amount is 110% of the previous year's snowfall. What was the previous year's snowfall?

33. A store advertised a scientific calculator for $55.80. This amount was 120% of the cost at a competing store. What is the price at the competitor's store?

34. A salesperson receives a commission of $820 for selling a car. This amount is 5% of the selling price of the car. What is the selling price of the car?

35. During a quality control test, a manufacturer of computer boards found that 24 boards were defective. This amount was 0.8% of the computer boards tested.
a. How many computer boards were tested?
b. How many computer boards tested were not defective?

36. Of the calls a directory assistance operator received, 441 were requests for telephone numbers listed in the current directory. This amount was 98% of the calls for assistance that the operator received.
a. How many calls did the operator receive?
b. How many telephone numbers requested were not listed in the current directory?

37. An orange grower produced 22,680 boxes of oranges this year. This amount was 108% of last year's harvest. What was the increase in the number of boxes of oranges over last year?

38. A steelworker's wage this year is $16.80 an hour. This amount is 112% of last year's wage. What was the increase in the hourly wage over last year?

39. A corporation earned $6,512,400 during the year. This amount was 120% of last year's earnings. Find the increase in the corporation's earnings for the year.

40. Defects were found in 5400 diodes. This amount was 0.12% of the total number of diodes produced in 1 month. Find the total number of diodes produced in that month.

SECTION 5.5 Percent Problems: Proportion Method

Objective A To solve percent problems using proportions

The percent problems in Sections 5.2 through 5.4 were solved using the basic percent equation. Those problems can also be solved using proportions.

The proportion method is based on writing two ratios. One ratio is the percent ratio, written as $\frac{\text{percent}}{100}$. The second ratio is the amount-to-base ratio, written as $\frac{\text{amount}}{\text{base}}$. These two ratios form the proportion

$$\frac{\textbf{percent}}{\textbf{100}} = \frac{\textbf{amount}}{\textbf{base}}$$

To use the proportion method, first identify the percent, the amount, and the base (the base usually follows the phrase "percent of").

What is 23% of 45?

$$\frac{23}{100} = \frac{n}{45}$$

$$\begin{aligned} 23 \times 45 &= 100 \times n \\ 1035 &= 100 \times n \\ 1035 \div 100 &= n \\ 10.35 &= n \end{aligned}$$

What percent of 25 is 4?

$$\frac{n}{100} = \frac{4}{25}$$

$$\begin{aligned} n \times 25 &= 100 \times 4 \\ n \times 25 &= 400 \\ n &= 400 \div 25 \\ n &= 16 \end{aligned}$$

12 is 60% of what number?

$$\frac{60}{100} = \frac{12}{n}$$

$$\begin{aligned} 60 \times n &= 100 \times 12 \\ 60 \times n &= 1200 \\ n &= 1200 \div 60 \\ n &= 20 \end{aligned}$$

Example 1 15% of what is 7? Round to the nearest hundredth.

Solution

$$\frac{15}{100} = \frac{7}{n}$$

$$\begin{aligned} 15 \times n &= 100 \times 7 \\ 15 \times n &= 700 \\ n &= 700 \div 15 \\ n &\approx 46.67 \end{aligned}$$

Example 2 26% of what is 22? Round to the nearest hundredth.

Your solution

Example 3 30% of 63 is what?

Solution

$$\frac{30}{100} = \frac{n}{63}$$

$$\begin{aligned} 30 \times 63 &= n \times 100 \\ 1890 &= n \times 100 \\ 1890 \div 100 &= n \\ 18.90 &= n \end{aligned}$$

Example 4 16% of 132 is what?

Your solution

Solutions on p. A23

Objective B To solve application problems

Example 5

An antique dealer found that 86% of the 250 items that were sold during 1 month sold for under $1000. How many items sold for under $1000?

Strategy

To find the number of items that sold for under $1000, write and solve a proportion using n to represent the number of items sold (amount) for less than $1000. The percent is 86% and the base is 250.

Solution

$$\frac{86}{100} = \frac{n}{250}$$
$$86 \times 250 = 100 \times n$$
$$21{,}500 = 100 \times n$$
$$21{,}500 \div 100 = n$$
$$215 = n$$

215 items sold for under $1000.

Example 6

Last year it snowed 64% of the 150 days of the ski season at a resort. How many days did it snow?

Your strategy

Your solution

Example 7

In a test of the strength of nylon rope, 5 pieces of the 25 pieces tested did not meet the test standards. What percent of the nylon ropes tested did meet the standards?

Strategy

To find the percent of ropes tested that met the standards:

- Find the number of ropes that met the test standards (25 − 5).
- Write and solve a proportion using n to represent the percent of ropes that met the test standards. The base is 25. The amount is the number of ropes that met the standards.

Solution $25 - 5 = 20$ ropes met test standards

$$\frac{n}{100} = \frac{20}{25}$$
$$n \times 25 = 100 \times 20$$
$$n \times 25 = 2000$$
$$n = 2000 \div 25$$
$$n = 80$$

80% of the ropes tested did meet the test standards.

Example 8

Five ballpoint pens in a box of 200 were found to be defective. What percent of the pens were not defective?

Your strategy

Your solution

Solutions on p. A23

5.5 EXERCISES

Objective A

Solve.

1. 26% of 250 is what?
2. What is 18% of 150?
3. 37 is what percent of 148?
4. What percent of 150 is 33?
5. 68% of what is 51?
6. 126 is 84% of what?
7. What percent of 344 is 43?
8. 750 is what percent of 50?
9. 82 is 20.5% of what?
10. 2.4% of what is 21?
11. What is 6.5% of 300?
12. 96% of 75 is what?
13. 7.4 is what percent of 50?
14. What percent of 1500 is 693?
15. 50.5% of 124 is what?
16. What is 87.4% of 255?
17. 120% of what is 6?
18. 14 is 175% of what?
19. What is 250% of 18?
20. 325% of 4.4 is what?
21. 33 is 220% of what?
22. 160% of what is 40?
23. 87 is what percent of 29?
24. What percent of 38 is 95?
25. 0.25% of what is 8?
26. 14 is 0.5% of what?
27. What percent of 250 is 6?
28. 15 is what percent of 5000?
29. Find 1.237% of 996,254.
30. 18,572 is what percent of 1,325,500?

Objective B *Application Problems*

31. A charity organization spent $2940 for administrative expenses. This amount is 12% of the money it collected. What is the total amount the charity organization collected?

32. A manufacturer of an anti-inflammatory drug claims that the drug will be effective for 6 hours. An independent testing service determined that the drug was effective only 80% of the length of time claimed by the manufacturer. Find the length of time the drug will be effective as determined by the testing service.

33. A calculator can be purchased for $28.50. This amount is 40% of the cost of the calculator 8 years ago. What was the cost of the calculator 8 years ago?

34. A produce market increased the price of lettuce to $.69 per pound, which is 115% of the original price. What was the original price?

35. A fire department received 24 false alarms out of a total of 200 alarms received. What percent of the alarms received were false alarms?

36. A typist made errors on 5 words on a typing test. The total length of the test was 250 words. What percent of the words were typed correctly?

37. It cost a mechanic $174 to repair a car. Of this cost, 55% was for labor and 45% was for parts.
    a. What was the labor cost?
    b. What was the total cost for parts?

38. In the last 500 hours of a computer operation, the computer malfunctioned 0.8% of the time.
    a. How many hours did the computer malfunction?
    b. How many hours did the computer operate properly?

39. In a test of a new antihistamine with 800 volunteers with allergy problems, 60 did not show any improvement. What percent of the volunteers did show improvement?

40. A monthly state income tax is 11% of the amount over $1700. What state income tax does a person pay on a salary of $2600?

41. A baseball player with a salary of $1,650,000 is offered a new contract that contains a 11.25% pay cut. What is the amount of money offered in the new contract?

42. You bought an automobile 1 year ago. Since then it has depreciated $932.09, which is 11.23% of the price you paid for it. How much did you pay for the automobile?

Calculators and Computers

The Percent Key on a Calculator

Finding the discount and sales price of an item that is on sale is one of the many uses of the percent key on a calculator.

A percent may be entered on a calculator in either of the following two ways:

1. The first way to enter a percent on a calculator is to enter it as a decimal. Remember, to write a percent as a decimal, remove the percent sign and multiply by 0.01. For example, to multiply a number by 12%, use the decimal 0.12.
2. The second way to enter a percent on a calculator is to use the [%] key.

The following two examples demonstrate the procedure for using this key.

Example 1 A car stereo that normally sells for $98.95 is on sale for 15% off the regular price. Find the discount and the sales price.

Enter	*Display*	*Comments*
98.95	98.95	Enter the regular price.
[×]	98.95	Multiply the regular price by the percent.
15	15	Enter the percent discount.
[%]	14.8425	The discount is $14.84 when rounded to the nearest cent.
[−]	84.1075	This step will work only if your calculator is a scientific calculator and programmed to use the Order of Operations Agreement.
		The sales price is $84.11.

Example 2 A new tire for a car costs $45.75. If the sales tax rate is 6%, find the sales tax and the total cost.

Enter	*Display*	*Comments*
45.75	45.75	Enter the cost.
[×]	45.75	Multiply the cost by the sales tax rate.
6	6.	Enter the sales tax rate.
[%]	2.745	The sales tax is $2.75.
[+]	48.495	Add the sales tax to the cost to get the total cost.
		The total cost is $48.50.

Chapter Summary

Key Words *Percent* means "parts of 100."

Essential Rules

To Write a Percent as a Fraction

To write a percent as a fraction, remove the percent sign and multiply by $\frac{1}{100}$.

To Write a Percent as a Decimal

To write a percent as a decimal, remove the percent sign and multiply by 0.01.

To Write a Decimal as a Percent

To write a decimal as a percent, multiply by 100%.

To Write a Fraction as a Percent

To write a fraction as a percent, multiply by 100%.

Chapter Review

SECTION 5.1

1. Write 12% as a fraction.

2. Write $16\frac{2}{3}\%$ as a fraction.

3. Write 42% as a decimal.

4. Write 7.6% as a decimal.

5. Write 0.38 as a percent.

6. Write $1\frac{1}{2}$ as a percent.

SECTION 5.2

7. What is 30% of 200?

8. What is 7.5% of 72?

9. Find 22% of 88.

10. Find 125% of 62.

11. A company uses 7.5% of its $60,000 expense budget for TV advertising. How much of the company's expense budget was spent for TV advertising?

12. A customer purchases a video camera for $980 and must pay a sales tax of 6.25% of the cost. What is the total cost of the video camera?

SECTION 5.3

13. What percent of 20 is 30?

14. 16 is what percent of 80?

15. What percent of 30 is 2.2? Round to the nearest tenth of a percent.

16. What percent of 15 is 92? Round to the nearest tenth of a percent.

17. A company's stock is trading at \$65 and paying a dividend of \$2.99. What percent of the stock price is the dividend?

18. A house in southern California had a value of \$125,000. In 2 years the house was valued at \$205,000. Find the percent increase in value during the 2 years.

SECTION 5.4

19. 20% of what is 15?

20. 78% of what is 8.5? Round to the nearest tenth.

21. $66\frac{2}{3}\%$ of what is 105?

22. $16\frac{2}{3}\%$ of what is 84?

23. A city's population this year is 157,500, which is 225% of what the city's population was 10 years ago. What was the city's population 10 years ago?

24. A baseball player has a batting average of 0.315. This average is an increase of 125% over the previous year's batting average. Find the previous year's batting average.

SECTION 5.5

25. What is 62% of 320?

26. What percent of 25 is 40?

27. A computer system can be purchased for \$1800. This price is 60% of what the computer cost 4 years ago. What was the cost of the computer 4 years ago?

28. A student misses 9 out of 60 questions on a history exam. What percent of the questions did the student answer correctly?

Chapter Test

1. Write 97.3% as a decimal.

2. Write $16\frac{2}{3}\%$ as a fraction.

3. Write 0.3 as a percent.

4. Write 1.63 as a percent.

5. Write $\frac{3}{2}$ as a percent.

6. Write $\frac{2}{3}$ as a percent.

7. What is 77% of 65? Round to the nearest hundredth.

8. 47.2% of 130 is what? Round to the nearest hundredth.

9. Which is larger:
7% of 120, or 76% of 13?

10. Which is smaller:
13% of 200, or 212% of 12?

11. A fast-food company uses 6% of its $75,000 budget for advertising. What amount of the budget is spent for advertising?

12. During the packaging process for vegetables, spoiled vegetables are discarded by an inspector. In 1 day an inspector found that 6.4% of the 1250 pounds of vegetables were spoiled. How many pounds of vegetables were not spoiled?

13. What percent of 120 is 30?

14. 26 is what percent of 12? Round to the nearest hundredth.

15. A department store has 125 permanent employees and must hire an additional 20 temporary employees for the holiday season. What percent of the permanent employees is hired as temporary employees for the holiday season?

16. A student missed 7 out of 80 questions on a math exam. What percent of the questions did the student answer correctly? (Round to the nearest tenth of a percent)

17. 12 is 15% of what?

18. 42.5 is 150% of what? Round to the nearest tenth.

19. A manufacturer of transistors found 384 defective transistors during a quality control study. This amount was 1.2% of the transistors tested. Find the number of transistors tested.

20. A new house was bought for $95,000; 5 years later the house sold for $152,000. The increase is what percent of the original price?

21. 123 is 86% of what number? Round to the nearest tenth.

22. What percent of 12 is 120?

23. A secretary receives a wage of $9.52 per hour. This amount is 112% of last year's salary. What is the dollar increase in the hourly wage over last year?

24. A city has a population of 71,500; 10 years ago the population was 32,500. The population now is what percent of what the population was 10 years ago?

25. The annual license fee on a car is 1.4% of the value of the car. If the license fee during a year was $91.00, what is the value of the car?

Cumulative Review

1. Simplify $18 \div (7 - 4)^2 + 2$.

2. Find the LCM of 16, 24, and 30.

3. Add: $2\frac{1}{3} + 3\frac{1}{2} + 4\frac{5}{8}$

4. Subtract: $27\frac{5}{12} - 14\frac{9}{16}$

5. Multiply: $7\frac{1}{3} \times 1\frac{5}{7}$

6. Divide: $\frac{14}{27} \div 1\frac{7}{9}$

7. Simplify: $\left(\frac{3}{4}\right)^3 \cdot \left(\frac{8}{9}\right)^2$

8. Simplify: $\left(\frac{2}{3}\right)^2 - \left(\frac{3}{8} - \frac{1}{3}\right) \div \frac{1}{2}$

9. Round 3.07973 to the nearest hundredth.

10. Subtract: $\begin{array}{r} 3.0902 \\ -1.9706 \\ \hline \end{array}$

11. Divide: $0.032\overline{)1.097}$
Round to the nearest ten-thousandth.

12. Convert $3\frac{5}{8}$ to a decimal.

13. Convert 1.75 to a fraction.

14. Place the correct symbol, < or >, between the two numbers.
$\frac{3}{8} \quad 0.87$

15. Solve the proportion $\frac{3}{8} = \frac{20}{n}$. Round to the nearest tenth.

16. Write "$76.80 earned in 8 hours" as a unit rate.

17. Write $18\frac{1}{3}\%$ as a fraction.

18. Write $\frac{5}{6}$ as a percent.

19. 16.3% of 120 is what? Round to the nearest hundredth.

20. 24 is what percent of 18?

21. 12.4 is 125% of what?

22. What percent of 35 is 120? Round to the nearest tenth.

23. A mechanic has an income of $740 per week. One-fifth of the income is deducted for income tax payments. Find the mechanic's take-home pay.

24. A student buys a car for $4321, with a down payment of $1000. The balance is paid in 36 equal monthly payments. Find the monthly payment.

25. Gasoline tax is $0.19 a gallon. Find the number of gallons of gasoline used during a month in which $79.80 was paid in taxes.

26. The real estate tax on a $72,000 home is $1440. At the same rate, find the real estate tax on a home valued at $150,000.

27. A customer purchases a stereo set for $490 and pays $29.40 in sales tax. What percent of the purchase price is the sales tax?

28. A survey of 300 people showed that 165 people favored a candidate for mayor. What percent of the people surveyed did not favor the candidate?

29. The value of a home in the northern part of the United States was $62,000 in 1985. The same home in 1990 had a value of $155,000. What percent of the 1985 value is the 1990 value?

30. The Environmental Protection Agency found that 990 out of 5500 children tested had levels of lead in their blood exceeding federal guidelines. What percent of the children tested had levels of lead in the blood that exceeded federal standards?

6 Applications for Business and Consumers: A Calculator Approach

OBJECTIVES

- To find unit cost
- To find the most economical purchase
- To find total cost
- To find percent increase
- To apply percent increase to business—markup
- To find percent decrease
- To apply percent decrease to business—discount
- To calculate simple interest
- To calculate compound interest
- To calculate the initial expenses of buying a home
- To calculate ongoing expenses of owning a home
- To calculate the initial expenses of buying a car
- To calculate ongoing expenses of owning a car
- To calculate commissions, total hourly wages, and salaries
- To calculate checkbook balances
- To balance a checkbook

A Penny a Day

A fictitious job offer in a newspaper claimed that it would pay a salary of 1¢ on the first of the month and 2¢ on the second of the month. Next month, the salary would be 4¢ on the first and 8¢ on the second of the month. The next month and each succeeding month for 12 months the procedure of doubling the previous payment would be continued the same way. Do you think you would want the job under these conditions?

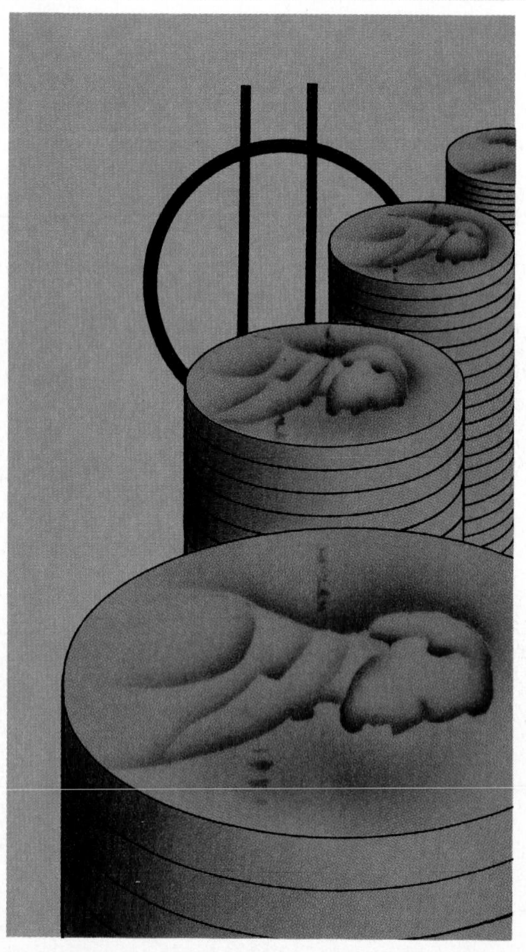

This problem is an example of compounding interest, which is similar to the type of interest that is earned on bank or savings deposits. The big difference is that banks and savings and loans do not compound the interest as quickly as in the given example.

The table below shows the amount of salary you would earn for each month and then gives the total annual salary.

Month	Salary
January	\$0.01 + \$0.02 = \$0.03
February	\$0.04 + \$0.08 = \$0.12
March	\$0.16 + \$0.32 = \$0.48
April	\$0.64 + \$1.28 = \$1.92
May	\$2.56 + \$5.12 = \$7.68
June	\$10.24 + \$20.48 = \$30.72
July	\$40.96 + \$81.92 = \$122.88
August	\$163.84 + \$327.68 = \$491.52
September	\$655.36 + \$1310.72 = \$1966.08
October	\$2621.44 + \$5242.88 = \$7864.32
November	\$10,485.76 + \$20,971.52 = \$31,457.28
December	\$41,943.04 + \$83,886.08 = \$125,829.12
	Total Annual Salary = \$167,772.15

Not a bad annual salary!

SECTION 6.1 Applications to Purchasing

Objective A To find unit cost

Frequently stores advertise items for purchase as

3 shirts for \$36 or

5 pounds of potatoes for 90¢.

The **unit cost** is the cost of 1 shirt or 1 pound of potatoes. To find the unit cost, divide the total cost by the number of units.

3 shirts for \$36

36 ÷ 3 = 12

\$12 is the cost of 1 shirt.

Unit cost: \$12 per shirt

5 pounds of potatoes for 90¢

90 ÷ 5 = 18

18¢ is the cost of 1 pound of potatoes.

Unit cost: 18¢ per pound

Example 1 Find the unit cost. Round to the nearest tenth of a cent.
a) 10 gallons of gasoline for \$15.49
b) 3 yards of material for \$10.

Strategy To find the unit cost, divide the total cost by the number of units.

Solution a) 15.49 ÷ 10 = 1.549
\$1.549 per gallon
b) 10 ÷ 3 ≈ 3.3333
\$3.333 per yard

Example 2 Find the unit cost. Round to the nearest tenth of a cent.
a) 5 quarts of oil for \$6.25
b) 4 ears of corn for 85¢

Your strategy

Your solution

Solution on p. A24

Objective B To find the most economical purchase

Comparison shoppers often find the most economical buy by comparing unit costs.

One store sells 6 cans of cola for \$2.04 and another store sells 24 cans of the same brand for \$7.92. To find the better buy, compare the unit costs.

2.04 ÷ 6 = 0.34
Unit cost: \$0.34 per can

7.92 ÷ 24 = 0.33
Unit cost: \$0.33 per can

Because \$0.33 < \$0.34, the better buy is 24 cans for \$7.92.

Example 3 Find the more economical purchase: 5 pounds of nails for \$3.25, or 4 pounds of nails for \$2.58.

Strategy To find the more economical purchase, compare the unit costs.

Solution 3.25 ÷ 5 = 0.65
2.58 ÷ 4 = 0.645
\$.645 < \$.65

The more economical purchase is 4 pounds for \$2.58.

Example 4 Find the more economical purchase: 6 cans of fruit for \$2.52, or 4 cans of fruit for \$1.66.

Your strategy

Your solution

Solution on p. A24

Objective C To find total cost

An installer of floor tile found the unit cost of identical floor tiles at three stores.

Store 1	*Store 2*	*Store 3*
\$1.22 per tile	\$1.18 per tile	\$1.28 per tile

By comparing the unit costs, the installer determined that Store 2 would provide the most economical purchase.

The installer also uses the unit cost to find the total cost of purchasing 300 floor tiles at Store 2. The **total cost** is found by multiplying the unit cost by the number of units purchased.

Unit cost	×	number of units	=	total cost
1.18	×	300	=	354

The total cost is \$354.

Example 5 Clear redwood lumber costs \$2.43 per foot. How much would 25 feet of clear redwood cost?

Strategy To find the total cost, multiply the unit cost (\$2.43) by the number of units (25).

Solution

Unit cost	×	number of units	=	total cost
2.43	×	25	=	60.75

The total cost is \$60.75.

Example 6 Pine saplings cost \$4.96 each. How much would 7 pine saplings cost?

Your strategy

Your solution

Solution on p. A24

6.1 EXERCISES

Objective A *Application Problems*

Find the unit cost. Round to the nearest tenth of a cent.

1. Syrup, 16 ounces for \$0.89

2. Wood, 8 feet for \$11.36

3. Lunch meat, 12 ounces for \$1.79

4. Light bulbs, 6 bulbs for \$2

5. Wood screws, 15 screws for \$0.50

6. Bay leaves, $\frac{1}{2}$ ounce for \$0.96

7. Antenna wire, 10 feet for \$1.95

8. Sponges, 4 for \$5.00

9. Catsup, 24 ounces for \$0.89

10. Strawberry jam, 18 ounces for \$1.23

11. Metal clamps, 2 for \$5.99

12. Coffee, 3 pounds for \$7.65

Objective B *Application Problems*

Find the more economical purchase.

13. Cheese, 12 ounces for \$1.79, or 16 ounces for \$2.40

14. Powder, 4 ounces for \$0.53, or 9 ounces for \$0.95

15. Shampoo, 8 ounces for \$1.89, or 12 ounces for \$2.64

16. Mayonnaise, 32 ounces for \$2.37, or 16 ounces for \$1.30

17. Fruit drink, 46 ounces for \$0.69, or 50 ounces for \$0.78

18. Detergent, 42 ounces for \$2.57, or 20 ounces for \$1.29

Find the more economical purchase.

19. Oil, 2 quarts for $4.79, or 5 quarts for $12.15

20. Potato chips, 8 ounces for $0.99, or 10 ounces for $1.23

21. Vitamins, 200 tablets for $9.98, or 150 tablets for $7.65

22. Wax, 16 ounces for $2.68, or 27 ounces for $4.39

23. Cereal, 18 ounces for $1.51, or 16 ounces for $1.32

24. Tires, 2 for $128.98, or 4 for $258.98

Objective C *Application Problems*

Solve.

25. Steak costs $2.79 per pound. Find the total cost of 3 pounds.

26. Red brick costs $0.56 per brick. Find the total cost of 50 bricks.

27. Flowering plants cost $0.49 each. Find the total cost of 6 plants.

28. Chicken costs $0.89 per pound. Find the total cost of 3.6 pounds. Round to the nearest cent.

29. Tea costs $0.54 per ounce. Find the total cost of 6.5 ounces. Round to the nearest cent.

30. Cheese costs $2.89 per pound. Find the total cost of 0.65 pounds. Round to the nearest cent.

31. Tomatoes cost $0.22 per pound. Find the total cost of 1.8 pounds. Round to the nearest cent.

32. Ham costs $1.49 per pound. Find the total cost of 3.4 pounds. Round to the nearest cent.

33. Candy costs $2.68 per pound. Find the total cost of $\frac{3}{4}$ pound.

34. Photocopying costs $0.05 per page. Find the total cost for photocopying 250 pages.

35. Paint costs $8.95 per gallon. How much change do you receive from $50 when purchasing 5 gallons of paint?

36. Crushed rock costs $15.75 per ton. How much change do you receive from $50 when purchasing 3 tons of crushed rock?

SECTION 6.2 Percent Increase and Percent Decrease

Objective A To find percent increase

Percent Increase is used to show how much a quantity has increased over its original value. The statements "car prices will show a 3.5% increase over last year's prices" and "employees were given an 11% pay increase" are illustrations of the use of percent increase.

A city's population increased from 50,000 to 51,500 in 1 year.

To find the percent increase,

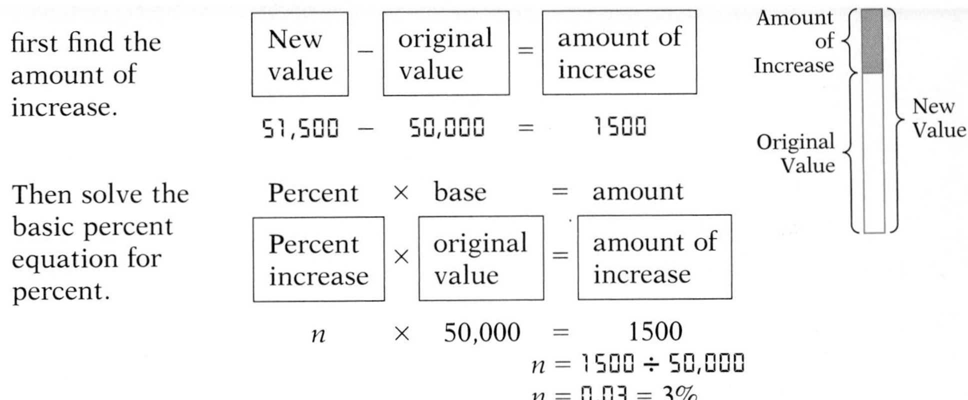

$n = 1500 \div 50{,}000$

$n = 0.03 = 3\%$

The population increased by 3% over the previous year.

Example 1 The average price of gasoline rose from \$0.92 to \$1.15 in 4 months. What was the percent increase in the price of gasoline?

Strategy To find the percent increase:

- Find the amount of increase.
- Solve the basic percent equation for *percent*.

Solution

New value − original value = amount of increase

1.15 − 0.92 = 0.23

Percent × base = amount

$n \times \$0.92 = \$0.23$

$n = 0.23 \div 0.92$

$n = 0.25 = 25\%$

The percent increase was 25%.

Example 2 An industrial plant increased its number of employees from 2000 to 2250. What was the percent increase in the number of employees?

Your strategy

Your solution

Solution on p. A24

Example 3 A seamstress was making a wage of \$4.80 an hour before a 10% increase in pay. What is the new hourly wage?

Strategy To find the new hourly wage:

- Solve the basic percent equation for *amount*.
- Add the amount of increase to the original wage.

Solution Percent × base = amount

0.10 × 4.80 = n

0.48 = n

4.80 + 0.48 = 5.28

The new hourly wage is \$5.28.

Example 4 A baker was making a wage of \$6.50 an hour before a 14% increase in pay. What is the new hourly wage?

Your strategy

Your solution

Solution on p. A24

Objective B To apply percent increase to business—markup

Many expenses are involved in operating a business. Employee salaries, rent of office space, and utilities are examples of operating expenses. To pay these expenses and earn a profit, a business must sell a product at a higher price than it paid for the product.

Cost is the price that a business pays for a product, and **selling price** is the price for which a business sells a product to a customer. The difference between selling price and cost is called **markup.**

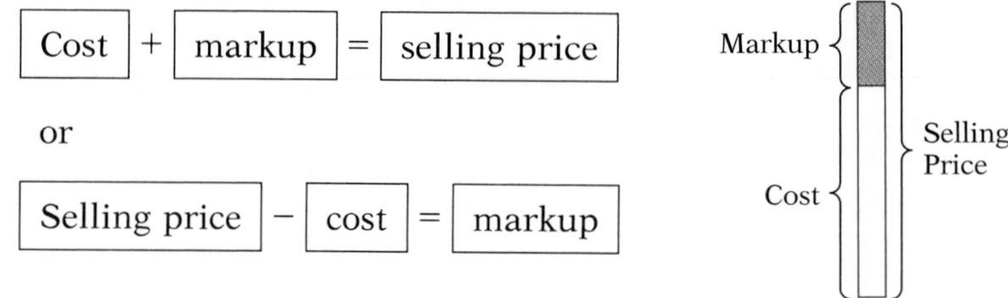

Markup is frequently expressed as a percent of a product's cost. This percent is called the **markup rate.**

Markup rate × cost = markup

A bicycle store owner purchases a bicycle for \$105 and sells it for \$147. To find the markup rate,

first find the markup. Selling price − cost = markup

147 − 105 = 42

Then solve the basic percent equation for *percent*.

Percent × base = amount

Markup rate	×	cost	=	markup

$n \times \$105 = \$42$

$n = 42 \div 105 = 0.4$

The markup rate is 40%

Example 5 The manager of a sporting goods store determines that a markup rate of 36% is necessary to make a profit. What is the markup on a pair of skis that costs the store $125?

Strategy To find the markup, solve the basic percent equation for *amount*.

Solution Percent × base = amount

Markup rate	×	cost	=	markup

$0.36 \times 125 = n$

$45 = n$

The markup is $45.

Example 6 A bookstore manager determines that a markup rate of 20% is necessary to make a profit. What is the markup on a book that costs the bookstore $8?

Your strategy

Your solution

Example 7 A plant nursery buys a citrus tree for $4.50 and uses a markup-rate of 46%. What is the selling price?

Strategy To find the selling price:

- Find the markup by solving the basic percent equation for *amount*.
- Add the markup to the cost.

Solution Percent × base = amount

Markup rate	×	cost	=	markup

$0.46 \times 4.50 = n$

$2.07 = n$

Cost	+	markup	=	selling price
4.50	+	2.07	=	6.57

The selling price is $6.57.

Example 8 A clothing store buys a suit for $72 and uses a markup rate of 55%. What is the selling price?

Your strategy

Your solution

Solutions on p. A24

Objective C To find percent decrease

Percent decrease is frequently used to show how much a quantity has decreased from its original value. The statements "the unemployment rate decreased by 0.4% over last month" and "there has been a 12% decrease in the number of industrial accidents" are illustrations of the use of percent decrease.

A family's electric bill decreased from $60 per month to $52.80 per month.

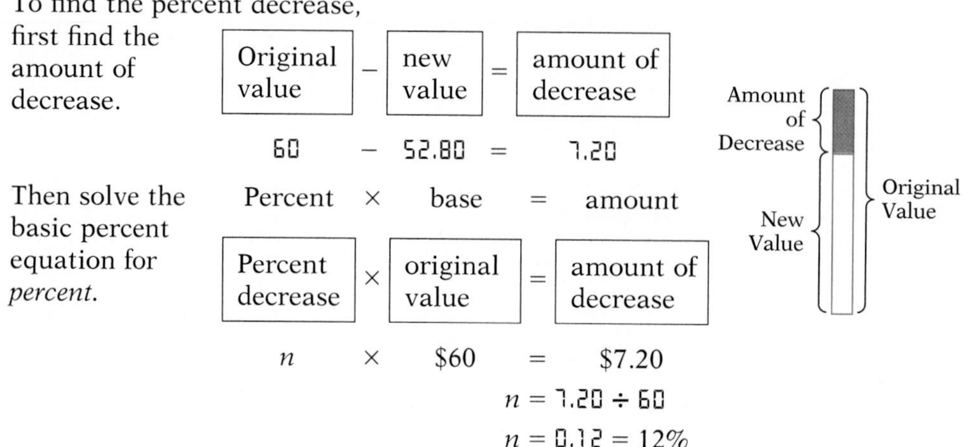

$n = 7.20 \div 60$

$n = 0.12 = 12\%$

The electric bill decreased by 12% per month.

Example 9 Because of unusually high temperatures, the number of people visiting a desert resort dropped from 650 to 525. Find the percent decrease in attendance. Round to the nearest tenth of a percent.

Strategy To find the percent decrease:

- Find the amount of decrease.
- Solve the basic percent equation for *percent*.

Solution

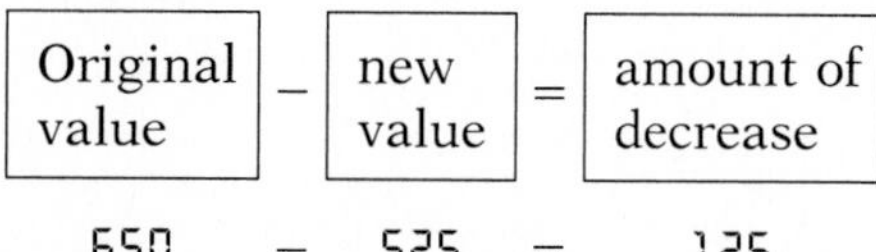

650 − 525 = 125

Percent × base = amount

$n \times 650 = 125$

$n = 125 \div 650$

$n \approx 0.192 = 19.2\%$

The percent decrease was 19.2%

Example 10 Car sales at an automobile dealership dropped from 150 in June to 120 in July. Find the percent decrease in car sales.

Your strategy

Your solution

Solution on pp. A24–A25

Example 11 The total sales for December for a stationery store were $16,000. For January, total sales showed an 8% decrease from December's sales. What were the total sales for January?

Strategy To find the total sales for January:

- Find the amount of decrease by solving the basic percent equation for *amount*.
- Subtract the amount of decrease from the December sales.

Solution Percent × base = amount

$0.08 \times 16{,}000 = n$

$1280 = n$

$16{,}000 - 1280 = 14{,}720$

The total sales for January were $14,720.

Example 12 Fog decreased the normal 5-mile visibility at an airport by 40%. What was the visibility in the fog?

Your strategy

Your solution

Solution on pp. A24–A25

Objective D To apply percent decrease to business—discount

To promote sales, a store may reduce the regular price of some of its products temporarily. The reduced price is called the **sale price.** The difference between the regular price and the sale price is called the **discount.**

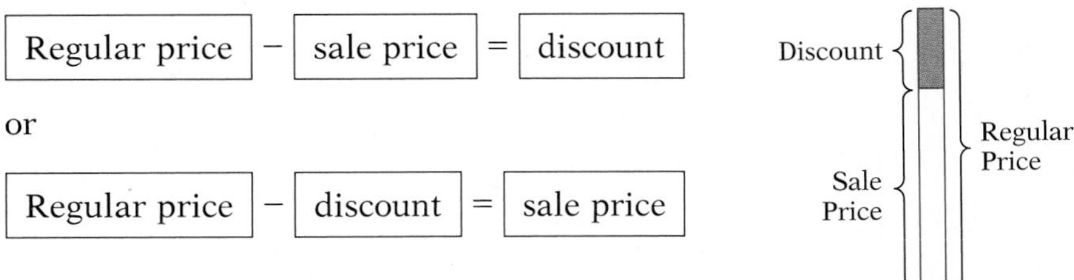

Discount is frequently stated as a percent of a product's regular price. This percent is called the **discount rate.**

Discount rate × regular price = discount

Example 13 An appliance store has a washing machine that regularly sells for \$350 on sale for \$297.50. What is the discount rate?

Strategy To find the discount rate:

- Find the discount.
- Solve the basic percent equation for *percent*.

Solution

Regular price	−	sale price	=	discount
350	−	297.50	=	52.50

Percent × base = amount

Discount rate	×	regular price	=	discount
n	×	\$350	=	\$52.50

$n = 52.50 \div 350$

$n = 0.15 = 15\%$

The discount rate is 15%.

Example 14 A lawn mower that regularly sells for \$125 is on sale for \$25 off the regular price. What is the discount rate?

Your strategy

Your solution

Example 15 A plumbing store is selling sinks that regularly sell for \$25 at 15% off the regular price. What is the sale price?

Strategy To find the sale price:

- Find the discount by solving the basic percent equation for *amount*.
- Subtract to find the sale price.

Solution

Percent × base = amount

Discount rate	×	regular price	=	discount
0.15	×	25	=	n

$3.75 = n$

Regular price	−	discount	=	sale price
25	−	3.75	=	21.25

The sale price is \$21.25.

Example 16 A florist is selling potted plants that regularly sell for \$10.25 at 20% off the regular price. What is the sale price?

Your strategy

Your solution

Solutions on p. A25

6.2 EXERCISES

Objective A *Application Problems*

1. A utility stock that sold for $30 per share increased in price by $1.65 in 1 day. What percent increase does this amount represent?

2. The value of a mutual fund investment of $2500 increased $500. What percent increase does this amount represent?

3. A professional baseball player had a batting average of 0.275. The next year the player had a batting average of 0.297. What is the percent increase in the player's batting average?

4. A running back for a high school football team had a rushing average of 3.80 yards per carry. The next year the back averaged 4.75 yards per carry. What percent increase does this amount represent?

5. You have increased your 60 words per minute typing speed 20 words per minute. What percent increase does this amount represent?

6. Your reading speed increased 70 words per minute from 280 words per minute. What is the percent increase in reading speed?

7. A teacher's union negotiated a new contract calling for a 9% increase from this year's contract.
   a. Find the amount of increase for a teacher who had a salary of $32,500.
   b. Find the new salary for the teacher who had a salary of $32,500.

8. A new labor contract called for an 8.5% increase in pay for all employees.
   a. What is the amount of the increase for an employee who makes $324 per week?
   b. What is the weekly wage of this employee after the wage increase?

9. A company whose stock is selling for $38 per share plans to raise its $1.60 dividend 15%.
   a. What is the increase in the dividend?
   b. Find the total dividend per share of the stock.

10. An amusement park increased its summer staff from 500 to 625 employees.
    a. How many employees were added to the staff?
    b. What percent increase does this amount represent?

11. A family increased its 1440-square foot home $16\frac{2}{3}$% by adding a family room. How much larger, in square feet, is the home now?

12. A restaurant increased its weekly hours of operation from 60 hours to 70 hours. What percent increase does this amount represent?

Objective B *Application Problems*

13. An air conditioner costs a retail store $285. Find the markup on an air conditioner that has a markup rate of 25%.

14. The owner of an electronics store uses a markup rate of 42%. What is the markup on a cassette player that costs the store $85?

15. A sporting goods store uses a markup rate of 30%. What is the markup on a tennis racket that costs the store $42?

16. The manager of a toy store determines that a markup rate of 38% is necessary for a profit to be made. What is the markup on a novelty item that costs the manager $5?

17. The markup on a computer system that costs $3250 is $975. What markup rate does this amount represent?

18. The markup on a calculator that costs an office supply store $24 is $12. What markup rate does this amount represent?

19. A camera shop uses a markup rate of 48% on its Model ZA cameras, which cost the shop $162.
    a. What is the markup?
    b. What is the selling price?

20. A golf pro shop uses a markup rate of 45% on a set of golf clubs that costs the shop $210.
    a. What is the markup?
    b. What is the selling price?

21. An outdoor furniture company uses a markup rate of 35% on a piece of lawn furniture that costs the store $62.
    a. What is the markup?
    b. What is the selling price?

22. The owner of a roadside fruit stand uses a 55% markup rate and pays $0.60 for a box of strawberries.
    a. What is the markup?
    b. What is the selling price for a box of strawberries?

23. The manager of a building supply store uses a markup rate of 42% for a table saw that costs the store $160. What is the selling price of the table saw?

24. The manager of a magic store uses a markup rate of 48%. What is the selling price of an item that costs $50?

Objective C *Application Problems*

25. A new bridge over a river reduced the normal 45-minute time between two cities by 18 minutes. What percent decrease does this amount represent?

26. In an election, 25,400 voters went to the polls to elect a mayor. Four years later, 1778 fewer voters participated in the election for mayor. What was the percent decrease in the number of voters who went to the polls?

27. By installing energy-saving equipment, a youth club reduced its normal $800 per month utility bill $320. What percent decrease does this amount represent?

28. During the last 40 years, the consumption of eggs in the United States has dropped from 400 eggs per person per year to 260 eggs per person per year. What percent decrease does this amount represent?

29. It is estimated that the value of a new car is reduced 30% after 1 year of ownership. Using this estimate, how much value does a $11,200 new car lose after 1 year?

30. A department store employs 1200 people during the holiday season. At the end of the holiday season, the department store reduces the number of employees 45%. What is the decrease in the number of employees?

31. Because of a decrease in demand for super-8 video cameras, a dealer reduced the orders for these models from 20 per month to 8 per month.
a. What is the amount of the decrease?
b. What percent decrease does this amount represent?

32. A new computer reduced the time for printing the payroll from 52 minutes to 39 minutes.
a. What is the amount of decrease?
b. What percent decrease does this amount represent?

33. An account executive's average expense for gasoline was $76. After joining a car pool, the executive was able to reduce this expense 20%.
a. What was the amount of decrease?
b. What is the average monthly gasoline bill now?

34. Last year a company paid a dividend of $3.50 per share. This year, because of unexpected losses, the company reduced the dividend 12%.
a. What was the amount of decrease?
b. What is the dividend this year?

35. Because of an improved traffic pattern at a sports stadium, the average amount of time a fan waits to park decreased from 3.5 minutes to 2.8 minutes. What percent decrease does this amount represent?

36. The price of gold dropped from approximately $600 per ounce to $375 per ounce in 7 months. What percent decrease does this amount represent?

Objective D *Application Problems*

37. A college bookstore is giving a discount of $8 on calculators that normally sell for $24. What is the discount rate?

38. A clothing store is selling a $72 sport jacket for $24 off the regular price. What is the discount rate?

39. A disk player that regularly sells for $340 is selling for 20% off the regular price. What is the discount?

40. An appliance store is selling its $450 washing machines for 15% off the regular price. What is the discount?

41. To promote business, a store manager offers an electric grill that regularly sells for $140 at $42 off the regular price. What is the discount rate?

42. A computer system that regularly sells for $1400 is on sale for $350 off the regular price. What is the discount rate?

43. A service station has its regularly priced $45 tune-up on sale for 16% off the regular price.
a. What is the discount?
b. What is the sale price?

44. Turkey that regularly sells for $0.85 per pound is on sale for 20% off the regular price.
a. What is the discount?
b. What is the sale price?

45. A department store has regularly priced $160 sleeping bags on sale for $120.
a. What is the discount?
b. What is the discount rate?

46. A nationally advertised brand of paint that regularly costs $16 per gallon is on sale for $12 per gallon.
a. What is the discount?
b. What is the discount rate?

47. A clothing store is offering 15% off its entire stock of children's clothing. What is the sale price of a jacket that regularly sells for $42?

48. During a sidwalk sale, all merchandise was reduced 20% off the regular price. What was the sale price of a sofa that normally sold for $650?

SECTION 6.3 Interest

Objective A

To calculate simple interest

When money is deposited in a bank account, the bank pays the depositor for the privilege of using that money. The amount paid to the depositor is called **interest.** When money is borrowed from a bank, the borrower pays for the privilege of using that money. The amount paid to the bank is also called interest.

The original amount deposited or borrowed is called the **principal.** The amount of interest is a percent of the principal. The percent used to determine the amount of interest is the **interest rate.** Interest rates are given for specific periods of time, usually months or years.

Interest computed on the original principal is called **simple interest.** To calculate simple interest, multiply the principal by the interest rate per period by the number of time periods.

Calculate the simple interest on \$1500 deposited for 2 years at an annual interest rate of 7.5%. As shown below, the interest earned is \$225.

Principal × annual interest rate × time (in years) = interest

1500 × 0.075 × 2 = 225

Example 1
A person borrows \$500 from a savings and loan association for 6 months at an annual interest rate of 7%. What is the simple interest due on the loan?

Strategy
To find the simple interest, multiply:

Principal × annual interest rate × time (in years)

Solution 500 × 0.07 × 0.5 = 17.5

The interest due is \$17.50.

Example 2
A company borrows \$15,000 from a bank for 18 months at an annual interest rate of 12%. What is the simple interest due on the loan?

Your strategy

Your solution

Example 3
A credit card company charges a customer 1.5% per month on the unpaid balance of charges on the credit card. What is the interest due in a month when the customer has an unpaid balance of \$54?

Strategy
To find the interest due, multiply the principal by the monthly interest rate by the time (in months).

Solution 54 × 0.015 × 1 = 0.81

The interest charge is \$0.81.

Example 4
A bank offers short-term loans at an interest rate of 2% per month. What is the interest due on a short-term loan of \$400 for 2 months?

Your strategy

Your solution

Solutions on p. A25

Objective B

To calculate compound interest

Usually the interest paid on money deposited or borrowed is compound interest. **Compound interest** is computed not only on the original principal but also on interest already earned. Compound interest is usually compounded annually (once a year), semiannually (twice a year), quarterly (four times a year), or daily.

$100 is invested for 3 years at an annual interest rate of 9%, compounded annually. The interest earned over the 3 years is calculated by first finding the interest earned each year.

1st year	Interest earned:	$0.09 \times \$100.00 = \$9.00$
	New principal:	$\$100.00 + \$9.00 = \underline{\$109.00}$
2nd year	Interest earned:	$0.09 \times \$109.00 = \$9.81$
	New principal:	$\$109.00 + \$9.81 = \underline{\$118.81}$
3rd year	Interest earned:	$0.09 \times \$118.81 \approx \$10.69$
	New principal:	$\$118.81 + \$10.69 = \$129.50$

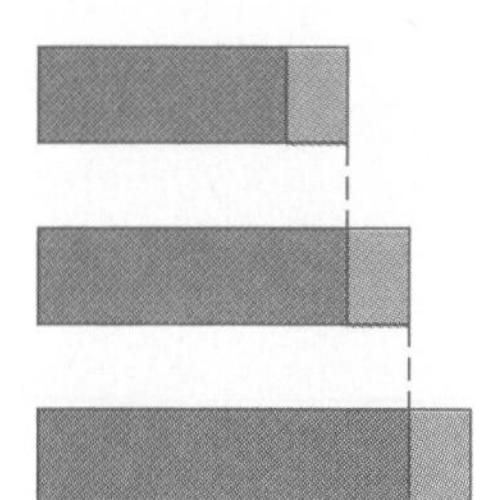

To find the interest earned, subtract the original principal from the new principal. As shown below, the interest earned is $29.50.

New principal	−	original principal	=	interest earned
129.50	−	100	=	29.5

Note that the compound interest earned is $29.50, and the simple interest earned on the investment would have been only $\$100 \times 0.09 \times 3 = \$27$.

Calculating compound interest can be very tedious, so there are tables that can be used to simplify these calculations. A portion of a compound interest table is given on pages A4 and A5.

Example 5
An investment of $650 pays 8% annual interest compounded semiannually. What is the interest earned in 5 years?

Strategy To find the interest earned:

- Find the new principal by multiplying the original principal by the factor found in the compound interest table.
- Subtract the original principal from the new principal.

Solution
650 × 1.48024 ≈ 962.16
(1.48024: from the table; 962.16: new principal)
962.16 − 650 = 312.16

The interest earned is $312.16.

Example 6
An investment of $1000 pays 10% annual interest compounded quarterly. What is the interest earned in 20 years?

Your strategy

Your solution

Solution on p. A25

6.3 EXERCISES

Objective A *Application Problems*

1. To finance the purchase of 15 new cars, a car rental agency borrows \$100,000 for 9 months at an annual interest rate of 12%. What is the simple interest due on the loan?

2. A home builder borrows \$50,000 for 8 months at an annual interest rate of 11.5%. What is the simple interest due on the loan? (Round to the nearest cent.)

3. A farmer obtained a \$25,000 loan at a 13.5% annual interest rate for 4 years. Find the simple interest due on the loan.

4. A nursery operator borrowed \$150,000 at a 9.5% annual interest rate for 4 years. What is the simple interest due on the loan?

5. A credit union charges its customers an interest rate of 2% per month for transferring money into an account that is overdrawn. Find the interest owed to the bank for 1 month when \$800 was transferred into an overdrawn account.

6. A credit card company charges a customer 1.6% per month on the customer's unpaid balance. Find the interest owed to the credit card company when the customer's unpaid balance for the month is \$1250.

7. A metalworking shop purchased a robot-controlled lathe for \$225,000 and financed the full amount at 8% simple annual interest for 4 years.
   a. Find the interest due on the loan.
   b. Find the monthly payment. (Monthly payment = $\frac{\text{loan amount + interest}}{\text{number of months}}$)

8. An entertainment center is purchased, and an \$1800 loan is obtained for 2 years at a simple interest rate of 12.4%.
   a. Find the interest due on the loan.
   b. Find the monthly payment. (Monthly payment = $\frac{\text{loan amount + interest}}{\text{number of months}}$)

9. To attract new customers, a car dealership is offering car loans at a simple interest rate of 4.5%.
   a. Find the interest charged to a customer who finances a car loan of \$12,000 for 2 years.
   b. Find the monthly payment.

10. A real estate syndicator purchases a small plane for \$57,000 and finances the full amount for 5 years at a simple annual interest rate of 10.5%.
    a. Find the interest due on the loan.
    b. Find the monthly payment.

11. A homeowner decided to build onto an existing structure instead of buying a new home. The homeowner borrowed \$42,000 for $3\frac{1}{2}$ years at a simple interest rate of 9.5%. Find the monthly payment.

12. An office has 15 new computers installed for a total cost of \$24,000. The entire amount was financed for $2\frac{1}{2}$ years at a simple interest rate of 11.2%. Find the monthly payment.

Objective B *Application Problems*

Solve. Use the table on pages A4 and A5. Round to the nearest cent.

13. A credit union pays 9% annual interest, compounded daily, on time savings deposits. Find the value of $750 deposited in this account after 1 year.

14. A high school student has an investment of $1000 that pays 9% annual interest, compounded quarterly. What is the value of the investment after 5 years?

15. An investment club invests $50,000 in a certificate of deposit that pays 10% annual interest, compounded quarterly. Find the value of this investment after 10 years.

16. A teacher invests $2500 in a tax-sheltered annuity that pays 8% annual interest, compounded daily. Find the value of this investment after 20 years.

17. A computer specialist invests $3000 in a corporate retirement account that pays 9% annual interest compounded semiannually. Find the value of this investment after 15 years.

18. To replace equipment, a farmer invests $20,000 in an account that pays 7% annual interest, compounded semiannually. What is the value of the investment after 5 years?

19. A corporation invests $75,000 in a trust account that pays 8% interest, compounded quarterly.
a. What will the value of the investment be in 5 years?
b. How much interest will be earned in the 5 years?

20. To save for retirement, a couple deposited $3000 in an account that pays 11% annual interest, compounded daily.
a. What will the value of the investment be in 10 years?
b. How much interest will be earned in the 10 years?

21. A fast-food restaurant has determined that it will need $50,000 in 5 years for expansion. In an attempt to meet that goal, the restaurant invests $35,000 today in an account that pays 8% annual interest, compounded quarterly.
a. Find the value of the $35,000 investment in 5 yrs.
b. Will the restaurant have enough money in the account for the planned expansion?

22. An investment of $5000 in an account that pays 10% annual interest, compounded quarterly, has a value of $8193.10 after 5 years.
a. What would the value of the investment have been if the 10% interest had been compounded daily over the 5 year period?
b. What is the difference in the values of the investment after the 5 years?

23. To save for a child's college education, a family deposits $2500 into an account that pays 10% annual interest, compounded daily. Find the amount of interest earned in this account over a 20-year period.

24. An account executive suggests buying a $5000 municipal bond that pays 8% annual interest, compounded quarterly. Find the amount of interest that would be earned in 5 years.

SECTION 6.4 Real Estate Expenses

Objective A To calculate the initial expenses of buying a home

One of the largest investments a person will make is the purchase of a home. The major initial expense in the purchase is the down payment. The amount of the down payment is normally a percent of the purchase price. This percent varies among banks, but it usually ranges from 5% to 25%.

The **mortgage** is the amount that is borrowed to buy real estate. The mortgage amount is the difference between the purchase price and the down payment.

A home is purchased for $85,000, and a down payment of $12,750 is made. Find the mortgage.

Purchase price	−	down payment	=	mortgage
85,000	−	12,750	=	72,250

The mortgage is $72,250.

Another large initial expense in buying a home is the loan origination fee, which is a fee the bank charges for processing the mortgage papers. The loan origination fee is usually a percent of the mortgage and is expressed in **points,** which is the term banks use to mean percent. For example, "5 points" means "5 percent."

Points	×	mortgage	=	loan origination fee

Example 1

A house is purchased for $87,000, and a down payment, which is 20% of the purchase price, is made. Find the mortgage.

Strategy

To find the mortgage:

- Find the down payment by solving the basic percent equation for *amount.*
- Subtract the down payment from the purchase price.

Solution

Percent × base = amount

Percent	×	purchase price	=	down payment
0.20	×	87,000	=	n

17,400 = n

Purchase price	−	down payment	=	mortgage
87,000	−	17,400	=	69,600

The mortgage is $69,600.

Example 2

An office building is purchased for $216,000, and a down payment, which is 25% of the purchase price, is made. Find the mortgage.

Your strategy

Your solution

Solution on p. A26

Example 3 A home is purchased with a mortgage of $65,000. The buyer pays a loan origination fee of $3\frac{1}{2}$ points. How much is the loan origination fee?

Strategy To find the loan origination fee, solve the basic percent equation for *amount*.

Solution

Percent × base = amount

Points	×	mortgage	=	fee

0.035 × 65,000 = n

2275 = n

The loan origination fee is $2275.

Example 4 The mortgage on a real estate investment is $80,000. The buyer paid a loan origination fee of $4\frac{1}{2}$ points. How much is the loan origination fee?

Your strategy

Your solution

Solution on p. A26

Objective B To calculate ongoing expenses of owning a home

Besides the initial expenses of buying a home, there are continuing monthly expenses involved in owning a home. The monthly mortgage payment, utilities, insurance, and taxes are some of these ongoing expenses. Of these expenses, the largest one is normally the monthly mortgage payment.

For a fixed rate mortgage, the monthly mortgage payment remains the same throughout the life of the loan. The calculation of the monthly mortgage payment is based on the amount of the loan, the interest rate on the loan, and the number of years required to pay back the loan. Calculating the monthly mortgage payment is fairly difficult, so tables such as the one on page A6 are used to simplify these calculations.

Find the monthly mortgage payment on a 30-year $60,000 mortgage at an interest rate of 9%. Use the monthly payment table on page A6.

60,000 × 0.0080462 = 482.77

↓ from the table

The monthly mortgage payment is $482.77.

The monthly mortgage payment includes the payment of both principal and interest on the mortgage. The interest charged during any one month is charged on the unpaid balance of the loan. Therefore, during the early years of the mortgage, when the unpaid balance is high, most of the monthly mortgage payment is interest charged on the loan. During the last few years of a mortgage, when the unpaid balance is low, most of the monthly mortgage payment goes toward paying off the loan.

Find the interest paid on a mortgage during a month when the monthly mortgage payment is $186.26 and $58.08 of that amount goes toward paying off the principal.

Monthly mortgage payment	−	principal	=	interest
186.26	−	58.08	=	128.18

The interest paid on the mortgage is $128.18.

Property tax is another ongoing expense of owning a house. Property tax is normally an annual expense that is paid on a monthly basis. The monthly property tax, which is determined by dividing the annual property tax by 12, is usually added to the monthly mortgage payment.

A home owner must pay $534 in property tax annually. Find the property tax that must be added each month to the home owner's monthly mortgage payment.

534 ÷ 12 = 44.5

$44.50 must be added to the monthly mortgage payment each month for property tax.

Example 5

A farmer purchases some land for $120,000 and makes a down payment of $25,000. The savings and loan association charges an annual interest rate of 11% on the farmer's 25-year mortgage. Find the monthly mortgage payment.

Strategy

To find the monthly mortgage payment:

- Subtract the down payment from the purchase price to find the mortgage.
- Multiply the mortgage by the factor found in the monthly payment table on page A6.

Solution

Purchase price	−	down payment	=	mortgage
120,000	−	25,000	=	95,000
95,000	×	0.0098011 (from the table)	=	931.10

The monthly mortgage payment is $931.10.

Example 6

A new condominium project is selling townhouses for $75,000. A down payment of $15,000 is required, and a 20-year mortgage at an annual interest rate of 12% is available. Find the monthly mortgage payment.

Your strategy

Your solution

Solution on p. A26

Example 7 A home has a mortgage of $55,000 for 25 years at an annual interest rate of 10%. During a month when $375.88 of the monthly mortgage payment is principal, how much of the payment is interest?

Strategy To find the interest:

- Multiply the mortgage by the factor found in the monthly payment table on page A6 to find the monthly mortgage payment.
- Subtract the principal from the monthly mortgage payment.

Solution

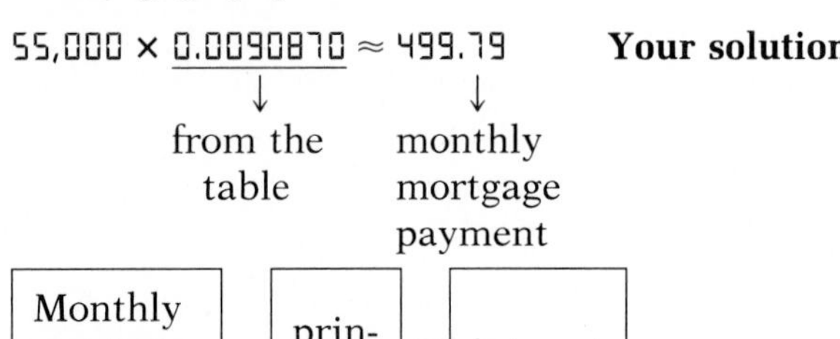

Monthly mortgage payment	−	principal	=	interest

499.79 − 375.88 = 123.91

$123.91 is interest on the mortgage.

Example 8 An office building has a mortgage of $125,000 for 25 years at an annual interest rate of 9%. During a month when $492.65 of the monthly mortgage payment is principal, how much of the payment is interest?

Your strategy

Your solution

Example 9 The monthly mortgage payment for a home is $598.75. The annual property tax is $900. Find the total monthly payment for the mortgage and property tax.

Strategy To find the monthly payment:

- Divide the annual property tax by 12 to find the monthly property tax.
- Add the monthly property tax to the monthly mortgage payment.

Solution 900 ÷ 12 = 75 monthly property tax

598.75 + 75 = 673.75

The total monthly payment is $673.75.

Example 10 The monthly mortgage payment for a home is $415.20. The annual property tax is $744. Find the total monthly payment for the mortgage and property tax.

Your strategy

Your solution

Solutions on p. A26

6.4 EXERCISES

Objective A *Application Problems*

1. A condominium at a ski resort is purchased for $97,000, and a down payment of $14,550 is made. Find the mortgage.

2. An insurance business is purchased for $173,000, and a down payment of $34,600 is made. Find the mortgage.

3. A building lot is purchased for $25,000. The lender requires a down payment of 30%. Find the down payment.

4. A new home buyer purchases a house for $88,500. The lender requires a down payment of 20%. Find the down payment.

5. An investor makes a down payment of 25% of the $250,000 purchase price of an apartment building. How much is the down payment?

6. A clothing store is purchased for $125,000, and a down payment, which is 25% of the purchase price, is made. How much is the down payment?

7. A loan of $150,000 is obtained to purchase a home. The loan origination fee is $2\frac{1}{2}$ points. Find the loan origination fee.

8. A savings and loan requires a borrower to pay $3\frac{1}{2}$ points for a loan. Find the loan origination fee for a loan of $90,000.

9. A real estate developer is selling developed homes for $95,000. A down payment of 5% is required.
   a. Find the down payment.
   b. Find the mortgage.

10. A cattle rancher purchases some land for $240,000. The bank requires a down payment of 15%.
    a. Find the down payment.
    b. Find the mortgage.

11. An electrician purchases a home for $96,000. Find the mortgage if the down payment is 10% of the purchase price.

12. A mortgage lender requires a down payment of 5% of the $56,000 purchase price of a condominium. How much is the mortgage?

Objective B *Application Problems*

Solve. Use the monthly payment table on page A6. Round to the nearest cent.

13. An investment club obtained a loan of $150,000 to buy a car wash business. The monthly mortgage payment was based on 25 years at 10%. Find the monthly mortgage payment.

14. A beautician obtains a 20-year mortgage of $90,000 to expand the business. The credit union charges an annual interest rate of 11%. Find the monthly mortgage payment.

15. A couple interested in buying a home determines that they can afford a monthly mortgage payment of $500. Can they afford to buy a home with a 30-year $65,000 mortgage at 9% interest?

16. A lawyer is considering purchasing a new office building with a 20-year $400,000 mortgage at 12% interest. The lawyer can afford a monthly mortgage payment of $6000. Can the lawyer afford the monthly mortgage payments on the new office building?

17. The county tax assessor has determined that the annual property tax on a $58,000 house is $645. Find the monthly property tax.

18. The annual property tax on an $86,000 home is $858. Find the monthly property tax.

19. An import company has a warehouse with a 25-year mortgage of $200,000 at an annual interest rate of 11%.
- **a.** Find the monthly mortgage payment.
- **b.** During a month when $941.72 of the monthly mortgage payment is principal, how much of the payment is interest?

20. A vacation home has a mortgage of $72,000 for 30 years at an annual interest rate of 10%.
- **a.** Find the monthly mortgage payment.
- **b.** During a month when $392.47 of the monthly mortgage payment is principal, how much of the payment is interest?

21. A duplex has a monthly mortgage payment of $452.73. The owner must pay an annual property tax of $948. Find the total monthly payment for the mortgage and property tax.

22. The monthly mortgage payment on a home is $356.48. The home owner must pay an annual property tax of $792. Find the total monthly payment for the mortgage and property tax.

23. A health inspector purchased a home for $86,000 and made a down payment of $15,000. The balance was financed for 30 years at an annual interest rate of 11%. Find the monthly mortgage payment.

24. A customer of a savings and loan purchased a $75,000 home and made a down payment of $20,000. The savings and loan charges its customers an annual interest rate of 9% for 30 years for a home mortgage. Find the monthly mortgage payment.

SECTION 6.5 Car Expenses

Objective A To calculate the initial expenses of buying a car

The initial expenses in the purchase of a car usually include the down payment, the license fees, and the sales tax. The down payment may be very small or as much as 25 or 30 percent of the purchase price of the car, depending on the lending institution. License fees and sales tax are regulated by each state, so these expenses vary from state to state.

Example 1
A car is purchased for \$8500, and the lender requires a down payment of 15% of the purchase price. Find the amount financed.

Strategy
To find the amount financed:

- Find the down payment by solving the basic percent equation for *amount*.
- Subtract the down payment from the purchase price.

Solution

Percent × base = amount

Percent	×	purchase price	=	down payment
0.15	×	8500	=	n

$1275 = n$

$8500 - 1275 = 7225$

The amount financed is \$7225.

Example 2
A down payment of 20% of the \$9200 purchase price of a new car is made. Find the amount financed.

Your strategy

Your solution

Example 3
A sales clerk purchases a car for \$6500 and pays a sales tax that is 5% of the purchase price. How much is the sales tax?

Strategy
To find the sales tax, solve the basic percent equation for *amount*.

Solution

Percent × base = amount

Percent	×	purchase price	=	sales tax
0.05	×	6500	=	n

$325 = n$

The sales tax is \$325.

Example 4
A car is purchased for \$7350. The car license fee is 1.5% of the purchase price. How much is the license fee?

Your strategy

Your solution

Solutions on pp. A26–A27

Objective B

To calculate ongoing expenses of owning a car

Besides the initial expenses of buying a car, there are continuing expenses involved in owning a car. These ongoing expenses include car insurance, gas and oil, general maintenance, and monthly car payment. The monthly car payment is calculated in the same manner as monthly mortgage payments on a home loan. A monthly payment table, such as the one on page A6, is used to simplify calculation of monthly car payments.

Example 5

At a cost of $0.27 per mile, how much does it cost to operate a car during a year the car is driven 15,000 miles?

Strategy

To find the cost, multiply the cost per mile by the number of miles driven.

Solution

15,000 × 0.27 = 4050 The cost is $4050.

Example 6

At a cost of $0.22 per mile, how much does it cost to operate a car during a year the car is driven 23,000 miles?

Your strategy

Your solution

Example 7

During 1 month the total gasoline bill was $84 and the car was driven 1200 miles. What was the cost per mile for gasoline?

Strategy

To find the cost per mile for gasoline, divide the cost for gasoline by the number of miles driven.

Solution 84 ÷ 1200 = 0.07

The cost per mile was $0.07.

Example 8

In a year when the total car insurance bill was $360 and the car was driven 15,000 miles, what was the cost per mile for car insurance?

Your strategy

Your solution

Example 9

A car is purchased for $8500 with a down payment of $1700. The balance is financed for 3 years at an annual interest rate of 12%. Find the monthly car payment.

Strategy

To find the monthly payment:

- Subtract the down payment from the purchase price to find the amount financed.
- Multiply the amount financed by the factor found in the monthly payment table on page A6.

Solution

8500 − 1700 = 6800

↓

amount financed

6800 × 0.0332143 ≈ 225.86

The monthly payment is $225.86.

Example 10

A truck is purchased for $15,900 with a down payment of $3975. The balance is financed for 4 years at an annual interest rate of 11%. Find the monthly payment.

Your strategy

Your solution

Solutions on p. A27

6.5 EXERCISES

Objective A *Application Problems*

1. A receptionist has saved $780 to make a down payment on a car. The bank requires a down payment of 12% of the purchase price of $7100. Does the receptionist have enough money to make the down payment on the car?

2. A jeep is purchased for $9288. A down payment of 15% is required. How much is the down payment?

3. A drapery installer buys a minivan to carry drapery samples. The purchase price of the van is $16,500, and a 4.5% sales tax is paid. How much is the sales tax?

4. A delivery truck for a lumber yard is purchased for $18,500. A sales tax of 4% of the purchase price is paid. Find the sales tax.

5. A license fee of 2% of the purchase price of a truck is to be paid on a pickup truck costing $12,500. How much is the license fee for the truck?

6. Your state charges a car license fee of 1.5% of the purchase price of a car. How much is the license fee for a car that costs $6998?

7. An electrician buys a $12,000 flat-bed truck. A state license fee of $175 and a sales tax of 3.5% of the purchase price are required.
   a. Find the sales tax.
   b. Find the total cost of the sales tax and the license fee.

8. A physical therapist buys a car for $9375 with a down payment of $1875 and a sales tax of 5% of the purchase price.
   a. Find the sales tax.
   b. Find the total cost of the sales tax and the down payment.

9. A student buys a motorcycle for $2200 and makes a down payment of 25% of the purchase price.
   a. Find the down payment.
   b. Find the amount financed.

10. A carpenter buys a utility van for $14,900 and makes a down payment that is 15% of the purchase price.
    a. Find the down payment.
    b. Find the amount financed.

11. An author buys a sports car for $35,000 and makes a down payment of 20% of the purchase price. Find the amount financed.

12. An executive secretary purchases a new car for $13,500. The secretary makes a down payment of 25% of the cost. Find the amount financed.

Objective B *Application Problems*

Solve. Use the monthly payment table on page A6. Round to the nearest cent.

13. A rancher purchases a truck, and $14,000 is financed through a credit union at 11% interest for 4 years. Find the monthly truck payment.

14. A car loan of $8000 is financed through a bank at an annual interest rate of 10% for 3 years. Find the monthly car payment.

15. An estimate of the cost and care of owning a compact car is $0.32 per mile. Using this estimate, how much does it cost to operate a car during a year the car is driven 16,000 miles?

16. An estimate of the cost of care and maintenance of automobile tires is $0.015 per mile. Using this estimate, how much would it cost for care and maintenance during a year the car is driven 14,000 miles?

17. A family spent $1600 on gas, oil, and car insurance during a period when the car was driven 14,000 miles. Find the cost per mile for gas, oil, and car insurance.

18. Last year, you spent $1050 for gasoline for your car. The car was driven 15,000 miles. What was the cost per mile for gasoline?

19. A car owner pays a monthly car payment of $143.50. During a month when $68.75 of the monthly payment is principal, how much of the payment is interest?

20. The cost for a pizza delivery truck for the year included $1870 in truck payments, $1200 on gasoline, and $675 on insurance. Find the total cost for truck payments, gasoline, and insurance for the year.

21. A city purchases a fire truck for $36,000, and a down payment of $5400 is made. The balance is financed for 5 years at an annual rate of 10%.
a. Find the amount financed.
b. Find the monthly truck payment.

22. A used car is purchased for $4995, and a down payment of $995 is made. The balance is financed for 3 years at an interest rate of 11%.
a. Find the amount financed.
b. Find the monthly car payment.

23. An artist purchases a new car costing $19,500, with a down payment of $3900. The balance is financed for 3 years at an annual interest rate of 10%. Find the monthly car payment.

24. A half-ton truck with a camper is purchased for $12,500, and a down payment of $1500 is made. The balance is financed for 4 years at an annual interest rate of 11%. Find the monthly payment.

SECTION 6.6 Wages

Objective A To calculate commissions, total hourly wages, and salaries

Commissions, hourly wage, and salary are three forms of receiving payment for doing work.

Commissions are usually paid to salespersons and are calculated as a percent of total sales.

A real estate broker receives a commission of 4.5% of the selling price of a house. Find the commission the broker earned for selling a home for $75,000.

To find the commission the broker earned, solve the basic percent equation for *amount*.

Percent × base = amount

Commission rate × total sales = commission

0.045 × 75,000 = 3375

The commission is $3375.

An employee who receives an **hourly wage** is paid a certain amount for each hour worked.

A plumber receives an hourly wage of $13.25. Find the plumber's total wages for working 37 hours.

To find the plumber's total wages, multiply the hourly wage by the number of hours worked.

Hourly wage × number of hours worked = total wages

13.25 × 37 = 490.25

The plumber's total wages for working 37 hours are $490.25.

An employee who is paid a **salary** receives payment based on a weekly, biweekly (every other week), monthly, or annual time schedule. Unlike the employee who receives an hourly wage, the salaried worker does not receive additional pay for working more than the regularly scheduled work day.

A computer operator receives a weekly salary of $295. Find the computer operator's salary for 1 month (4 weeks).

To find the computer operator's salary for 1 month, multiply the salary per pay period by the number of pay periods.

Salary per pay period × number of pay periods = total salary

295 × 4 = 1180

The computer operator's total salary for 1 month is $1180.

Example 1
A pharmacist's hourly wage is $18. On Saturday, the pharmacist earns time and a half ($1\frac{1}{2}$ times the regular hourly wage). How much does the pharmacist earn for working 6 hours on Saturday?

Strategy
To find the pharmacist's earnings:

- Find the hourly wage for working on Saturday by multiplying the hourly wage by $1\frac{1}{2}$.
- Multiply the product by the number of hours worked.

Solution
18 × 1.5 = 27   27 × 6 = 162

The pharmacist earns $162.

Example 2
A construction worker's hourly wage is $8.50. The worker earns double time (2 times the regular hourly wage) for working overtime. How much does the worker earn for working 8 hours of overtime?

Your strategy

Your solution

Example 3
An efficiency expert received a contract for $3000. The consultant spent 75 hours on the project. Find the consultant's hourly wage.

Strategy
To find the hourly wage, divide the total earnings by the number of hours worked.

Solution
3000 ÷ 75 = 40 The hourly wage was $40.

Example 4
A contractor for a bridge project receives an annual salary of $28,224. What is the contractor's salary per month?

Your strategy

Your solution

Example 5
A computer salesperson receives $18,500 per year plus a $5\frac{1}{2}$% commission on sales over $100,000. During the year, the salesperson sold $150,000 worth of computers. Find the salesperson's total earnings for the year.

Strategy
To find the total earnings:

- Find the commission earned by multiplying the commission rate by sales over $100,000.
- Add the commission to the annual pay.

Solution
150,000 − 100,000 = 50,000
50,000 × 0.055 = 2750 commission

18,500 + 2750 = 21,250

The salesperson earned $21,250.

Example 6
An insurance agent receives $12,000 per year plus a $9\frac{1}{2}$% commission on sales over $50,000. During 1 year, the agent's sales totaled $175,000. Find the agent's total earnings for the year.

Your strategy

Your solution

Solutions on p. A27

6.6 EXERCISES

Objective A *Application Problems*

1. A salesclerk in a clothing store earns \$5.50 per hour. How much does the clerk earn in a 40-hour week?

2. A homeowner pays a gardener an hourly wage of \$5.25. How much does the homeowner pay the gardener for working 25 hours?

3. A real estate agent receives a 3% commission for selling a house. Find the commission the real estate agent earned for selling a house for \$98,500.

4. An insurance agent receives a commission of 40% of the first year's premium. Find the agent's commission for selling a life insurance policy with a first year premium of \$1050.

5. A stock broker receives a commission of 1.5% of the price of stock that is bought or sold. Find the commission on 100 shares of stock that were bought for \$5600.

6. The owner of a gallery receives a commission of 20% of paintings that are sold on consignment. Find the commission on a painting sold for \$22,500.

7. A teacher receives an annual salary of \$38,928. How much does the teacher receive per month?

8. An apprentice plumber receives an annual salary of \$19,680. How much does the plumber receive per month?

9. A car salesperson receives a 4% commission for selling a new car. Find the commission the salesperson earned for selling a new car for \$10,400.

10. An electrician's hourly wage is \$15.80. For working overtime, the electrician earns double time. What is the electrician's hourly wage for working overtime?

11. A sales representative for a shoe company receives a commission of 12% of the weekly sales. Find the commission earned during the week when sales were \$4500.

12. A golf pro receives a commission of 25% for selling a golf set. Find the commission the pro earned for selling a golf set that cost \$450.

13. A carpet installer receives $1.75 per square yard to install carpet. How much does the installer receive for installing 160 square yards of carpet?

14. A typist charges $1.75 per page for typing technical material. How much does the typist earn for typing a 225-page book?

15. A nuclear chemist receives $15,000 for consulting fees while working on a nuclear power plant. The chemist worked 120 hours on the project. Find the consultant's hourly wage.

16. A computer consultant received a contract for $3400. The consultant spent 40 hours working on the project. Find the consultant's hourly wage.

17. A physical therapist's hourly wage is $10.78. For working overtime, the therapist receives double time.
- **a.** What is the therapist's hourly wage for working overtime?
- **b.** How much does the therapist earn for working 16 hours of overtime?

18. A lathe operator receives an hourly wage of $8.50. When working on Saturday, the operator receives time and a half.
- **a.** What is the lathe operator's hourly wage on Saturday?
- **b.** How much does the lathe operator earn for working 8 hours on Saturday?

19. A stock clerk at a supermarket earns an hourly wage of $6.20. For working the late night shift, the clerk's wage is increased by 15%.
- **a.** What is the increase in pay for working the late night shift?
- **b.** What is the clerk's hourly wage for working the late night shift?

20. A nurse's hourly wage is $8.50. For working the night shift, the nurse's wage is increased 10%.
- **a.** What is the increase in pay for working the night shift?
- **b.** What is the nurse's hourly wage for working the night shift?

21. A sales executive sells heavy equipment to a highway construction company. The executive receives 14% of all sales over $100,000. During 1 year, the sales executive sold $350,000 worth of heavy equipment.
- **a.** Find the amount of sales over $100,000.
- **b.** Find the commission the sales executive earned.

22. A sales engineer receives a salary of $15,000 per year plus a commission of 20% of the total annual sales. During 1 year, the total sales were $125,000.
- **a.** Find the commission the sales engineer earned.
- **b.** Find the sales engineer's total earnings for the year.

23. A service station attendant's hourly wage is $6.40. For working the night shift, the attendant's wage is increased 15%. What is the attendant's hourly wage for working the night shift?

24. A door-to-door salesperson receives a salary of $150 per week plus a commission of 15% on all sales over $1500. Find the salesperson's earnings during a week that sales totaled $3000.

SECTION 6.7 Bank Statements

Objective A To calculate checkbook balances

A checking account can be opened at most banks or savings and loan associations by depositing an amount of money in the bank. A checkbook contains checks and deposit slips and a checkbook register in which to record checks written and amounts deposited in the checking account. Each time a check is written, the amount of the check is subtracted from the amount in the account. When a deposit is made, the amount deposited is added to the amount in the account.

A portion of a checkbook register is shown below. The account holder had a balance of $587.93 before writing two checks, one for $286.87 and the other for $102.38, and making one deposit for $345.00.

RECORD ALL CHARGES OR CREDITS THAT AFFECT YOUR ACCOUNT

NUMBER	DATE	DESCRIPTION OF TRANSACTION	PAYMENT/DEBIT (-)		√ T	FEE (IF ANY) (-)	DEPOSIT/CREDIT (+)		BALANCE $	
									587	93
108	8/4	Plumber	$286	87		$	$		301	06
109	8/10	Car Payment	102	38					198	68
	8/14	Deposit					345	00	543	68

To find the current checking account balance, subtract the amount of each check from the previous balance. Then add the amount of the deposit.

The current checking account balance is $543.68.

Example 1 A mail carrier had a checking account balance of \$485.93 before writing two checks, one for \$18.98 and another for \$35.72, and making a deposit of \$250. Find the current checking account balance.

Strategy To find the current balance:

- Subtract the amounts of each check from the old balance.
- Add the amount of the deposit.

Solution

```
  485.93
-  18.98   first check
  466.95
-  35.72   second check
  431.23
+250.00    deposit
  681.23
```

The current checking account balance is \$681.23.

Example 2 A cement mason had a checking account balance of \$302.46 before writing a check for \$20.59 and making two deposits, one in the amount of \$176.86 and another in the amount of \$94.73. Find the current checking account balance.

Your strategy

Your solution

Solution on p. A28

Objective B To balance a checkbook

Each month a bank statement is sent to the account holder. The bank statement shows the checks that the bank has paid, the deposits received, and the current bank balance.

A bank statement and checkbook register are shown on the next page.

Balancing the checkbook, or determining if the checking account balance is accurate, requires a number of steps.

1. In the checkbook register, put a check (√) by each check paid by the bank and each deposit recorded by the bank.

RECORD ALL CHARGES OR CREDITS THAT AFFECT YOUR ACCOUNT

NUMBER	DATE	DESCRIPTION OF TRANSACTION	PAYMENT/DEBIT (-)	√ T	FEE (IF ANY) (-)	DEPOSIT/CREDIT (+)	BALANCE $ 840 27
263	5/20	Dentist	$ 25 00	✓	$	$	815 27
264	5/22	Meat market	33 61	✓			781 66
265	5/22	Gas company	67 14				714 52
	5/29	Deposit		✓		192 00	906 52
266	5/29	Pharmacy	18 95	✓			887 57
267	5/30	Telephone	43 85				843 72
268	6/2	Groceries	43 19	✓			800 53
	6/3	Deposit		✓		215 00	1015 53
269	6/7	Insurance	103 00	✓			912 53
	6/10	Deposit				225 00	1137 53
270	6/15	Clothing store	16 63	✓			1120 90
271	6/18	Newspaper	7 00				1113 90

CHECKING ACCOUNT Monthly Statement — Account Number: 924-297-8

Date	Transaction	Amount	Balance
5/20	OPENING BALANCE		840.27
5/21	CHECK	25.00	815.27
5/23	CHECK	33.61	781.66
5/29	DEPOSIT	192.00	973.66
6/1	CHECK	18.95	954.71
6/1	INTEREST	4.47	959.18
6/3	CHECK	43.19	915.99
6/3	DEPOSIT	215.00	1130.99
6/9	CHECK	103.00	1027.99
6/16	CHECK	16.63	1011.36
6/20	SERVICE CHARGE	3.00	1008.36
6/20	CLOSING BALANCE		1008.36

2. Add to the current checkbook balance all checks that have been written but not yet paid by the bank and any interest paid on the account.

3. Subtract any service charges and any deposits not yet recorded by the bank. This is the checkbook balance.

Current checkbook balance:	1113.90
Checks: 265	67.14
267	43.85
271	7.00
Interest:	+ 4.47
	1236.36
Service charge:	− 3.00
	1233.36
Deposit:	− 225.00
Checkbook balance:	1008.36

4. Compare the balance with the bank balance listed on the bank statement. If the two numbers are equal, the bank statement and checkbook balance.

Closing bank balance from bank statement		Checkbook balance
$1008.36	=	$1008.36

The bank statement and checkbook balance.

RECORD ALL CHARGES OR CREDITS THAT AFFECT YOUR ACCOUNT

NUMBER	DATE	DESCRIPTION OF TRANSACTION	PAYMENT/DEBIT (-)		√ T	FEE (IF ANY) (-)	DEPOSIT/CREDIT (+)		BALANCE $ 1620	42
413	3/2	Car Payment	$132	15	√	$	$		1488	27
414	3/2	Utility	67	14	√				1421	13
415	3/5	Restaurant - Dinner for 4	78	14					1342	99
	3/8	Deposit			√		1842	66	3185	65
416	3/10	House Payment	672	14	√				2513	51
417	3/14	Insurance	177	10					2336	41

CHECKING ACCOUNT Monthly Statement — Account Number: 924-297-8

Date	Transaction	Amount	Balance
3/1	OPENING BALANCE		1620.42
3/4	CHECK	132.15	1488.27
3/5	CHECK	67.14	1421.13
3/8	DEPOSIT	1842.66	3263.79
3/10	INTEREST	6.77	3270.56
3/12	CHECK	672.14	2598.42
3/25	SERVICE CHARGE	2.00	2596.42
3/30	CLOSING BALANCE		2596.42

Balance the bank statement shown above.

1. In the checkbook register, put a check (√) by each check paid by the bank and each deposit recorded by the bank.

2. Add to the current checkbook balance all checks that have been written but not yet paid by the bank and any interest paid on the account.

Current checkbook balance:	2336.41
Checks: 415	78.14
417	177.10
Interest:	+ 6.77
	2598.42

3. Subtract any service charges and any deposits not yet recorded by the bank. This is the checkbook balance.

Service charge:	− 2.00
	2596.42
Checkbook Balance:	2596.42

4. Compare the balance with the bank balance listed on the bank statement. If the two numbers are equal, the bank statement and checkbook balance.

Closing bank balance from bank statement		Checkbook balance:
$2596.42	=	$2596.42

The bank statement and checkbook balance.

RECORD ALL CHARGES OR CREDITS THAT AFFECT YOUR ACCOUNT											
NUMBER	DATE	DESCRIPTION OF TRANSACTION	PAYMENT/DEBIT (-)		√ T	FEE (IF ANY) (-)	DEPOSIT/CREDIT (+)		BALANCE	
									$ 412	64
345	1/14	Phone bill	$ 34	75		$	$		377	89
346	1/19	Magazine	8	98					368	91
347	1/23	Theatre tickets	45	00					323	91
	1/31	Deposit					947	00	1270	91
348	2/5	Cash	250	00					1020	91
349	2/12	Rent	440	00					580	91

CHECKING ACCOUNT Monthly Statement — Account Number: 924-297-8

Date	Transaction	Amount	Balance
1/10	OPENING BALANCE		412.64
1/18	CHECK	34.75	377.89
1/23	CHECK	8.98	368.91
1/31	DEPOSIT	947.00	1315.91
2/1	INTEREST	4.52	1320.43
2/10	CHECK	250.00	1070.43
2/10	CLOSING BALANCE		1070.43

Example 3

Balance the bank statement shown above.

Solution

Current checkbook balance:	580.91
Checks: 347	45.00
349	440.00
Interest:	+ 4.52
	1070.43
Service charge:	− 0.00
	1070.43
Deposit:	− 0.00
Checkbook balance:	1070.43

Closing bank balance from bank statement: $1070.43

Checkbook balance: $1070.43

The bank statement and checkbook balance.

RECORD ALL CHARGES OR CREDITS THAT AFFECT YOUR ACCOUNT

NUMBER	DATE	DESCRIPTION OF TRANSACTION	PAYMENT/DEBIT (-)		√ T	FEE (IF ANY) (-)	DEPOSIT/CREDIT (+)		BALANCE	
									$ 603	17
	2/15	Deposit	$			$	$ 523	84	1127	01
234	2/20	Mortgage	473	21					653	80
235	2/27	Cash	200	00					453	80
	3/1	Deposit					523	84	977	64
236	3/12	Insurance	275	50					702	14
237	3/12	Telephone	48	73					653	41

CHECKING ACCOUNT Monthly Statement — Account Number: 314-271-4

Date	Transaction	Amount	Balance
2/14	OPENING BALANCE		603.17
2/15	DEPOSIT	523.84	1127.01
2/21	CHECK	473.21	653.80
2/28	CHECK	200.00	453.80
3/1	INTEREST	2.11	455.91
3/14	CHECK	275.50	180.41
3/14	CLOSING BALANCE		180.41

Example 4

Balance the bank statement shown above.

Your solution

Solution on p. A28

6.7 EXERCISES

Objective A *Application Problems*

1. You had a checking account balance of $342.51 before making a deposit of $143.81. What is your new checking account balance?

2. A student had a checking account balance of $493.26 before writing a check for $48.39. What is the current checking account balance?

3. A real estate firm had a balance of $2431.76 in its rental property checking account. What is the balance in this account after a check for $1209.29 has been written?

4. The business checking account for a tire store showed a balance of $1536.97. What is the balance in this account after a deposit of $439.21 has been made?

5. A nutritionist had a checking account balance of $1204.63 before writing one check for $119.27 and another check for $260.09. Find the current checkbook balance.

6. A farmer had a checking account balance of $3046.93 before writing a check for $1027.33 and making a deposit of $150.00. Find the current checkbook balance.

7. The business checking account for a dry cleaning store had a balance of $3476.85 before a deposit of $1048.53 was made. The store manager then wrote checks, one for $848.37 and another for $676.19. Find the current checkbook balance.

8. A pottery store manager had a checking account balance of $427.38 before a deposit of $127.29 was made. The manager then wrote two checks, one for $43.52 and one for $249.78. Find the current checkbook balance.

9. A carpenter had a checkbook balance of $104.96 before making a deposit of $350 and writing a check for $71.29. Is there enough money in the account to purchase a refrigerator for $375?

10. A taxidriver had a checkbook balance of $149.85 before making a deposit of $245 and writing a check for $387.68. Is there enough money in the account for the bank to pay the check?

11. A sporting goods store has the opportunity to buy downhill skis and cross-country skis at a manufacturer's close-out sale. The downhill skis will cost $3500, and the cross-country skis will cost $2050. There is currently $5625.42 in the sporting goods store checking account. Is there enough money in the account to make both purchases by check?

12. A lathe operator's current checkbook balance is $643.42. The operator wants to purchase a utility trailer for $225 and a used piano for $450. Is there enough money in the account to make the two purchases?

Objective B *Application Problems*

13. Balance the checkbook.

RECORD ALL CHARGES OR CREDITS THAT AFFECT YOUR ACCOUNT

NUMBER	DATE	DESCRIPTION OF TRANSACTION	PAYMENT/DEBIT (-)		√ T	FEE (IF ANY) (-)	DEPOSIT/CREDIT (+)		BALANCE	
									$ 1035	18
218	7/2	Mortgage	$ 284	60		$	$		750	58
219	7/4	Telephone	23	36					727	22
×220	7/7	Cash	200	00					527	22
	7/12	Deposit					792	60	1319	82
221	7/15	Insurance	192	30					1127	52
222	7/18	Investment	100	00					1027	52
×223	7/20	Credit card	214	83					812	69
	7/26	Deposit					792	60	1605	29
224	7/27	Department store	113	37					1491	92

CHECKING ACCOUNT Monthly Statement — Account Number: 122-345-1

Date	Transaction	Amount	Balance
7/1	OPENING BALANCE		1035.18
7/1	INTEREST	5.15	1040.33
7/4	CHECK	284.60	755.73
7/6	CHECK	23.36	732.37
7/12	DEPOSIT	792.60	1524.97
7/20	CHECK	192.30	1332.67
7/24	CHECK	100.00	1232.67
7/26	DEPOSIT	792.60	2025.27
7/28	CHECK	200.00	1825.27
7/30	CLOSING BALANCE		1825.27

14. Balance the checkbook.

RECORD ALL CHARGES OR CREDITS THAT AFFECT YOUR ACCOUNT

NUMBER	DATE	DESCRIPTION OF TRANSACTION	PAYMENT/DEBIT (-) $		√ T	FEE (IF ANY) (-) $	DEPOSIT/CREDIT (+) $		BALANCE $	
									466	79
223	3/2	Groceries	67	32					399	47
	3/5	Deposit					560	70	960	17
224	3/5	Rent	460	00					500	17
225	3/7	Gas & electric	42	35					457	82
226	3/7	Cash	100	00					357	82
227	3/7	Insurance	118	44					239	38
228	3/7	Credit card	119	32					120	06
229	3/12	Dentist	42	00					78	06
230	3/13	Drug store	17	03					61	03
	3/19	Deposit					560	70	621	73
231	3/22	Car payment	141	35					480	38
232	3/25	Cash	100	00					380	38
233	3/25	Oil company	66	40					313	98
234	3/28	Plumber	55	73					258	25
235	3/29	Department store	88	39					169	86
	3/30									

CHECKING ACCOUNT Monthly Statement — Account Number: 122-345-1

Date	Transaction	Amount	Balance
3/1	OPENING BALANCE		466.79
3/5	DEPOSIT	560.70	1027.49
3/7	CHECK	67.32	960.17
3/8	CHECK	460.00	500.17
3/8	CHECK	100.00	400.17
3/9	CHECK	42.35	357.82
3/12	CHECK	118.44	239.38
3/14	CHECK	42.00	197.38
3/18	CHECK	17.03	180.35
3/19	DEPOSIT	560.70	741.05
3/25	CHECK	141.35	599.70
3/27	CHECK	100.00	499.70
3/29	CHECK	55.73	443.97
3/30	INTEREST	13.22	457.19
4/1	CLOSING BALANCE		457.19

15. Balance the checkbook.

RECORD ALL CHARGES OR CREDITS THAT AFFECT YOUR ACCOUNT

NUMBER	DATE	DESCRIPTION OF TRANSACTION	PAYMENT/DEBIT (-)		√ T	FEE (IF ANY) (-)	DEPOSIT/CREDIT (+)		BALANCE	
									$ 219	43
	5/1	Deposit	$			$	$ 219	14	438	57
515	5/2	Electric Bill	22	35					416	22
516	5/2	Groceries	55	14					361	08
517	5/4	Insurance	122	17					238	91
518	5/5	Theatre tickets	24	50					214	41
	5/8	Deposit					219	14	433	55
519	5/10	Telephone	17	39					416	16
520	5/12	Newspaper	12	50					403	66
	5/15	Interest					7	82	411	48
	5/15	Deposit					219	14	630	62
521	5/20	Hotel	172	90					457	72
522	5/21	Credit card	113	44					344	28
523	5/22	Eye Exam	42	00					302	28
524	5/24	Groceries	77	14					225	14
525	5/24	Deposit					219	14	444	28
526	5/25	Oil company	44	16					400	12
527	5/30	Car payment	88	62					311	50
528	5/30	Doctor	37	42					274	08

CHECKING ACCOUNT Monthly Statement — Account Number: 122-345-1

Date	Transaction	Amount	Balance
5/1	OPENING BALANCE		219.43
5/1	DEPOSIT	219.14	438.57
5/3	CHECK	55.14	383.43
5/4	CHECK	22.35	361.08
5/6	CHECK	24.50	336.58
5/8	CHECK	122.17	214.41
5/8	DEPOSIT	219.14	433.55
5/15	INTEREST	7.82	441.37
5/15	CHECK	17.39	423.98
5/15	DEPOSIT	219.14	643.12
5/23	CHECK	42.00	601.12
5/23	CHECK	172.90	428.22
5/24	CHECK	77.14	351.08
5/24	DEPOSIT	219.14	570.22
5/30	CHECK	88.62	481.60
6/1	CLOSING BALANCE		481.60

Calculators and Computers

Amortization Did you ever wonder how your car payments are calculated? Maybe you own a house and make a monthly mortgage payment. In both these cases the amount of your payment is calculated so that you repay the amount you borrowed (the principal) and the interest you were charged to borrow the money. In the words of a banker, your monthly payments "amortize" the loan.

The series of monthly payments you make is called an amortization schedule. The program AMORTIZATION on the Math ACE disk can be used to calculate your monthly payment on any fixed interest loan. The program will also tell you the amount of interest you will repay.

Here are three guidelines that will make using the program a little easier:

1. Enter the annual interest rate as a percent in decimal form. For example, if the interest rate is $9\frac{1}{4}\%$, enter the interest rate as 9.25.
2. Because the payments are *monthly,* you must enter the number of months of the loan. For example, if the loan is for 5 years, then the number of months is 60 (5×12).
3. Enter the amount of money borrowed. Do not use dollar signs ($) or commas in your number.

Chapter Summary

Key Words The *unit cost* is the cost of one item.

Percent increase is used to show how much a quantity has increased over its original value.

Cost is the price that a business pays for a product.

Selling price is the price at which a business sells a product to a customer.

Markup is the difference between selling price and cost.

Markup rate is frequently expressed as a percent of a product's cost.

Percent decrease is used to show how much a quantity has decreased from its original value.

Sale price is the reduced price from the regular price.

Discount is the difference between the regular price and the sale price.

Discount rate is frequently stated as a percent of a product's regular price.

Interest is the amount of money paid for the privilege of using someone else's money.

Principal is the amount of money originally deposited or borrowed.

The percent used to determine the amount of interest is the *interest rate.*

Interest computed on the original amount is called *simple interest.*

Compound interest is computed not only on the original principal but also on interest already earned.

The *mortgage* is the amount that is borrowed to buy real estate.

The loan origination fee is usually a percent of the mortgage and is expressed in *points.*

Commissions are usually paid to salespersons and are calculated as a percent of total sales.

An employee who receives an *hourly wage* is paid a certain amount for each hour worked.

An employee who is paid a *salary* receives payment based on a weekly, biweekly, monthly, or annual time schedule.

Essential Rules

To Find Unit Cost To find the unit cost, divide the total cost by the number of units.

To Find Total Cost To find the total cost, multiply the unit cost by the number of units.

Basic Markup Equations:

Selling Price = cost + markup

$$\mathbf{S} = \mathbf{C} + \mathbf{M}$$

Markup = markup rate × cost

$$\mathbf{M} = \mathbf{r} \times \mathbf{C}$$

Basic Discount Equations:

Sale Price = regular price − discount

$$\mathbf{S} = \mathbf{R} - \mathbf{D}$$

Discount = discount rate × regular price

$$\mathbf{D} = \mathbf{r} \times \mathbf{R}$$

Annual Simple Interest Equation:

$$\text{Principal} \times \begin{array}{c}\text{annual}\\\text{interest rate}\end{array} \times \begin{array}{c}\text{time}\\\text{in years}\end{array} = \text{Interest}$$

$$\mathbf{P} \times \mathbf{r} \times \mathbf{t} = \mathbf{I}$$

Chapter Review

SECTION 6.1

1. A 20-ounce box of cereal costs \$2.90. Find the unit cost.

2. Twenty-four ounces of a mouthwash cost \$3.49. A 60-ounce container of the same kind of mouthwash costs \$8.40. Which is the better buy?

SECTION 6.2

3. An oil stock was bought for $\$42\frac{3}{8}$ per share; 6 months later the stock was selling for $\$55\frac{1}{4}$ per share. Find the percent increase in the price of the stock for 6 months. Round to the nearest tenth of a percent.

4. A professional baseball player received a salary of \$1 million last year. This year the player signed a contract paying \$12 million over 4 years. Find the yearly percent increase in the player's salary.

5. A computer store uses a markup rate of 35% on all computer systems. Find the selling price of a computer system that costs the store \$1540.

6. Last year an oil company had earnings of \$4.12 per share. This year the earnings are \$4.73 per share. What is the percent increase in earnings per share? Round to the nearest percent.

7. A sporting goods store uses a markup rate of 40%. What is the markup on a ski suit that costs the store \$180?

8. A suit that regularly costs \$235 is on sale for 40% off the regular price. Find the sale price.

SECTION 6.3

9. A contractor borrows \$100,000 from a credit union for 9 months at an annual interest rate of 11%. What is the simple interest due on the loan?

10. A sporting goods store borrowed \$30,000 at an annual interest rate of 11% for 6 months. Find the simple interest due on the loan.

11. A computer programmer invests \$25,000 in a retirement account that pays 9% interest, compounded daily. What is the value of the investment in 10 years? Use the table on page A4. Round to the nearest cent.

12. A fast food restaurant invests \$50,000 in an account that pays 11% annual interest, compounded quarterly. What will be the value of the investment in 1 year? Use the table on page A4.

SECTION 6.4

13. A home buyer purchases a home for $125,000. The lender requires a down payment of 15%. Find the amount of the down payment.

14. A credit union requires a borrower to pay $2\frac{1}{2}$ points for a loan. Find the origination fee for a loan of $75,000.

15. The monthly mortgage payment for a condominium is $523.67. The owner must pay an annual property tax of $658.32. Find the total monthly payment for the mortgage and property tax.

16. A family bought a home for $156,000. The family made a 10% down payment and financed the remainder with a 30-year loan with an annual interest rate of 10%. Find the monthly mortgage payment. Use the monthly payment table on page A6. Round to the nearest cent.

SECTION 6.5

17. A plumber buys a truck for $13,500. A state license of $315 and a sales tax of 6.25% of the purchase price are required. Find the total cost of the sales tax and the license fee.

18. An account executive spent $1025.58 for insurance, $605.82 for gas, $37.92 for oil, and $188.27 for maintenance during a year in which 15,320 miles were driven. Find the cost per mile for these four items. Round to the nearest tenth of a cent.

19. A car owner pays a monthly car payment of $122.78. During a month when $25.45 is principal, how much of the payment is interest?

20. A pickup truck with a slide-in camper is purchased for $14,450. A down payment of 8% is made, and the remaining cost is financed for 4 years at an annual interest rate of 11%. Find the monthly payment. Use the monthly payment schedule on page A6. Round to the nearest cent.

SECTION 6.6

21. The manager of the retail store at a ski resort receives a commission of 3% on all sales at the alpine shop. Find the total commission received during a month in which the shop had $108,000 of sales.

22. A nurse receives $12.60 per hour for working 40 hours a week and time and a half for working over 40 hours. Find the nurse's total income during a week in which the nurse worked 48 hours.

SECTION 6.7

23. A student had a checking account balance of $1568.45 before writing checks for $123.76, $756.45, and $88.77. The student then deposited a check for $344.21. Find the student's current checkbook balance.

24. The business checking account of a donut shop showed a balance of $9567.44 before checks of $1023.55, $345.44, and $23.67 were written and checks of $555.89 and $135.91 were deposited. Find the current checkbook balance.

Chapter Test

1. Twenty feet of lumber cost \$138.40. What is the cost per foot?

2. Find the more economical purchase: 5 pounds of tomatoes for \$1.65, or 8 pounds for \$2.72.

3. Red snapper cost \$4.15 per pound. Find the cost of $3\frac{1}{2}$ pounds. (Round to the nearest cent.)

4. An exercise bicycle increased in price from \$415 to \$498. Find the percent increase in the cost of the exercise bicycle.

5. Fifteen years ago a painting was priced at \$6000. Today the same painting has a value of \$15,000. Find the percent increase in the price of the painting during the fifteen years.

6. A department store uses a 40% markup rate. Find the selling price of a compact disc player that the store purchased for \$215.

7. A bookstore buys a paperback book for \$5 and uses a markup rate of 25%. Find the selling price of the book.

8. The price of gold dropped from \$850 per ounce to \$360 per ounce. What percent decrease does this amount represent? (Round to the nearest tenth of a percent.)

9. The price of a video camera drops from \$1120 to \$896. What percent decrease does this price drop represent?

10. A corner hutch with a regular price of \$299 is on sale for 30% off the regular price. Find the sale price.

11. A box of stationery that regularly sells for \$4.50 is on sale for \$2.70. Find the discount rate.

12. A construction company borrowed \$75,000 at an annual interest rate of 11% for 4 months. Find the simple interest due on the loan.

13. A self-employed individual placed \$30,000 in an account that pays 9% annual interest, compounded quarterly. How much interest was earned in 10 years? Use the table on page A4.

14. A savings and loan institution is giving mortgage loans that have a loan origination fee of $2\frac{1}{2}$ points. Find the loan origination fee on a home purchased with a loan of \$134,000.

15. A new housing development offers homes with a mortgage of $122,000 for 25 years at an annual interest rate of 10%. Find the monthly mortgage payment. Use the table on page A6.

16. A $17,500 minivan is purchased with an 18% down payment. Find the amount financed.

17. A rancher buys a pickup for $11,500, with a down payment of 15% of the cost. The balance is financed for 3 years at an annual interest rate of 11%. Find the monthly car payment. Use the table on page A6.

18. An emergency room nurse receives an hourly wage of $13.40 an hour. When called in at night, the nurse receives time and a half. How much does the nurse earn in a week when working 30 hours at normal rates and 15 hours during the night?

19. The business checking account for a pottery store had a balance of $7349.44 before checks for $1349.67 and $344.12 were written. The store manager then made a deposit of $956.60. Find the current checkbook balance.

20. Balance the checkbook shown.

RECORD ALL CHARGES OR CREDITS THAT AFFECT YOUR ACCOUNT

NUMBER	DATE	DESCRIPTION OF TRANSACTION	PAYMENT/DEBIT (-)	√ T	FEE (IF ANY) (-)	DEPOSIT/CREDIT (+)	BALANCE
							$ 422.13
	8/1	House Payment	$ 213.72		$	$	208.41
	8/4	Deposit				552.60	761.01
	8/5	Plane Tickets	162.40				598.61
	8/6	Groceries	66.44				532.17
x	8/10	Car Payment	122.37				409.80
	8/15	Deposit				552.60	962.40
x	8/16	Credit card	213.45				748.95
	8/18	Doctor	92.14				656.81
	8/22	Utilities	72.30				584.51
x	8/28	T.V. Repair	78.20				506.31

CHECKING ACCOUNT Monthly Statement — Account Number: 122-345-1

Date	Transaction	Amount	Balance
8/1	OPENING BALANCE		422.13
8/3	CHECK	213.72	208.41
8/4	DEPOSIT	552.60	761.01
8/8	CHECK	66.44	694.57
8/8	CHECK	162.40	532.17
8/15	DEPOSIT	552.60	1084.77
8/23	CHECK	72.30	1012.47
8/24	CHECK	92.14	920.33
9/1	CLOSING BALANCE		920.33

Cumulative Review

1. Simplify $12 - (10 - 8)^2 \div 2 + 3$.

2. Add: $3\frac{1}{3} + 4\frac{1}{8} + 1\frac{1}{12}$

3. Subtract: $12\frac{3}{16} - 9\frac{5}{12}$

4. Multiply: $5\frac{5}{8} \times 1\frac{9}{15}$

5. Divide: $3\frac{1}{2} \div 1\frac{3}{4}$

6. Simplify $\left(\frac{3}{4}\right)^2 \div \left(\frac{3}{8} - \frac{1}{4}\right) + \frac{1}{2}$.

7. Divide: $0.059\overline{)3.0792}$.
Round to the nearest tenth.

8. Convert $\frac{17}{12}$ to a decimal. Round to the nearest thousandth.

9. Write "$410 in 8 hours" as a unit rate.

10. Solve the proportion $\frac{5}{n} = \frac{16}{35}$.
Round to the nearest hundredth.

11. Write $\frac{5}{8}$ as a percent.

12. Find 6.5% of 420.

13. Write 18.2% as a decimal.

14. What percent of 20 is 8.4?

15. 30 is 12% of what?

16. 65 is 42% of what? Round to the nearest hundredth.

Cumulative Review

17. A series of late summer storms produced rainfall of $3\frac{3}{4}$, $8\frac{1}{2}$, and $1\frac{2}{3}$ inches during a 3-week period. Find the total rainfall during the 3 weeks.

18. A family pays $\frac{1}{5}$ of its total monthly income for taxes. The family has a total monthly income of $2850. Find the amount of the monthly income that the family pays in taxes.

19. In 5 years, the cost of a scientific calculator went from $75 to $30. What is the ratio of the decrease in price to the original price?

20. A compact car drove 417.5 miles on 12.5 gallons of gasoline. Find the number of miles driven per gallon of gasoline.

21. A 14 pound turkey costs $12.96. Find the unit cost. (Round to the nearest cent.)

22. Eighty shares of a stock has a dividend of $112. At the same rate, find the dividend on 200 shares of the stock.

23. A video camera that regularly sells for $900 is on sale for 20% off the regular price. What is the sale price?

24. A department store buys a portable disc player for $85 and uses a markup rate of 40%. Find the selling price of the disc player.

25. An elementary school teacher received an increase in salary from $2800 per month to $3024 per month. Find the percent increase in the teacher's salary.

26. A contractor borrowed $120,000 for 6 months at an annual interest rate of 10%. How much simple interest is due on the loan?

27. A red sports car is purchased for $14,000, and a down payment of $2000 is made. The balance is financed for 3 years at an annual interest rate of 9%. Find the monthly payment. Use the table on page A6. (Round to the nearest cent.)

28. A family had a checking account balance of $1846.78. A check of $568.30 was deposited into the account, and checks of $123.98 and $47.33 were written. Find the new checking account balance.

29. A car owner spent $840 on gasoline and oil, $520 on insurance, $185 on tires, and $432 on repairs. Find the cost per mile to drive the car 10,000 miles during the year. (Round to the nearest cent.)

30. A house has a mortgage of $72,000 for 20 years at an annual interest rate of 11%. Find the monthly mortgage payment. Use the table on page A6. (Round to the nearest cent.)

7 Statistics

OBJECTIVES

- To read a pictograph
- To read a circle graph
- To read a bar graph
- To read a broken-line graph
- To read a histogram
- To read a frequency polygon
- To find the mean of a set of numbers
- To find the median of a set of numbers

Frequencies of Letters

ZH WKH SHRSOH

The above phrase is a cryptogram. For this phrase to be read, the cryptogram has to be decoded.

Cryptology is the study of encrypting and decrypting messages. Encrypting means to write the message in code; decrypting means breaking a secret code. One of the methods the cryptologist uses in breaking a code is statistics.

Statistics is a study of the organization and analysis of data. A cryptologist uses statistics by analyzing ordinary text, like that in a novel or a newspaper, and determining how frequently different letters of the alphabet occur. For example, in English, the letter "e" is the most frequently occurring letter. A table of the approximate frequencies of each letter is given below:

A—7.3%	J—0.2%	S—6.3%
B—0.9%	K—0.3%	T—9.3%
C—3.0%	L—3.6%	U—2.7%
D—4.3%	M—2.5%	V—1.3%
E—13.0%	N—7.8%	W—1.6%
F—2.7%	O—7.4%	X—0.6%
G—1.7%	P—2.7%	Y—1.8%
H—3.4%	Q—0.3%	Z—0.1%
I—7.5%	R—7.3%	

Knowing these frequencies, the cryptologist reasons that the most frequently occurring letters in a coded message corresponds to the most frequently occurring letters in an ordinary message. Thus, to decode the phrase above, a cryptologist might guess that the letter H in the coded message corresponds to the letter E in the ordinary message. This guess may not be correct, but it is a good first choice.

See if you can decode the above phrase.

Answer: WE THE PEOPLE. The phrase was coded by taking the letter three spaces beyond the original letter. For example, A gets coded as D, B gets coded as E, C gets coded as F, and so on. This method of coding a message is called the Caesar Cipher, after Julius Caesar, who used this method.

SECTION 7.1 Pictographs and Circle Graphs

Objective A To read a pictograph

Statistics is the branch of mathematics concerned with **data,** or numerical information. A graph may be used to present data in a form that is easily read.

A **pictograph** uses symbols to represent information. A pictograph is usually used to represent comparisons when the exact numbers are not too important. The pictograph in Figure 1 shows the number of cars sold on a Friday, Saturday, and Sunday. Each picture represents two cars.

Fig. 1 Number of cars sold.

The pictograph shows that 8 cars were sold on Friday, 15 on Saturday, and 7 on Sunday.

The pictograph can be used to determine ratios among the number of cars sold or the ratio of the number of cars sold on 1 day to the total number of cars sold on the weekend.

The ratio of the number of cars sold on Saturday to the total number of cars sold on the weekend is

$$\frac{15 \text{ cars}}{30 \text{ cars}} = \frac{1}{2}$$

The pictograph in Figure 2 shows the number of students receiving an A, B, C, D, or F in a class of 30 students. Each figure represents 2 students.

Use the pictograph in Figure 2 to find the percent *(N)* of the total number of students (30) receiving A's.

Percent × base = amount

$$N \times 30 = 5$$

$$N = \frac{5}{30} = \frac{1}{6}$$

$$N = 16\frac{2}{3}\%$$

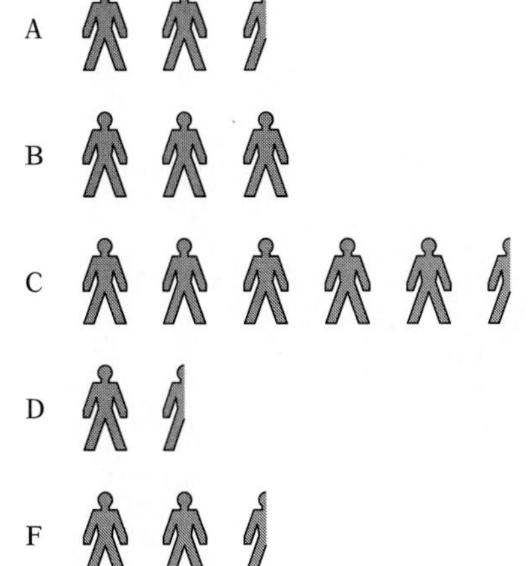

Fig. 2 Number of students receiving grades.

The pictograph in Figure 3 shows the housing starts in a midwestern town during a 4-month period. Each picture represents four housing starts.

The ratio of the number of housing starts in May to the number of housing starts in August is

$$\frac{4 \text{ starts}}{8 \text{ starts}} = \frac{1}{2}$$

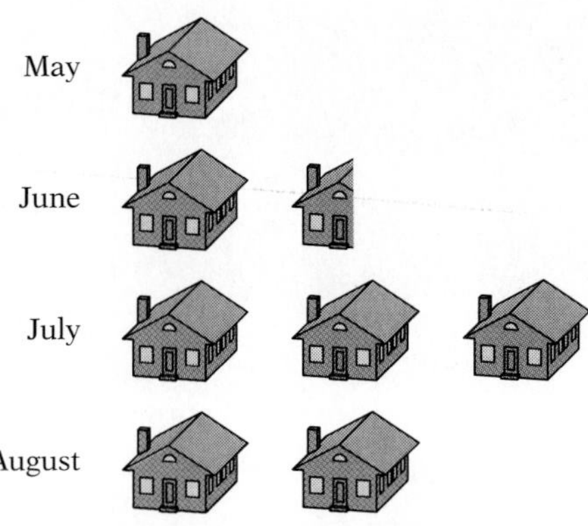

Fig. 3 Number of housing starts.

Example 1

Use Figure 3 to find the total number of housing starts for the 4 months.

Strategy

To find the total number of housing starts:

- Read the pictograph to determine the number of housing starts each month.
- Add the four numbers.

Solution

Starts for May—4
Starts for June—6
Starts for July—12
Starts for August—8

Total starts—30

The total number of housing starts is 30.

Example 2

Use Figure 3 to find what percent of the total number of housing starts the number of July housing starts is.

Your strategy

Your solution

Solution on p. A28

Objective B To read a circle graph

The circle graph in Figure 4 shows the activities of an airplane mechanic in a 24-hour day. The complete circle represents all 24 hours of the day. Each pie-shaped section, or **sector,** shows the number of hours spent on one activity.

One use of the circle graph is to find the ratio of the number of hours spent on one activity to the total number of hours in the day.

The working-to-total ratio is

$$\frac{8 \text{ hours}}{24 \text{ hours}} = \frac{1}{3}$$

The circle graph can also be used to find the ratio of the time spent on one activity to the time spent on a second activity.

The recreation-to-working ratio is

$$\frac{4 \text{ hours}}{8 \text{ hours}} = \frac{1}{2}$$

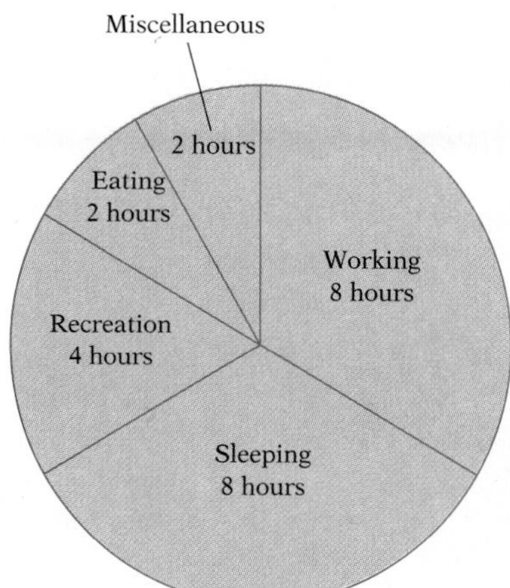

Fig. 4 Airplane mechanic's activities in a 24-hour day.

The circle graph in Figure 5 shows how a family's monthly income of $1600 is budgeted. The complete circle represents 100% of the budget. Each sector of the graph expresses one item of the budget as a percent of the total monthly income.

Use the circle graph to find the amount of money budgeted for a particular expense by solving the basic percent equation for *amount.*

How much money is budgeted for the car payment?

Percent ×		*base*	=	*amount*
15%	of	total income	=	amount for car payment
0.15	×	$1600	=	$240

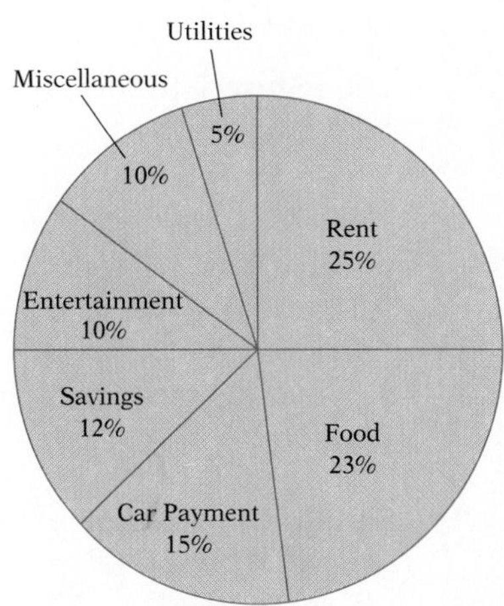

Fig. 5 Budget for a monthly income of $1600.

The circle graph in Figure 6 shows the annual expenses of owning and operating a car.

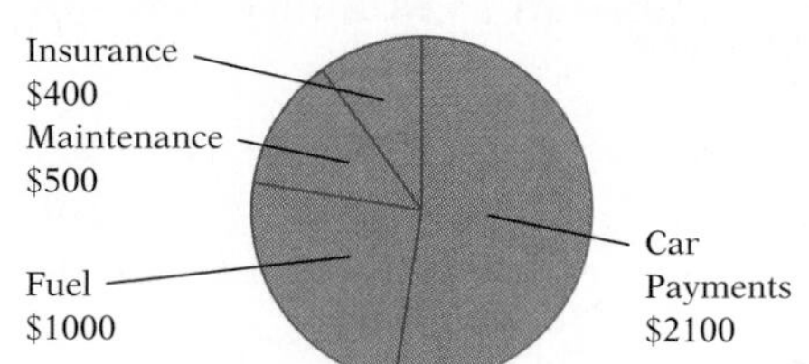

Fig. 6 Annual car expenses totaling $4000.

Example 3

Use Figure 6 to find the ratio of the annual fuel expense to the total annual cost of operating a car.

Strategy To find the ratio:

- Locate the annual fuel expense in the circle graph.
- Write the ratio of the annual fuel expense to the total annual cost of operating a car in simplest form.

Solution Annual fuel expense: $1000

$\frac{\$1000}{\$4000} = \frac{1}{4}$

The ratio is $\frac{1}{4}$.

Example 4

Use Figure 6 to find the ratio of the annual cost of insurance to the annual cost of maintenance.

Your strategy

Your solution

The circle graph in Figure 7 shows the distribution of an employee's gross monthly income.

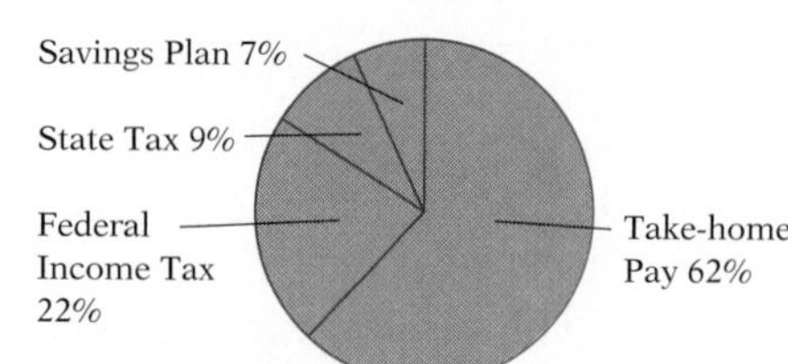

Fig. 7 Distribution of gross monthly income of $2000.

Example 5

Use Figure 7 to find the employee's take-home pay.

Strategy To find the take-home pay:

- Locate the percent of the distribution that is take-home pay in the circle graph.
- Solve the basic percent equation for *amount*.

Solution Take-home pay: 62%

Percent × *base* = *amount*

0.62 × $2000 = $1240

The employee's take-home pay is $1240.

Example 6

Use Figure 7 to find the federal income tax the employee paid.

Your strategy

Your solution

Solutions on pp. A28–A29

7.1 EXERCISES

Objective A

The pictograph in Figure 8 shows the amount of gasoline a service station sold during a 4-week period. Each barrel in the pictograph represents 1000 gallons of gasoline.

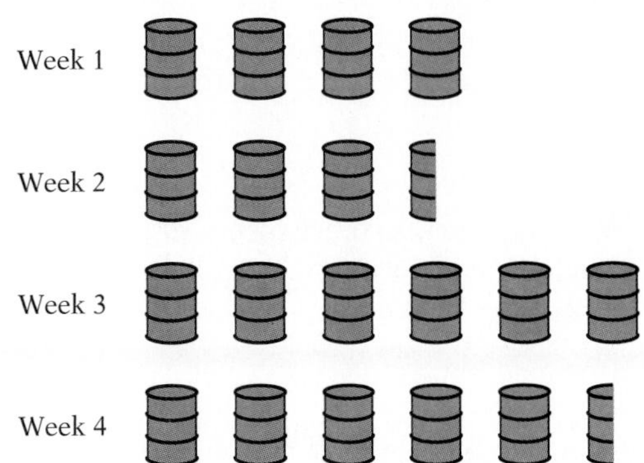

Fig. 8 Barrels of gasoline sold.

1. Find the total number of gallons of gasoline sold during the month.
2. Find the ratio of the amount of gasoline sold in week 2 to the amount sold in week 4.
3. Find the percent of the amount of gasoline sold during the month that was sold in week 1. (Round to the nearest tenth of a percent.)

Of each dollar that a school receives (see Figure 9), 55¢ comes from the state government, 30¢ from local sources, and 15¢ from the federal government.

Fig. 9 Sources of school's budget.

4. Find the ratio of the amount of money that comes from local sources to the amount that comes from the state.
5. Find the percent of the budget that comes from the federal government.
6. If the total budget is $15,000,000, find the amount of the budget that comes from local sources.

The pictograph in Figure 10 shows the number of fish caught in Lost Lake during four seasons. Each picture represents 200 fish.

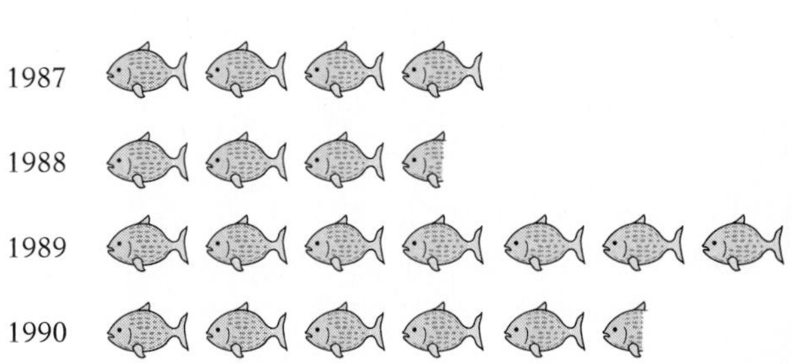

Fig. 10 Number of fish caught.

7. Find the total number of fish caught during the 4 years.
8. Find the ratio of the number of fish caught in 1987 to the number caught in 1988.
9. Find the percent of the number of fish caught in 1990 to the total number of fish caught. (Round to the nearest tenth of a percent.)

Objective B

The circle graph in Figure 11 shows the number of students in each class in a small midwestern liberal arts college.

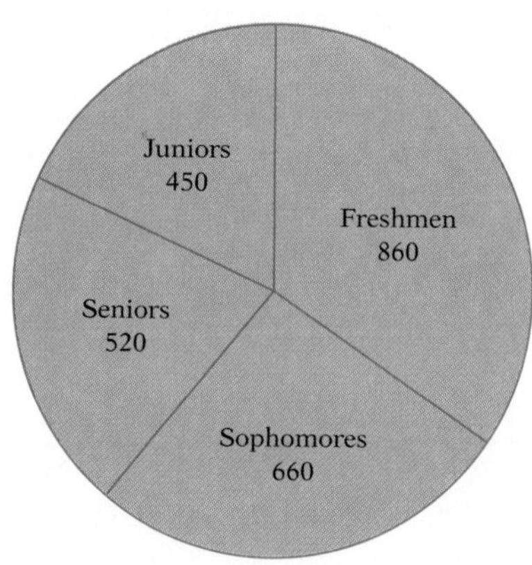

Fig. 11 Enrollment in a small college.

10. Find the total number of students attending the college.

11. What is the ratio of the number of students in the junior class to the total number of students?

12. What is the ratio of the number of senior students to the number of freshmen students?

13. What is the ratio of the number of students in the freshmen class to the number of junior students?

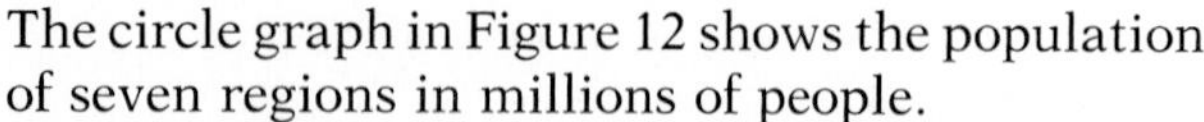

The circle graph in Figure 12 shows the population of seven regions in millions of people.

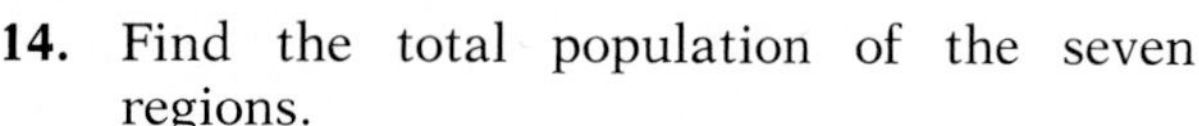

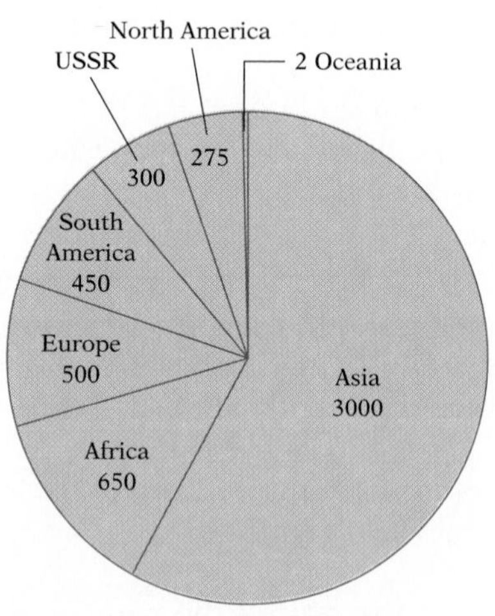

Fig. 12 Population in millions of people.

14. Find the total population of the seven regions.

15. What is the ratio of the population of Asia to the population of Africa?

16. What is the ratio of the population of North America to the population of Asia?

17. What is the ratio of the population of North America to the population of South America?

The circle graph in Figure 13 shows the land area of each of the seven continents in square miles.

18. Find the total land area of the seven continents.

19. Find the ratio of the land area of North America to the land area of South America.

20. Find the ratio of the land area of Asia to the total land area of the seven continents.

21. Find the ratio of the land area of Australia to the total land area of the seven continents.

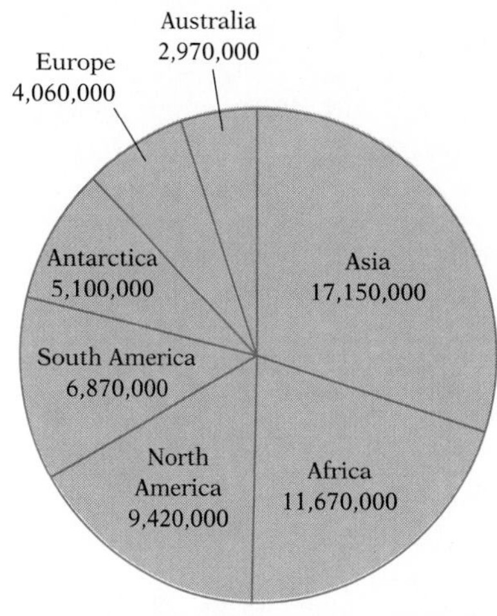

Fig. 13 Land area of the seven continents in square miles.

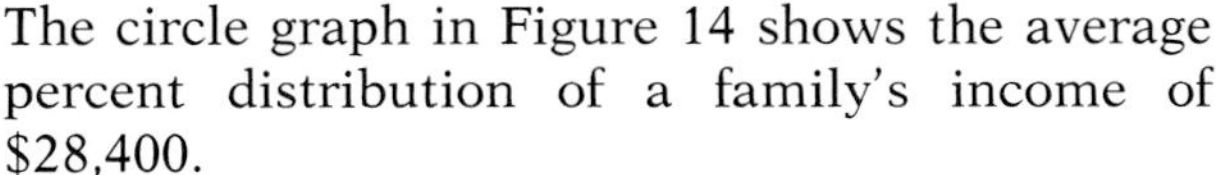

The circle graph in Figure 14 shows the average percent distribution of a family's income of $28,400.

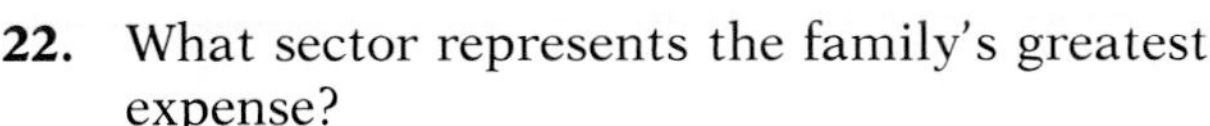

22. What sector represents the family's greatest expense?

23. How much of the family's income was spent on food?

24. How much of the family's income was spent on medical purposes?

25. How much of the family's income was spent on clothing and entertainment?

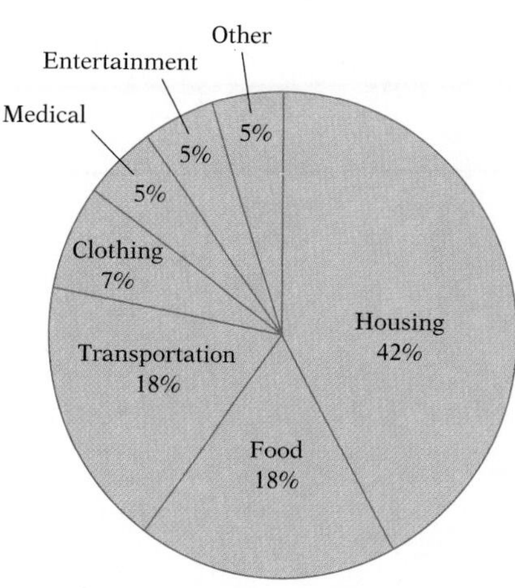

Fig. 14 Percent distribution of a family's income.

The circle graph in Figure 15 shows the age distribution in a city of 240,000 people.

26. Find the number of people in the city older than 65.

27. How many people are between the ages of 31 and 45?

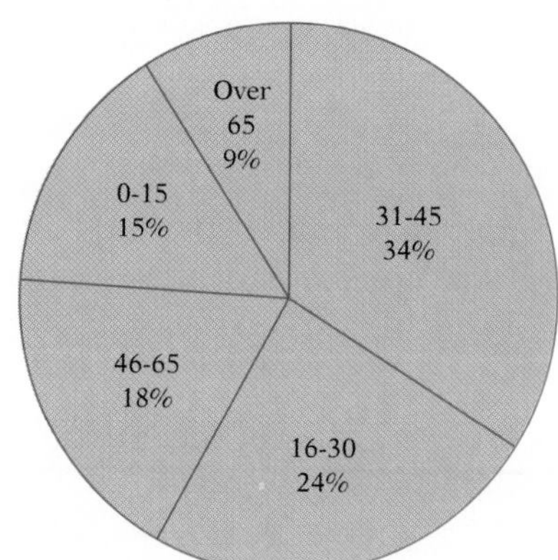

Fig. 15 Age distribution in a city of 240,000 people.

28. What fractional part of the population of the city is older than 46?

29. How many people are between the ages of 31 and 65?

The circle graph in Figure 16 shows how a college's annual income of $22,450,000 is budgeted.

30. Find the amount of money budgeted for supplies.

31. Find the amount budgeted for administrative costs.

32. Find the total amount of money budgeted for supplies and utilities.

33. Find the total amount of money budgeted for administrative costs and teacher salaries.

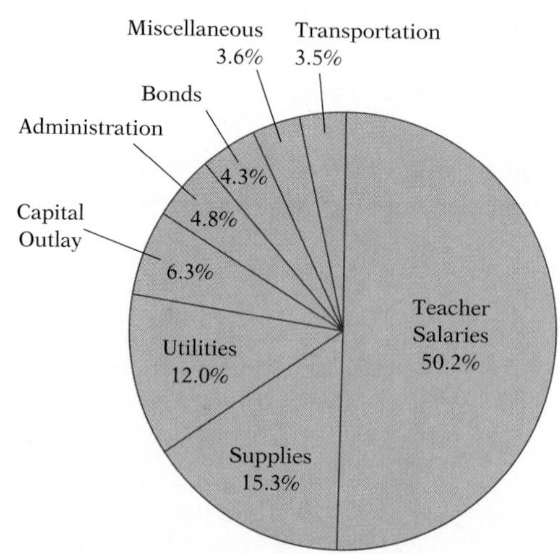

Fig. 16 College budget of $22,450,000.

SECTION 7.2 Bar Graphs and Broken-Line Graphs

Objective A

To read a bar graph

The bar graph in Figure 17 shows the market value of a home for each of 5 years. For each year, the height of the bar indicates the market value of the home.

The value of the home in 1991 was 210,000.

The value of the home in 1988 was 120,000.

Since the 1987 bar is the lowest bar in the graph, the home had the lowest market value in 1987.

Market Value (in thousands of dollars)
220
210
200
180
160
140
120
100
90
80
60
40
20
0
1987
1988
1989
1990
1991

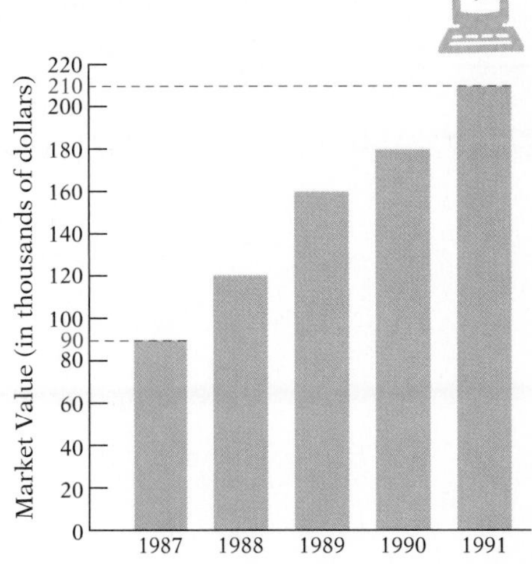

Fig. 17

A double-bar graph is used to display data for purposes of comparison. The double-bar graph in Figure 18 shows the quarterly profits for a company for the years 1990 and 1991.

The profit in the second quarter of 1990 was $14,000.

The profit in the second quarter of 1991 was $16,000.

In 1990 the company had the largest quarterly profit during the third quarter.

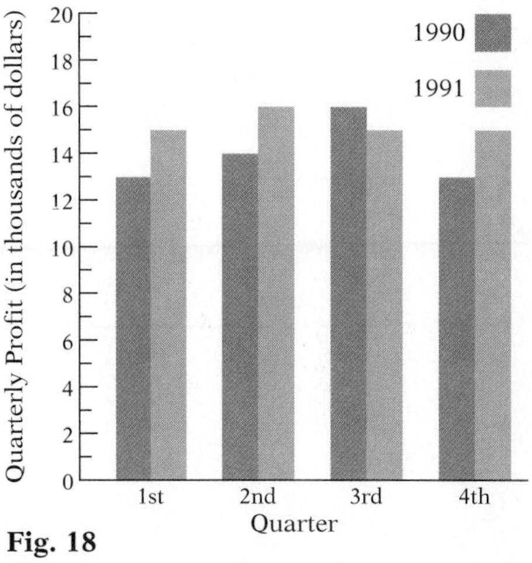

Quarterly Profit (in thousands of dollars)
20
18
16
14
12
10
8
6
4
2
0
1990
1991
1st
2nd
3rd
4th
Quarter

Fig. 18

Example 1

Use Figure 18 to find the difference between the company's fourth quarter profits for 1991 and 1990.

Strategy To find the difference:

- Read the bar graph to find the fourth quarter profit for each year.
- Subtract to find the difference between the profits.

Solution

1991 fourth quarter profit: $15,000
1990 fourth quarter profit: $13,000

$15,000 − $13,000 = $2000
The difference is $2000.

Example 2

Use Figure 18 to find the difference between the company's first quarter profits for 1991 and 1990.

Your strategy

Your solution

Solution on p. A29

Objective B To read a broken-line graph

The broken-line graph in Figure 19 shows the number of new car sales by an automobile manufacturer for a 5-week period. The height of each dot indicates the number of new car sales each week.

The number of cars sold during week 1 was 20,000.

The number of cars sold during week 3 was 15,000.

Since the highest dot in the graph is above week 4, the greatest number of cars was sold during week 4.

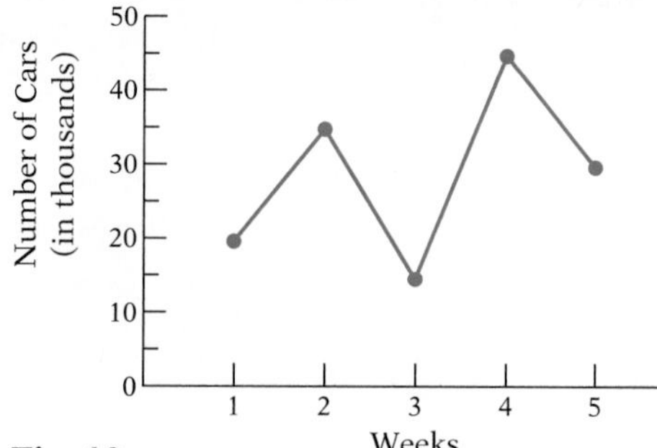

Fig. 19

Two broken-line graphs are often shown in the same figure for comparison. Figure 20 shows the number of landings of private planes and commercial planes at an airport for a 5-year period.

The number of private plane landings in 1990 was 55,000.

The number of commercial plane landings in 1990 was 80,000.

The greatest number of private plane landings occurred in 1989.

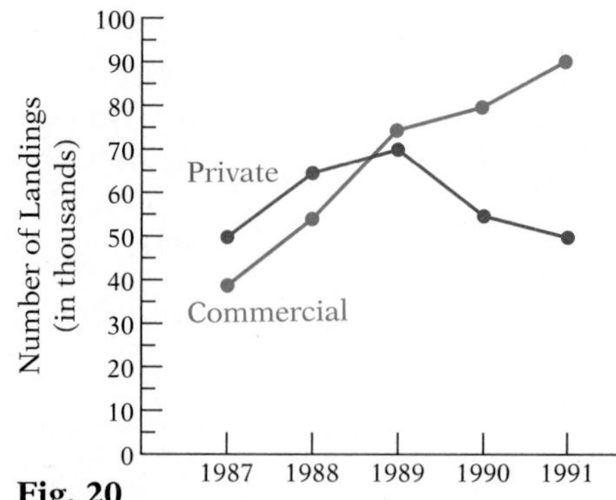

Fig. 20

Example 3

Use Figure 20 to find the difference between the number of private plane landings and commercial plane landings in 1988.

Strategy To find the difference:

- Read the line graphs to find the number of landings for each type of aircraft in 1988.
- Subtract to find the difference in landings.

Solution

Private plane landings: 65,000
Commercial plane landings: 55,000

65,000 − 55,000 = 10,000.
The difference is 10,000.

Example 4

Use Figure 20 to find the difference between the number of commercial and private plane landings in 1991.

Your strategy

Your solution

Solution on p. A29

7.2 EXERCISES

Objective A

The bar graph in Figure 21 shows the number of cars a corporation sold during the first 6 months of the year.

1. How many cars were sold in May?
2. How many cars were sold in February and March?
3. Find the ratio of the number of cars sold in February to the number of cars sold in June.
4. In which month was the greatest number of cars sold?

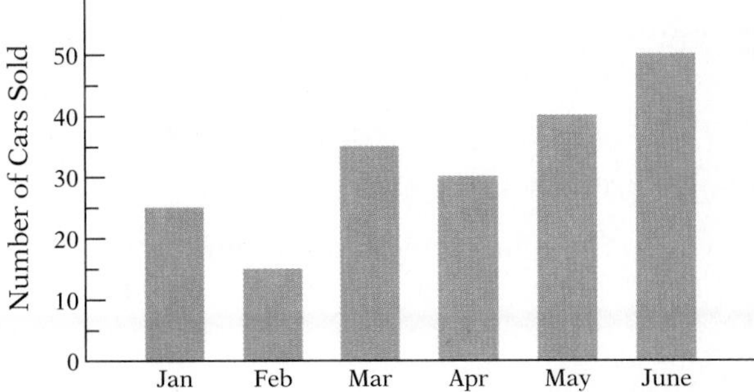

Fig. 21

The double-bar graph in Figure 22 shows the premiums earned and the benefits an insurance company paid during a 5-year period.

5. Find the amount of premiums earned in 1987.
6. Find the amount of benefits paid in 1989.
7. In what year did the amount of benefits exceed the premiums earned?
8. Find the difference in premiums earned and benefits paid in 1984.

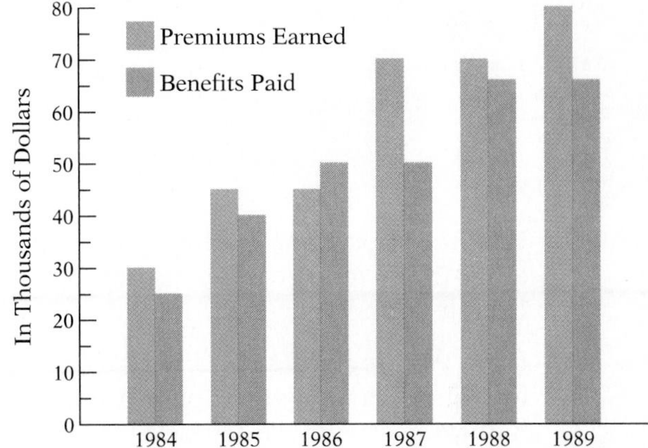

Fig. 22

The double-bar graph in Figure 23 shows some crime statistics per 100,000 population in an eastern city.

9. How many burglaries were committed per 100,000 population in 1985?
10. What crimes decreased during the period 1985 to 1989?
11. Find the ratio of auto thefts that occurred in 1985 to the number of auto thefts that occurred in 1989.
12. Find the percent increase in armed robbery during the years 1985 to 1989. (Round to the nearest tenth of a percent.)

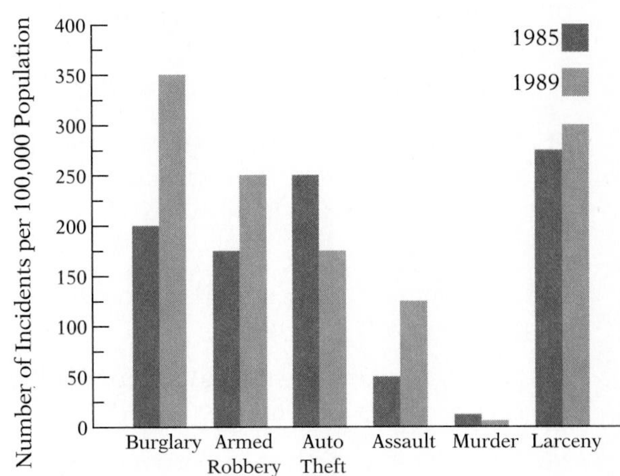

Fig. 23

Objective B

The broken-line graph in Figure 24 shows the snowfall at a ski resort during the ski season.

13. What was the amount of snowfall during January?

14. During which month was the snowfall the lowest?

15. What was the total snowfall during November and December?

16. Find the ratio of the amount of snowfall in December to the amount of snowfall during January.

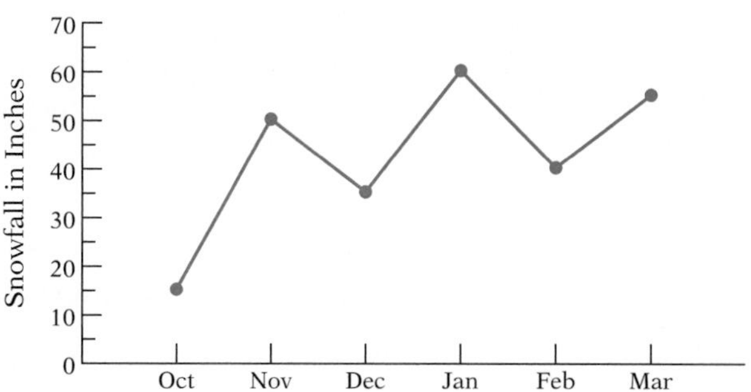

Fig. 24

The double-broken-line graph in Figure 25 shows the average high temperature in Honolulu and the average high temperature in New York.

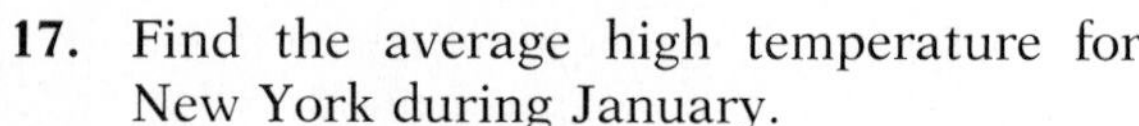

17. Find the average high temperature for New York during January.

18. Find the average high temperature for Honolulu during January.

19. Find the approximate difference in the average high temperature during July for Honolulu and New York.

20. Find the approximate difference in the lowest average high temperature and the highest average high temperature for Honolulu.

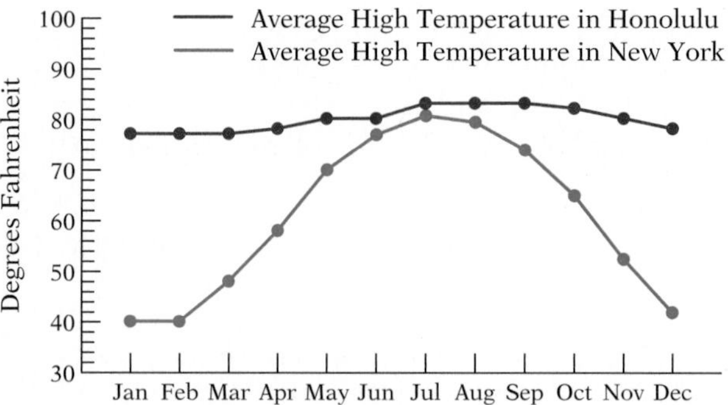

Fig. 25

The double-broken-line graph in Figure 26 shows the number of business calls and the number of residential calls made each hour from 9 A.M. to 5 P.M. during an 8-hour business day in a small city.

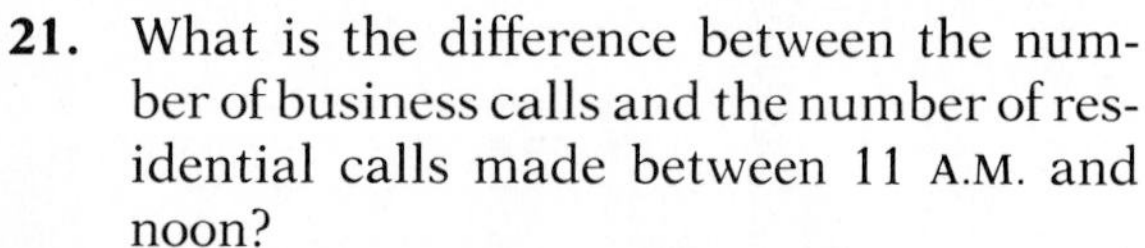

21. What is the difference between the number of business calls and the number of residential calls made between 11 A.M. and noon?

22. How many business calls were made between 9 A.M. and noon?

23. How many residential calls were made between 9 A.M. and noon?

24. What is the difference between the number of business calls and the number of residential calls made between 2 and 5 P.M.?

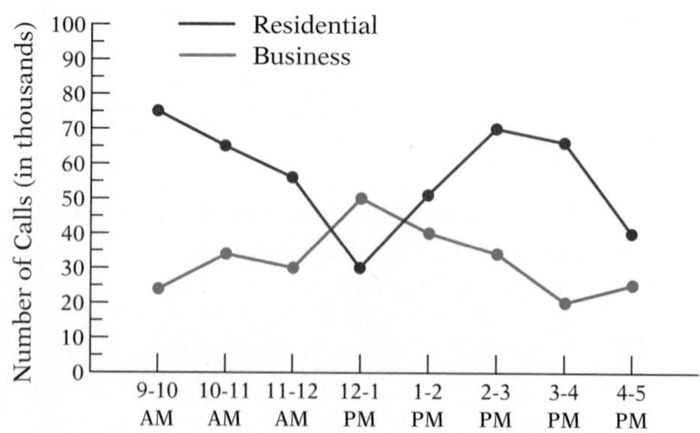

Fig. 26

SECTION 7.3 Histograms and Frequency Polygons

Objective A To read a histogram

7

A research group measured the fuel usage of 92 cars. The results are recorded in the histogram in Figure 27. A **histogram** is a special type of bar graph. The width of each bar corresponds to a range of numbers called a **class interval.** The height of each bar corresponds to the number of occurrences of data in each class interval and is called the **class frequency.**

Class Intervals (Miles per Gallon)	*Class Frequencies (Number of Cars)*
18–20	12
20–22	19
22–24	24
24–26	17
26–28	15
28–30	5

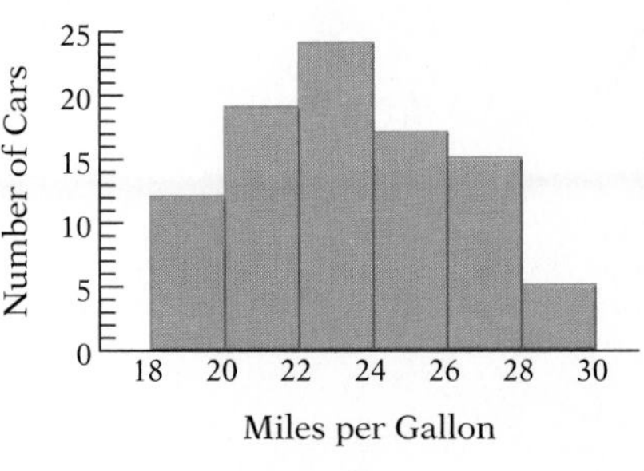

Fig. 27

Twenty-four cars get between 22 and 24 miles per gallon.

A precision tool company has 85 employees. Their hourly wages are recorded in the histogram in Figure 28.

The ratio of the number of employees whose hourly wage is between \$10 and \$12 to the total number of employees is $\frac{17 \text{ employees}}{85 \text{ employees}} = \frac{1}{5}$.

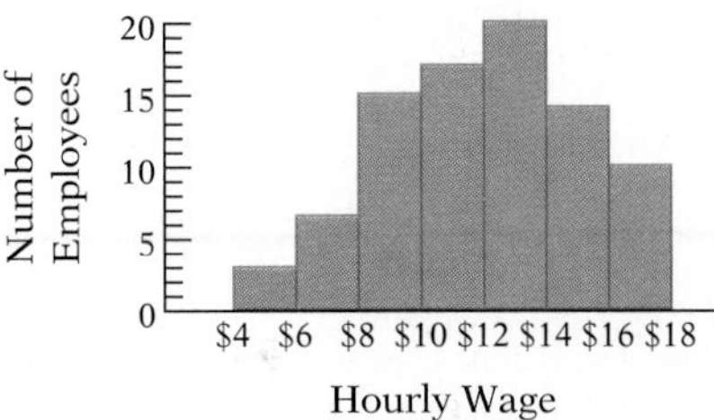

Fig. 28

Example 1

Use Figure 28 to find the number of employees whose hourly wage is between \$14 and \$18.

Strategy

To find the number of employees:

- Read the histogram to find the number of employees whose hourly wage is between \$14 and \$16 and the number whose wage is between \$16 and \$18.
- Add the two numbers.

Solution Number whose wage is between \$14 and \$16: 14; between \$16 and \$18: 10.

14 + 10 = 24

24 employees have an hourly wage between \$14 and \$18.

Example 2

Use Figure 28 to find the number of employees whose hourly wage is between \$8 and \$12.

Your strategy

Your solution

Solution on p. A29

Objective B To read a frequency polygon

The speeds of 70 cars on a highway were measured by radar. The results are recorded in the frequency polygon in Figure 29. A **frequency polygon** is a graph that displays information in a manner similar to a histogram. A dot is placed above the center of each class interval at a height corresponding to that class's frequency; the dots are then connected to form a broken-line graph. The center of a class interval is called the **class midpoint.**

Class Interval (Miles per Hour)	*Class Midpoint*	*Class Frequency*
30–40	35	7
40–50	45	13
50–60	55	25
60–70	65	21
70–80	75	4

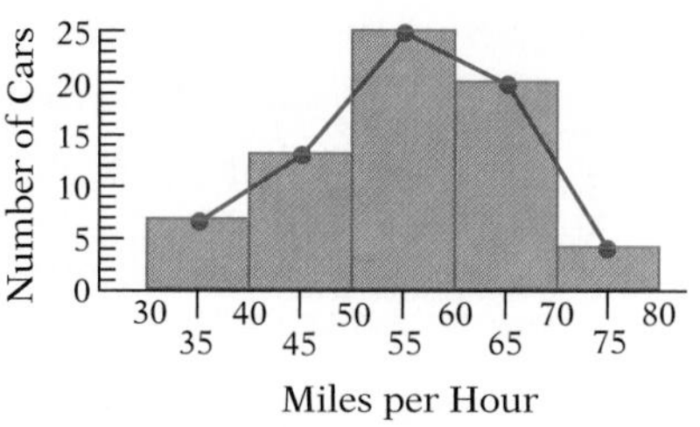

Fig. 29

Twenty-five cars were traveling between 50 and 60 miles per hour.

Sixty people took an exam for a real estate license. The scores on the exam are recorded in the frequency polygon in Figure 30.

The ratio of the number of people scoring between 70 and 80 to the total number of people tested is $\frac{24 \text{ people}}{60 \text{ people}} = \frac{2}{5}$.

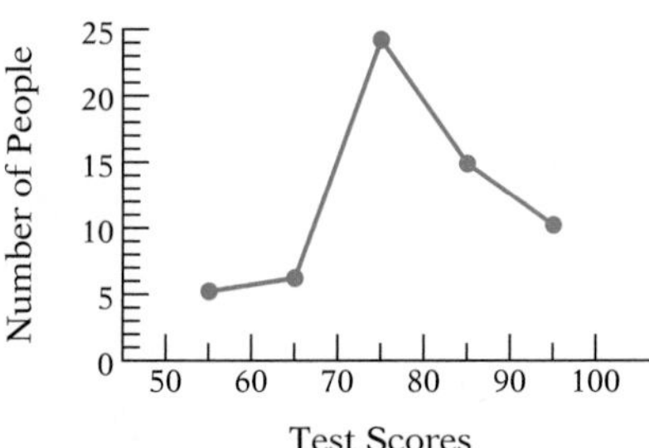

Fig. 30

Example 3

Use Figure 30 to find the number of people who scored between 80 and 100 on the real estate exam.

Strategy

To find the number of people who scored between 80 and 100 on the exam:

- Read the frequency polygon to find the number of people who scored between 80 and 90 and the number who scored between 90 and 100.
- Add the two numbers.

Solution

The number who scored between 80 and 90: 15; between 90 and 100: 10.

$15 + 10 = 25$

25 people scored between 80 and 100.

Example 4

Use Figure 30 to find the number of people who scored between 50 and 70 on the real estate exam.

Your strategy

Your solution

Solution on p. A29

7.3 EXERCISES

Objective A

The test scores of 34 students are recorded in the histogram in Figure 31.

1. How many students scored between 60 and 80?
2. Find the ratio of the number of students who scored between 50 and 60 to the total number of students.
3. Find the number of students who scored above 80.
4. Find the percent of the students who scored below 60. (Round to the nearest tenth of a percent.)

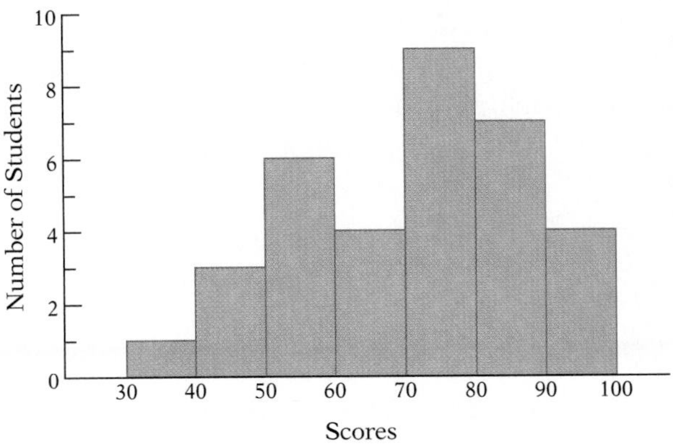

Fig. 31

A department store keeps records of the amounts its customers spend. The histogram in Figure 32 records the dollar amounts 153 customers spent.

5. How many customers made purchases between $20 and $30?
6. What is the ratio of the number of customers whose purchases were between $30 and $40 to the total number of customers?
7. How many customers made purchases of more than $30?
8. What percent of the total number of customers spent more than $50? Round to the nearest tenth of a percent.

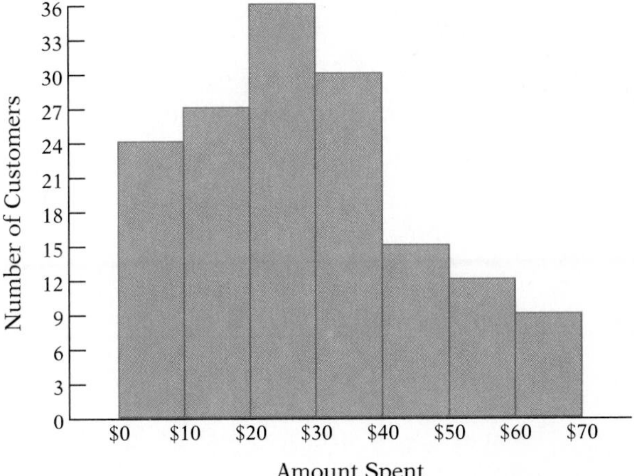

Fig. 32

The histogram in Figure 33 shows the number of cars sold in different price ranges.

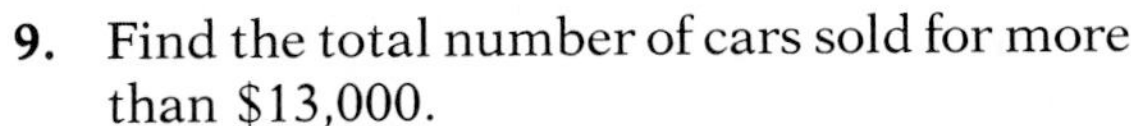

9. Find the total number of cars sold for more than $13,000.
10. Find the number of cars sold whose price was between $5000 and $11,000.
11. Find the ratio of the number of cars sold that were priced between $7000 and $9000 and the number sold that were priced between $11,000 and $13,000.
12. What percent of the cars sold were priced over $15,000? (Round to the nearest tenth of a percent.)

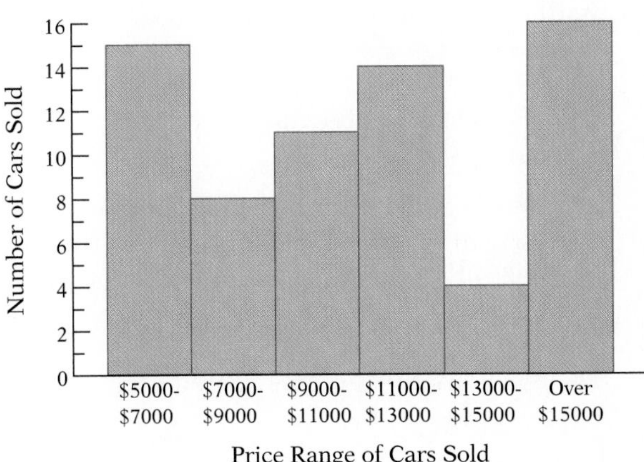

Fig. 33

Objective B

A total of 34 runners ran the 100-yard dash. The results are recorded in Figure 34.

13. How many runners ran the 100-yard dash in less than 11 seconds?

14. Find the ratio of the number of runners who ran the race between 10 and 11 seconds to the number who ran between 12 and 13 seconds.

15. How many runners ran the race between 11 and 12 seconds?

16. What percent of the runners ran the race in less than 11 seconds? (Round to the nearest tenth of a percent.)

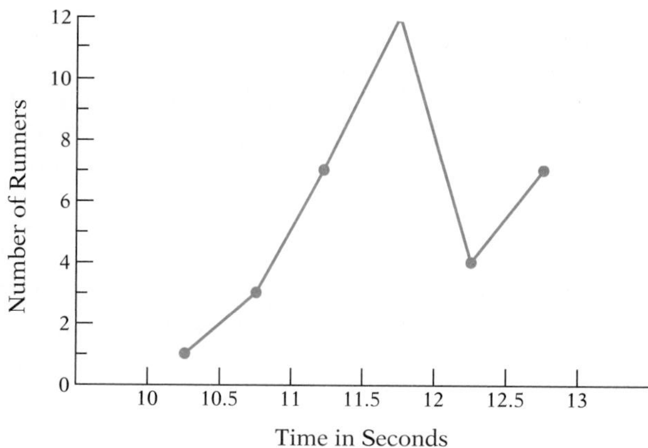

Fig. 34

A survey of 50 families was made to determine the percent of their income that was spent on vacations. The results are recorded in Figure 35.

17. How many families spent between 4% and 8% of their incomes on vacations?

18. Find the number of families who spent more than 8% of their income on their vacations.

19. How many families spent less than 4% of their income on vacations?

20. Find the ratio of the number of families who spent less than 4% of their income on vacations to the number of families who spent more than 8%.

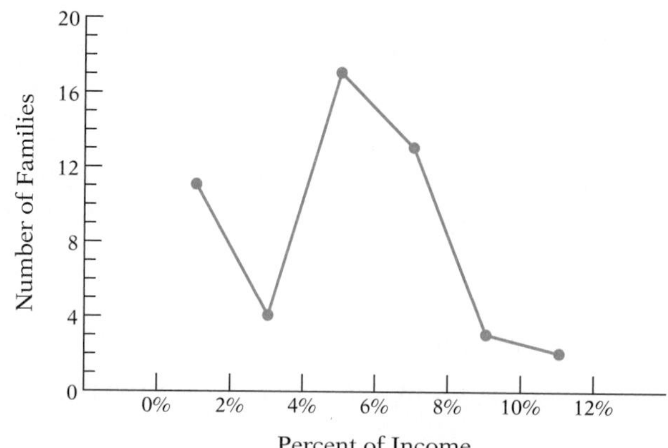

Fig. 35

Each member of a university's entering freshman class of 600 students was given a mathematics placement exam. The scores are recorded in Figure 36.

21. How many students scored between 600 and 700?

22. What is the ratio of the number of students scoring between 500 and 600 to the number of students tested?

23. How many students scored below 600?

24. How many students scored above 400?

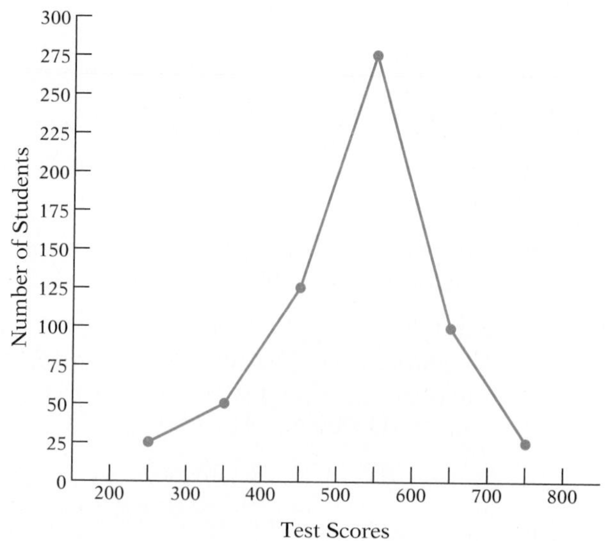

Fig. 36

SECTION 7.4 Means and Medians

Objective A To find the mean of a set of numbers

A student's test scores for five tests are listed below.

Test 1	*Test 2*	*Test 3*	*Test 4*	*Test 5*
86	95	94	87	93

The **average** or **mean** score is the sum of all the test scores divided by the number of tests:

$$\frac{\text{Sum of scores}}{\text{Number of tests}} = \frac{86 + 95 + 94 + 87 + 93}{5} = \frac{455}{5} = 91$$

Notice that a student who scored 91 on each of the five tests would have the same grade average as the student whose test scores are given in the example above.

Example 1

For a price comparison of six supermarkets, identical items were purchased in each store. The results are listed below:

Store	*Cost*	*Store*	*Cost*
1	\$40.74	4	\$39.05
2	\$39.45	5	\$38.86
3	\$38.57	6	\$39.25

Find the mean cost of the items purchased.

Strategy To find the mean cost of the items purchased:

- Find the sum of the costs.
- Divide the sum of the costs by the number of stores (6).

Solution

$$\begin{array}{r} \$40.74 \\ 39.45 \\ 38.57 \\ 39.05 \\ 38.86 \\ +39.25 \\ \hline \$235.92 \end{array} \text{ sum of costs} \qquad 6\overline{)\$235.92} = \$39.32$$

The mean cost of the items purchased was \$39.32.

Example 2

The amounts that a sales representative spent each month for gasoline are listed below:

Month	*Cost*	*Month*	*Cost*
Jan	\$118	Apr	\$141
Feb	\$130	May	\$134
Mar	\$109	June	\$136

Find the monthly mean cost for gasoline.

Your Strategy

Your Solution

Solution on p. A30

Objective B To find the median of a set of numbers

The ages of the presidents of nine corporations are 54, 38, 62, 45, 56, 60, 59, 39, 64.

The **median** age of the presidents is the middle age when the ages are arranged from smallest to largest.

To find the median, arrange the numbers from the smallest to largest. Locate the middle number.

38, 39, 45, 54, 56, 59, 60, 62, 64

4 numbers | middle number | 4 numbers

56 is the median.

The median separates a list of numbers so that there are the same number of values below the median as there are above the median.

A student received test scores of 89, 90, 58, 92, 84, and 96.

To find the median, arrange the numbers from the smallest to largest. When there are two middle numbers, the median is the average of those two numbers.

58, 84, 89, 90, 92, 96

2 numbers | middle numbers | 2 numbers

$$\frac{89 + 90}{2} = \frac{179}{2} = 89.5$$

89.5 is the median.

Example 3

In a neighborhood survey it was found that seven houses sold during the past year had been sold for \$90,000, \$87,000, \$86,000, \$92,000, \$88,000, \$91,500, and \$96,500. Find the median price of a house in this neighborhood.

Strategy

To find the median price, arrange the prices in order from smallest to largest. The median is the middle number.

Solution

\$86,000
\$87,000 } 3 numbers
\$88,000
\$90,000 ← middle number
\$91,500
\$92,000 } 3 numbers
\$96,500

The median price of a house in this neighborhood is \$90,000.

Example 4

A library's records show the number of books loaned on each day last week. Find the median number of books loaned.

Mon	Tue	Wed	Thu	Fri	Sat
450	375	420	500	480	490

Your strategy

Your solution

Solution on p. A30

7.4 EXERCISES

Objective A

1. You received grades of 75, 88, 94, 76, and 82 on five mathematics exams. Find your mean grade.

2. The closing prices of a utility stock for 5 days were $36.25, $35.75, $36.50, $36, and $37.20. Find the mean closing price of the stock.

3. The starting five players on a basketball team had the following heights: 70 inches, 85 inches, 80 inches, 74 inches, and 78 inches. Find the mean height of the basketball team in inches. (Round off to tenths of an inch.)

4. The prices of 1 pound of T-bone steak at six different grocery stores were $2.58, $2.62, $2.49, $2.75, $2.66, and $2.68. Find the mean price of T-bone steak at the six grocery stores.

5. The low temperatures in degrees Fahrenheit for 5 days at a ski resort were 28°, 21°, 18°, 9°, and 11°. Find the mean low temperature. (Round off to tenths of a degree.)

6. The eight everyday players on a baseball team hit 8, 15, 42, 18, 1, 24, 6, and 13 home runs during the season. Find the mean number of home runs the eight players hit.

7. A nurse's pay checks for 4 months are given in the table below. Find the mean monthly check.

Jan	*Feb*	*Mar*	*Apr*
$1200	$1350	$1190	$1220

8. A taxi driver's records in the table below show the number of gallons of gasoline purchased each day on the job last week. Find the mean number of gallons of gasoline purchased.

Wed	*Thu*	*Fri*	*Sat*	*Sun*
9.4	9.3	11.8	10.3	9.7

9. A test to determine the braking distance of a car yielded the following results: 220 feet, 208 feet, 216 feet, 219 feet, 227 feet. Find the mean braking distance of the car.

10. During the past year, six houses in a small town sold for the following prices: $75,400, $89,450, $116,295, $247,600, $82,500, and $176,300. Find the mean price of a house in this town.

11. The annual rainfall for each of 6 years is recorded in the table below. Find the mean annual rainfall.

Year	*Number of Inches*
1	16.2
2	19.3
3	17.4
4	16.5
5	20.1
6	18.5

12. The number of hours of television 10 families watched during 1 day is recorded in the table below. Find the mean number of hours of television the 10 families watched.

Family	*Hours*	*Family*	*Hours*
1	3.4	6	3.8
2	2.5	7	4.1
3	3.0	8	2.7
4	2.6	9	5.0
5	4.2	10	2.8

Objective B

13. The daily maximum temperature in degrees Fahrenheit for 7 days in the summer were 88°, 94°, 99°, 96°, 102°, 95°, and 87°. Find the median temperature for the week.

14. The prices of identical calculators at each of five department stores were \$38.75, \$44.50, \$42.95, \$41.00, and \$52.00. Find the median price of the calculator.

15. The times for running the seven most popular programs at a community college requires 17, 102, 32, 87, 56, 44, and 160 seconds. Find the median time for running the most popular programs.

16. The hourly wages for seven job classifications at a company are \$6.42, \$9.24, \$8.98, \$6.38, \$7.24, \$6.26, and \$7.16. Find the median hourly wage.

17. The ages of the seven most recently hired employees at a company are 26, 45, 22, 25, 24, 30, and 34. Find the median age.

18. The number of hours the custodian of an office building worked each day last week are shown at the right. Find the median number of hours worked.

Mon	*Tue*	*Wed*	*Thu*	*Fri*	*Sat*	*Sun*
$6\frac{3}{4}$	$7\frac{1}{2}$	$8\frac{1}{4}$	$7\frac{3}{4}$	$8\frac{1}{2}$	0	0

19. The number of tickets eight police officers gave out during 1 day are 17, 5, 9, 10, 23, 5, 13, and 20. Find the median number of tickets given out.

20. The number of requests for a conference room at a hotel during a 6-day period were 46, 18, 29, 48, 38, and 24. Find the median number of requests.

21. During a study of the amount of coffee a vending machine dispensed, the following amounts of coffee were obtained: 3.88 ounces, 4.07 ounces, 3.89 ounces, 4.15 ounces, 4.01 ounces, 3.92 ounces, 4.11 ounces, and 3.85 ounces. Find the median amount of coffee obtained from the vending machine.

22. The monthly utility bills for eight homes are \$86.48, \$92.81, \$48.92, \$74.16, \$112.53, \$61.92, \$86.48, and \$97.92. Find the median utility bill.

23. The populations of the ten largest cities in the world are shown below. Find the median population.

Buenos Aires: 11,600,000
Calcutta: 13,700,000
Bombay: 13,100,000
Mexico City: 22,000,000
New York: 15,700,000

Sao Paulo: 18,400,000
Seoul: 12,300,000
Shanghai: 13,400,000
Teheran: 11,300,000
Tokyo: 21,000,000

Calculators and Computers

Calculation of Grade Point Average

Grade point average (GPA) is an example of a mean. In this case the mean is a **weighted mean** because each course may have a different number of units. For example, courses that are 5 units have more weight on your GPA than 2-unit courses.

The program on the Math ACE disk to calculate your GPA uses a 4 point scale: $A = 4$, $B = 3$, $C = 2$, $D = 1$, and $F = 0$. The program will compute your GPA for as many as 7 courses.

To illustrate how the program computes your GPA, an example using 4 classes will be used.

Class	*Units*	*Grade*
Math	3	B
English	3	C
Biology	4	A
PE	2	A

First, multiply the number of units for each class by the grade points for that class. Second, add up these numbers. Third, divide by the total number of units.

Here is what it would look like:

$$\begin{aligned} \text{GPA} &= \frac{3 \cdot 3 + 3 \cdot 2 + 4 \cdot 4 + 2 \cdot 4}{12} \\ &= \frac{9 + 6 + 16 + 8}{12} \\ &= \frac{39}{12} \\ &= 3.25 \end{aligned}$$

Chapter Summary

Key Words *Statistics* is the branch of mathematics concerned with *data,* or numerical information.

The *circle graph* represents data by the size of the sectors.

The *bar graph* represents data by the height of the bars.

The *broken-line graph* represents data by the position of the lines and shows trends and comparisons.

A *histogram* is a special kind of bar graph.

In a histogram, the width of each bar corresponds to a range of numbers called a *class interval.*

In a histogram, the height of each bar corresponds to the number of occurrences of data in each class interval and is called the *class frequency.*

A *frequency polygon* is a graph that displays information in a manner similar to a histogram. A dot is placed above the center of each class interval at a height corresponding to that class' frequency.

The *median* separates a list of numbers so that there are the same number of values below the median as there are above the median.

The *average* or *mean* score is the sum of all the test scores divided by the number of tests.

Essential Rules ***To Find the Average or Mean***

To find the average or mean of a set of numbers, divide the sum of the numbers by the number of addends.

$$\textbf{Mean} = \frac{\textbf{Sum of Numbers}}{\textbf{Number of Addends}}$$

To Find the Median

To find the median, arrange the numbers from smallest to largest and locate the middle number. When there are two middle numbers, the median is the average of those two numbers.

Chapter Review

SECTION 7.1

The pictograph in Figure 37 shows the number of students earning A, B, C, D, and F grades in a geology class. Each figure represents two students.

1. Find the total number of students receiving grades.
2. Find the ratio of the number of students receiving B grades to the number of students receiving D grades.
3. Find the percent of the total number of students receiving C grades. Round to the nearest tenth of a percent.

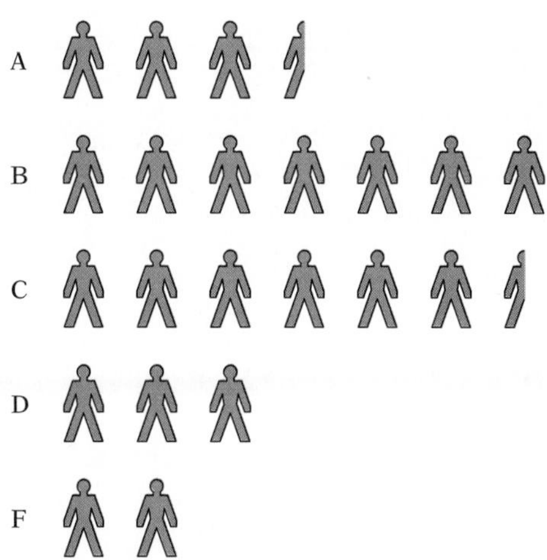

Fig. 37 Number of students receiving grades.

The circle graph in Figure 38 shows the income received from all the major league baseball teams.

4. Find the baseball teams' total income.
5. What is the ratio of the income received from tickets sold at the gate to the income received from local broadcasting?
6. Find the percent of the total income received from national broadcasting. Round to the nearest tenth of a percent.

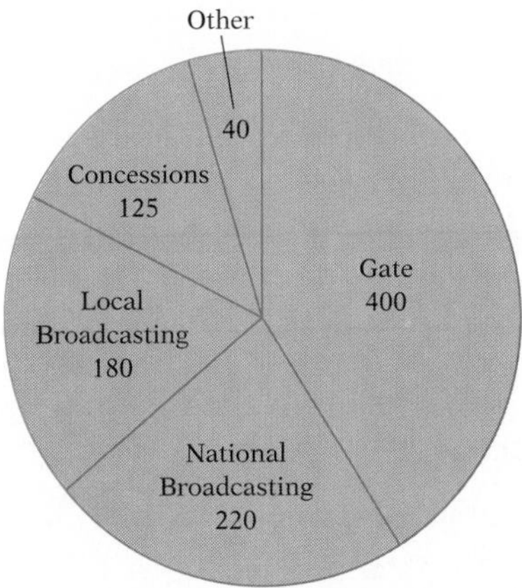

Fig. 38 Income for major league baseball (in millions of dollars).

SECTION 7.2

The double-bar graph in Figure 39 shows the maximum and minimum temperatures during 5 days.

7. Find the difference in the maximum and minimum temperatures for Tuesday.
8. Find the ratio of the maximum to minimum temperatures for Friday.
9. Which day had the lowest temperature? What was this temperature?

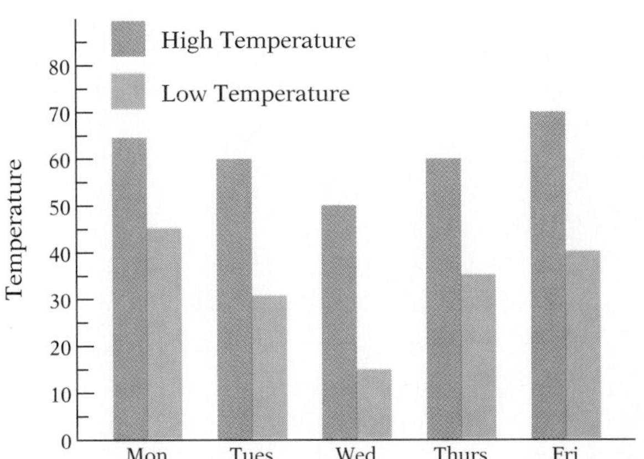

Fig. 39

The double-broken-line graph in Figure 40 shows the profit for a corporation during the four quarters of 1989 and 1990.

10. Find the difference in profit for the second quarters of 1989 and 1990.

11. Find the total profit made in the third and fourth quarters of 1990.

12. Find the ratio of the profit made in the third quarter of 1989 to the profit made in the third quarter of 1990.

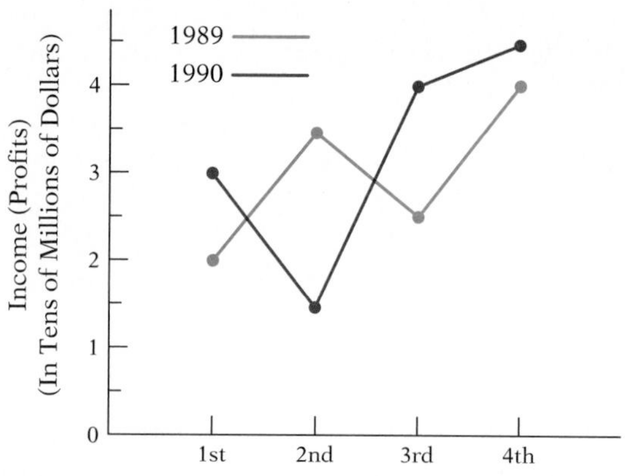

Fig. 40

SECTION 7.3

The histogram in Figure 41 shows the heights of 41 trees in a nursery.

13. How many of the trees were over 72 inches tall?

14. Find the ratio of the number of trees under 66 inches to the number of trees that were between 69 to 72 inches tall.

15. Find the percent of the trees that had a height between 66 to 69 inches.

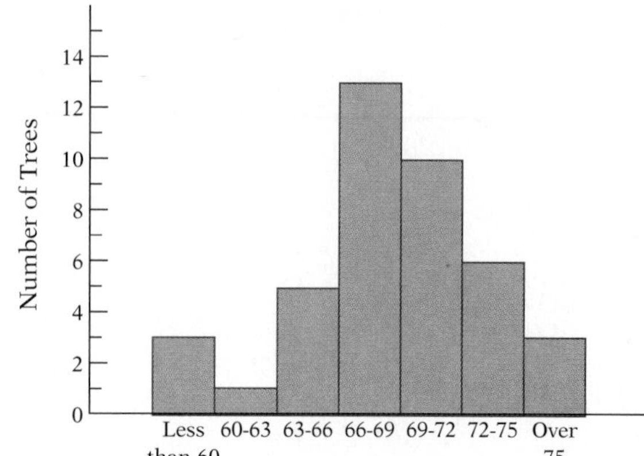

Fig. 41

The frequency polygon in Figure 42 shows the range of the scores of a team in the national basketball association.

16. Find the number of games in which fewer than 100 points were scored.

17. Find the ratio of the number of games in which 90 to 100 points were scored to the number in which 110 to 120 points were scored.

18. Find the percent of the total number of games in which over 120 points were scored. (Round to the nearest tenth of a percent.)

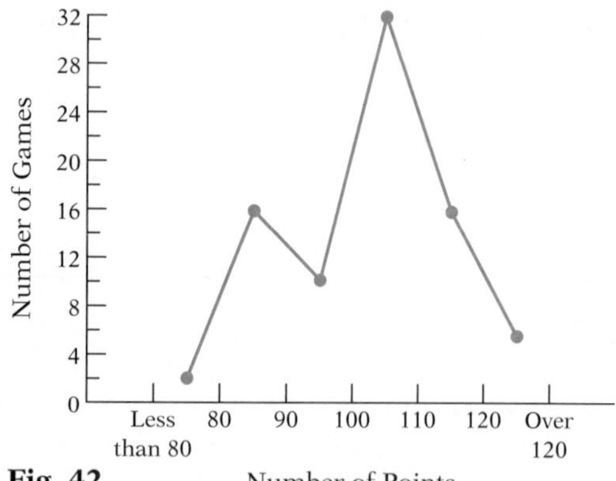

Fig. 42 Number of Points

SECTION 7.4

19. You received grades of 67, 87, 98, 83, and 77 on five history exams. What is your mean score?

20. The number of students in eight classes of an elementary school are 23, 29, 25, 31, 27, 26, 19, and 28. Find the median number of students in the eight classes.

Chapter Test

The pictograph in Figure 43 shows the number of students receiving grades in a geology class. Each picture in the pictograph represents two students.

1. Find the total number of students in the geology class.
2. Find the ratio of the number of students receiving an A grade to the number of students receiving a D grade.
3. Find the percent of the total number of students who received a B grade. Round to the nearest tenth of a percent.

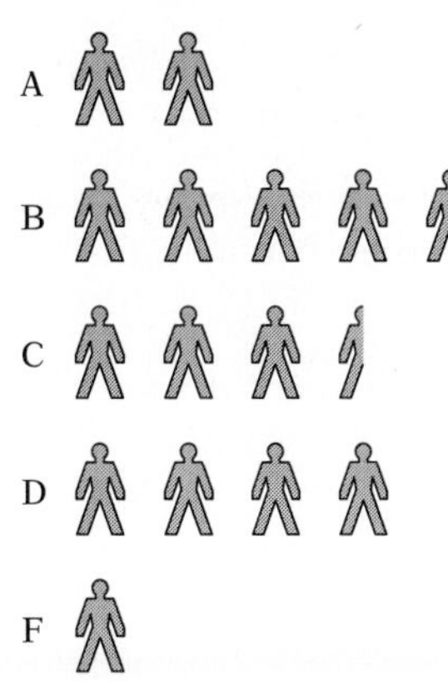

Fig. 43 Number of students receiving grades.

The circle graph in Figure 44 shows sources of income for a community college that has a total budget of $32,000,000.

4. Find the ratio of the amount of federal funds to the amount in the total budget.
5. Find the percent of the total budget that comes from state funds.
6. What is the ratio of the amount of state funds to the amount of federal funds?

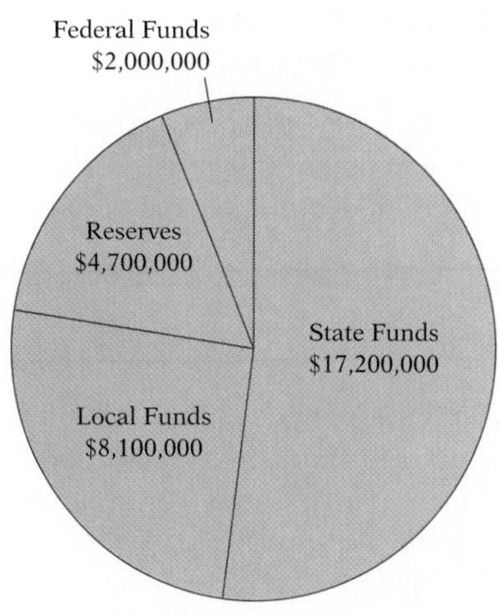

Fig. 44 Sources of income for college budget of $32,000,000.

The double bar graph in Figure 45 shows the number of economy cars an automobile manufacturer sold during the last 6 months of 1990 and 1991.

7. Find the number of cars sold in November 1990.
8. Find the difference between the number of cars sold in October 1990 and the number of cars sold in October 1991.
9. What is the difference between the first 3 months' sales for 1990 and 1991?

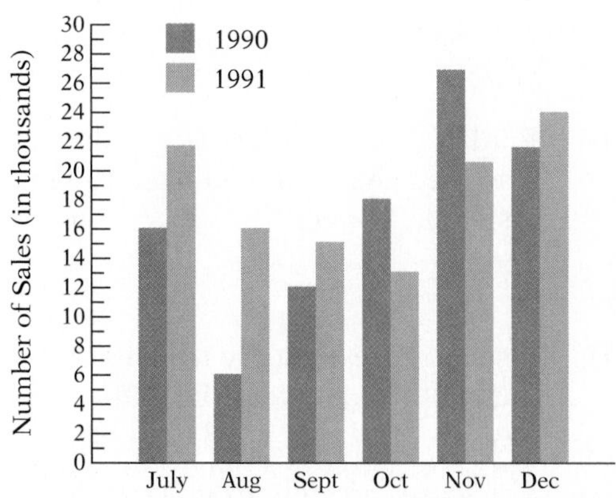

Fig. 45

The double-broken-line graph in Figure 46 shows the quarterly income for a company for the years 1990 and 1991.

10. What is the company's third quarter income for 1991?

11. Find the ratio of the second quarter income for 1991 to the second quarter income for 1990.

12. Find the difference between the company's fourth quarter incomes for 1990 and 1991.

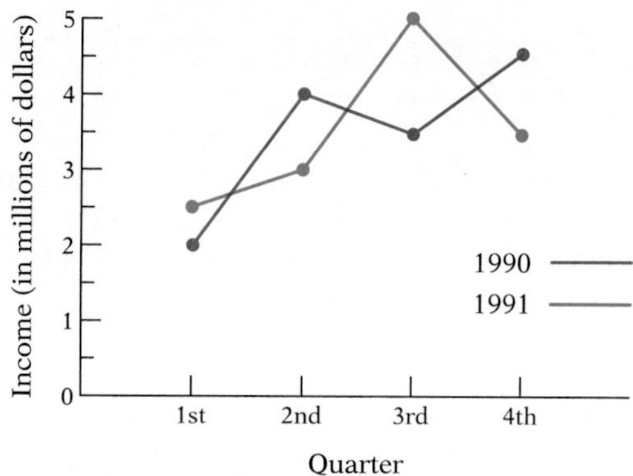

Fig. 46

The histogram in Figure 47 shows the salaries for 51 employees in a small computer company.

13. Find the number of employees that receive a salary over $25,000.

14. Find the ratio of the number of employees whose salary is above $30,000 to the total number of employees.

15. Find the percent of the total number of employees whose salaries are between $20,000 and $30,000. Round to the nearest percent.

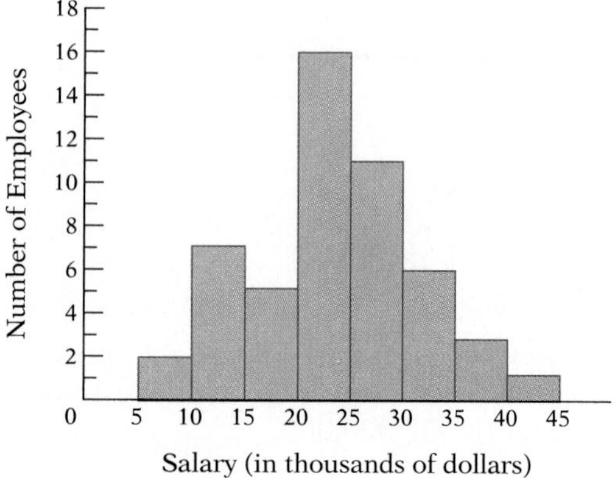

Fig. 47

A television rating service surveyed 100 families to find the number of hours they watched television. The results are recorded in a frequency polygon as shown in Figure 48.

16. How many families watched between 15 and 25 hours of television a week?

17. Find the ratio of the number of families that watched between 5 and 10 hours of television each week to the number of families that watched between 20 and 25 hours of television.

18. Find the percent of the number of families surveyed that watched over 15 hours of television each week.

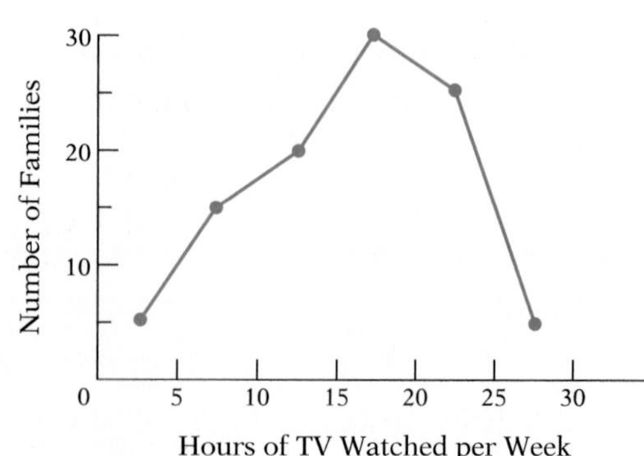

Fig. 48

19. A sales representative traveled 125, 212, 188, 231, and 189 miles for 5 days during the past week. Find the mean number of miles driven during the 5 days.

20. The scores on eight exams for obtaining an insurance license are 68, 76, 98, 88, 73, 56, 78, and 83. Find the median score.

Cumulative Review

1. Simplify $2^2 \cdot 3^3 \cdot 5$.

2. Simplify $3^2 \cdot (5 - 2) \div 3 + 5$.

3. Find the LCM of 24 and 40.

4. Write $\frac{60}{144}$ in simplest form.

5. Add: $4\frac{1}{2} + 2\frac{3}{8} + 5\frac{3}{15}$

6. Subtract: $12\frac{5}{8} - 7\frac{11}{12}$

7. Multiply: $\frac{5}{8} \times 3\frac{1}{5}$

8. Divide: $3\frac{1}{5} \div 4\frac{1}{4}$

9. Simplify $\frac{5}{8} \div \left(\frac{3}{4} - \frac{2}{3}\right) + \frac{3}{4}$.

10. Write two hundred nine and three hundred five thousandths in standard form.

11. Multiply: 4.092×0.69

12. Convert $16\frac{2}{3}$ to a decimal. Round to the nearest hundredth.

13. Write "330 miles on 12.5 gallons of gas" as a unit rate.

14. Solve the proportion. $\frac{n}{5} = \frac{16}{25}$

15. Write $\frac{4}{5}$ as a percent.

16. 8 is 10% of what?

17. What is 38% of 43? Round to the nearest hundredth.

18. What percent of 75 is 30?

19. A salesperson at a department store receives \$100 per week plus 2% commission on sales. Find the income for the week in which the salesperson had \$27,500 in sales.

20. A life insurance policy costs \$4.15 for every \$1000 of insurance. At this rate, what is the cost for \$50,000 of life insurance?

21. A contractor borrowed \$125,000 for 6 months at an annual simple interest rate of 11%. Find the interest due on the loan.

22. A compact disc player with a cost of \$180 is sold for \$279. Find the markup rate.

23. The circle graph in Figure 49 shows how a family's monthly income of \$3000 is budgeted. How much is budgeted for food?

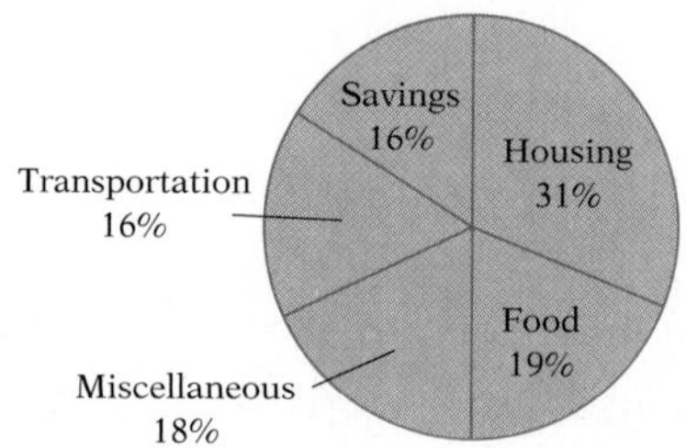

Fig. 49 Budget for a monthly income of \$3000.

24. The double-broken-line graph in Figure 50 shows two student's scores on five math tests of 30 problems each. Find the difference between the number of problems the two students answered correctly on Test 1.

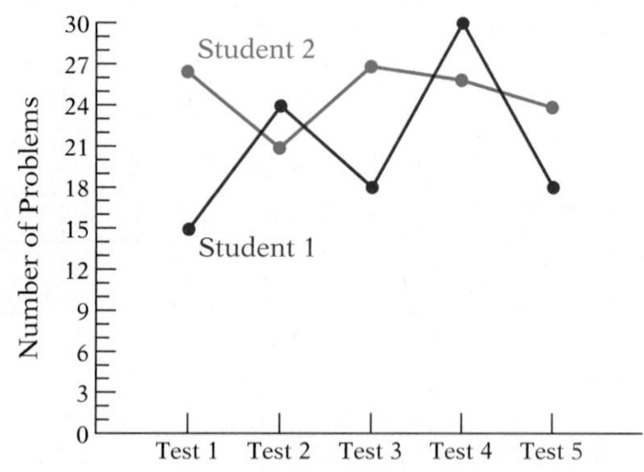

Fig. 50

25. The average high temperatures for a week in a mountain town were 56°, 72°, 80°, 75°, 68°, 62°, and 74°. Find the mean high temperature for the week. (Round to the nearest tenth of a degree.)

26. The salaries for six teachers in a small school are \$17,000, \$21,000, \$28,500, \$22,300, \$20,200, and \$25,000. Find the median salary.

8 U.S. Customary Units of Measurement

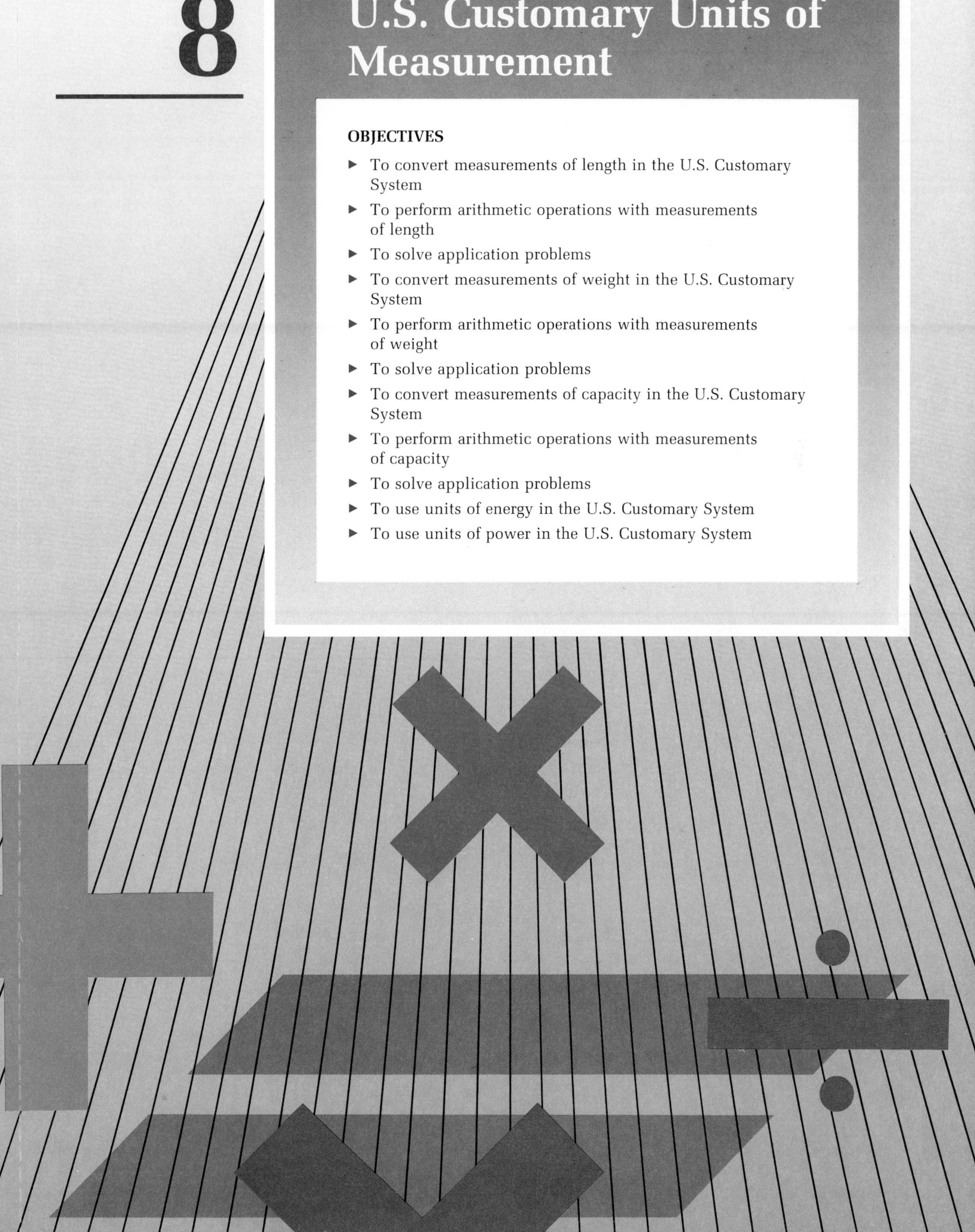

OBJECTIVES

- To convert measurements of length in the U.S. Customary System
- To perform arithmetic operations with measurements of length
- To solve application problems
- To convert measurements of weight in the U.S. Customary System
- To perform arithmetic operations with measurements of weight
- To solve application problems
- To convert measurements of capacity in the U.S. Customary System
- To perform arithmetic operations with measurements of capacity
- To solve application problems
- To use units of energy in the U.S. Customary System
- To use units of power in the U.S. Customary System

Light Years and Astronomical Units

The measurement of a short distance between two objects normally can be made by using some kind of ruler or tape measure. For example, the distance across a room could be found with a tape measure. This distance most likely would be given in feet.

On the other hand, a tape measure would be quite useless for finding the distance between Chicago and New York City. Even measuring that distance in feet would not be practical. Instead, an odometer is used, and the results are recorded in miles. It is approximately 1700 miles between New York City and Chicago.

But even miles become impractical for measuring the distance between two objects that are separated by a large distance. For example, the distance between the earth and Alpha Centauri, a relatively nearby star, is approximately 25 trillion miles, which is 25,000,000,000,000 miles.

Astronomers have units of measurement that are useful for measuring the vast difference in space. A unit that is used frequently is the light year. One light year is the distance light could travel in 1 year. Because light travels at 186,000 miles per second, in 1 year, light could travel about 5,880,000,000,000 miles, which is close to 6 trillion miles.

Using this measure, the distance between the earth and Alpha Centauri is approximately 4.2 light years, a more convenient number for measuring that distance. On the other hand, the distance between Chicago and New York is 0.0000000029 light years—not a very convenient measure.

Another measure astronomers use is the astronomical unit (AU). 1 AU is approximately 92,900,000 miles and is the approximate distance of the earth from the sun. Using this measure, Pluto, the farthest planet from the sun, is 39.4 AU from the sun.

SECTION 8.1 Length

Objective A

To convert measurements of length in the U.S. Customary System

A **measurement** includes a number and a unit.

number	unit
3	feet
7	miles
12	yards

Standard units of measurement have been established to simplify trade and commerce.

The unit of length, or distance, called the **yard,** was originally defined as the length of a specified bronze bar located in London.

The standard U.S. Customary System units of length are **inch, foot, yard,** and **mile.** Equivalences between units of length in the U.S. Customary System are:

12 inches (in.) = 1 foot (ft)
3 ft = 1 yard (yd)
36 in. = 1 yard (yd)
5280 ft = 1 mile (mi)

These equivalences can be used to form conversion rates to change one unit of measurement to another. For example, because 3 ft = 1 yd, the following conversion rates, equivalent to 1, can be written:

$$\frac{3 \text{ ft}}{1 \text{ yd}} = \frac{1 \text{ yd}}{3 \text{ ft}} = 1$$

Convert 27 ft to yards.

$$27 \text{ ft} = 27 \text{ ft} \times \boxed{\frac{1 \text{ yd}}{3 \text{ ft}}}$$

$$= 27 \cancel{\text{ft}} \times \frac{1 \text{ yd}}{3 \cancel{\text{ft}}}$$

$$= \frac{27 \text{ yd}}{3}$$

$$= 9 \text{ yd}$$

Convert 5 yd to feet.

$$5 \text{ yd} = 5 \text{ yd} \times \boxed{\frac{3 \text{ ft}}{1 \text{ yd}}}$$

$$= 5 \cancel{\text{yd}} \times \frac{3 \text{ ft}}{1 \cancel{\text{yd}}}$$

$$= \frac{15 \text{ ft}}{1}$$

$$= 15 \text{ ft}$$

Notice that in the conversion rate chosen, the unit in the numerator is the same as the unit desired in the answer.

The unit in the denominator is the same as the unit in the given measurement. Therefore the units in the denominator and in the given measurement can be eliminated.

Example 1 Convert 40 in. to feet.

Solution $40 \text{ in.} = 40 \cancel{\text{in}} \times \frac{1 \text{ ft}}{12 \cancel{\text{in}}} = 3\frac{1}{3} \text{ ft}$

Example 2 Convert 60 in. to feet.

Your solution

Example 3 Convert 36 ft to yards.

Solution $36 \text{ ft} = 36 \cancel{\text{ft}} \times \frac{1 \text{ yd}}{3 \cancel{\text{ft}}} = 12 \text{ yd}$

Example 4 Convert 14 ft to yards.

Your solution

Example 5 Convert $3\frac{1}{4}$ yd to feet.

Solution $3\frac{1}{4} \text{ yd} = 3\frac{1}{4} \cancel{\text{yd}} \times \frac{3 \text{ ft}}{1 \cancel{\text{yd}}} = 9\frac{3}{4} \text{ ft}$

Example 6 Convert 9800 ft to miles.

Your solution

Solutions on p. A30

Objective B To perform arithmetic operations with measurements of length

When performing arithmetic operations with measurements of length, write the answer in simplest form. For example, 1 ft 14 in. should be written as 2 ft 2 in.

Convert: 50 in. = ____ ft ____ in.

Because 12 in. = 1 ft, divide 50 in. by 12. The whole number part of the quotient is the number of feet. The remainder is the number of inches.

$$\begin{array}{r} 4 \text{ ft } 2 \text{ in.} \\ 12\overline{)\ 50} \\ -48 \\ \hline 2 \end{array}$$

50 in. = 4 ft 2 in.

Example 7 Convert:
17 in. = ____ ft ____ in.

Solution

$$\begin{array}{r} 1 \text{ ft } 5 \text{ in} \\ 12\overline{)\ 17} \\ -12 \\ \hline 5 \end{array}$$

17 in. = 1 ft 5 in.

Example 8 Convert:
42 in. = ____ ft ____ in.

Your solution

Example 9 Convert:
31 ft = ____ yd ____ ft

Solution

$$\begin{array}{r} 10 \text{ yd } 1 \text{ ft} \\ 3\overline{)\ 31} \\ -30 \\ \hline 1 \end{array}$$

31 ft = 10 yd 1 ft

Example 10 Convert:
14 ft = ____ yd ____ ft

Your solution

Solutions on p. A30

Example 11 Add: 4 ft 4 in. + 1 ft 11 in.

Solution

$$\begin{array}{r} 4\text{ ft }\ 4\text{ in.} \\ \underline{+1\text{ ft }11\text{ in.}} \\ 5\text{ ft }15\text{ in.} \end{array} = 6\text{ ft }3\text{ in.}$$

Example 12 Add: 3 ft 5 in. + 4 ft 9 in.

Your solution

Example 13 Subtract: 9 ft 6 in. − 3 ft 8 in.

Solution

$$\begin{array}{r} \overset{8\text{ ft.}}{\cancel{9\text{ ft}}}\ \ \overset{18\text{ in.}}{\cancel{6\text{ in.}}} \\ \underline{-3\text{ ft }\ \ 8\text{ in.}} \\ 5\text{ ft }10\text{ in.} \end{array}$$

Borrow 1 ft (12 in.) from 9 ft and add to 6 in.

Example 14 Subtract: 4 ft 2 in. − 1 ft 8 in.

Your solution

Example 15 Multiply: 3 yd 2 ft × 4

Solution

$$\begin{array}{r} 3\text{ yd }2\text{ ft} \\ \underline{\times \qquad 4} \\ 12\text{ yd }8\text{ ft} \end{array} = 14\text{ yd }2\text{ ft}$$

Example 16 Multiply: 4 yd 1 ft × 8

Your solution

Example 17 Divide: 4 ft 3 in. ÷ 3

Solution

$$\begin{array}{rrr} & 1\text{ ft} & 5\text{ in.} \\ 3\,) & 4\text{ ft} & 3\text{ in.} \\ & \underline{-3\text{ ft}} & \\ & 1\text{ ft} = & \underline{12\text{ in.}} \\ & & 15\text{ in.} \\ & & \underline{-15\text{ in.}} \\ & & 0 \end{array}$$

Example 18 Divide: 7 yd 1 ft ÷ 2

Your solution

Example 19 Multiply: $2\frac{3}{4}$ ft × 3

Solution

$$\begin{aligned} 2\frac{3}{4}\text{ ft} \times 3 &= \frac{11}{4}\text{ ft} \times 3 \\ &= \frac{33}{4}\text{ft} \\ &= 8\frac{1}{4}\text{ ft} \end{aligned}$$

Example 20 Subtract: $6\frac{1}{4}$ ft − $3\frac{2}{3}$ ft

Your solution

Solutions on p. A30

Objective C To solve application problems

Example 21

One shrub extends 15 ft along a freeway. How many shrubs are to be planted along 6450 ft of the freeway?

Strategy

To find the number of shrubs to be planted along the freeway, divide the total distance (6450 ft) by the length of one shrub (15 ft).

Solution

$$\frac{6450 \text{ ft}}{15 \text{ ft}} = 430$$

430 shrubs are to be planted along the freeway.

Example 22

The floor of a storage room is being tiled. Eight tiles, each a 9-inch square, fit across the width of the floor. Find the width in feet of the storage room.

Your strategy

Your solution

Example 23

A plumber used 3 ft 9 in., 2 ft 6 in., and 11 in. of copper tubing to install a sink. Find the total length of copper tubing used.

Strategy

To find the total length of copper tubing used, add the three lengths of copper tubing (3 ft 9 in., 2 ft 6 in., and 11 in.).

Solution

$$\begin{array}{rr} & 3 \text{ ft} \;\; 9 \text{ in.} \\ & 2 \text{ ft} \;\; 6 \text{ in.} \\ + & 11 \text{ in.} \\ \hline & 5 \text{ ft } 26 \text{ in.} \end{array} = 7 \text{ ft } 2 \text{ in.}$$

The plumber used 7 ft 2 in. of copper tubing.

Example 24

A board 9 ft 8 in. is cut into four pieces of equal length. How long is each piece?

Your strategy

Your solution

Solutions on p. A31

8.1 EXERCISES

Objective A

Convert.

1. 6 ft = ______ in.
2. 9 ft = ______ in.
3. 30 in. = ______ ft
4. 64 in. = ______ ft
5. 13 yd = ______ ft
6. $4\frac{1}{2}$ yd = ______ ft
7. 16 ft = ______ yd
8. $4\frac{1}{2}$ ft = ______ yd
9. $2\frac{1}{3}$ yd = ______ in.
10. 5 yd = ______ in.
11. 120 in. = ______ yd
12. 66 in. = ______ yd
13. 2 mi = ______ ft
14. $1\frac{1}{2}$ mi = ______ ft
15. $7\frac{1}{2}$ in. = ______ ft
16. $2\frac{1}{4}$ ft = ______ in.
17. $4\frac{3}{4}$ ft = ______ in.
18. $5\frac{1}{3}$ yd = ______ ft

Objective B

Perform the arithmetic operation.

19. 100 in. = ____ ft ____ in.
20. 6400 ft = ____ mi ____ ft
21. 15 in. = ____ ft ____ in.
22. 6 ft 7 in. + 3 ft 4 in.
23. 9 ft 11 in. + 3 ft 6 in.
24. 5 ft 3 in. − 2 ft 6 in.
25. 9 yd 1 ft − 3 yd 2 ft
26. 2 ft 5 in. × 6
27. $3\frac{2}{3}$ ft × 4
28. $2\overline{)5 \text{ ft } 4 \text{ in.}}$
29. $12\frac{1}{2}$ in. ÷ 3
30. $4\frac{2}{3}$ ft + $6\frac{1}{2}$ ft

Perform the arithmetic operations.

31. 3 yd 2 ft
+6 yd 2 ft

32. 1 mi 4200 ft
+2 mi 3600 ft

33. 5 yd 1 ft
−2 yd 2 ft

34. 12 ft 6 in.
− 7 ft 9 in.

35. $3\frac{5}{8}$ yd − $1\frac{3}{4}$ yd

36. 3 ft 7 in.
× 4

Objective C *Application Problems*

37. A kitchen counter is to be covered with tile that is 4 inches square. How many tiles can be placed along one row of a counter top that is 4 ft 8 in. long?

38. Thirty two yards of material were used for making pleated draperies. How many feet of material were used?

39. Find the missing dimension.

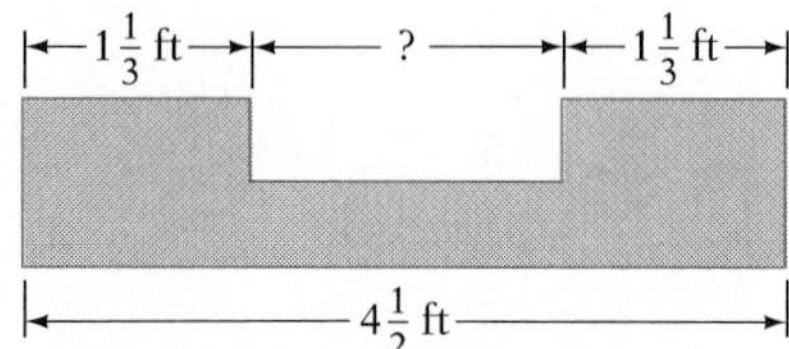

40. Find the total length of the shaft.

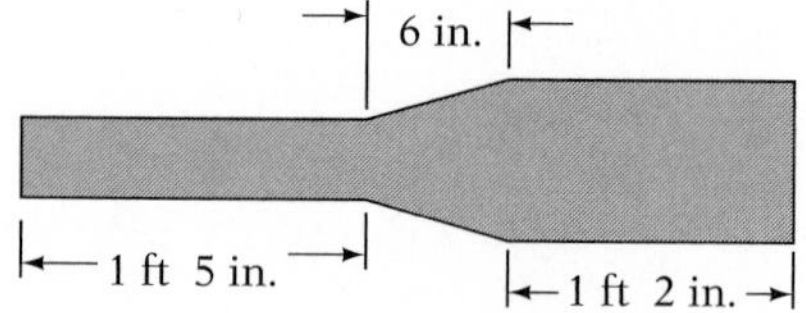

41. What length of material is needed to drill two holes 3 in. in diameter and leave $\frac{1}{2}$ inch between the holes and on either side as shown in the diagram?

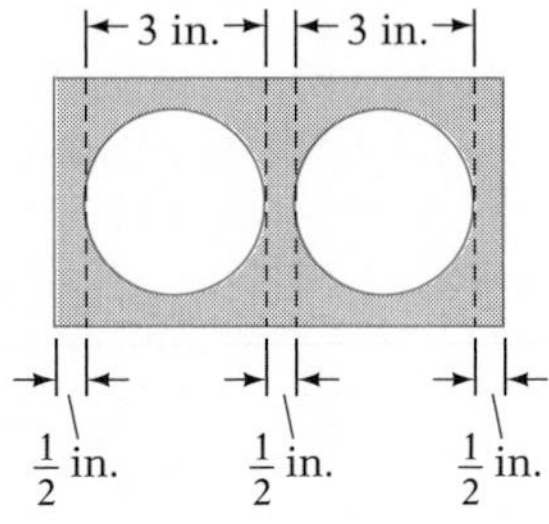

42. Find the missing dimension in the figure.

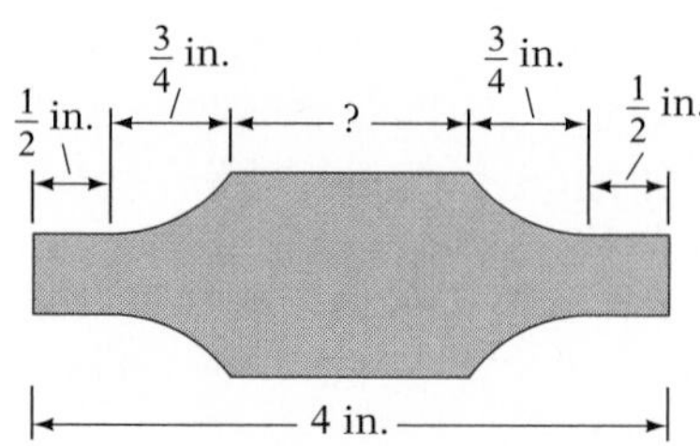

43. A board $6\frac{2}{3}$ ft is cut into four equal pieces. How long is each piece?

44. How long must a board be if four pieces, each 3 ft 4 in. long, are to be cut from it?

45. A picture is 1 ft 9 in. high and 1 ft 6 in. wide. Find the length of framing needed to frame the picture.

46. You bought 32 feet of baseboard to install in the kitchen of your house. How many inches of baseboard did you purchase?

47. Forty-five bricks, each 9 in. long, are laid end to end to make the base for a wall. Find the length of the wall in feet.

48. A roof is constructed with nine rafters, each rafter 8 ft 4 in. long. Find the total number of feet of material needed to build the rafters.

SECTION 8.2 Weight

Objective A

To convert measurements of weight in the U.S. Customary System

Weight is a measure of how strongly the earth is pulling on an object. The unit of weight called the **pound** is defined as the weight of a standard solid kept at the Bureau of Standards in Washington, D.C. The U.S. Customary System units of weight are **ounce, pound,** and **ton.**

Equivalences between units of weight in the U.S. Customary System are:

$$16 \text{ ounces (oz)} = 1 \text{ pound (lb)}$$
$$2000 \text{ lb} = 1 \text{ ton}$$

These equivalences can be used to form conversion rates to change one unit of measurement to another. For example, because 16 oz = 1 lb, the following conversion rates equivalent to 1 can be written:

$$\frac{16 \text{ oz}}{1 \text{ lb}} = \frac{1 \text{ lb}}{16 \text{ oz}} = 1$$

Convert 62 oz to pounds

$$62 \text{ oz} = 62 \text{ oz} \times \boxed{\frac{1 \text{ lb}}{16 \text{ oz}}}$$
$$= \frac{62 \not{\text{oz}}}{1} \times \frac{1 \text{ lb}}{16 \not{\text{oz}}}$$
$$= \frac{62 \text{ lb}}{16}$$
$$= 3\frac{7}{8} \text{ lb}$$

The conversion rate must contain lb (the unit desired in the answer) in the numerator and oz (the original unit) in the denominator.

Example 1 Convert $3\frac{1}{2}$ tons to pounds.

Solution $3\frac{1}{2} \text{ tons} = 3\frac{1}{2} \not{\text{tons}} \times \frac{2000 \text{ lb}}{1 \not{\text{ton}}}$

$= 7000 \text{ lb}$

Example 2 Convert 3 lb to ounces.

Your solution

Example 3 Convert 42 oz to pounds.

Solution $42 \text{ oz} = 42 \not{\text{oz}} \times \frac{1 \text{ lb}}{16 \not{\text{oz}}}$

$= \frac{42 \text{ lb}}{16} = 2\frac{5}{8} \text{ lb}$

Example 4 Convert 4200 lb to tons.

Your solution

Solutions on p. A31

Objective B To perform arithmetic operations with measurements of weight

When performing arithmetic operations with measurements of weight, write the answer in simplest form. For example, 1 lb 22 oz should be written 2 lb 6 oz.

Example 5 Subtract: 14 lb 5 oz − 8 lb 14 oz

Solution

```
  13 lb 21 oz
  14 lb  5 oz
−  8 lb 14 oz
   5 lb  7 oz
```

Borrow 1 lb (16 oz) from 14 lb and add to 5 oz.

Example 6 Subtract: 7 lb 1 oz − 3 lb 4 oz

Your solution

Example 7 Divide: 7 lb 14 oz ÷ 3

Solution

```
    2 lb   10 oz
3)  7 lb   14 oz
   −6 lb
    1 lb = 16 oz
           30 oz
          −30 oz
            0
```

Example 8 Multiply: 3 lb 6 oz × 4

Your solution

Solutions on p. A31

Objective C To solve application problems

Example 9

Four teachers spent their summer vacation panning for gold. How much money did each teacher receive if they found 1 lb 9 oz of gold and the price of gold was $525.80 per ounce?

Strategy To find the amount of money each teacher received:

- Convert 1 lb 9 oz to ounces.
- Multiply the number of ounces by the price per ounce ($525.80) to find the total income.
- Divide the total income by the number of teachers (4).

Solution 1 lb 9 oz = 25 oz

```
  $525.80           $3,286.25
×      25        4)$13,145.00
$13,145.00 total income
```

Each teacher received $3286.25.

Example 10

Find the weight in pounds of 12 bars of soap. Each bar weighs 9 oz.

Your strategy

Your solution

Solution on p. A31

8.2 EXERCISES

Objective A

Convert.

1. 64 oz = ______ lb
2. 36 oz = ______ lb
3. 4 lb = ______ oz
4. 7 lb = ______ oz
5. 3200 lb = ______ tons
6. 7000 lb = ______ tons
7. 6 tons = ______ lb
8. $1\frac{1}{4}$ tons = ______ lb
9. 66 oz = ______ lb
10. 90 oz = ______ lb
11. $1\frac{1}{2}$ lb = ______ oz
12. $2\frac{5}{8}$ lb = ______ oz
13. $1\frac{3}{10}$ tons = ______ lb
14. $\frac{4}{5}$ tons = ______ lb
15. 500 lb = ______ tons
16. 5000 lb = ______ tons
17. 180 oz = ______ lb
18. 12 oz = ______ lb

Objective B

Perform the arithmetic operation.

19. 9000 lb = ______ tons ______ lb
20. 85 oz = ______ lb ______ oz
21. 40 oz = ______ lb ______ oz
22. 4 lb 7 oz + 3 lb 12 oz
23. 1 ton 800 lb + 3 tons 1600 lb
24. 7 lb 5 oz − 3 lb 8 oz
25. 3 tons 500 lb − 1 ton 800 lb
26. 3 lb 6 oz × 4
27. $5\frac{1}{2}$ lb × 6
28. 2)3 lb 8 oz
29. $4\frac{2}{3}$ lb × 3
30. 7 lb 7 oz + 6 lb 9 oz

Perform the arithmetic operation.

31. $6\frac{1}{2}$ oz
$+2\frac{1}{2}$ oz

32. $6\frac{3}{8}$ lb
$-2\frac{5}{6}$ lb

33. 5 lb 12 oz ÷ 4

Objective C *Application Problems*

34. A machinist has 25 iron rods to mill. Each rod weighs 20 oz. Find the total weight of the rods in pounds.

35. A fireplace brick weighs 2 1/2 lb. What is the weight of a load of 800 bricks?

36. A college bookstore received 1200 textbooks, each book weighing 9 oz. Find the total weight of the 1200 textbooks in pounds.

37. A 4 × 4-in. tile weighs 7 oz. Find the weight in pounds of a package of 144 tiles.

38. A farmer orders 20 tons of feed for 100 cattle. After 15 days, the farmer has 5 tons of feed left. How many pounds of food has each cow eaten per day?

39. A case of soft drink contains 24 cans, each weighing 6 oz. Find the weight, in pounds, of the case of soft drink.

40. A baby weighed 7 lb 8 oz at birth. At 6 months of age, the baby weighed 15 lb 13 oz. Find the baby's increase in weight during the 3 months.

41. Shampoo weighing 5 lb 4 oz is divided equally and poured into four containers. How much shampoo is in each container?

42. A steel rod weighing 16 lb 11 oz is cut into three pieces. Find the weight of each piece of steel rod.

43. Find the cost of a ham roast weighing 5 lb 10 oz if the price per pound is $2.40.

44. A candy store buys candy weighing 12 lb for $14.40. The candy is repackaged and sold in 6-oz packages for $1.15 each. Find the markup on the 12 lb of candy.

45. A manuscript weighing 2 lb 3 oz is mailed at the postage rate of $0.25 per ounce. Find the cost of mailing the manuscript.

SECTION 8.3 Capacity

Objective A To convert measurements of capacity in the U.S. Customary System

Liquid substances are measured in units of **capacity.** The standard U.S. Customary units of capacity are the **fluid ounce, cup, pint, quart,** and **gallon.** Equivalences between units of capacity in the U.S. Customary System are:

8 fluid ounces (fl oz) = 1 cup (c)
2 c = 1 pint (pt)
2 pt = 1 quart (qt)
4 qt = 1 gallon (gal)

These equivalences can be used to form conversion rates to change one unit of measurement to another. For example, because 8 fl oz = 1 c, the following conversion rates equivalent to 1 can be written:

$$\frac{8 \text{ fl oz}}{1 \text{ c}} = \frac{1 \text{ c}}{8 \text{ fl oz}} = 1$$

Convert 36 fl oz to cups

$$36 \text{ fl oz} = 36 \text{ fl oz} \times \boxed{\frac{1 \text{ c}}{8 \text{ fl oz}}}$$

$$= \frac{36 \cancel{\text{fl oz}}}{1} \times \frac{1 \text{ c}}{8 \cancel{\text{fl oz}}}$$

$$= \frac{36 \text{ c}}{8}$$

$$= 4\frac{1}{2}\text{c}$$

The conversion rate must contain c in the numerator and fl oz in the denominator.

Convert 3 qt to cups.

$$3 \text{ qt} = 3 \text{ qt} \times \boxed{\frac{2 \text{ pt}}{1 \text{ qt}}} \times \boxed{\frac{2 \text{ c}}{1 \text{ pt}}}$$

$$= \frac{3 \cancel{\text{qt}}}{1} \times \frac{2 \cancel{\text{pt}}}{1 \cancel{\text{qt}}} \times \frac{2 \text{ c}}{1 \cancel{\text{pt}}}$$

$$= \frac{12 \text{ c}}{1}$$

$$= 12 \text{ c}$$

The direct equivalence is not given above. Use two conversion rates. First convert quarts to pints and then convert pints to cups. The unit in the denominator of the second conversion rate and the unit in the numerator of the first conversion rate must be the same in order to cancel.

Example 1 Convert 42 c to quarts.

Solution $42 \text{ c} = 42 \cancel{\text{c}} \times \frac{1 \cancel{\text{pt}}}{2 \cancel{\text{c}}} \times \frac{1 \text{ qt}}{2 \cancel{\text{pt}}}$

$= \frac{42 \text{ qt}}{4} = 10\frac{1}{2}\text{qt}$

Example 2 Convert 18 pt to gallons.

Your solution

Solution on p. A32

Objective B To perform arithmetic operations with measurements of capacity

When performing arithmetic operations with measurements of capacity, write the answer in simplest form. For example, 1 c 12 fl oz should be written as 2 c 4 fl oz.

Example 3 Subtract: 4 gal 1 qt − 2 gal 3 qt

Solution

$$\begin{array}{r} \overset{3\text{ gal}}{\cancel{4\text{ gal}}}\ \overset{5\text{ qt}}{\cancel{1\text{ qt}}} \\ -2\text{ gal } 3\text{ qt} \\ \hline 1\text{ gal } 2\text{ qt} \end{array}$$

Borrow 1 gal (4 qt) from 4 gal and add to 1 qt.

Example 4 Divide: 4 gal 2 qt ÷ 3

Your solution

Solution on p. A32

Objective C To solve application problems

Example 5

A can of apple juice contains 25 fl oz. Find the number of quarts of apple juice in a case of 24 cans.

Strategy

To find the number of quarts of apple juice in one case:

- Multiply the number of cans (24) by the number of fluid ounces per can (25) to find the total number of fluid ounces in the case.
- Convert the number of fluid ounces in the case to quarts.

Solution

$$\begin{array}{l} 25\text{ fl oz} \\ \underline{\times 24} \\ 600\text{ fl oz in the case} \end{array}$$

$$600\text{ fl oz} = \frac{600\ \cancel{\text{fl oz}}}{1} \cdot \frac{1\ \cancel{\text{c}}}{8\ \cancel{\text{fl oz}}} \cdot \frac{1\ \cancel{\text{pt}}}{2\ \cancel{\text{c}}} \cdot \frac{1\text{ qt}}{2\ \cancel{\text{pt}}}$$

$$= \frac{600\text{ qt}}{32} = 18\frac{3}{4}\text{qt}$$

One case of apple juice contains $18\frac{3}{4}$ qt.

Example 6

Five students are going backpacking in the desert. Each student requires 1 qt of water per day. How many gallons of water should they take for a 3-day trip?

Your strategy

Your solution

Solution on p. A32

8.3 EXERCISES

Objective A

Convert.

1. 60 fl oz = ______ c
2. 48 fl oz = ______ c
3. 3 c = ______ fl oz
4. $2\frac{1}{2}$ c = ______ fl oz
5. 8 c = ______ pt
6. 5 c = ______ pt
7. $3\frac{1}{2}$ pt = ______ c
8. 12 pt = ______ qt
9. 22 qt = ______ gal
10. 10 qt = ______ gal
11. $2\frac{1}{4}$ gal = ______ qt
12. 7 gal = ______ qt
13. $7\frac{1}{2}$ pt = ______ qt
14. $3\frac{1}{2}$ qt = ______ pt
15. 20 fl oz = ______ pt
16. $1\frac{1}{2}$ pt = ______ fl oz
17. 17 c = ______ qt
18. $1\frac{1}{2}$ qt = ______ c

Objective B

Perform the arithmetic operation.

19. 14 qt = ______ gal ______ qt
20. 9 pt = ______ qt ______ pt
21. 5 pt = ______ qt ______ pt
22. 3 gal 2 qt + 4 gal 3 qt
23. 4 qt 1 pt + 2 qt 1 pt
24. 3 gal 1 qt − 1 gal 2 qt
25. 3 c 3 fl oz − 2 c 5 fl oz
26. 2 qt 1 pt × 5
27. $3\frac{1}{2}$ pt × 5
28. 5)6 gal 1 qt
29. $3\frac{1}{2}$ gal ÷ 4
30. 5 c 3 fl oz + 3 c 6 fl oz

Perform the arithmetic operation.

31. $\begin{array}{r} 3 \text{ gal } 3 \text{ qt} \\ +1 \text{ gal } 2 \text{ qt} \\ \hline \end{array}$

32. $\begin{array}{r} 4 \text{ c } 6 \text{ fl oz} \\ -2 \text{ c } 7 \text{ fl oz} \\ \hline \end{array}$

33. $\begin{array}{r} 3 \text{ gal} \\ -1 \text{ gal } 2 \text{ qt} \\ \hline \end{array}$

34. $1\frac{1}{2}$ pt + $2\frac{2}{3}$ pt

35. $4\frac{1}{2}$ gal − $1\frac{3}{4}$ gal

36. $2\overline{)3 \text{ gal } 2 \text{ qt}}$

Objective C *Application Problems*

37. It is estimated that 60 adults will attend a church social. Assume that each adult will drink 2 c of coffee. How many gallons of coffee should be prepared?

38. A local playhouse serves punch during intermission. Assume that 200 people will each drink 1 c of punch. How many gallons of punch should be ordered?

39. A solution needed for a class of 30 chemistry students required 72 oz of water, 16 oz of one solution, and 48 oz of another solution. Find the number of quarts of the final solution.

40. A cafeteria sold 124 cartons of milk in 1 day. Each carton contained 1 c of milk. How many quarts of milk were sold that day?

41. A farmer changed the oil in the tractor seven times during the year. Each oil change required 5 qt of oil. How many gallons of oil did the farmer use in the seven oil changes?

42. The gasoline tank on a compact car holds $10\frac{1}{2}$ gal of gas. How many quarts of gasoline does the gasoline tank hold?

43. A can of grape juice contains 24 oz. Find the number of cups of grape juice in a case of 12 cans.

44. There are 24 cans in a case of tomato juice. Each can contains 10 oz of tomato juice. Find the number of 1-cup servings in the case of tomato juice.

45. One brand of orange juice costs $1.20 for 1 qt. Another brand of orange juice costs 96¢ for 24 oz. Find the more economical purchase.

46. A backpacker is carrying 6 qt of water for 3 days of desert camping. Water weighs 8 1/3 lb per gallon. Find the weight of water the backpacker carries.

47. A department store buys hand-lotion in 5-qt containers and then repackages the hand lotion in 8-fl oz bottles. The hand lotion cost $41.50 and each 8-fl oz bottle is sold for $4.25. How much profit is made on each 5-qt package of hand lotion?

48. A garage mechanic buys oil in 50-gal containers for changing the oil in customer's cars. The mechanic pays $80 for the 50 gal of oil and charges customers $1.05 per quart. Find the profit the mechanic makes on one 50-gal container of oil.

SECTION 8.4 Energy and Power

Objective A To use units of energy in the U.S. Customary System

Energy can be defined as the ability to do work. Energy is stored in coal, in gasoline, in water behind a dam, and in one's own body.

One **foot pound** (ft·lb) of energy from your body is required to lift 1 pound a distance of 1 foot.

To lift 50 lb a distance of 5 ft requires
$50 \times 5 = 250$ ft·lb of energy

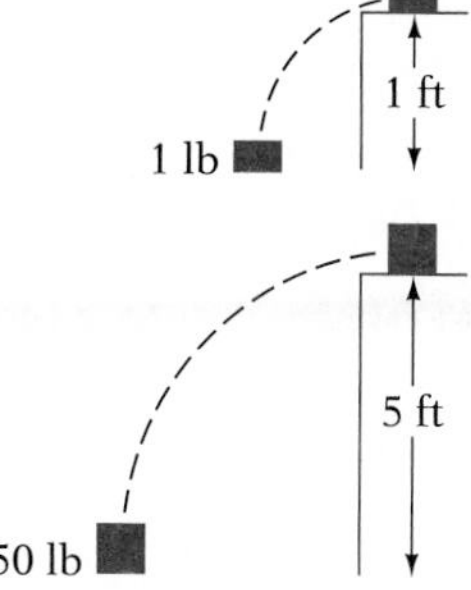

Consumer items, such as furnaces, stoves, and air conditioners, which use energy, are rated in **British Thermal Units** (BTU).

A furnace with a rating of 35,000 BTU per hour releases 35,000 BTU of energy in one hour (1 h).

$$1 \text{ BTU} = 778 \text{ ft·lb}$$

Therefore the following conversion rate equivalent to 1 can be written:

$$\frac{778 \text{ ft·lb}}{1 \text{ BTU}} = 1$$

Example 1

Convert 250 BTU to foot pounds.

Solution

$$250 \text{ BTU} = 250 \cancel{\text{BTU}} \times \frac{778 \text{ ft·lb}}{1 \cancel{\text{BTU}}}$$
$$= 194{,}500 \text{ ft·lb}$$

Example 2

Convert 4.5 BTU to foot pounds.

Your solution

Example 3

Find the energy required for a 125-lb person to climb a mile-high mountain.

Solution

In climbing the mountain, the person is lifting 125 lb a distance of 5280 ft.

$$\text{Energy} = 125 \text{ lb} \times 5280 \text{ ft}$$
$$= 660{,}000 \text{ ft·lb}$$

Example 4

Find the energy required for a motor to lift 800 lb through a distance of 16 ft.

Your solution

Solutions on p. A32

Example 5

A furnace is rated at 80,000 BTU per hour. How many foot pounds of energy are released in 1 h?

Solution

$$80{,}000 \text{ BTU} = 80{,}000 \cancel{\text{BTU}} \times \frac{778 \text{ ft}\cdot\text{lb}}{1 \cancel{\text{BTU}}} = 62{,}240{,}000 \text{ ft}\cdot\text{lb}$$

Example 6

A furnace is rated at 56,000 BTU per hour. How many foot pounds of energy are released in 1 h?

Your solution

Solution on p. A32

Objective B

To use units of power in the U.S. Customary System

Power is the rate at which work is done, or the rate at which energy is released.

Power is measured in **foot pounds per second** $\left(\frac{\text{ft}\cdot\text{lb}}{\text{s}}\right)$. In each of the following examples, the amount of energy released is the same, but the time taken to release the energy is different; thus the power is different.

100 lb is lifted 10 ft in 10 s.

$$\text{Power} = \frac{10 \text{ ft} \times 100 \text{ lb}}{10 \text{ s}} = 100 \frac{\text{ft}\cdot\text{lb}}{\text{s}}$$

100 lb is lifted 10 ft in 5s.

$$\text{Power} = \frac{10 \text{ ft} \times 100 \text{ lb}}{5 \text{ s}} = 200 \frac{\text{ft}\cdot\text{lb}}{\text{s}}$$

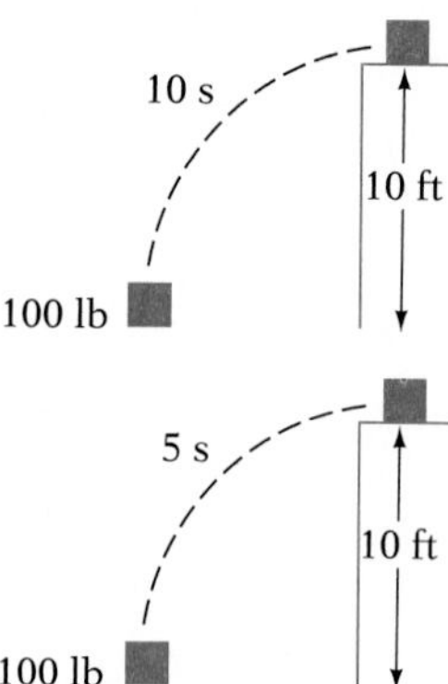

The U.S. Customary unit of power is the **horsepower.** A horse doing average work can pull 550 lb a distance of 1 ft in 1 s and can continue this work all day.

$$1 \text{ horsepower (hp)} = 550 \frac{\text{ft}\cdot\text{lb}}{\text{s}}$$

Example 7

Find the power needed to raise 300 lb a distance of 30 ft in 15 s.

Solution $\text{Power} = \frac{30 \text{ ft} \times 300 \text{ lb}}{15 \text{ s}} = 600 \frac{\text{ft}\cdot\text{lb}}{\text{s}}$

Example 8

Find the power needed to raise 1200 lb a distance of 90 ft in 24 s.

Your solution

Example 9

A motor has a power of 2750 $\frac{\text{ft}\cdot\text{lb}}{\text{s}}$. Find the horsepower of the motor.

Solution $\frac{2750}{550} = 5 \text{ hp}$

Example 10

A motor has a power of 3300 $\frac{\text{ft}\cdot\text{lb}}{\text{s}}$. Find the horsepower of the motor.

Your solution

Solutions on p. A32

8.4 EXERCISES

Objective A

1. Convert 25 BTU to foot pounds.

2. Convert 6000 BTU to foot pounds.

3. Convert 25,000 BTU to foot pounds.

4. Convert 40,000 BTU to foot pounds.

5. Find the energy required to lift 150 lb a distance of 10 ft.

6. Find the energy required to lift 300 lb a distance of 16 ft.

7. Find the energy required to lift a 3300-lb car a distance of 9 ft.

8. Find the energy required to lift a 3680-lb elevator a distance of 325 ft.

9. Three tons are lifted 5 ft. Find the energy required in foot pounds.

10. Seven tons are lifted 12 ft. Find the energy required in foot pounds.

11. A construction worker carries 3-lb blocks up a 10-ft flight of stairs. How many foot pounds of energy are required to carry 850 blocks up the stairs?

12. A crane lifts an 1800-lb steel beam to the roof of a building 36 ft high. Find the amount of energy the crane requires in lifting the beam.

13. A furnace is rated at 45,000 BTU per hour. How many foot pounds of energy are released by the furnace in 1 h?

14. A furnace is rated at 22,500 BTU per hour. How many foot pounds of energy does the furnace release in 1 h?

15. Find the amount of energy in foot pounds given off when 1 lb of coal is burned. 1 lb of coal gives off 12,000 BTU of energy when burned.

16. Find the amount of energy in foot pounds given off when 1 lb of gasoline is burned. 1 lb of gasoline gives off 21,000 BTU of energy when burned.

Objective B

17. Convert 1100 $\frac{\text{ft} \cdot \text{lb}}{\text{s}}$ to horsepower.

18. Convert 6050 $\frac{\text{ft} \cdot \text{lb}}{\text{s}}$ to horsepower.

19. Convert 4400 $\frac{\text{ft} \cdot \text{lb}}{\text{s}}$ to horsepower.

20. Convert 1650 $\frac{\text{ft} \cdot \text{lb}}{\text{s}}$ to horsepower.

21. Convert 5 hp to foot pounds per second.

22. Convert 3 hp to foot pounds per second.

23. Convert 7 hp to foot pounds per second.

24. Convert 2 hp to foot pounds per second.

25. Find the power in foot pounds per second needed to raise 125 lb a distance of 12 ft in 3 s.

26. Find the power in foot pounds per second needed to raise 500 lb a distance of 60 ft in 8 s.

27. Find the power in foot pounds per second needed to raise 3000 lb a distance of 40 ft in 25 s.

28. Find the power in foot pounds per second needed to raise 12,000 lb a distance of 40 ft in 60 s.

29. Find the power in foot pounds per second of an engine that can raise 180 lb to a height of 40 ft in 5 s.

30. Find the power in foot pounds per second of an engine that can raise 1200 lb to a height of 18 ft in 30 s.

31. A motor has a power of 1650 $\frac{\text{ft} \cdot \text{lb}}{\text{s}}$. Find the horsepower of the motor.

32. A motor has a power of 16,500 $\frac{\text{ft} \cdot \text{lb}}{\text{s}}$. Find the horsepower of the motor.

33. A motor has a power of 6600 $\frac{\text{ft} \cdot \text{lb}}{\text{s}}$. Find the horsepower of the motor.

34. A motor has a power of 4400 $\frac{\text{ft} \cdot \text{lb}}{\text{s}}$. Find the horsepower of the motor.

Calculators and Computers

Conversions The program U.S. CUSTOMARY on the Math ACE disk allows you to practice changing from one unit of measure in the system to another unit of measure in the system. The following conversions will help you while you practice.

Distance

12 inches = 1 foot
3 feet = 1 yard
5280 feet = 1 mile

Weight

16 ounces = 1 pound
2000 pounds = 1 ton

Capacity

8 fluid ounces = 1 cup
2 cups = 1 pint
2 pints = 1 quart
4 quarts = 1 gallon

When using the program, remember to enter any fractional answers as decimals. For example:

Change 21 quarts to gallons.

Enter the answer $5\frac{1}{4}$ as 5.25.

Chapter Summary

Key Words A *measurement* includes a number and a unit.

The U.S. Customary units of length are *inch, foot, yard,* and *mile.*

Weight is a measure of how strongly the earth is pulling on an object.

The U.S. Customary units of weight are *ounce, pound,* and *ton.*

Liquid substances are measured in units of *capacity.*

The U.S. Customary units of capacity are *fluid ounce, cup, pint, quart,* and *gallon.*

Energy is the ability to do work.

One *foot pound* of energy is the energy necessary to lift 1 pound a distance of 1 foot.

Furnaces, stoves, and air conditioners are rated in energy units called *British Thermal Units.*

Power is the rate at which work is done.

The U.S. Customary unit of power is the *horsepower.*

Essential Rules A conversion unit is used to convert one unit of measurement to another. For example, the conversion unit 12 in./1 ft is used to convert feet to inches.

Chapter Review

SECTION 8.1

1. Convert 4 ft to inches.

2. Convert 14 ft to yards.

3. Add: $\begin{array}{r} 3\text{ ft } 9\text{ in.} \\ +5\text{ ft } 6\text{ in.} \\ \hline \end{array}$

4. Subtract: $\begin{array}{r} 5\text{ yd } 1\text{ ft} \\ -3\text{ yd } 2\text{ ft} \\ \hline \end{array}$

5. Multiply: $\begin{array}{r} 2\text{ ft } 8\text{ in.} \\ \times \quad 5 \\ \hline \end{array}$

6. Divide: $3\overline{)7\text{ ft } 6\text{ in.}}$

7. A board 6 ft 11 in. is cut from a board 10 ft 5 in. long. Find the length of the remaining piece of board.

SECTION 8.2

8. Convert $3\frac{3}{8}$ lb to ounces.

9. Convert 2400 lb to tons.

10. Add: $\begin{array}{r} 5\text{ lb } 11\text{ oz} \\ +3\text{ lb } 8\text{ oz} \\ \hline \end{array}$

11. Subtract: $\begin{array}{r} 3\text{ tons } 500\text{ lb} \\ -1\text{ ton } 1500\text{ lb} \\ \hline \end{array}$

12. Multiply: $\begin{array}{r} 5\text{ lb } 6\text{ oz} \\ \times \quad 8 \\ \hline \end{array}$

13. Divide: $3\overline{)7\text{ lb } 5\text{ oz}}$

14. A book weighing 2 lb 3 oz is mailed at the postage rate of \$0.18 per ounce. Find the cost of mailing the book.

SECTION 8.3

15. Convert 12 c to quarts.

16. Convert $2\frac{1}{2}$ pt to fluid ounces.

17. A can of pineapple juice contains 18 fl oz. Find the number of quarts in a case of 24 cans.

18. A cafeteria sold 256 cartons of milk in one school day. Each carton contains 1 c of milk. How many gallons of milk were sold that day?

SECTION 8.4

19. Convert 50 BTU to foot pounds. (1 BTU = 778 ft · lb)

20. Find the energy needed to lift 200 lb a distance of 8 ft.

21. A furnace is rated at 35,000 BTU per hour. How many foot pounds of energy does the furnace release in 1 h? (1 BTU = 778 ft · lb)

22. Convert 3850 $\frac{\text{ft} \cdot \text{lb}}{\text{s}}$ to horsepower. (1 hp = 550 $\frac{\text{ft} \cdot \text{lb}}{\text{s}}$)

23. Convert 2.5 hp to foot pounds per second.

24. Find the power in foot pounds per second of an engine that can raise 800 lb to a height of 15 ft in 25 s.

Chapter Test

1. Convert $2\frac{1}{2}$ ft to inches.

2. Subtract: 4 ft 2 in. − 1 ft 9 in.

3. A board $6\frac{2}{3}$ ft long is cut into five equal pieces. How long is each piece?

4. Seventy-two bricks, each 8 in. long, are laid end to end to make the base for a wall. Find the length of the wall in feet.

5. Convert $2\frac{7}{8}$ lb to ounces.

6. Convert: 40 oz = ___ lb ___ oz [8.2A]

7. Add: 9 lb 6 oz + 7 lb 11 oz

8. Divide: 6 lb 12 oz ÷ 4

9. A college bookstore received 1000 workbooks, each workbook weighing 12 oz. Find the total weight of the 1000 workbooks in pounds.

10. An elementary school class gathered 800 aluminum cans for recycling. Four aluminum cans weigh 3 oz. Find the amount the class received if the rate of pay was $0.75 per pound for the aluminum cans. (Round to the nearest cent.)

11. Convert 13 qt to gallons.

12. Convert $3\frac{1}{2}$ gal to pints.

13. Multiply: $1\frac{3}{4}$ gal × 7

14. Add: 5 gal 2 qt + 2 gal 3 qt

15. A can of grapefruit juice contains 20 oz. Find the number of cups of grapefruit juice in a case of 24 cans.

16. A mechanic buys oil in 40-gal containers for changing the oil in customer's cars. The mechanic pays $90 for the 40 gal of oil and charges customers $1.35 per quart. Find the profit the mechanic makes on one 40-gal container of oil.

17. Find the energy required to lift 250 lb a distance of 15 ft.

18. A furnace is rated at 40,000 BTU per hour. How many foot pounds of energy are released by the furnace in 1 hr? (1 BTU = 778 ft · lb)

19. Find the power needed to lift 200 lb a distance of 20 ft in 25 s.

20. A motor has a power of 2200 $\frac{\text{ft} \cdot \text{lb}}{\text{s}}$. Find the motor's horsepower. (1 hp = 550 $\frac{\text{ft} \cdot \text{lb}}{\text{s}}$)

Cumulative Review

1. Find the LCM of 9, 12, and 15.

2. Write $\frac{43}{8}$ as a mixed number.

3. Subtract: $5\frac{7}{8} - 2\frac{7}{12}$

4. Divide: $5\frac{1}{3} \div 2\frac{2}{3}$

5. Simplify $\frac{5}{8} \div \left(\frac{3}{8} - \frac{1}{4}\right) - \frac{10}{16}$.

6. Round 2.0972 to the nearest hundredth.

7. Multiply: $\begin{array}{r} 0.0792 \\ \times\ 0.49 \\ \hline \end{array}$

8. Solve the proportion. $\frac{n}{12} = \frac{44}{60}$

9. Find $2\frac{1}{2}\%$ of 50.

10. 18 is 42% of what? Round to the nearest hundredth.

11. A 7.2-lb roast costs \$15.48. Find the unit cost.

12. Add: $3\frac{2}{5}$ in. + $5\frac{1}{3}$ in.

13. Convert: 24 oz = _____ lb _____ oz

14. Multiply: 3 lb 8 oz × 9

15. Subtract: $4\frac{1}{3}$ qt − $1\frac{5}{6}$ qt

16. Subtract: 4 lb 6 oz − 2 lb 10 oz

17. An investor receives a dividend of \$56 from 40 shares of stock. At the same rate, find the dividend that would be received from 200 shares of stock.

18. You have a balance of \$578.56 in your checkbook. You write checks of \$216.98 and \$34.12 and make a deposit of \$315.33. What is your new checking balance?

19. An account executive receives a salary of \$800 per month plus a commission of 2% on all sales over \$25,000. Find the total monthly income of an account executive who has monthly sales of \$140,000.

20. A health inspector found that 3% of a shipment of carrots were spoiled and could not be sold. Find the amount of carrots from a shipment of 2500 pounds that could be sold.

21. The scores on the final exam of a trigonometry class are recorded in the histogram in the figure. What percent of the class received a score between 80% and 90%? (Round to the nearest percent.)

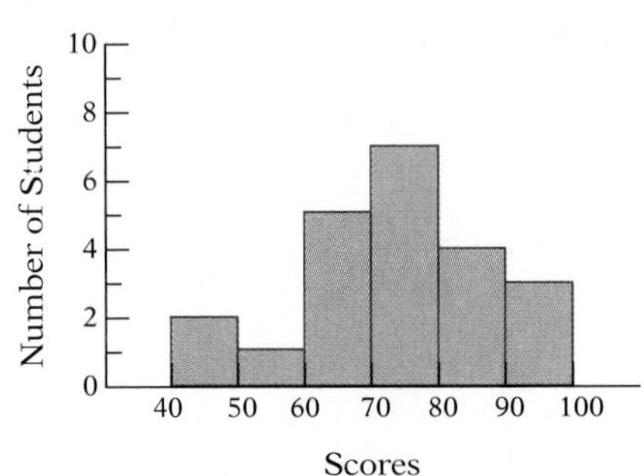

22. A department store uses a markup rate of 40% on all merchandise. What is the selling price of a compact disc player that cost the store \$220?

23. A construction firm received a loan of \$200,000 for 8 months at a simple interest rate of 11%. Find the interest paid on the loan.

24. Six college students spent several weeks panning for gold during their summer vacation. The students obtained 1 lb 3 oz of gold. After selling the gold for \$400 per ounce, how much money did each student receive? Round to the nearest dollar.

25. Four books were mailed at the rate of \$0.15 per ounce. The books weighed 1 lb 3 oz, 13 oz, 1 lb 8 oz, and 1 lb. Find the cost of mailing the books.

26. One brand of yogurt cost \$0.61 for 8 oz; 36 oz of another brand can be bought for \$2.70. Which purchase is the better buy?

27. A contractor can buy a 4000-gal tank of gasoline for \$3600. The pump price of the gasoline is \$1.17 per gallon. How much does the contractor save by buying the 4000 gal of gasoline?

28. Find the energy required to lift 400 lb a distance of 8 ft.

29. Find the power in foot pounds per second needed to raise 600 lb a distance of 8 ft in 12 s.

9 The Metric System of Measurement

OBJECTIVES

- To convert units of length in the metric system of measurement
- To perform arithmetic operations with measurements of length
- To solve application problems
- To convert units of mass in the metric system of measurement
- To perform arithmetic operations with measurements of mass
- To solve application problems
- To convert units of capacity in the metric system of measurement
- To perform arithmetic operations with measurements of capacity
- To solve application problems
- To use units of energy in the metric system of measurement
- To convert U.S. Customary units to metric units
- To convert metric units to U.S. Customary units

Is a Pound a Pound?

Which is heavier, a pound of feathers or a pound of gold? It would seem that a pound is a pound, whether it is gold or feathers. However, this is not the case.

Metals, like gold and silver, are measured by using the Troy Weight System, whereas the weight of feathers, meat, people, and other nonmetal quantities are measured using the Avoirdupois Weight System. Each system is part of the U.S. Customary System of Measurement.

One grain is a small weight in the Customary System and is approximately 0.02 ounces. A pound of feathers weighs 7000 grains, but a pound of gold weighs only 5760 grains. Therefore a pound of feathers weighs more than a pound of gold.

To complicate matters, 1 avoirdupois pound contains 16 ounces, but 1 troy pound contains 12 ounces, which means that there are 437.5 grains in 1 avoirdupois ounce but 480 grains in a troy ounce. Thus an ounce of feathers weighs less than an ounce of gold.

To summarize, a pound of feathers weighs more than a pound of gold, but an ounce of feathers weighs less than an ounce of gold. This kind of confusion is one reason the metric system of measurement was developed.

SECTION 9.1 Length

Objective A To convert units of length in the metric system of measurement

In 1789, an attempt was made to standardize units of measurement internationally in order to simplify trade and commerce between nations. A commission in France developed a system of measurement known as the **metric system.** The basic unit of length in the metric system is the **meter.** One meter is approximately the distance from a doorknob to the floor. Originally the meter was defined as $\frac{1}{10{,}000{,}000}$ of the distance from the equator to the north pole. The meter is now defined as 1,650,763.73 wavelengths of a particular orange-red wavelength given off by atoms of krypton.

All units of length in the metric system are derived from the meter. Prefixes to the basic unit denote the length of each unit. For example, the prefix "centi-" means one-hundredth; therefore 1 centimeter is 1 one-hundredth of a meter.

kilo- = 1000	1 kilometer (km) = 1000 meters (m)
hecto- = 100	1 hectometer (hm) = 100 m
deca- = 10	1 decameter (dam) = 10 m
	1 meter (m) = 1 m
deci- = 0.1	1 decimeter (dm) = 0.1 m
centi- = 0.01	1 centimeter (cm) = 0.01 m
milli- = 0.001	1 millimeter (mm) = 0.001 m

Conversion between units of length in the metric system involves moving the decimal point to the right or to the left. Listing the units in order from largest to smallest will indicate how many places to move the decimal point and in which direction.

To convert 4200 cm to m, write the units in order from largest to smallest.

km hm dam m dm cm mm — Converting cm to m requires moving 2 positions to the left.

2 positions

4200 cm = 42.00 m — Move the decimal point the same number of places and in the same direction.

2 places

A metric measurement involving two units is customarily written in terms of one unit. Convert the smaller unit to the larger unit and then add.

To convert 8 km 32 m to km, first convert 32 m to km.

km hm dam m dm cm mm

32 m = 0.032 km

Then add the result to 8 km. 8 km 32 m = 8 km + 0.032 km = 8.032 km

Example 1 Convert 0.38 m to millimeters.

Solution 0.38 m = 380 mm

Example 2 Convert 3.07 m to centimeters.

Your solution

Solution on p. A33

Example 3 Convert 4 m 62 cm to centimeters.

Solution 4 m = 400 cm
4 m 62 cm
= 400 cm + 62 cm
= 462 cm

Example 4 Convert 3 m 7 cm to meters.

Your solution

Solution on p. A33

Objective B To perform arithmetic operations with measurements of length

Arithmetic operations can be performed with measurements of length in the metric system. A measurement involving two units should be written in terms of a single unit before adding, subtracting, multiplying, or dividing.

Add:
6 m 42 cm + 7 m 98 cm

$$\begin{array}{r} 6\text{ m }42\text{ cm} = 6.42\text{ m} \\ +7\text{ m }98\text{ cm} = 7.98\text{ m} \\ \hline 14.40\text{ m} \end{array} \quad \text{or} \quad \begin{array}{r} 6\text{ m }42\text{ cm} = 642\text{ cm} \\ +7\text{ m }98\text{ cm} = 798\text{ cm} \\ \hline 1440\text{ cm} \end{array}$$

Note that the measurements can be changed to either meters or centimeters before adding. In this textbook, unless otherwise stated, the units should be changed to the larger unit before the arithmetic operation is performed.

Example 5 Divide: 42 km 765 m ÷ 14 Round to the nearest thousandth.

Solution 42 km 765 m = 42.765 km

$$14\overline{)42.765}\text{ km} = 3.0546\text{ km} \approx 3.055\text{ km}$$

Example 6 Subtract: 3 m − 42 cm

Your solution

Solution on p. A33

Objective C To solve application problems

Example 7

A piece measuring 1 m 42 cm is cut from a board 4 m 20 cm long. Find the length of the remaining piece.

Strategy

To find the length of the remaining piece, subtract the length of the piece cut (1 m 42 cm) from the original length (4 m 20 cm).

Solution

$$\begin{array}{r} 4\text{ m }20\text{ cm} = 4.20\text{ m} \\ -1\text{ m }42\text{ cm} = 1.42\text{ m} \\ \hline 2.78\text{ m} \end{array}$$

The length of the piece remaining is 2.78 m.

Example 8

A bookcase 1 m 75 cm long has four shelves. Find the cost of the shelves when the price of lumber is $11.75 per meter.

Your strategy

Your solution

Solution on p. A33

9.1 EXERCISES

Objective A

Convert.

1. 42 cm = ____ mm
2. 62 cm = ____ mm
3. 81 mm = ____ cm
4. 68.2 mm = ____ cm
5. 6804 m = ____ km
6. 3750 m = ____ km
7. 2.109 km = ____ m
8. 32.5 km = ____ m
9. 432 cm = ____ m
10. 61.7 cm = ____ m
11. 0.88 m = ____ cm
12. 3.21 m = ____ cm
13. 6 m 42 cm = ____ m
    = ____ cm
14. 62 km 482 m = ____ km
    = ____ m
15. 42 cm 6 mm = ____ cm
    = ____ mm
16. 62 m 7 cm = ____ m
    = ____ cm

Objective B

Perform the arithmetic operation.

17. 42 cm + 8 m
18. 69 km 432 m + 7 km 921 m
19. 98 m 67 cm − 19 m 82 cm
20. 2 km − 435 m
21. 7 cm 4 mm × 4
22. 16 km 691 m × 12
23. 3 km 726 m ÷ 9
24. 42 m 60 cm ÷ 12
25. 8 km 312 m + 9 km 814 m
26. 729 km 467 m + 837 km 942 m
27. 954.62 km × 944
28. 165 km 429 cm ÷ 27

Objective C *Application Problems*

29. A carpenter needs 15 ceiling joists, each joist 4 m 60 cm long. Find the total length of ceiling joists needed in meters.

30. A person can walk 1 km 600 m in $\frac{1}{2}$ hour. How far can the person walk in 2 hours?

31. Find the missing dimension in centimeters.

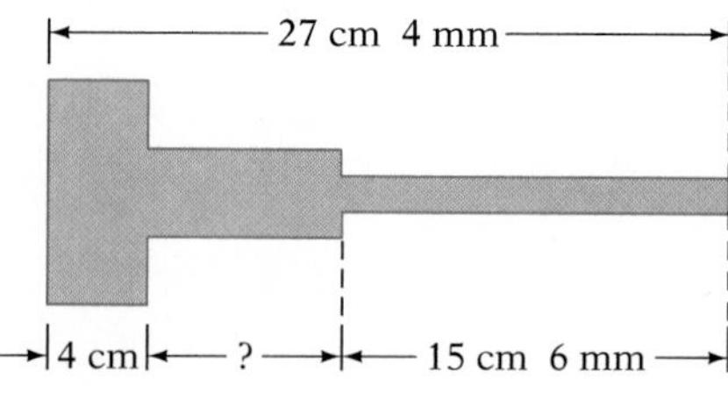

32. Find the total length in centimeters.

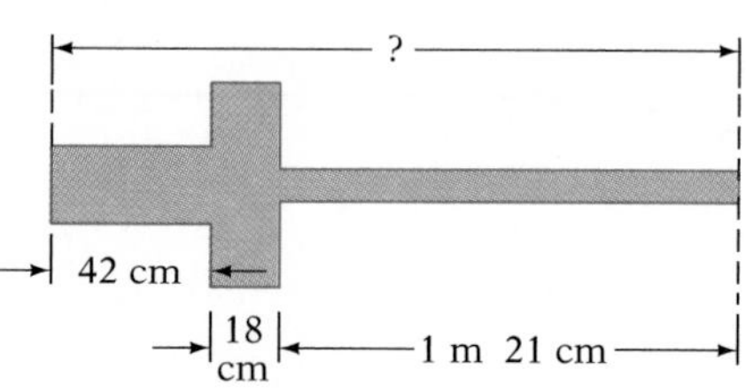

33. A bicycle race had two checkpoints. One checkpoint was 12 km 400 m from the starting point. The second checkpoint was 9 km 300 m from the first checkpoint. The second checkpoint was 8 km 800 m from the finish line. How long was the race?

34. How many shelves, 1 m 40 cm long, can be cut from a board that is 4 m 20 cm in length? Find the length of board remaining after the shelves are cut.

35. Twenty rivets are used to fasten two steel plates together. The plates are 3 m 40 cm long, and the rivets are equally spaced, with a rivet at each end. Find the distance between the rivets. (Round to the nearest tenth of a centimeter.)

36. A frame for a picture is 1 m 10 cm long and 80 cm wide. Find the total length of framing needed to frame the picture.

37. Four pieces of fencing are cut from a 50-m roll to build a dog run. The dog run is 3 m 40 cm wide and 13 m 80 cm long. How much of the fencing is left on the roll after the dog run is completed?

38. During the week, a cross-country runner ran 12 km 500 m, 15 km 800 m, 12 km 500 m, 13 km 200 m, and 18 km 400 m during the week. Find the average distance run each of the 5 days.

39. Light travels 3×10^8 m in 1 s. How far does the light travel in 1 year? (Astronomers refer to this distance as 1 light year.)

40. The distance around the earth is 41,000 km. Light travels 3×10^8 m in 1 s. How many times will light travel around the earth in 1 s?

SECTION 9.2 Mass

Objective A

To convert units of mass in the metric system of measurement

Mass and weight are closely related. Weight is a measure of how strongly the earth is pulling on an object. Therefore, an object's weight will be less in space than on the earth's surface. However, the amount of material in the object, its **mass**, remains the same. On the surface of the earth, mass and weight can be used interchangeably.

The basic unit of mass in the metric system is the **gram.** If a box that is 1 cm long on a side is filled with water, then the mass of that water is 1 gram.

1 cm
1 cm
1 cm

1 gram = the mass of water in the box

The gram is a very small unit of mass. A paperclip weighs about 1 gram. The kilogram (1000 grams) is a more useful unit of mass in consumer applications. This textbook weighs about 1 kilogram.

The units of mass in the metric system have the same prefixes as the units of length:

1 kilogram (kg) = 1000 grams (g)
1 hectogram (hg) = 100 g
1 decagram (dag) = 10 g
1 gram (g) = 1 g
1 decigram (dg) = 0.1 g
1 centigram (cg) = 0.01 g
1 milligram (mg) = 0.001 g

Conversion between units of mass in the metric system involves moving the decimal point to the right or to the left. Listing the units in order from largest to smallest will indicate how many places to move the decimal point and in which direction.

To convert 324 g to kg, first write the units in order from largest to smallest.

kg hg dag g dg cg mg
3 positions

Converting g to kg requires moving 3 positions to the left.

324 g = 0.324 kg
3 places

Move the decimal point the same number of places and in the same direction.

Example 1 Convert 4.23 g to milligrams.

Solution 4.23 g = 4230 mg

Example 2 Convert 42.3 mg to grams.

Your solution

Solution on p. A33

Example 3 Convert 2 kg 564 g to kilograms.

Solution 564 g = 0.564 kg
2 kg 564 g =
2 kg + 0.564 kg = 2.564 kg

Example 4 Convert 3 g 54 mg to milligrams.

Your solution

Solution on p. A33

Objective B To perform arithmetic operations with measurements of mass

Arithmetic operations can be performed with measurements of mass. A measurement involving two units should be written in terms of a single unit before adding, subtracting, multiplying, or dividing.

Add:
3 g 42 mg + 5 g 690 mg

3 g 42 mg = 3.042 g		3 g 42 mg = 3042 mg
+5 g 690 mg = 5.690 g	or	+5 g 690 mg = 5690 mg
8.732 g		8732 mg

Note that the units can be changed to either grams or milligrams before adding. In this textbook, unless otherwise stated, the units should be changed to the larger unit before the arithmetic operation is performed.

Example 5 Add: 3 kg + 62 g

Solution 3 kg + 62 g =
3 kg + 0.062 kg = 3.062 kg

Example 6 Multiply: 4 g 620 mg × 8

Your solution

Solution on p. A33

Objective C To solve application problems

Example 7

Find the cost of a roast weighing 3 kg 320 g if the price per kilogram is $4.17. Round to the nearest cent.

Strategy To find the cost of the roast:

- Convert 3 kg 320 g to kilograms.
- Multiply the weight by the cost per kilogram ($4.17).

Solution 3 kg 320 g = 3.320 kg

$$\begin{array}{r} 3.320 \\ \times 4.17 \\ \hline 13.84440 \end{array}$$

The cost of the roast is $13.84.

Example 8

Three hundred grams of fertilizer are used to fertilize each tree in an apple orchard. How many kilograms of fertilizer are required to fertilize 400 trees?

Your strategy

Your solution

Solution on p. A33

9.2 EXERCISES

Objective A

Convert.

1. 420 g = ______ kg
2. 7421 g = ______ kg
3. 127 mg = ______ g
4. 43 mg = ______ g
5. 4.2 kg = ______ g
6. 0.027 kg = ______ g
7. 0.45 g = ______ mg
8. 325 g = ______ mg
9. 1856 g = ______ kg
10. 1.37 kg = ______ g
11. 4057 mg = ______ g
12. 0.0456 g = ______ mg
13. 3 kg 922 g = ______ kg
    = ______ g
14. 1 kg 47 g = ______ kg
    = ______ g
15. 7 g 891 mg = ______ g
    = ______ mg
16. 209 g 42 mg = ______ g
    = ______ mg

Objective B

Perform the arithmetic operation. Round to the nearest thousandth.

17. 4 g − 692 mg

18.
```
  4 kg  67 g
−1 kg 296 g
```

19. 4 kg + 692 g

20.
```
  362 g 419 mg
+192 g  97 mg
```

21. 46 g × 16

22.
```
  3 kg 496 g
×          9
```

Perform the arithmetic operation. Round to the nearest thousandth.

23. 456 g ÷ 14

24. 3 kg 119 g ÷ 7

25. 4 kg 319 g − 2 kg 599 g

26. 7432 kg 89 g − 5196 kg 72 g

27. 832 g 498 mg × 967

28. 325 kg 72 g ÷ 167

Objective C *Application Problems*

29. A baby weighs 3 kg 800 g at birth. At the end of 6 months, the baby weighs 8 kg 650 g. Find the increase in weight in kilograms during the 6 months.

30. A doctor advises a patient weighing 108 kg 400 g to lose 20 kg of weight. How much more does the patient need to lose after losing 13 kg 800 g?

31. A concrete block weighs 2 kg 350 g. Find the weight of a load of 1200 concrete blocks.

32. A 15 × 15-cm tile weighs 220 g. Find the weight of a box of 72 tiles in kilograms.

33. Five scouts are taking 104 kg 600 g of supplies on a 5-day backpacking trip. How much weight in kilograms will each scout carry if the gear is divided equally?

34. A doctor recommends that a patient take 3.5 g of vitamin C a week. Find the daily dosage of the vitamin in milligrams.

35. Find the cost of a ham weighing 4 kg 700 g if the price per kilogram is $4.20.

36. Eighty grams of grass seed are used for every 100 m^2 of lawn. Find the amount of seed needed to cover 2000 m^2.

37. An overseas flight charges $5.40 for each kilogram or part of a kilogram over 50 kg of luggage weight. How much extra must be paid for three pieces of luggage weighing 20 kg 400 g, 18 kg 300 g, and 15 kg?

38. A variety goods store buys nuts in 10-kg containers and repackages the nuts for resale. The store packages the nuts in 200-g bags and sells them for $1.27 per bag. Find the profit on a 10-kg container of nuts costing $30.

39. A train car is loaded with 15 automobiles weighing 1405 kg each. Find the total weight of the automobiles.

40. During 1 year there were 2076 million kg of cotton fiber produced, 237 million kg of wood fibers produced, and 2678 million kg of synthetic fibers produced. What percent of the total fiber production is the cotton fiber production? (Round to the nearest hundredth of a percent.)

SECTION 9.3 Capacity

Objective A To convert units of capacity in the metric system of measurement

The basic unit of capacity in the metric system is the liter. One **liter** is defined as the capacity of a box that is 10 cm long on each side.

10 cm
10 cm
10 cm

The units of capacity in the metric system have the same prefixes as the units of length:

1 kiloliter (kl) = 1000 L
1 hectoliter (hl) = 100 L
1 decaliter (dal) = 10 L
1 liter (L) = 1 L

1 deciliter (dl) = 0.1 L
1 centiliter (cl) = 0.01 L
1 milliliter (ml) = 0.001 L

The milliliter is equal to 1 **cubic centimeter** (cm^3).

1 cm
1 cm
1 cm
1 ml = 1 cm^3

Conversion between units of capacity in the metric system involves moving the decimal point to the right or to the left. Listing the units in order from largest to smallest will indicate how many places to move the decimal point and in which direction.

To convert 824 ml to liters, first write the units in order from largest to smallest.

kl hl dal L dl cl ml
3 positions

Converting ml to L requires moving 3 positions to the left.

824 ml = 0.824 L
3 places

Move the decimal point the same number of places and in the same direction.

Example 1 Convert 4 L 32 ml to liters.

Solution 32 ml = 0.032 L
4 L 32 ml = 4 L + 0.032 L
= 4.032 L

Example 2 Convert 2 kl 167 L to liters.

Your solution

Example 3 Convert 1.23 L to cubic centimeters.

Solution 1.23 L = 1230 ml = 1230 cm^3

Example 4 Convert 325 cm^3 to liters.

Your solution

Solutions on p. A34

Objective B To perform arithmetic operations with measurements of capacity

Arithmetic operations can be performed with measurements of capacity. A measurement involving two units should be written in terms of a single unit before adding, subtracting, multiplying, or dividing.

Subtract: 8 L 27 ml − 5 L 635 ml

8 L 27 ml = 8.027 L		8 L 27 ml = 8027 ml
−5 L 635 ml = 5.635 L	or	−5 L 635 ml = 5635 ml
2.392 L		2392 ml

Note that the units can be changed to either liters or milliliters before subtracting. In this textbook, unless otherwise stated, the units should be changed to the larger unit before the arithmetic operation is performed.

Example 5 Multiply: 4 L 147 ml × 9

Solution 4 L 147 ml = 4.147 L
× 9 = 9
37.323 L

Example 6 Divide: 22 kl 992 L ÷ 12

Your solution

Solution on p. A34

Objective C To solve application problems

Example 7

A laboratory assistant is in charge of ordering acid for three chemistry classes of 30 students each. Each student requires 80 ml of acid. How many liters of acid should be ordered? (The assistant must order by the whole liter.)

Strategy

To find the number of liters of acid to be ordered:

- Find the amount of acid needed by multiplying the number of classes by the number of students per class by the amount of acid required by each student.
- Convert to liters.

Solution 3 × 30 × 80 = 7200 ml
7200 ml = 7.2 L

The laboratory assistant should order 8 L of acid.

Example 8

For $42.50 a druggist buys 5 L of hand lotion and repackages it in 125 ml containers. Each container is sold for $2.79. Find the profit on the 5 L of hand lotion.

Your strategy

Your solution

Solution on p. A34

9.3 EXERCISES

Objective A

Convert.

1. 4200 ml = _____ L
2. 7.5 ml = _____ L
3. 3.42 L = _____ ml
4. 0.037 L = _____ ml
5. 423 ml = _____ cm^3
6. 0.32 ml = _____ cm^3
7. 642 cm^3 = _____ ml
8. 0.083 cm^3 = _____ ml
9. 42 cm^3 = _____ L
10. 3075 cm^3 = _____ L
11. 0.435 L = _____ cm^3
12. 2.57 L = _____ cm^3
13. 1 L 267 ml = _____ cm^3
14. 4 L 105 ml = _____ cm^3
15. 3 L 23 ml = _____ cm^3
16. 4.62 kl = _____ L
17. 0.035 kl = _____ L
18. 1423 L = _____ kl
19. 3 L 42 ml = _____ L
    = _____ ml
20. 1 L 127 ml = _____ L
    = _____ ml
21. 3 kl 4 L = _____ kl
    = _____ L
22. 6 kl 32 L = _____ kl
    = _____ L

Objective B

Perform the arithmetic operation.

23. 8 L 163 ml
    +4 L 275 ml
24. 3 L 123 ml
    +5 L 892 ml
25. 3 L 162 ml
    × 12
26. 4 L 792 ml ÷ 4
27. 10 L 72 ml
    − 3 L 818 ml
28. 437 ml + 3 L
29. 198 L 625 ml
    +932 L 747 ml
30. 62 L 47 ml × 3928
31. 423 L 89 ml ÷ 763

Objective C *Application Problems*

32. A can of tomato juice contains 1260 ml. How many 180-ml servings are in one can of tomato juice?

33. A Little League auxiliary serves punch to the ballplayers after the game. How many liters of punch are needed for 30 servings? Each serving contains 220 ml of punch.

34. A youth club uses 800 ml of chlorine each day for their swimming pool. How many liters of chlorine are used in a month of 30 days?

35. A flu vaccine is being given for the coming winter season. A medical corporation buys 12 L of flu vaccine. How many patients can be immunized if each patient receives 3 cm^3 of the vaccine?

36. There are 24 bottles in a case of shampoo. Each bottle of shampoo contains 320 ml. Find the number of liters in one case of shampoo.

37. A chemistry experiment requires 12 ml of an acid solution. How many liters of acid are used when 90 students perform the experiment?

38. A case of 12 1-L cans of apple juice costs $11.40. A case of 24 cans, each can containing 300 ml, of apple juice costs $6.90. Which case of apple juice is the better buy?

39. Nineteen percent of air is oxygen. Find the amount of oxygen in 50 L of air.

40. Six L of liquid soap are bought for $11.40 per liter. The soap is repackaged in 150-ml bottles and sold for $3.29 per bottle. Find the profit on the 6 L of liquid soap.

41. A pharmacy buys cough syrup in 5-L containers and then repackages the cough syrup in 250 ml bottles. Thirteen bottles of the cough syrup have been sold. How many bottles of the cough syrup are still in stock?

42. A wholesale distributer purchased 32 kl of cooking oil for $34,880. The wholesaler repackaged the cooking oil in 1.25-L bottles and sold each bottle for $2.17. Find the profit on the 32 kl of cooking oil.

43. A service station operator bought 85 kl of gasoline for $19,250. The gasoline was sold for $0.329 per liter. Find the profit on the 85 kl of gasoline.

SECTION 9.4 Energy

Objective A To use units of energy in the metric system of measurement

Two commonly used units of energy in the metric system are the calorie and the watthour.

The **calorie** (cal) is used for measuring heat energy. One calorie is the amount of energy that will raise the temperature of 1 kg of water 1 degree Celsius. One calorie is the energy required to lift 1 kg a distance of 427 m.

Swimming uses 480 cal per hour. How many calories would be used by swimming $\frac{1}{2}$ hour each day for 30 days?

To find the number of calories used: first find the number of hours spent swimming. $\frac{1}{2} \times 30 = 15$ h

Multiply the number of hours spent swimming by the calories used per hour. $15 \times 480 = 7200$ cal

7200 cal would be used by swimming $\frac{1}{2}$ hour each day for 30 days.

The **watthour** is used for measuring electrical energy. One watthour is the amount of energy required to lift 1 kg a distance of 370 m. A light bulb rated at 100 watts (W) will emit 100 watthours of energy each hour.

1000 watthours (Wh) = 1 kilowatthour (kWh)

A 150-W light bulb is on for 8 h. At 8¢ per kWh, find the cost of the energy used.

To find the cost: first find the number of watthours used. $150 \times 8 = 1200$ Wh

Convert to kilowatthours. 1200 Wh = 1.2 kWh

Multiply the number of kilowatthours used by the cost per kilowatthour. The cost of the energy used is \$0.096. $1.2 \times \$0.08 = \$0.096$

Example 1

Walking uses 180 cal per hour. How many calories will you burn off by walking $\frac{3}{4}$ h each day for 1 week?

Strategy To find the number of calories:

- Multiply to find the number of hours spent walking per week.
- Multiply the product by the calories burned off per hour.

Example 2

Housework requires 240 cal per hour. How many calories are used in 5 days by doing $1\frac{1}{2}$ h of active housework per day?

Your strategy

Solution on p. A34

Solution

$\frac{3}{4} \times 7 = \frac{21}{4} = 5\frac{1}{4}$ h

$5\frac{1}{4} \times 180 = 945$ cal

You will burn off 945 cal.

Your solution

Example 3

An iron is rated at 1200 W. If the iron is used for 1.5 h, how much energy is used in kilowatthours?

Strategy

To find the energy used:

- Multiply to find the number of watthours used.
- Convert to kilowatthours.

Solution

1200 W × 1.5 h = 1800 Wh
1800 Wh = 1.8 kWh
1.8 kWh of energy are used.

Example 4

Find the number of kilowatthours of energy used when a 150-W light bulb burns for 200 h.

Your strategy

Your solution

Example 5

A TV set rated at 1800 W is operated an average of 3.5 h per day. At 7.2¢ per kilowatthour, find the cost of operating the set for 1 week. Round to the nearest cent.

Strategy

To find the cost:

- Multiply to find the total number of hours the set is used.
- Multiply the product by the number of watts to find the watthours.
- Convert to kilowatthours.
- Multiply the number of kilowatthours by the cost per kilowatthour.

Solution

3.5 × 7 = 24.5 h
24.5 × 1800 = 44,100 Wh
44,100 Wh = 44.1 kWh
44.1 × $0.072 = $3.1752

The cost is approximately $3.18.

Example 6

A microwave oven rated at 500 W is used an average of 20 min per day. At 8.7¢ per kilowatthour, find the cost of operating the oven 30 days.

Your strategy

Your solution

Solutions on pp. A34–A35

9.4 EXERCISES

Objective A *Application Problems*

1. How many calories can be eliminated from your diet for 30 days if you omit one slice of bread per day? One slice of bread contains 110 cal.

2. You omit one egg containing 75 cal from your usual breakfast. If you continue this practice for 90 days, how many calories will you omit from your diet?

3. Moderately active people need 20 cal per pound of weight to maintain their weight. How many calories would a 150-lb moderately active person need per day to maintain weight?

4. A person whose daily activities would be characterized as light activity needs 15 cal per pound of body weight to maintain weight. How many calories would a 135-lb person with light activity need per day to maintain body weight?

5. You consume 350 cal for breakfast and 650 cal for lunch. Find the percent of the total daily intake of calories left for the evening meal if you limit your daily intake to 1800 cal. (Round to the nearest tenth of a percent.)

6. For a healthy diet, it is recommended that 55% of the daily intake of calories come from carbohydrates. Find the daily intake of calories from carbohydrates if you want to limit your calorie intake to 1600 cal.

7. Playing tennis singles requires 450 cal per hour. How many calories do you burn up in 30 days playing 45 min per day?

8. Cycling at 8 mi per hour requires 320 cal per hour. If you ride a bicycle for $1\frac{1}{2}$ h per day for 5 days a week, how many calories do you burn up in 4 weeks?

9. After playing golf for 3 h, you had a banana split containing 550 cal. Playing golf uses 320 cal per hour. How many calories did you gain or lose from these two activities?

10. Playing the violin requires 185 cal per hour. How many calories are consumed playing the violin for 45 min per day for 30 days?

11. Hiking requires approximately 315 cal per hour. How many hours would you have to hike to burn off the calories in a 375-cal sandwich, a 150-cal soda, and a 280-cal ice cream cone? (Round to the nearest tenth.)

12. Riding a bicycle requires 265 cal per hour. How many hours would you have to ride a bicycle to burn off the calories in a 320-cal milkshake, a 310-cal cheeseburger, and a 150-cal apple? (Round to the nearest tenth.)

13. An oven uses 500 W of energy. How many watthours of energy are used to cook a 5-kg roast for $2\frac{1}{2}$ h?

14. Find the cost of 560 kWh of electricity at $0.092 per kilowatthour.

15. A 21-in. color TV set is rated at 90 W. The TV is used an average of $3\frac{1}{2}$ h each day for a week. How many kilowatthours of energy are used during the week?

16. A 120-W bulb is kept burning 24 h a day. How many kilowatthours of electrical energy are used in 1 day?

17. A 120-W stereo set is on an average of 2 h a day. Find the cost of listening to the stereo for 2 weeks if the cost is 9.4¢ per kilowatthour. (Round to the nearest cent.)

18. How much does it cost to run a 2200-W air conditioner for 8 h at 9¢ per kilowatthour? (Round to the nearest cent.)

19. A 1200-W hair dryer is used an average of 5 h per week. How many kilowatthours of energy are used each week?

20. A space heater is used for 3 h. The heater uses 1400 W per hour. Electricity cost 11.1¢ per kilowatthour. Find the cost of using the electric heater. (Round to the nearest cent.)

21. A house is insulated to save energy. The house used 265 kWh of electrical energy per month before insulation and saved 45 kWh of energy per month after insulation. What percent decrease does this amount represent? (Round to the nearest tenth.)

22. A household uses an average of 16.3 kWh of electrical energy each day. Electrical energy costs 10.2¢ per kilowatthour. Find the cost of using electrical energy for 31 days for the household. (Round to the nearest cent.)

23. A welder used 6.5 kw of energy each hour. Find the cost of using the welder for 6 h a day for 30 days. The cost is 9.4¢ per kilowatthour.

24. A manufacturing company uses 3250 kWh of energy each day. At 9.5¢ per kilowatthour, find the annual cost of electrical energy.

SECTION 9.5 Conversion Between the U.S. Customary and the Metric System of Measurement

Objective A To convert U.S. Customary units to metric units

More than 90% of the world's population use the metric system of measurement. Therefore, converting U.S. Customary units to metric units is essential in trade and commerce, for example, in importing foreign goods and exporting domestic goods. Approximate equivalences between the two systems are:

Units of Length	Units of Weight	Units of Capacity
1 m ≈ 3.28 ft	28.35 g ≈ 1 oz	1 L ≈ 1.06 qt
1 cm ≈ 0.39 in.	454 g ≈ 1 lb	
1.61 km ≈ 1 mi	0.454 kg ≈ 1 lb	
0.91 m ≈ 1 yd	1 kg ≈ 2.2 lb	
0.305 m ≈ 1 ft		
2.54 cm ≈ 1 in.		
1 m ≈ 1.09 yd		

These equivalences can be used to form conversion rates to change one unit of measurement to another. For example, because 1.61 km ≈ 1 mi, the following conversion rates equivalent to 1 can be written:

$$\frac{1.61 \text{ km}}{1 \text{ mi}} = \frac{1 \text{ mi}}{1.61 \text{ km}} = 1$$

Convert 55 mi to kilometers.

$$55 \text{ mi} \approx 55 \text{ mi} \times \boxed{\frac{1.61 \text{ km}}{1 \text{ mi}}}$$

The conversion rate must contain km in the numerator and mi in the denominator.

$$= \frac{55 \cancel{\text{mi}}}{1} \times \frac{1.61 \text{ km}}{1 \cancel{\text{mi}}}$$

$$= \frac{88.55 \text{ km}}{1}$$

55 mi ≈ 88.55 km

Example 1 Convert 4 gal to liters. Round to the nearest tenth.

Solution Because the conversion factor for gallons to liters is not given, convert gallons to quarts and then convert quarts to liters.

$$4 \text{ gal} \approx 4 \cancel{\text{gal}} \times \frac{4 \cancel{\text{qt.}}}{1 \cancel{\text{gal}}} \times \frac{1 \text{ L}}{1.06 \cancel{\text{qt.}}} = \frac{16 \text{ L}}{1.06}$$

4 gal ≈ 15.1 L

Example 2 Convert 10 c to liters. Round to the nearest hundredth.

Your solution

Solution on p. A35

Example 3 Convert 45 mi/h to kilometers per hour.

Solution $\frac{45 \text{ mi}}{\text{h}} \approx \frac{45 \not{\text{mi}}}{\text{h}} \times \frac{1.61 \text{ km}}{1 \not{\text{mi}}}$

≈ 72.45 km/h

Example 4 Convert 60 ft/s to meters per second. Round to the nearest hundredth.

Your solution

Example 5 The price of gasoline is \$1.52/gal. Find the cost per liter to the nearest cent.

Solution $\frac{\$1.52}{\text{gal}} \approx \frac{\$1.52}{\not{\text{gal}}} \times \frac{1 \not{\text{gal}}}{4 \not{\text{qt}}} \times \frac{1.06 \not{\text{qt}}}{\text{L}}$

$\approx \$.40$/L

Example 6 The price of milk is \$1.86/gal. Find the cost per liter to the nearest cent.

Your solution

Solutions on p. A35

Objective B To convert metric units to U.S. Customary units

Metric units are already being used in the United States today. Gasoline may be sold by the liter, 35-mm film is available, cereal is sold by the gram, and 100-mm-long cigarettes are common. The same conversion rates used in Objective A are used for converting metric units to U.S. customary units.

Example 7 Convert 200 m to feet.

Solution $200 \text{ m} \approx 200 \not{\text{m}} \times \frac{3.28 \text{ ft}}{1 \not{\text{m}}}$
$200 \text{ m} \approx 656$ ft

Example 8 Convert 45 cm to inches. Round to the nearest hundredth.

Your solution

Example 9 Convert 90 km/h to miles per hour. Round to the nearest hundredth.

Solution $\frac{90 \text{ km}}{\text{h}} \approx \frac{90 \not{\text{km}}}{\text{h}} \times \frac{1 \text{ mi}}{1.61 \not{\text{km}}}$

≈ 55.90 mi/h

Example 10 Express 75 km/h in miles per hour. Round to the nearest hundredth.

Your solution

Example 11 The price of gasoline is \$0.372/L. Find the cost per gallon to the nearest cent.

Solution

$\frac{\$0.372}{1 \text{ L}} \approx \frac{\$0.372}{1 \not{\text{L}}} \times \frac{1 \not{\text{L}}}{1.06 \not{\text{qt}}} \times \frac{4 \not{\text{qt}}}{1 \text{ gal}}$

$\approx \$1.40$/gal

Example 12 The price of ice cream is \$1.50/L. Find the cost per gallon to the nearest cent.

Your solution

Solutions on p. A35

9.5 EXERCISES

Objective A

Convert. Round to the nearest hundredth.

1. Convert the 100 yd dash to meters.

2. Find the weight in kilograms of a 135-lb person.

3. Find the height in meters of a person 6 ft 4 in.

4. Find the number of liters in 1 gal of punch.

5. How many kilograms does a 12-lb ham weigh?

6. Find the number of liters in 14.3 gal of gasoline.

7. Find the number of milliliters in 1 c.

8. The winning long jump at a track meet was 29 ft 2 in. Convert this distance to meters.

9. Express 65 mi/h in kilometers per hour.

10. Express 30 mi/h in kilometers per hour.

11. Bacon costs $1.69/lb. Find the cost per kilogram.

12. Peaches cost $0.69/lb. Find the cost per kilogram.

13. The cost of gasoline is $1.47/gal. Find the cost per liter.

14. Paint costs $9.80/gal. Find the cost per liter.

15. A hiker planning a 5-day backpacking trip decides to hike an average of 5 h each day. Hiking requires an extra 320 cal per hour. How many pounds will the hiker lose during the trip if an extra 900 cal is consumed each day? (3500 cal are equivalent to 1 lb.)

16. Swimming requires 550 cal per hour. How many pounds could be lost by swimming $1\frac{1}{2}$ h each day for 5 days if no extra calories were consumed? (3500 cal are equivalent to 1 lb.)

17. The distance around the earth is 24,887 mi. Convert this distance to kilometers.

18. The distance from the earth to the sun is 93,000,000 mi. Convert this distance to kilometers.

Objective B

Convert. Round to the nearest hundredth.

19. Convert the 100-m dash to feet.

20. Find the weight in pounds of a 98-kg person.

21. Find the number of gallons in 6 L of antifreeze.

22. Your height is 1.65 m. Find your height in inches.

23. Find the distance of the 1500-m race in feet.

24. Find the weight of 327 g of cereal in ounces.

25. How many gallons does a 48-L tank hold?

26. Find the width of 35-mm film in inches.

27. A bottle of syrup contains 876 ml. Find the number of pints in 876 ml.

28. A backpack tent weighs 2.1 kg. Find the weight in pounds.

29. Express 80 km/h in miles per hour.

30. Express 30 m/s in feet per second.

31. Gasoline costs 38.5¢/L. Find the cost per gallon.

32. A turkey roast costs $3.15/kg. Find the cost per pound.

33. A pat of butter contains 50 cal. How many pounds can be lost in 30 days by omitting one pat of butter from a daily diet? (3500 cal are equivalent to 1 lb.)

34. Hiking uses approximately 300 cal per hour. How many hours of hiking are necessary to lose 1 lb? (3500 cal are equivalent to 1 lb.)

35. A milk truck has a capacity of 3400 L. Convert this capacity to gallons.

36. A shipment of steel weighs 114,000 kg. Convert this weight to tons.

Calculators and Computers

Metric System The metric system was established in France in 1789. The basic unit of length in the metric system is the **meter,** the basic unit of capacity is the **liter,** and the basic unit of mass is the **gram.**

One of the most appealing ideas behind the metric system is the ease of a conversion from one unit in the system to another unit. Because the metric system uses a system of prefixes for its units, the name of a unit gives its relation to the standard unit. The same prefixes are used for length, capacity, and weight. The prefixes indicate multiples or divisions of the standard units. For example, the prefix "milli-" means "one thousandth of," so 1 millimeter (mm) denotes 1 one thousandth of a meter. The prefixes are:

kilo- = 1000	1 kilometer (km) = 1000 meters (m)
hecto- = 100	1 hectometer (hm) = 100 m
deca- = 10	1 decameter (dam) = 10 m
	1 meter (m) = 1 m
deci- = 0.1	1 decimeter (dm) = 0.1 m
centi- = 0.01	1 centimeter (cm) = 0.01 m
milli- = 0.001	1 millimeter (mm) = 0.001 m

As the prefixes indicate, the metric system is based on 10. Thus conversion from one unit to another is a matter of just moving the decimal point. Although this procedure seems quite simple, many people have a difficult time making the conversions. Thus, to help you learn the metric system, a program has been written to provide you with additional practice problems.

The program METRIC CONVERSIONS on the Math ACE disk will allow you to practice changing from one unit to another. Each of the primary units—**meter, liter,** and **gram**—will be practiced, along with the prefixes.

The above table is helpful when using this program.

Chapter Summary

Key Words The *metric system of measurement* is a system of measurement based on the decimal system.

The basic unit of measurement of length in the metric system is the *meter.*

The basic unit of mass in the metric system is the *gram.*

The basic unit of capacity in the metric system is the *liter.*

The basic unit of energy in the metric system is the *calorie.*

The *watthour* is also used in the metric system for measuring electrical energy.

Essential Rules Prefixes to the basic unit denote the magnitude of each unit in the metric system.

kilo-	1000
hecto-	100
deca-	10
deci-	0.1
centi-	0.01
milli-	0.001

Conversion between units in the metric system involves moving the decimal point:

1. When converting from a larger unit to a smaller unit, move the decimal point to the **right.**
2. When converting from a smaller unit to a larger unit, move the decimal point to the **left.**

Chapter Review

SECTION 9.1

1. Convert 0.37 cm to millimeters.

2. Convert 1.25 km to meters.

3. Add: $\begin{array}{r} 5 \text{ m } 67 \text{ cm} \\ +3 \text{ m } 88 \text{ cm} \\ \hline \end{array}$

4. Subtract: $\begin{array}{r} 56 \text{ cm } 3 \text{ mm} \\ -35 \text{ cm } 8 \text{ mm} \\ \hline \end{array}$

5. Multiply: $\begin{array}{r} 4 \text{ m } 55 \text{ cm} \\ \times \quad 8 \\ \hline \end{array}$

6. Divide: $15\overline{)52 \text{ km } 500 \text{ m}}$

7. Three pieces of wire fence are cut from a 50-m roll. The three pieces measure 2 m 40 cm, 5 m 60 cm, and 4 m 80 cm. How much wire fence is left on the roll after the three pieces are cut?

SECTION 9.2

8. Convert 0.450 g to milligrams.

9. Convert 4.050 kg to grams.

10. Add: $\begin{array}{r} 4 \text{ g } 677 \text{ mg} \\ +\ 9 \text{ g } 566 \text{ mg} \\ \hline \end{array}$

11. Subtract: $\begin{array}{r} 45 \text{ kg } \ \ 45 \text{ g} \\ -\ 32 \text{ kg } 585 \text{ g} \\ \hline \end{array}$

12. Multiply: $\begin{array}{r} 3 \text{ kg } 450 \text{ g} \\ \times \quad 11 \\ \hline \end{array}$

13. Divide: $18\overline{)45 \text{ g } 340 \text{ mg}}$
Round to the nearest thousandths.

14. Find the total cost of a 7-kg 300-g turkey costing $2.79 per kilogram. Round to the nearest cent.

SECTION 9.3

15. Convert 0.0056 L to milliliters.

16. Convert 1.2 L to cubic centimeters.

17. Add:
$$\begin{array}{r} 3\text{ L } \ 45\text{ ml} \\ +\ 7\text{ L } 568\text{ ml} \\ \hline \end{array}$$

18. Subtract:
$$\begin{array}{r} 15\text{ L } 569\text{ ml} \\ -\ \ 8\text{ L } 972\text{ ml} \\ \hline \end{array}$$

19. Multiply:
$$\begin{array}{r} 5\text{ L } 122\text{ ml} \\ \times \qquad 20 \\ \hline \end{array}$$

20. Divide: $25\overline{)45\text{ L } 250\text{ ml}}$

21. One-hundred twenty-five guests are expected to attend a reception. Assuming that each person drinks 400 ml of coffee, how many liters of coffee should be prepared?

SECTION 9.4

22. A large egg contains approximately 90 cal. How many calories can be eliminated from your diet in 1 month by eliminating one large egg per day from your usual breakfast?

23. A TV uses 240 W of energy. The set is on an average of 5 h a day in a 30-day month. At a cost of 9.5¢ per kilowatthour, how much does it cost to run the set 30 days?

SECTION 9.5

24. Convert the 1000-m run to yards. Round to the nearest tenth.

25. A backpack tent weighs 1.90 kg. Find the weight in pounds. Round to the nearest hundredth.

26. Ham costs $3.40 per pound. Find the cost per kilogram.

27. Cycling burns up approximately 400 cal per hour. How many hours of cycling is necessary to lose 1 lb? (3500 cal is equivalent to 1 lb.)

Chapter Test

1. Convert 42.6 mm to centimeters.

2. Convert 5 km 38 m to meters.

3. Subtract:
7 m 63 cm − 2 m 98 cm

4. Add: 4 m 29 cm + 17 m 87 cm

5. A carpenter needs 30 rafters, each rafter 3 m 80 cm long. Find the total length of rafters needed in meters.

6. Twenty-five rivets are used to fasten two steel plates together. The plates are 4 m 20 cm long, and the rivets are equally spaced, with a rivet at each end. Find the distance between the rivets.

7. Convert 3.29 kg to grams.

8. Convert 3 g 89 mg to grams.

9. Multiply: 3 kg 480 g × 7

10. Subtract:
17 g 164 mg − 9 g 867 mg

11. A 20 × 20-cm tile weighs 250 g. Find the weight of a box of 144 tiles in kilograms.

12. Two hundred grams of fertilizer are used for each tree in apple orchard containing 1200 trees. At \$2.25 per kilogram of fertilizer, how much does it cost to fertilize the apple orchard?

13. Convert 3.25 L to milliliters.

14. Convert 1.6 L to cubic centimeters.

15. Divide: 3 L 750 ml ÷ 5

16. Subtract:
7 L 180 ml − 3 L 249 ml

17. The community health clinic is giving flu shots for the coming flu season. Each flu shot contains 2 cm^3 of vaccine. How many liters of vaccine are needed to inoculate 2600 people?

18. A TV set rated at 1600 W is operated an average of 4 h per day. Electrical energy costs 8.5¢ per kilowatthour. How much does it cost to operate the TV set 30 days?

19. The record ski jump for men is 636 ft. Convert this value to meters. Round to the nearest hundredth. (1 m = 3.28 ft)

20. The record ski jump for women is 110 m. Convert this value to feet. Round to the nearest hundredth. (1 m = 3.28 ft)

Cumulative Review

1. Simplify:
$12 - 8 \div (6 - 4)^2 \cdot 3$

2. Add: $5\frac{3}{4} + 1\frac{5}{6} + 4\frac{7}{9}$

3. Subtract: $4\frac{2}{9} - 3\frac{5}{12}$

4. Divide: $5\frac{3}{8} \div 1\frac{3}{4}$

5. Simplify: $\left(\frac{2}{3}\right)^4 \cdot \left(\frac{9}{4}\right)^2$

6. Subtract: 12.0072 − 9.937

7. Solve the proportion $\frac{5}{8} = \frac{n}{50}$. Round to the nearest tenth.

8. Write $1\frac{3}{4}$ as a percent.

9. 6.09 is 4.2% of what number?

10. Convert 18 pt to gallons.

11. Convert: 18 m 75 cm = ____ m

12. Subtract: 4 km 420 m − 1 km 892 m

13. Convert: 5 kg 50 g = ____ kg

14. Add: 3 g + 672 mg

15. Divide: 12 kg 450 g ÷ 15

16. Subtract: 6 L − 452 ml

17. A family has a monthly income of $2244 per month. The family spends one-fourth of its monthly income on rent. How much money is left after the rent is paid?

18. The state income tax on a business is \$620 plus 0.08 times the profit the business makes. The business made a profit of \$82,340.00 last year. Find the amount of income tax the business paid.

19. The property tax on a \$45,000 is \$900. At the same rate, what is the property tax on a home worth \$75,000?

20. A car dealer offers new car buyers a 12% rebate on some new car models. What rebate would a new car buyer receive on a car that cost \$13,500?

21. An investor received a dividend of \$533 on an investment of \$8200. What percent of the investment is the dividend?

22. You received grades of 78, 92, 45, 80, and 85 on five English exams. Find your average grade.

23. A ski instructor receives a salary of \$22,500. The ski instructor's salary will increase by 12% next year. What will the salary be next year?

24. A sporting goods store has regularly priced \$80 fishing rods on sale for \$62.40. What is the discount rate?

25. Forty-eight blocks, each block 9 in. long, are laid end to end to make the base for a wall. Find the length of the wall in feet.

26. A jar of apple juice contains 24 oz. Find the number of quarts of apple juice in a case of 16 jars.

27. A garage mechanic buys oil in 40-gal containers. The mechanic buys the oil for \$2.88 per gallon and sells the oil for \$1.09 per quart. Find the profit on one 40-gal container of oil.

28. A school swimming pool uses 1 L 200 ml of chlorine each school day. How much chlorine is used for 20 days during the month?

29. A 1200-W hair dryer is used an average of 30 min a day. At a cost of 10.5¢ per kilowatthour, how much does it cost to operate the hair dryer 30 days?

30. Convert 60 mi/h to kilometers per hour. Round to the nearest tenth. (1.61 km = 1 mi)

10

Geometry

OBJECTIVES

- To define and describe lines and angles
- To define and describe geometric figures
- To solve problems involving angles formed by intersecting lines
- To find the perimeter of geometric figures
- To find the perimeter of composite geometric figures
- To solve application problems
- To find the area of geometric figures
- To find the area of composite geometric figures
- To solve application problems
- To find the volume of geometric solids
- To find the volume of composite geometric solids
- To solve application problems
- To use a table to find the square root of a number
- To find the unknown side of a right triangle using the Pythagorean Theorem
- To solve application problems
- To solve similar and congruent triangles
- To solve application problems

Mobius Strips and Klein Bottles

Some geometric shapes have some very unusual characteristics. Among these figures are a Mobius strip and a Klein bottle.

A Mobius strip is formed by taking a long strip of paper and twisting it one-half turn. The resulting figure is called a "one-sided" surface. It is one-sided in the sense that if you tried to paint the strip in one continuous motion beginning at one spot the entire surface would be painted the same color, unlike a strip that has not been twisted.

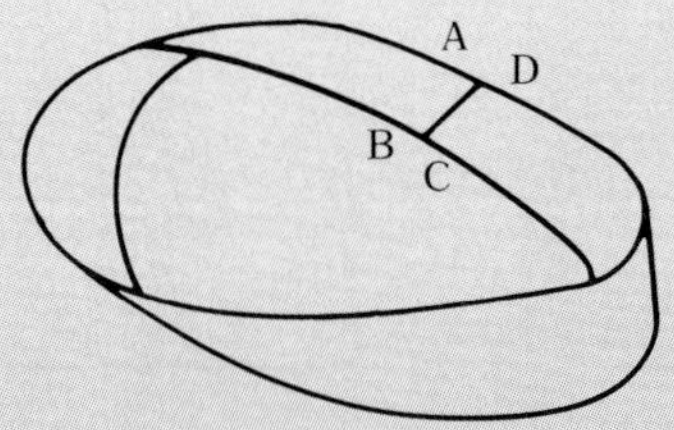

Another remarkable fact of being one sided can be demonstrated by making a Mobius strip and sealing the junction with tape. Now cut the strip by cutting along the center of the strip. Try this; you will be amazed at the result.

A second interesting surface is called a Klein bottle, which is a one-sided surface with no edges and no "inside" or "outside." A Klein bottle is formed by pulling the small open end of a tapering tube through the side of the tube and joining the ends of the small open end to the ends of the larger open end.

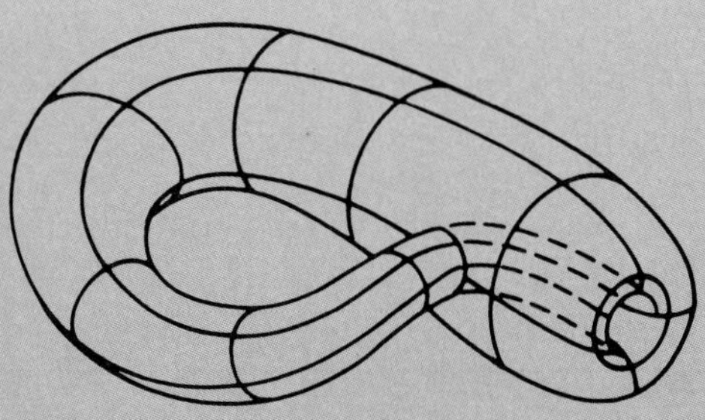

SECTION 10.1 Angles, Lines, and Geometric Figures

Objective A To define and describe lines and angles

The word "geometry" comes from the Greek words for "earth" *(geo)* and "measure." The original purpose of geometry was to measure land. Today geometry is used in many sciences, such as physics, chemistry, and geology, and in applied fields such as mechanical drawing and astronomy. Geometric form is used in art and design.

Two basic geometric concepts are plane and space.

A **plane** is a flat surface, such as a tabletop or a blackboard. Figures that can lie totally in a plane are called **plane figures.**

Space extends in all directions. Objects in space, such as trees, ice cubes, or doors, are called **solids.**

A **line** extends indefinitely in two directions in a plane. A line has no width.

A **line segment** is part of a line and has two end points. The line segment AB is shown in the figure.

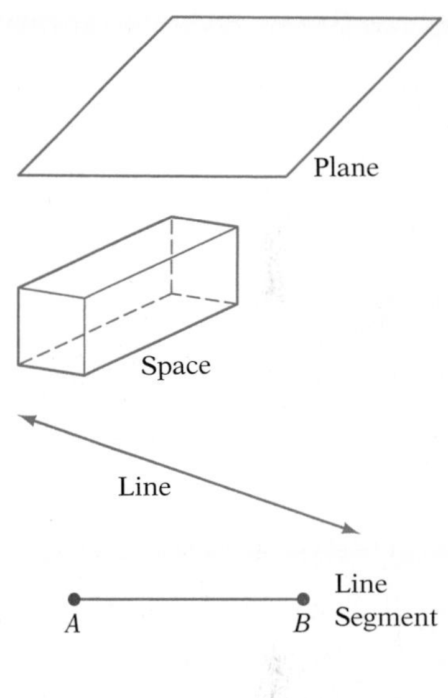

The length of a line segment is the distance between the endpoints of the line segment. The length of a line segment may be expressed as the sum of two or more shorter line segments, as shown. For this example, $AB = 5$, $BC = 3$, and $AC = AB + BC = 5 + 3 = 8$. When no units, such as feet or meters, are given for the length, the distances are assumed to be in the same units of length.

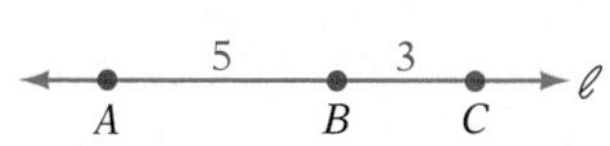

Given that $AB = 22$ and $AC = 31$, find the length of BC.

BC is the difference between AC and AB. $BC = AC - AB$

Substitute 22 for AB and 31 for AC and solve for BC.

$BC = 31 - 22$
$BC = 9$

Lines in a plane can be parallel or intersecting. **Parallel lines** never meet; the distance between them is always the same. **Intersecting lines** cross at a point in the plane.

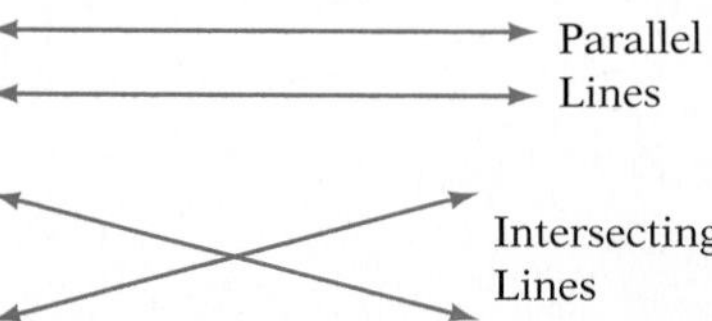

The symbol ∥ means "is parallel to." In the accompanying figure, $AB \parallel CD$ and $p \parallel q$. Note that line p contains line segment AB and line q contains line segment CD. Parallel lines contain parallel line segments.

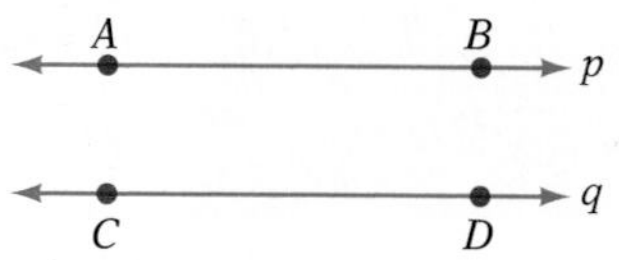

A **ray** starts at a point and extends indefinitely in one direction.

An **angle** is formed when two rays start from the same point. Rays r_1 and r_2 start from point B. The common endpoint is called the **vertex** of the angle.

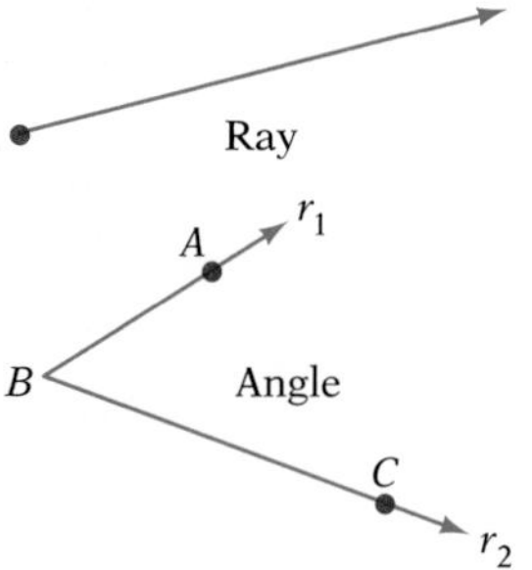

If A and C are points on rays r_1 and r_2, respectively, then the angle is called $\angle B$ or $\angle ABC$, where $\angle$ is the symbol for angle. Note that the angle is named by the vertex, or the vertex is the second point listed when the angle is named by giving three points.

An angle can also be named by a variable written between the rays close to the vertex. In the figure, $\angle x = \angle QRS = \angle SRQ$ and $\angle y = \angle SRT = \angle TRS$. Note that in this figure more than two rays meet at the vertex. In this case, the vertex cannot be used to name the angle.

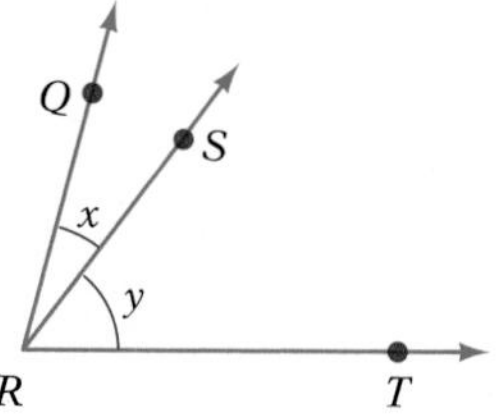

An angle is measured in **degrees.** The symbol for degree is °.

One complete revolution is 360° (360 degrees).

$\frac{1}{4}$ of a revolution is 90°. A 90° angle is called a **right angle.** The symbol ∟ represents a right angle.

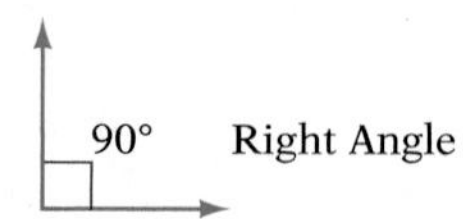

Perpendicular lines are intersecting lines that form right angles.

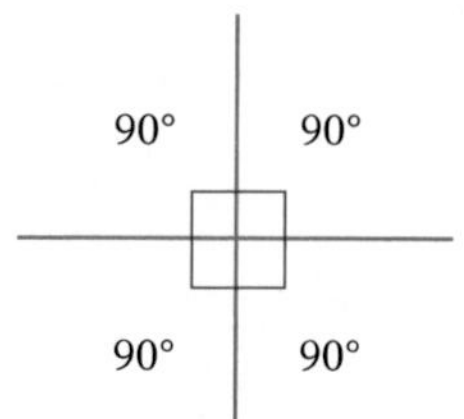

The symbol ⊥ means "is perpendicular to." In the accompanying figure, $AB \perp CD$ and $p \perp q$. Note that line p contains line segment AB and line q contains line segment CD. Perpendicular lines contain perpendicular line segments.

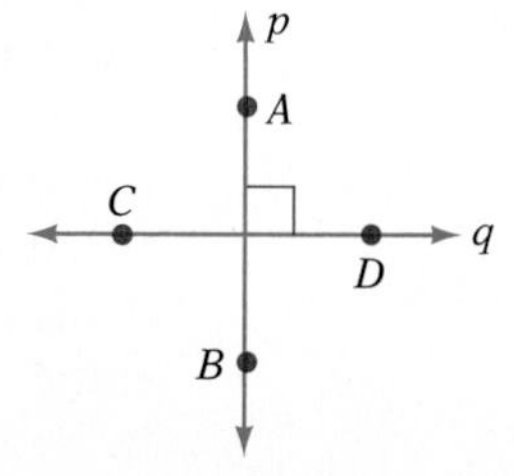

Complementary angles are two angles whose sum is 90°.

$\angle A + \angle B = 70° + 20° = 90°$
$\angle A$ and $\angle B$ are complementary angles.

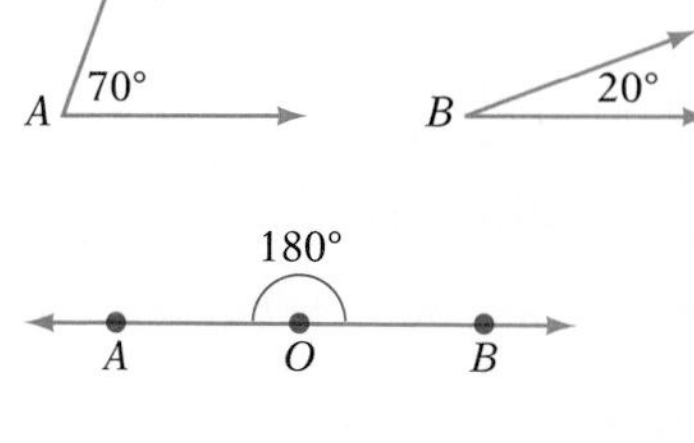

$\frac{1}{2}$ of a revolution is 180°. A 180° angle is called a **straight angle.** $\angle AOB$ in the figure is a straight angle.

Supplementary angles are two angles whose sum is 180°.

$\angle A + \angle B = 130° + 50° = 180°$
$\angle A$ and $\angle B$ are supplementary angles.

An **acute angle** is an angle whose measure is between 0° and 90°. $\angle B$ in the figure above is an acute angle. An **obtuse angle** is an angle whose measure is between 90° and 180°. $\angle A$ in the figure above is an obtuse angle.

In the accompanying figure, $\angle DAC = 45°$ and $\angle CAB = 55°$.

$\angle DAB = \angle DAC + \angle CAB = 45° + 55° = 100°$

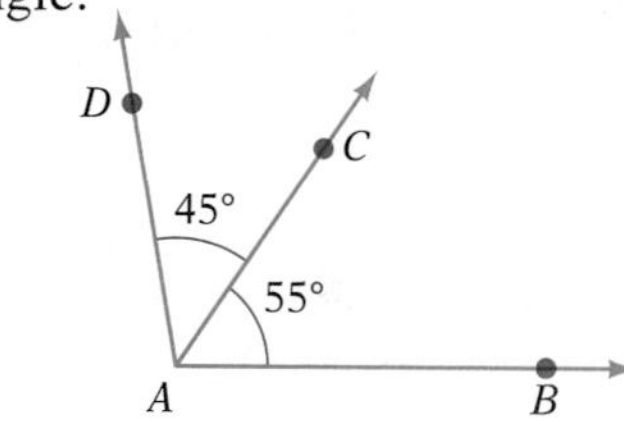

Example 1

Given that $MN = 15$, $NO = 18$, and $MP = 48$, find the length of OP.

Solution
$OP = MP - MN - NO$
$OP = 48 - 15 - 18$
$OP = 48 - 33$
$OP = 15$

Example 2

Given that $QR = 24$, $ST = 17$, and $QT = 62$, find the length of RS.

Your solution

Example 3

Find the complement of a 32° angle.

Solution To find an angle that, when added to 32°, will equal 90°, subtract 32° from 90°.

$90° - 32° = 58°$

58° is the complement of 32°.

Example 4

Find the complement of a 67° angle.

Your solution

Example 5

Find the supplement of a 105° angle.

Solution Subtract 105° from 180°.

$180° - 105° = 75°$

75° is the supplement of 105°.

Example 6

Find the supplement of a 32° angle.

Your solution

Solutions on p. A35

Example 7

Find the measure of $\angle x$ in the figure.

Solution

$\angle x = 90° - 47°$
$\angle x = 43°$

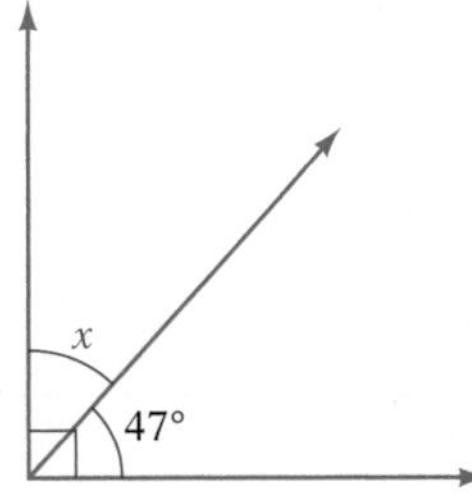

Example 8

Find the measure of $\angle a$ in the figure.

Your solution

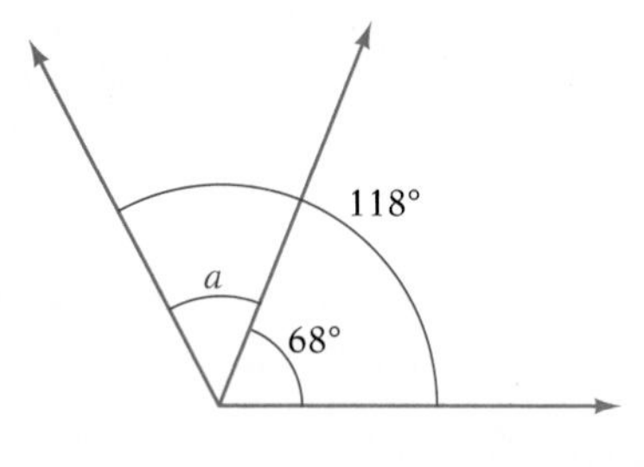

Solution on p. A35

Objective B To define and describe geometric figures

10

A **triangle** is a three-sided plane figure. Figure ABC is a triangle. AB is called the **base.** The line CD, perpendicular to the base, is called the **height.**

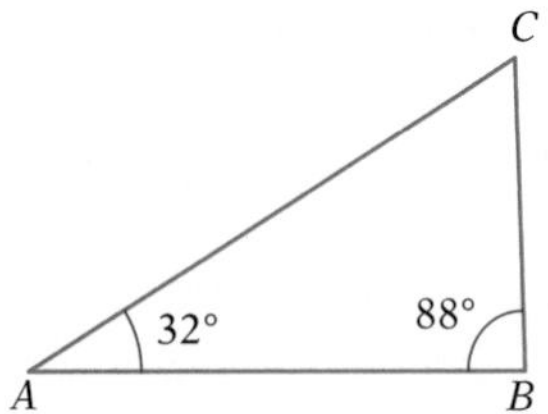

The sum of the three angles in a triangle is 180°.

Angle A + Angle B + Angle C = 180°

In the figure, $\angle A = 32°$ and $\angle B = 88°$.

The sum of the three angles is 180°; therefore

$\angle C = 180° - \angle A - \angle B = 180° - 32° - 88° = 60°$

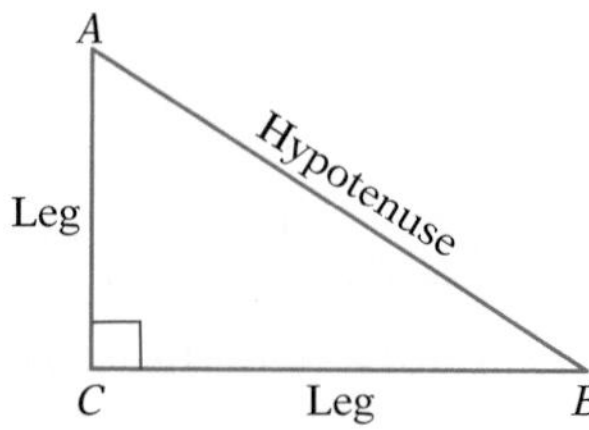

A **right triangle** contains one right angle. The side opposite the right angle is called the **hypotenuse.** The other two sides are called **legs.**

Angle A + Angle B = 90°

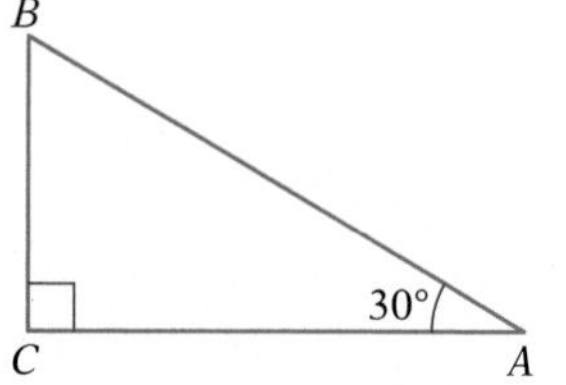

In right triangle ACB, angle $A = 30°$.

The sum of angles A and B equals 90°. Therefore

$\angle B = 90° - \angle A = 90° - 30° = 60°$

A **quadrilateral** is a four-sided plane figure. Three quadrilaterals with special characteristics are described here.

A **parallelogram** has opposite sides parallel and equal. The distance AE between the parallel sides is called the **height.**

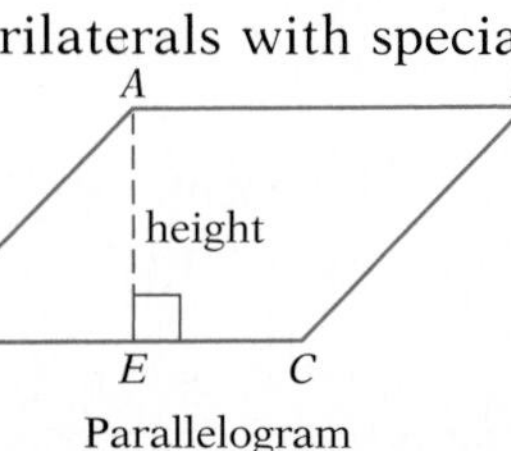

Parallelogram

A **rectangle** is a parallelogram that has four right angles.

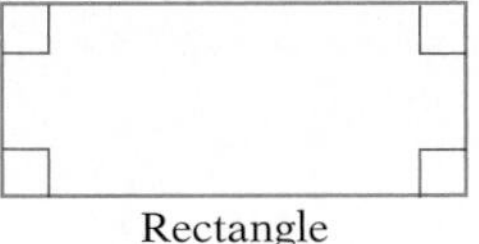
Rectangle

A **square** is a rectangle that has four equal sides.

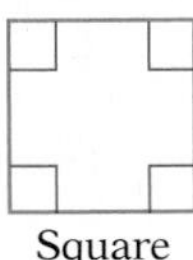
Square

A **circle** is a plane figure in which all points are the same distance from point O, called the **center** of the circle.

The **diameter** is a line segment across the circle going through point O. AB is a diameter of the circle.

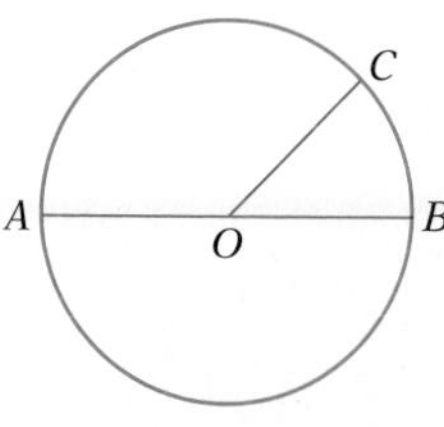

Circle

The **radius** is a line segment from the center of a circle to a point on the circle. OC is a radius of the circle.

Diameter $= 2 \times$ radius or Radius $= \frac{1}{2} \times$ diameter

The diameter in the circle shown is 8 in.

The radius is one-half the diameter. Therefore

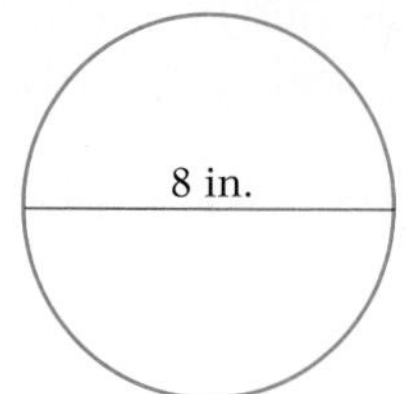

Radius $= \frac{1}{2} \times$ diameter $= \frac{1}{2} \times 8$ in. $= 4$ in.

Geometric solids are figures in space, or space figures. Four common space figures are the rectangular solid, cube, sphere, and cylinder.

A **rectangular solid** is a solid in which all six faces are rectangles.

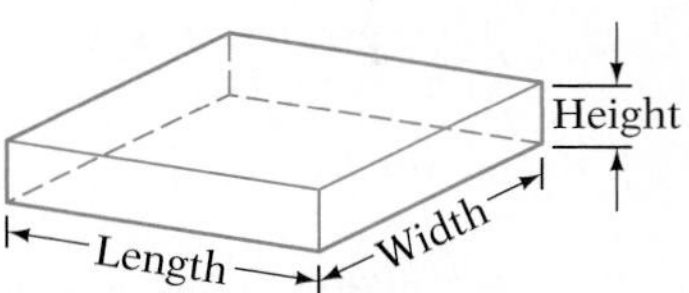

Rectangular Solid

A **cube** is a rectangular solid in which all six faces are squares.

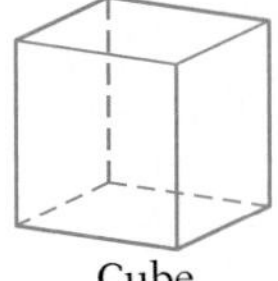
Cube

A **sphere** is a solid in which all points are the same distance from point O, called the **center** of the sphere.

The **diameter** of the sphere is a line segment across the sphere going through point O. AB is a diameter of the sphere.

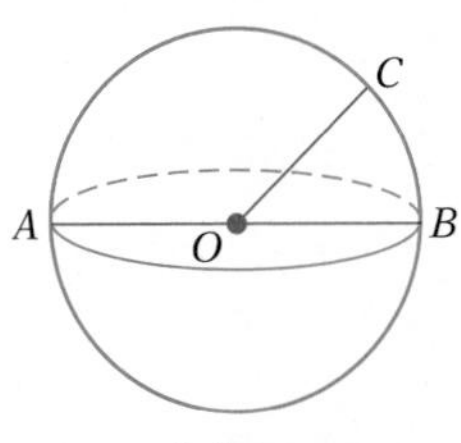

Sphere

The **radius** of the sphere is a line segment from the center to a point on the sphere. OC is a radius of the sphere.

Radius $= \frac{1}{2} \times$ diameter or Diameter $= 2 \times$ radius

The radius of the sphere shown is 5 cm.

The diameter is equal to twice the radius.

Diameter = 2 × radius = 2 × 5 cm = 10 cm

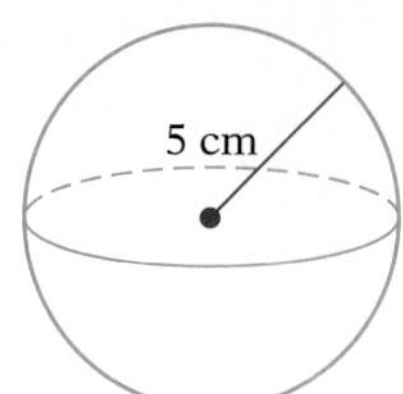

The most common **cylinder** is one in which the bases are circles and perpendicular to the height.

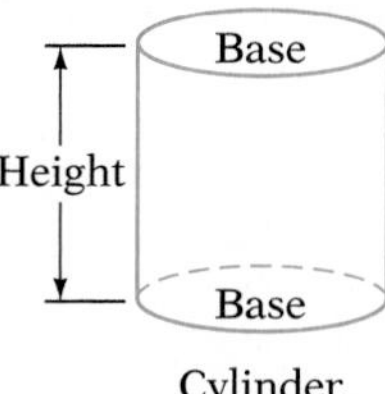

Cylinder

Example 9

One angle in a right triangle is equal to 30°. Find the measure of the other angles.

Solution A right triangle has one 90° angle. The sum of the other two angles equals 90°. Subtract 30° from 90°.

90° − 30° = 60°

The other angles of the triangle are 90° and 60°.

Example 10

A right triangle has one angle equal to 7°. Find the measure of the other angles.

Your solution

90° and 83°

Example 11

Two angles of a triangle are 42° and 103°. Find the measure of the third angle of the triangle.

Solution The sum of the three angles of a triangle is 180°. Add 42° + 103° and subtract the sum from 180° to find the third angle.

42° + 103° = 145°
180° − 145° = 35°

The third angle is 35°.

Example 12

Two angles of a triangle are 62° and 45°. Find the measure of the other angle.

Your solution

73°

Example 13

A circle has a radius of 8 cm. Find the diameter.

Solution Diameter = 2 · radius
= 2 · 8 cm = 16 cm

Example 14

A circle has a diameter of 8 in. Find the radius.

Your solution

4 in.

Solutions on p. A35

Objective C

To solve problems involving angles formed by intersecting lines

Four angles are formed by the intersection of two lines. If the two lines are perpendicular, each of the four angles is a right angle. If the two lines are not perpendicular, then two of the angles formed are acute angles and two of the angles are obtuse angles. The two acute angles are always opposite each other, and the two obtuse angles are always opposite each other.

In the figure, $\angle w$ and $\angle y$ are acute angles. $\angle x$ and $\angle z$ are obtuse angles. Two angles that are on opposite sides of the intersection of two lines are called **vertical angles.** Vertical angles have the same measure. $\angle w$ and $\angle y$ are vertical angles. $\angle x$ and $\angle z$ are vertical angles.

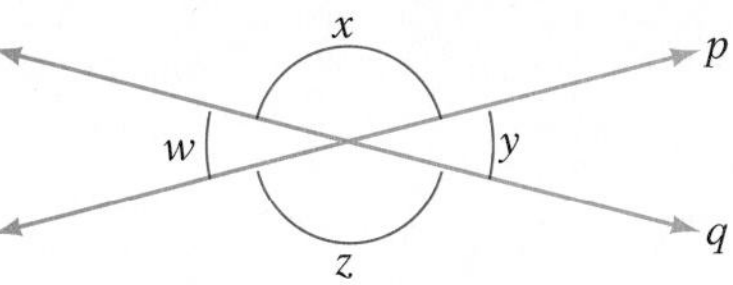

$\angle w = \angle y$
$\angle x = \angle z$

Two angles that share a common side are called **adjacent angles.** In the previous figure, $\angle x$ and $\angle y$ are adjacent angles, as are $\angle y$ and $\angle z$, $\angle w$ and $\angle x$, and $\angle w$ and $\angle z$. Adjacent angles of intersecting lines are supplementary angles.

$\angle x + \angle y = 180°$
$\angle y + \angle z = 180°$
$\angle z + \angle w = 180°$
$\angle w + \angle x = 180°$

In the accompanying figure, given that $\angle c = 65°$, find the measure of angles a, b, and d.

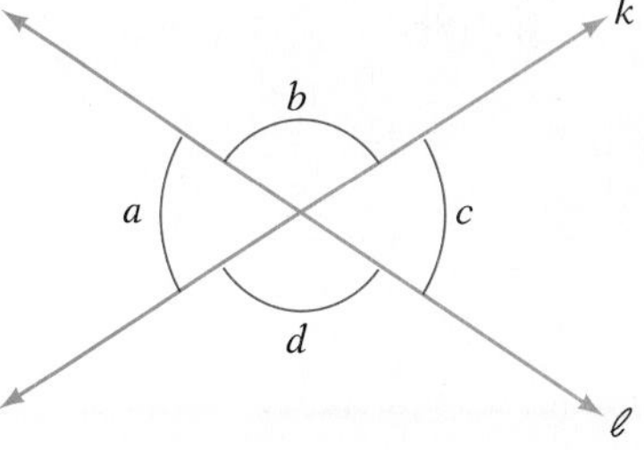

$\angle c = \angle a$ because $\angle c$ and $\angle a$ are vertical angles. $\angle a = 65°$

$\angle c$ is supplementary to $\angle b$ because $\angle c$ and $\angle b$ are adjacent angles. Subtract 65° from 180° to find $\angle b$.

$\angle b = 180° - 65°$
$\angle b = 115°$

$\angle b = \angle d$ because $\angle b$ and $\angle d$ are vertical angles. $\angle d = 115°$

A line intersecting two other lines at two different points is called a **transversal.**

If the lines cut by a transversal are parallel lines and the transversal is perpendicular to the parallel lines, all eight angles formed are right angles.

Transversal
ℓ_1
ℓ_2

If the lines cut by a transversal are parallel lines and the transversal is not perpendicular to the parallel lines, all four acute angles have the same measure and all four obtuse angles have the same measure. For the accompanying figure:

Transversal
a b d c w x z y
ℓ_1
ℓ_2

$\angle a = \angle c = \angle w = \angle y$ and $\angle b = \angle d = \angle x = \angle z$.

Alternate interior angles are two angles that are on opposite sides of the transversal and between the parallel lines. For the previous figure, $\angle c$ and $\angle w$ are alternate interior angles. $\angle d$ and $\angle x$ are alternate interior angles. Alternate interior angles have the same measure.

Alternate exterior angles are two angles that are on opposite sides of the transversal and outside the parallel lines. For the previous figure, $\angle a$ and $\angle y$ are alternate exterior angles. $\angle b$ and $\angle z$ are alternate exterior angles. Alternate exterior angles have the same measure.

Corresponding angles are two angles that are on the same side of the transversal and are both acute angles or are both obtuse angles. For the previous figure, the following pairs of angles are corresponding angles: $\angle a$ and $\angle w$, $\angle d$ and $\angle z$, $\angle b$ and $\angle x$, $\angle c$ and $\angle y$. Corresponding angles have the same measure.

In the accompanying figure, given that $\ell_1 \parallel \ell_2$ and $\angle c = 58°$, find the measures of $\angle f$, $\angle h$, and $\angle g$.

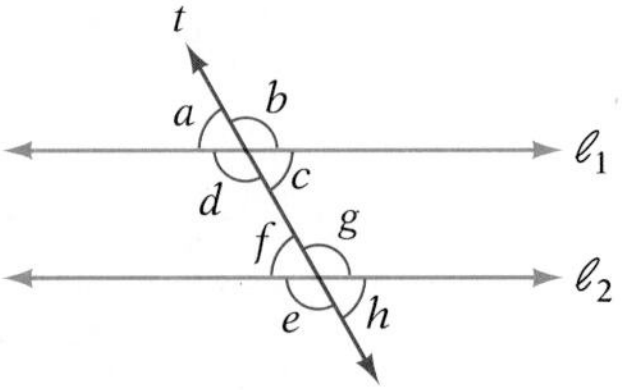

$\angle c$ and $\angle f$ are alternate interior angles. $\angle f = \angle c = 58°$

$\angle c$ and $\angle h$ are corresponding angles. $\angle h = \angle c = 58°$

$\angle g$ is supplementary to $\angle h$. Subtract 58° from 180° to find $\angle g$.

$\angle g = 180° - \angle h$
$\angle g = 180° - 58°$
$\angle g = 122°$

Example 15

In the figure, angle $a = 75°$. Find angle b.

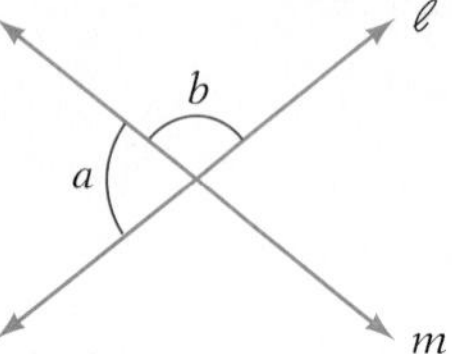

Solution Angles a and b are supplementary angles. Subtract 75° from 180°.

Angle $b = 180° - 75° = 105°$

Example 16

In the figure, angle $a = 125°$. Find angle b.

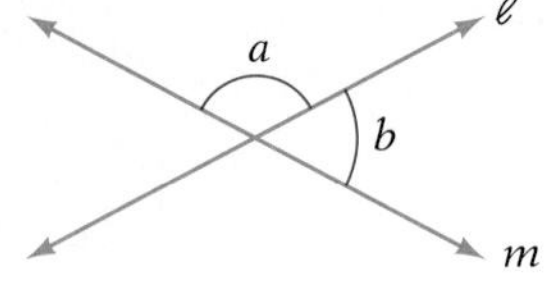

Your solution

Example 17

In the figure, $\ell_1 \parallel \ell_2$ and angle $a = 70°$. Find angle b.

Solution

Angles a and c are corresponding angles; therefore angle $c = 70°$.

Angles b and c are supplementary angles; therefore $\angle b = 180° - 70° = 110°$.

Example 18

In the figure, $\ell_1 \parallel \ell_2$ and angle $a = 120°$. Find angle b.

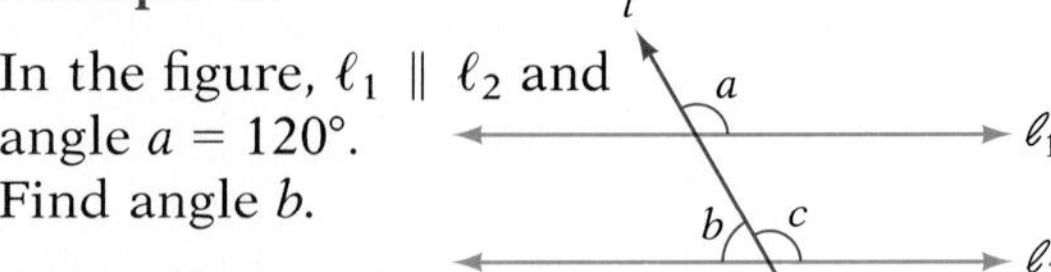

Your solution

Solutions on p. A36

10.1 EXERCISES

Objective

1. Name the geometric figure measured in degrees.

2. How many degrees are in one complete revolution?

3. How many degrees are in a straight angle?

4. Give the name of two lines that intersect at right angles.

5. In the figure, $EF = 20$ and $FG = 10$. Find the length of EG.

6. In the figure, $EF = 18$ and $FG = 6$. Find the length of EG.

7. In the figure, it is given that $QR = 7$ and $QS = 28$. Find the length of RS.

8. In the figure, it is given that $QR = 15$ and $QS = 45$. Find the length of RS.

9. In the figure, it is given that $AB = 12$, $CD = 9$, and $AD = 35$. Find the length of BC.

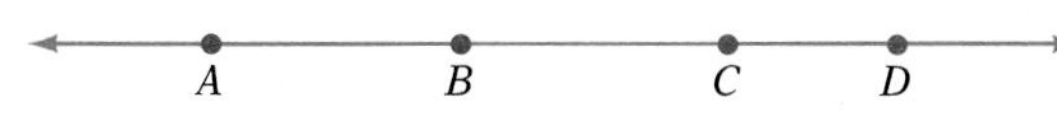

10. In the figure, it is given that $AB = 21$, $BC = 14$, and $AD = 54$. Find the length of CD.

11. Find the complement of a 31° angle.

12. Find the complement of a 62° angle.

13. Find the supplement of a 72° angle.

14. Find the supplement of a 162° angle.

15. Find the complement of a 13° angle.

16. Find the complement of a 88° angle.

17. Find the supplement of a 127° angle.

18. Find the supplement of a 7° angle.

19. In the figure, find the measure of angle AOB.

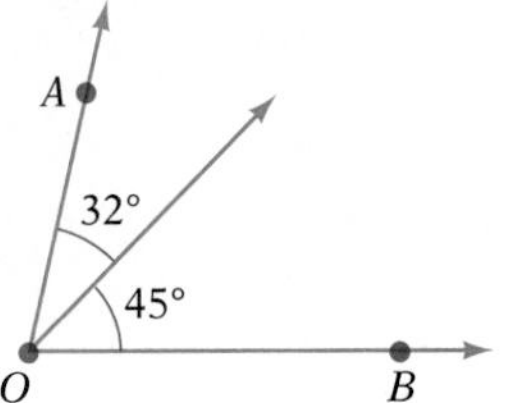

20. In the figure, find the measure of angle AOB.

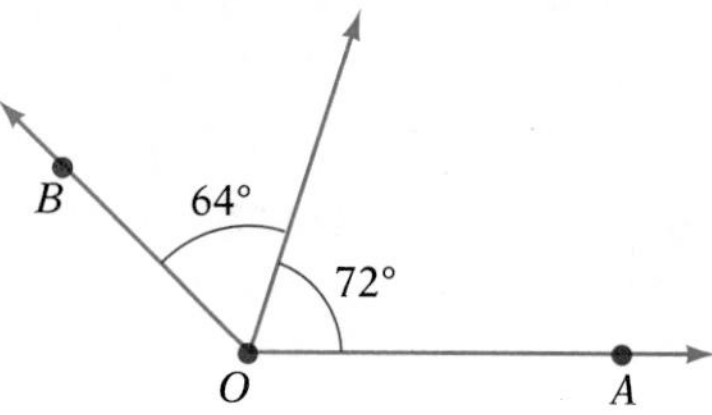

21. Find the measure of angle a in the figure.

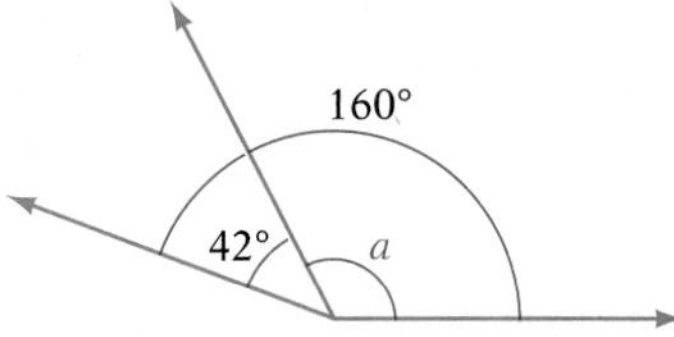

22. Find the measure of angle a in the figure.

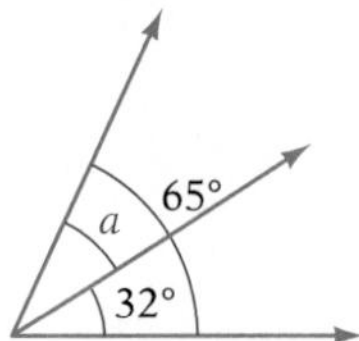

23. Find the measure of angle a in the figure.

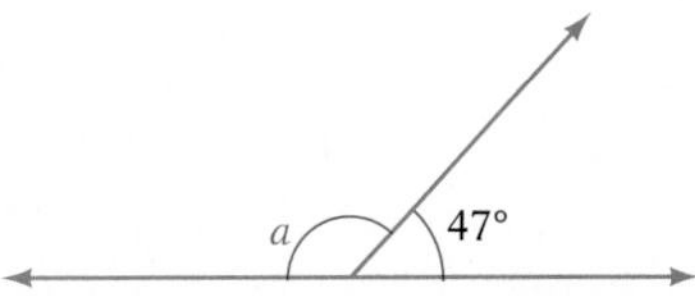

24. Find the measure of angle a in the figure.

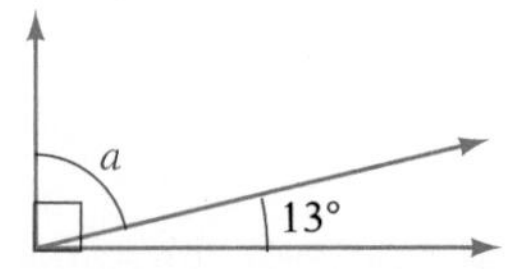

25. In the figure, it is given that $\angle LOM = 53°$ and $\angle LON = 139°$. Find the measure of $\angle MON$.

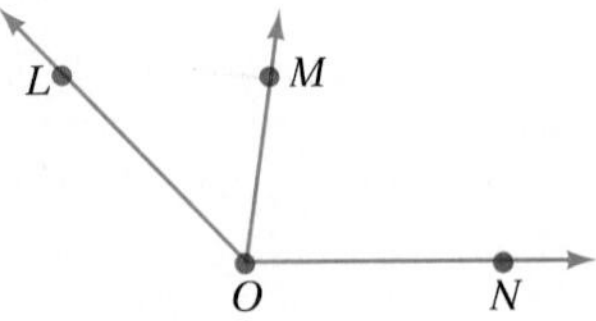

26. In the figure, it is given that $\angle MON = 38°$ and $\angle LON = 85°$. Find the measure of $\angle LOM$.

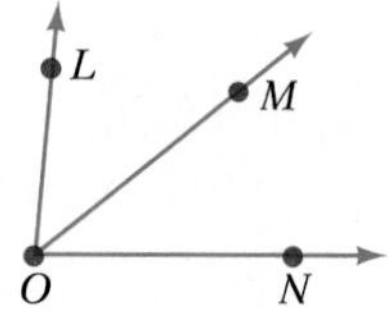

Objective B

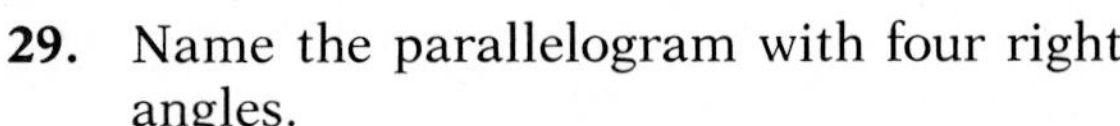

27. What is the sum of the three angles of a triangle?

28. Name the side opposite the right angle in a right triangle.

29. Name the parallelogram with four right angles.

30. Name the rectangle with four equal sides.

31. Name the rectangular solid in which all six faces are squares.

32. Name the solid in which all points are the same distance from the center.

33. Name the quadrilateral in which opposite sides are parallel and equal.

34. Name the plane figure in which all points are the same distance from the center.

35. Name the solid in which the bases are circular and perpendicular to the height.

36. Name the solid in which all the faces are rectangles.

37. A triangle has a 13° angle and a 65° angle. Find the measure of the other angle.

38. A triangle has a 105° angle and a 32° angle. Find the measure of the other angle.

39. A right triangle has a 45° angle. Find the measure of the other two angles.

40. A right triangle has a 62° angle. Find the measure of the other two angles.

41. A triangle has a 62° angle and a 104° angle. Find the measure of the other angle.

42. A triangle has a 30° angle and a 45° angle. Find the measure of the other angle.

43. A right triangle has a 25° angle. Find the measure of the other two angles.

44. Two angles of a triangle are 42° and 105°. Find the measure of the other angle.

45. Find the radius of a circle with a diameter of 16 in.

46. Find the radius of a circle with a diameter of 9 ft.

47. Find the diameter of a circle with a radius of $2\frac{1}{3}$ ft.

48. Find the diameter of a circle with a radius of 24 cm.

49. The radius of a sphere is 3.5 cm. Find the diameter.

50. The radius of a sphere is $1\frac{1}{2}$ ft. Find the diameter.

51. The diameter of a sphere is 4 ft 8 in. Find the radius.

52. The diameter of a sphere is 1.2 m. Find the radius.

Objective C

53. Find the measures of angles a and b in the figure.

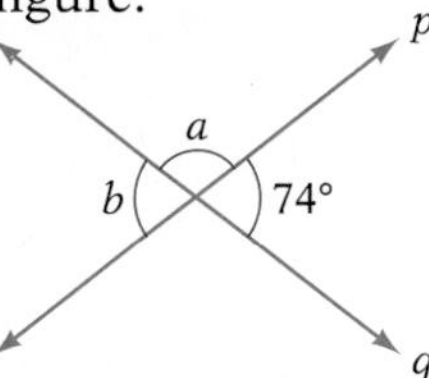

54. Find the measures of angles a and b in the figure.

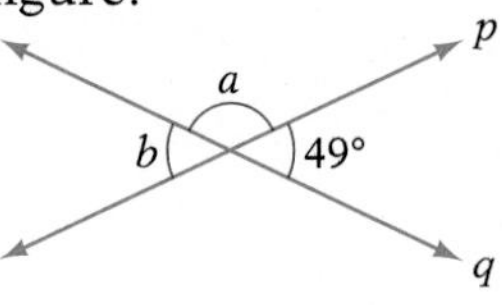

55. Find the measures of angles a and b in the figure.

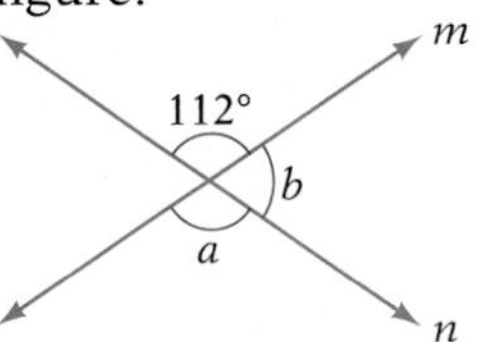

56. Find the measures of angles a and b in the figure.

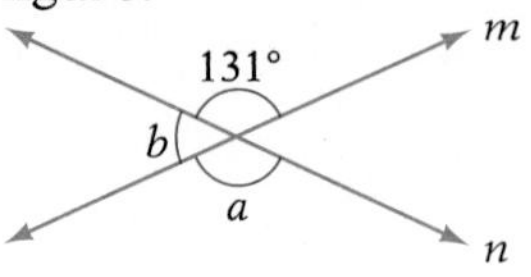

57. In the figure, it is given that $\ell_1 \parallel \ell_2$. Find the measures of angles a and b.

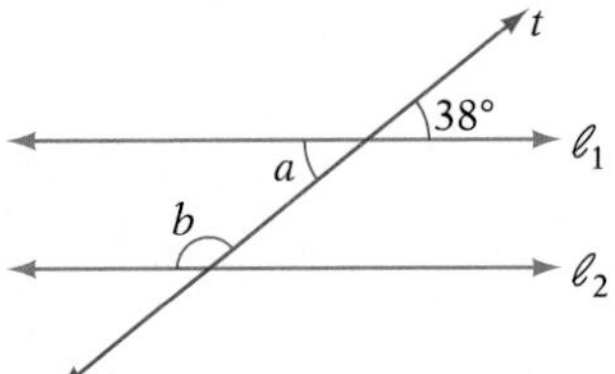

58. In the figure, it is given that $\ell_1 \parallel \ell_2$. Find the measures of angles a and b.

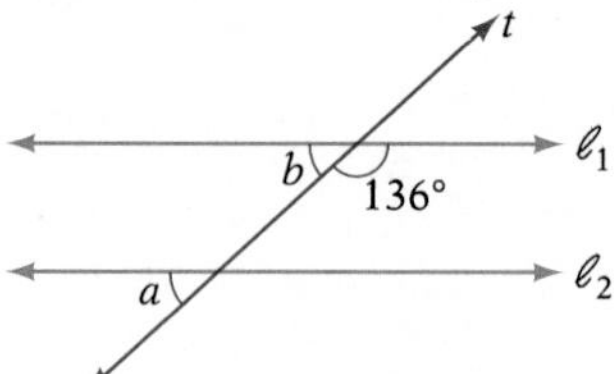

59. In the figure, it is given that $\ell_1 \parallel \ell_2$. Find the measures of angles a and b.

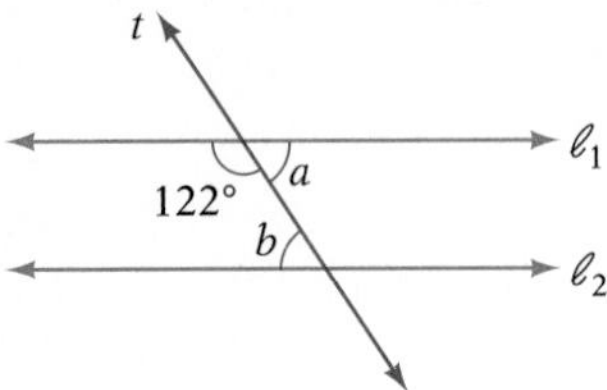

60. In the figure, it is given that $\ell_1 \parallel \ell_2$. Find the measures of angles a and b.

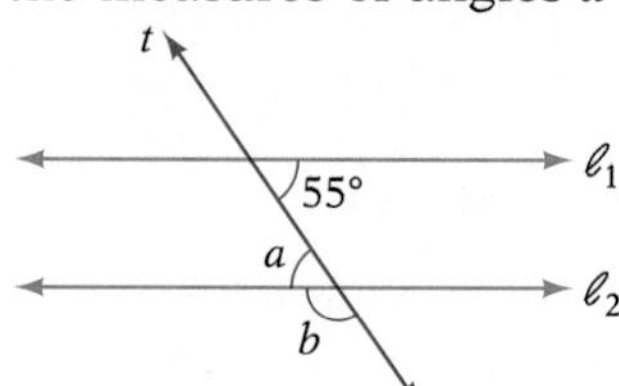

61. In the figure, it is given that $\ell_1 \parallel \ell_2$. Find the measures of angles a and b.

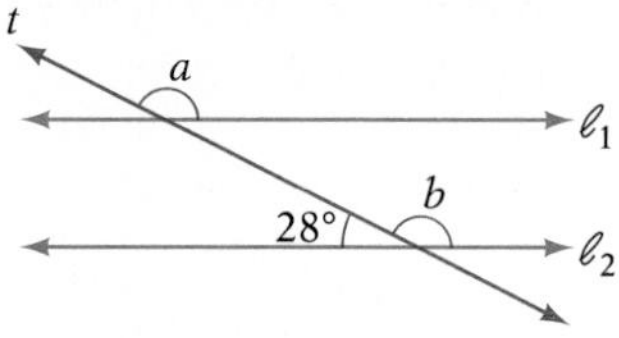

62. In the figure, it is given that $\ell_1 \parallel \ell_2$. Find the measures of angles a and b.

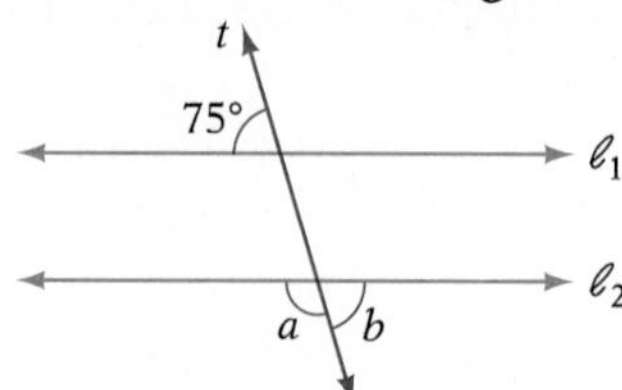

63. In the figure, it is given that $\ell_1 \parallel \ell_2$. Find the measures of angles a and b.

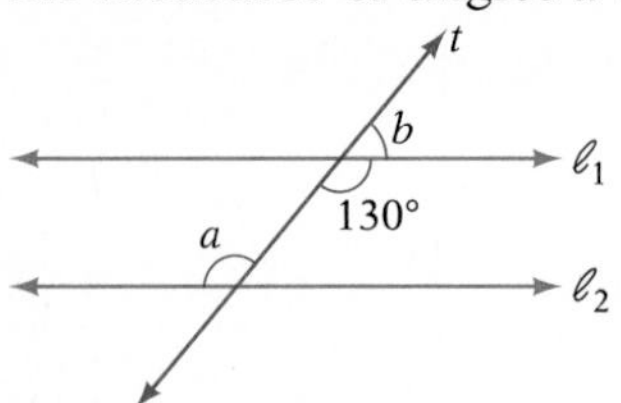

64. In the figure, it is given that $\ell_1 \parallel \ell_2$. Find the measures of angles a and b.

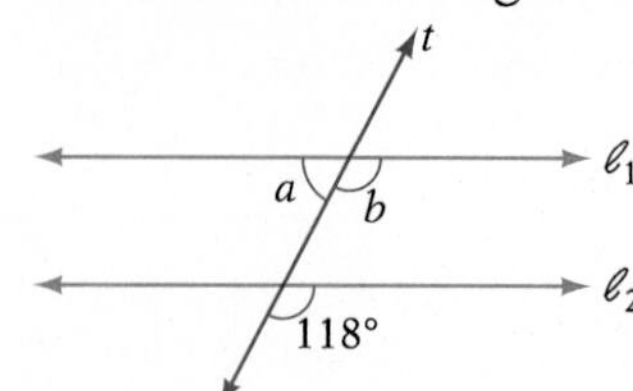

SECTION 10.2 Perimeter

Objective A To find the perimeter of geometric figures

Perimeter is the distance around a plane figure. Perimeter is used in buying fencing for a lawn or determining how much baseboard is needed for a room. The perimeter formulas for four common geometric figures are given here.

The perimeter of a triangle or quadrilateral is the sum of the length of each side.

Triangle

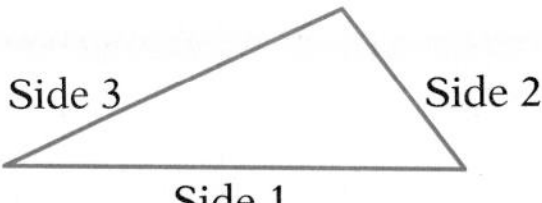

Perimeter = side 1 + side 2 + side 3

The perimeter of the triangle shown is the sum of three sides.

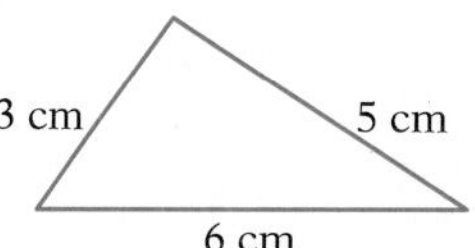

Perimeter = 3 cm + 5 cm + 6 cm = 14 cm

Square

Perimeter = side + side + side + side
= $4 \cdot$ side

The perimeter of the square is four times the length of one side:

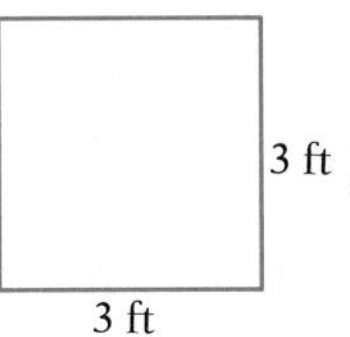

Perimeter = 4×3 ft = 12 ft

Rectangle

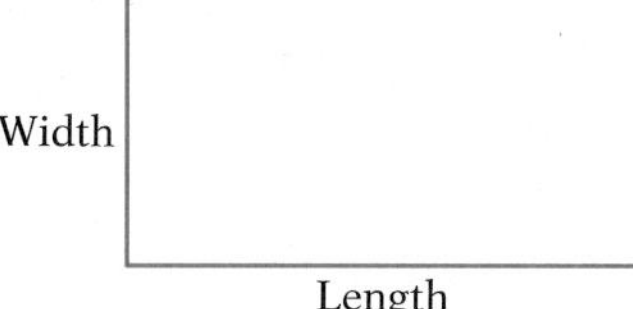

Perimeter = ($2 \cdot$ length) + ($2 \cdot$ width)

The perimeter of the rectangle shown is the sum of two times the width and two times the length:

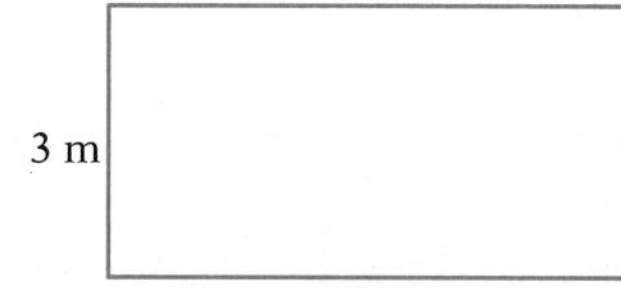

Perimeter = (2×3 m) + (2×6 m)
= 6 m + 12 m
= 18 m

Circle

The distance around a circle is called the **circumference.**

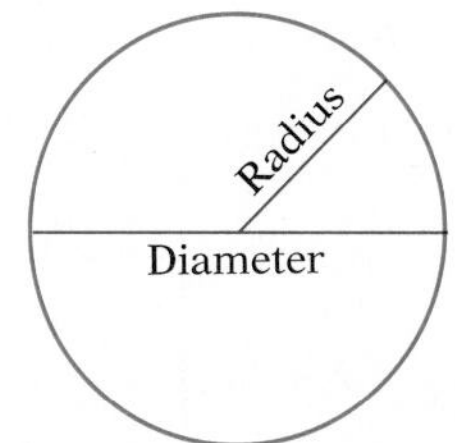

Circumference = $2 \cdot \pi \cdot$ radius
or
Circumference = $\pi \cdot$ diameter

The formula for circumference uses the number π (pi). The value of π can be approximated by a fraction or a decimal.

$$\pi \approx \frac{22}{7} \qquad \pi \approx 3.14$$

The circumference of the circle shown is two times π times the radius.

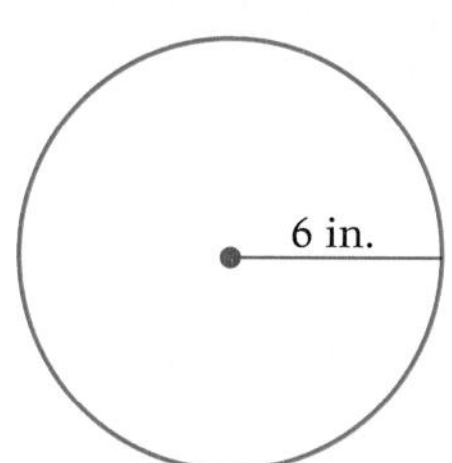

Circumference $= 2 \cdot \pi \cdot 6$ in. $\approx 2 \cdot 3.14 \cdot 6$ in.
Circumference $\approx$ 37.68 in

In this last example 37.68 in. is the approximate circumference of the circle. The answer to this example could also be written in terms of π.

Circumference $= 2 \cdot \pi \cdot 6$ in.
$= 12\pi$ in

The exact circumference of the circle is 12π in.

The approximate circumference of the circle is $12 \times 3.14 = 37.68$ in.

Example 1

Find the perimeter of a rectangle with a width of $\frac{2}{3}$ ft and a length of 2 ft.

Solution

Perimeter $= (2 \cdot \text{length}) + (2 \cdot \text{width})$

$= (2 \cdot 2 \text{ ft}) + (2 \cdot \frac{2}{3} \text{ ft})$

$= 4 \text{ ft} + \frac{4}{3} \text{ ft}$

$= 5\frac{1}{3} \text{ ft}$

Example 2

Find the perimeter of a rectangle with a length of 2 m and a width of 0.85 m.

Your solution

Example 3

Find the circumference of a circle with a radius of 18 cm. Use $\pi \approx 3.14$.

Solution

Circumference $= 2 \cdot \pi \cdot \text{radius}$
$\approx 2 \cdot 3.14 \cdot 18$ cm
Circumference $\approx$ 113.04 cm

Example 4

Find the circumference of a circle with a diameter of 6 in. Use $\pi \approx 3.14$.

Your solution

Solutions on p. A36

Objective B

To find the perimeter of composite geometric figures

Composite geometric figures are figures made from two or more geometric figures. The following composite is made from part of a rectangle and part of a circle:

Composite figure = 3 sides of a rectangle + $\frac{1}{2}$ the circumference of a circle

Perimeter of the composite figure $= (2 \cdot \text{length}) + \text{width} + \frac{1}{2} \cdot \pi \cdot \text{diameter}$

The perimeter of the composite figure below is found by adding the measures of twice the length plus the width plus one-half the circumference of the circle.

12 m, 4 m = 12 m, 4 m, 12 m + $\frac{1}{2}\pi \times 4$ m

$$\text{Perimeter} = 2 \cdot \text{length} + \text{width} + \frac{1}{2} \cdot \pi \cdot \text{diameter}$$

$$\approx 2 \cdot 12 \text{ m} + 4 \text{ m} + \frac{1}{2} \cdot 3.14 \cdot 4 \text{ m}$$

$$= 24 \text{ m} + 4 \text{ m} + 6.28 \text{ m}$$

Perimeter ≈ 34.28 m

Example 5

Find the perimeter of the composite figure. Use $\pi \approx \frac{22}{7}$.

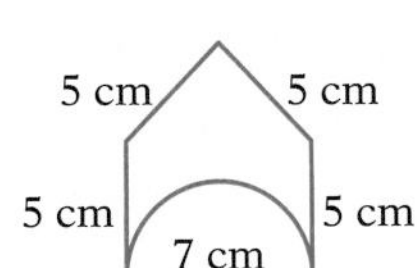

Solution

Perimeter of composite figure = sum of lengths of the 4 sides + $\frac{1}{2}$ the circumference of the circle

$$\approx (4 \cdot 5 \text{ cm}) + \frac{1}{2}\left(\frac{22}{7} \cdot 7 \text{ cm}\right)$$

$$= 20 \text{ cm} + 11 \text{ cm}$$

Perimeter ≈ 31 cm

Example 6

Find the perimeter of the composite figure. Use $\pi \approx 3.14$.

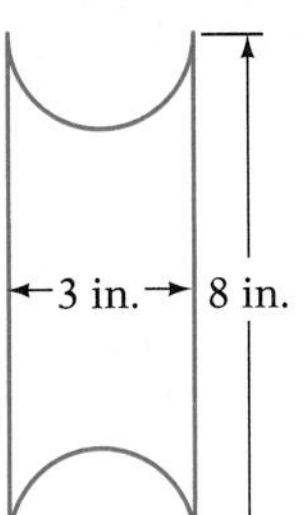

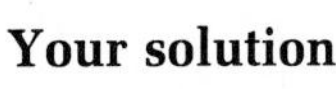

Your solution

Solutions on p. A36

Objective C To solve application problems

Example 7

If fencing costs \$2.75 per foot, how much will it cost to fence a rectangular lot that is 108 ft wide and 240 ft long?

Strategy

To find the cost of the fence:

- Find the perimeter of the lot.
- Multiply the perimeter by the per-foot cost of fencing.

Solution

$$\begin{aligned} \text{Perimeter} &= (2 \cdot \text{length}) + (2 \cdot \text{width}) \\ &= (2 \cdot 240 \text{ ft}) + (2 \cdot 108 \text{ ft}) \\ &= 480 \text{ ft} + 216 \text{ ft} \\ &= 696 \text{ ft} \\ \text{Cost} &= 696 \times \$2.75 = \$1914 \end{aligned}$$

The cost is \$1914.

Example 8

A metal strip is being installed around a workbench that is 0.74 m wide and 3 m long. At \$1.76 per meter, find the cost of the metal stripping. Round to the nearest cent.

Your strategy

Your solution

Solution on p. A36

10.2 EXERCISES

Objective A

In Exercises 1 to 8, find the perimeter or circumference of the given figures.

1. 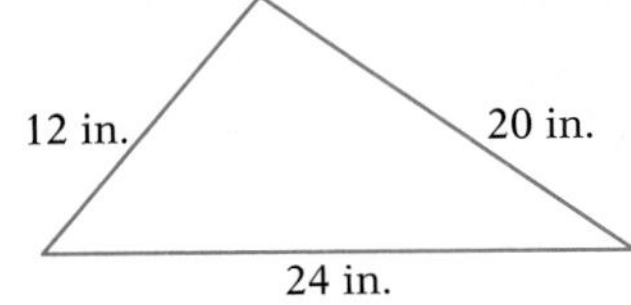

2. 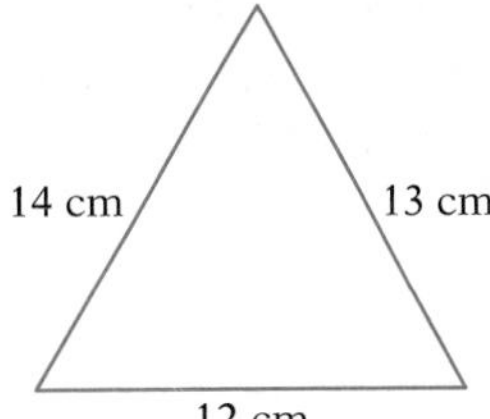

3. 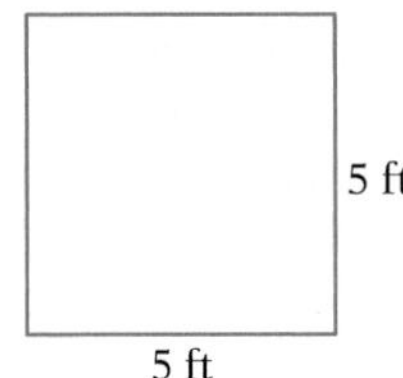

4. 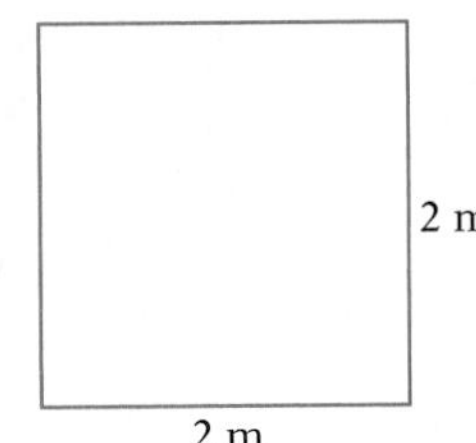

5. 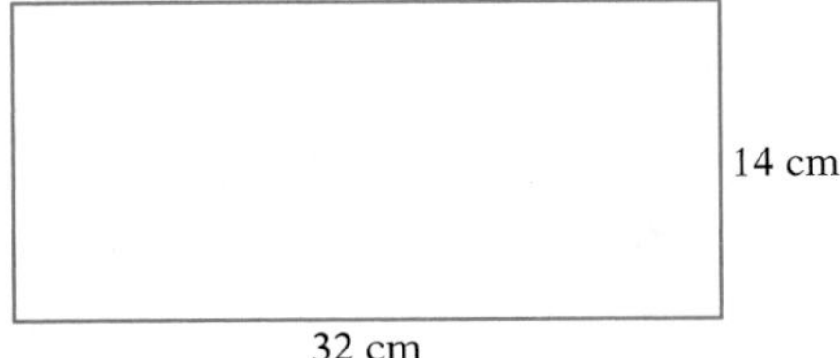

6. 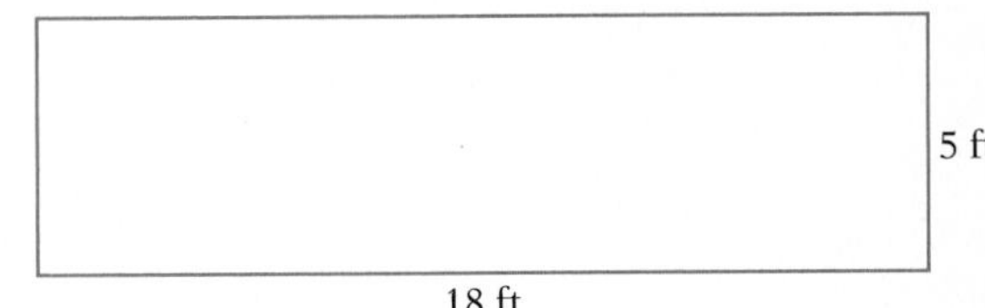

7. 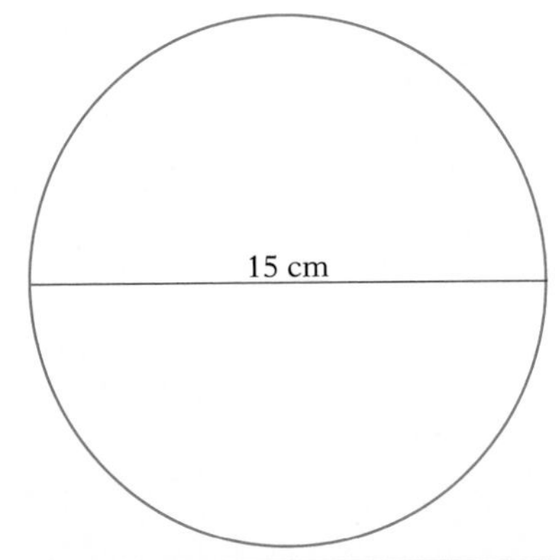

8. 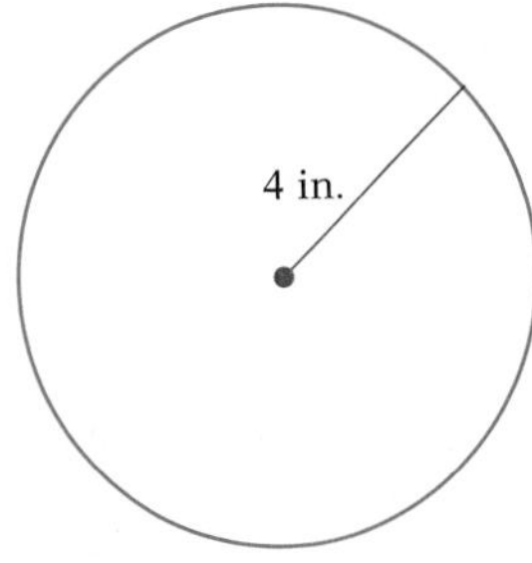

9. Find the perimeter of a triangle with sides 2 ft 4 in., 3 ft, and 4 ft 6 in.

10. Find the perimeter of a rectangle with a length of 2 m and a width of 0.8 m.

11. Find the circumference of a circle with a radius of 8 cm. Use $\pi \approx 3.14$.

12. Find the circumference of a circle with a diameter of 14 in. Use $\pi \approx \frac{22}{7}$.

13. Find the perimeter of a square in which the sides are equal to 60 m.

14. Find the perimeter of a triangle in which each side is $1\frac{2}{3}$ ft.

15. Find the perimeter of a five-sided figure with sides of 22 cm, 47 cm, 29 cm, 42 cm, and 17 cm.

16. Find the perimeter of a rectangular farm that is $\frac{1}{2}$ mi wide and $\frac{3}{4}$ mi long.

Objective B

In the figures accompanying Exercises 17 to 24, find the perimeter. Use $\pi \approx 3.14$.

17.

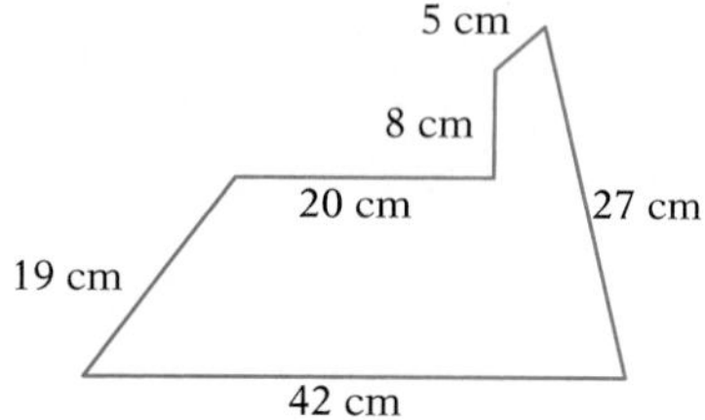

18.

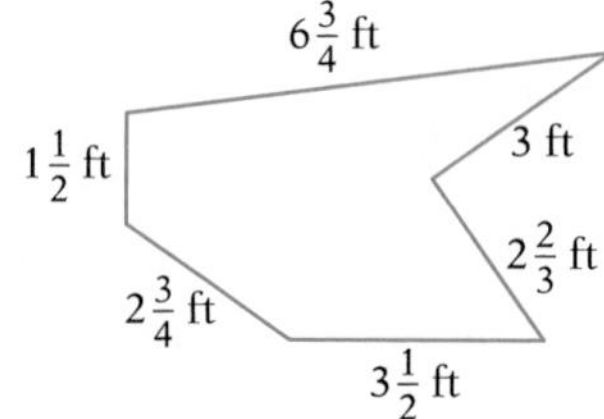

19.

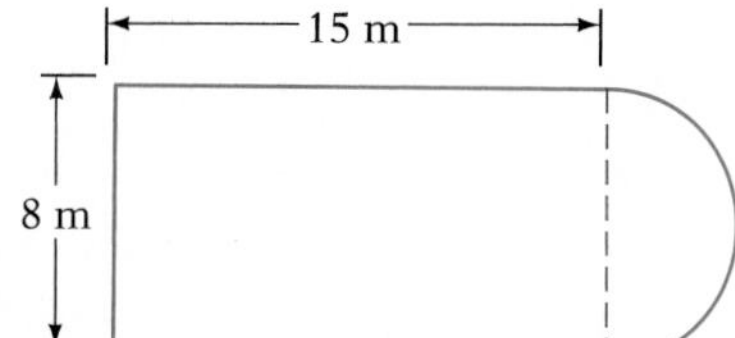

20.

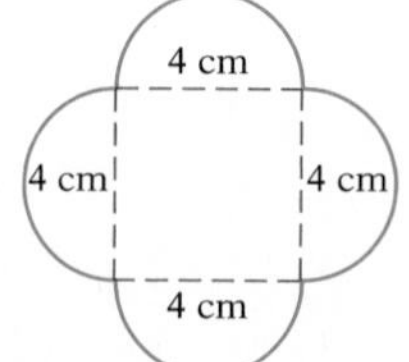

21.

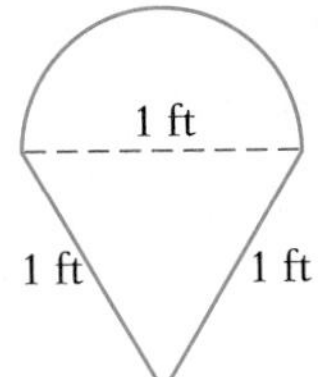

22.

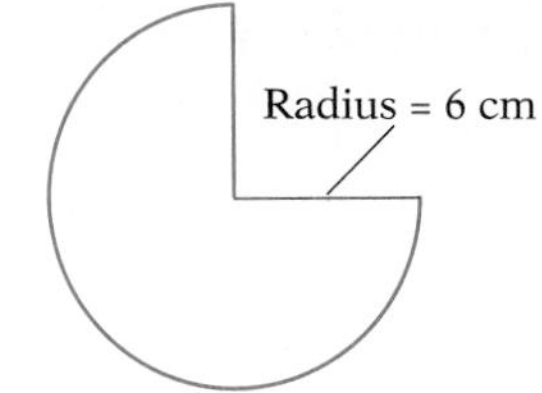

23.

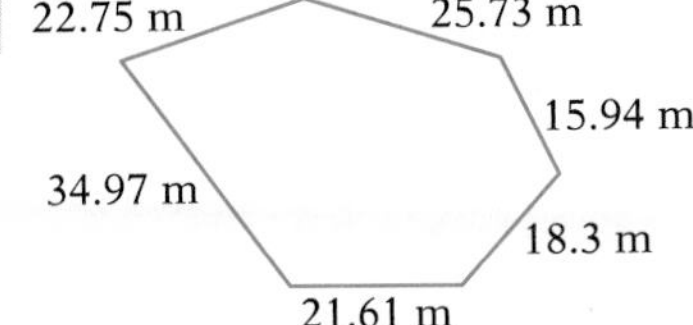

24.

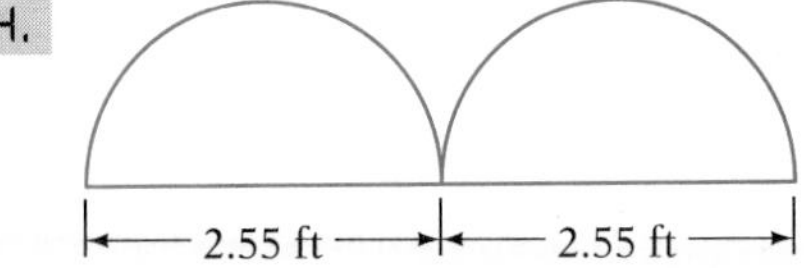

Objective C *Application Problems*

25. Find the amount of fencing needed to fence a farm that is $1\frac{1}{2}$ mi long and $\frac{3}{4}$ mi wide.

26. Find the number of feet of framing needed to frame a picture that is $2\frac{1}{2}$ ft by $1\frac{2}{3}$ ft.

27. How much binding is needed to bind the outside of a circular rug that is 5 m in diameter? Use $\pi \approx 3.14$.

28. Find the length of molding needed to put around a circular table that is 3.8 ft in diameter. Use $\pi \approx 3.14$. (Round to hundredths.)

29. A bicycle tire has a diameter of 24 in. How many feet does the bicycle travel if the wheel makes 5 revolutions? Leave the answer in terms of π.

30. A tricycle tire has a diameter of 12 in. How many feet does the tricycle travel if the wheel makes 8 revolutions? Leave the answer in terms of π.

31. Find the length of weather stripping installed around the door shown in the figure below. Use $\pi \approx 3.14$.

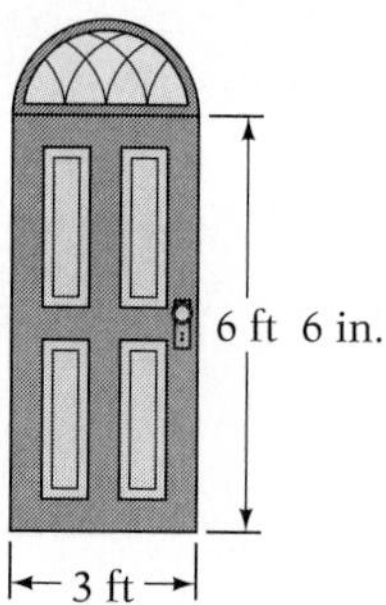

32. Find the perimeter of a roller rink with the dimensions shown in the figure. Use $\pi \approx 3.14$.

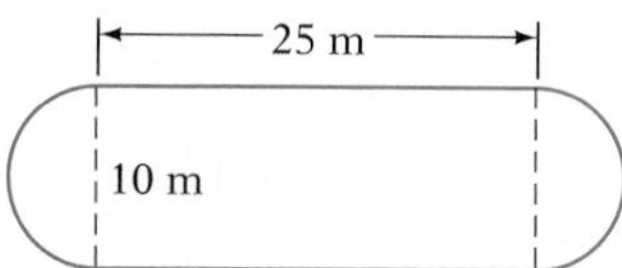

33. The rectangular lot shown in the figure is being fenced. The fencing along the road is to cost \$2.20 per foot. The rest of the fencing costs \$1.85 per foot. Find the total cost to fence the lot.

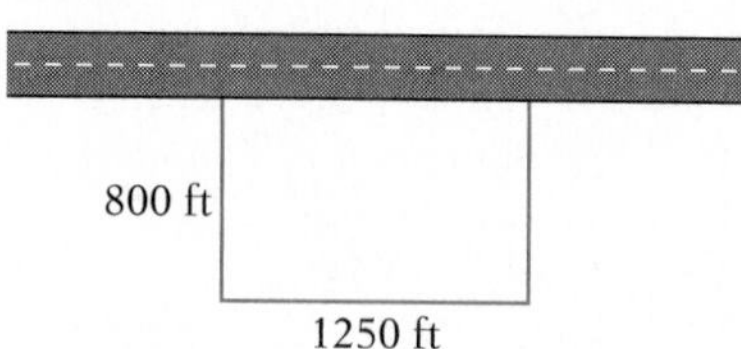

34. A rain gutter is being installed on a home that has the dimensions shown below. At \$8.29 per meter, how much will it cost to install the rain gutter?

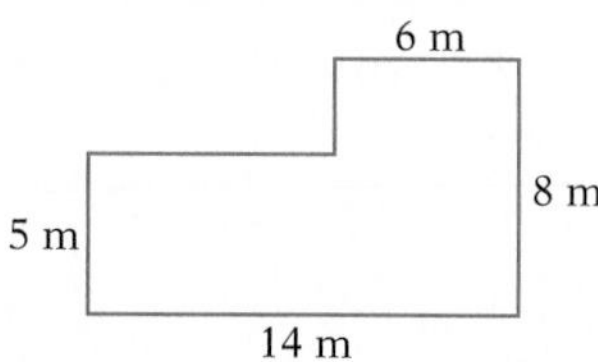

35. The distance from the earth to the sun is 93,000,000 mi. Approximate the distance the earth travels in making 1 revolution about the sun. Use $\pi \approx 3.14$.

36. The distance from the surface to the center of the earth is 6356 km. Find the circumference of the earth. Use $\pi \approx 3.14$.

SECTION 10.3 Area

Objective A

To find the area of geometric figures

Area is a measure of the amount of surface in a region. Area can be used to describe the size of a rug, a parking lot, a farm, or a national park.

Area is measured in square units.

A square that is 1 ft on each side has an area of 1 square foot, written 1 ft^2.

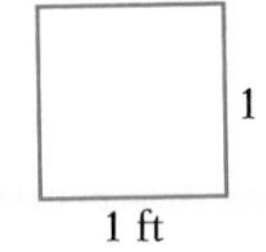

A square that is 1 cm on each side has an area of 1 square centimeter, written 1 cm^2.

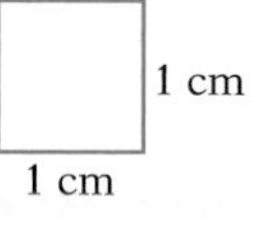

Areas of common geometric figures are given by the following formulas.

Rectangle

Area = length · width
= 3 cm · 2 cm = 6 cm^2

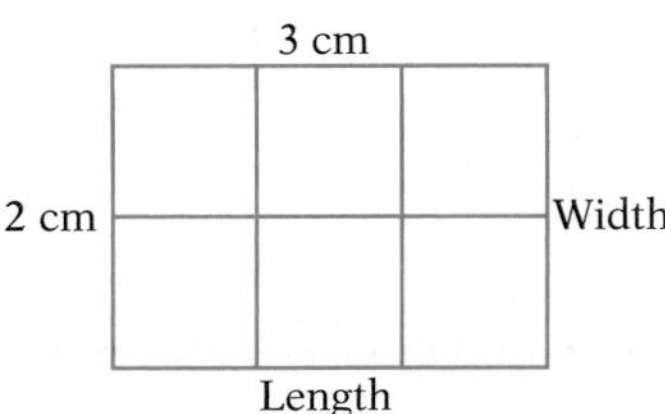

The area of the accompanying rectangle is found by multiplying the length times the width.

Area = length · width = 8 ft · 5 ft = 40 ft^2

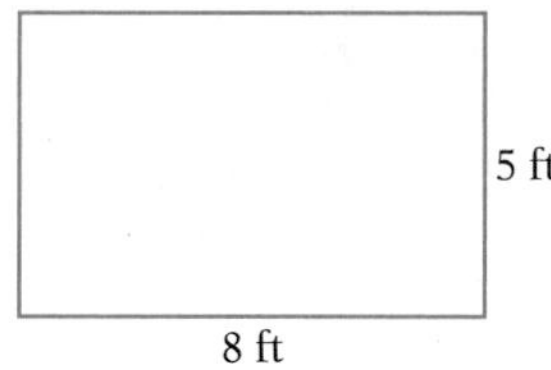

Square

Area = side · side = $(\text{side})^2$
= 2 cm · 2 cm = 4 cm^2

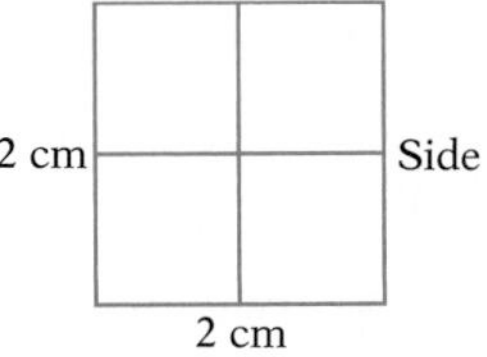

The area of the accompanying square is found by multiplying the length of a side by itself or side squared.

Area = side · side = 14 cm · 14 cm
= 196 cm^2

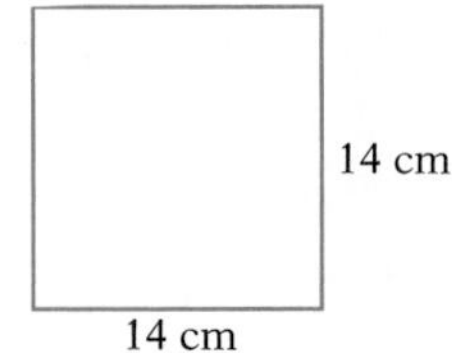

Circle

Area = $\pi \cdot$ radius $\cdot$ radius = $\pi(\text{radius})^2$
$\approx 3.14(5 \text{ in.})^2$
Area $\approx 78.50 \text{ in.}^2$

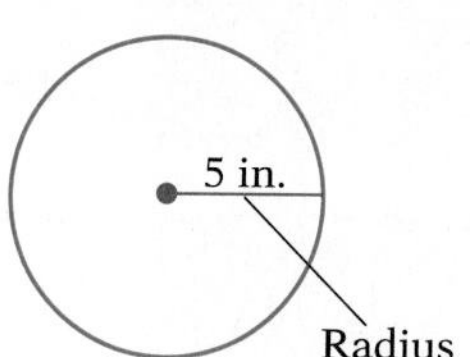

The area of the accompanying circle is found by multiplying pi times the radius squared.

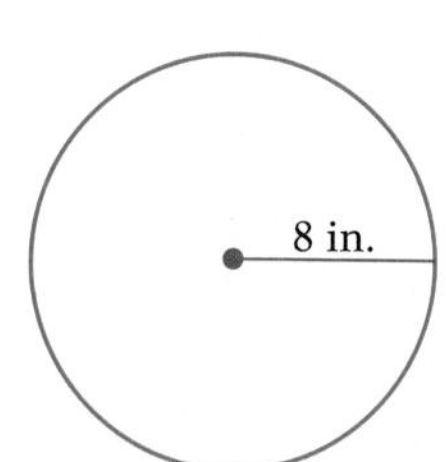

Area = $\pi \cdot (\text{radius})^2 \approx 3.14 \cdot (8 \text{ in})^2 \approx 200.96 \text{ in.}^2$

The answer can also be written in terms of π as
Area = $64\ \pi \text{ in.}^2$

Triangle

Area = $\frac{1}{2} \cdot$ base $\cdot$ height
$= \frac{1}{2} \cdot 6 \text{ in.} \cdot 4 \text{ in.} = 12 \text{ in.}^2$

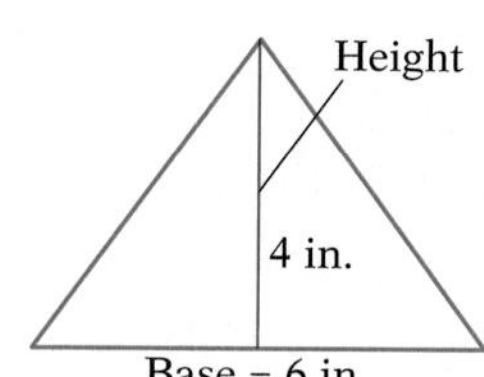

The area of the accompanying triangle is found by multiplying $\frac{1}{2}$ times the base times the height.

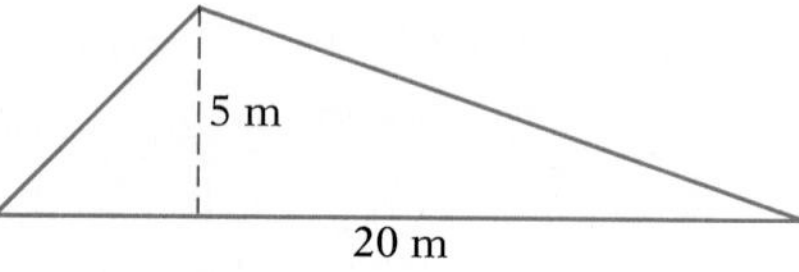

Area = $\frac{1}{2} \cdot$ base $\cdot$ height
$= \frac{1}{2} \cdot 20 \text{ m} \cdot 5 \text{ m} = 50 \text{ m}^2$

Example 1

Find the area of a circle with a diameter of 9 cm. Use $\pi \approx 3.14$.

Solution

Radius = $\frac{1}{2} \cdot$ diameter
$= \frac{1}{2} \cdot 9 \text{ cm} = 4.5 \text{ cm}$

Area = $\pi(\text{radius})^2$
$\approx 3.14(4.5 \text{ cm})^2$
Area $\approx 63.585 \text{ cm}^2$

Example 2

Find the area of a triangle with a base of 24 in. and a height of 14 in.

Your solution

Solution on p. A36

Objective B To find the area of composite geometric figures

The area of the composite figure in the figure below is found by calculating the area of the rectangle and then subtracting the area of the triangle.

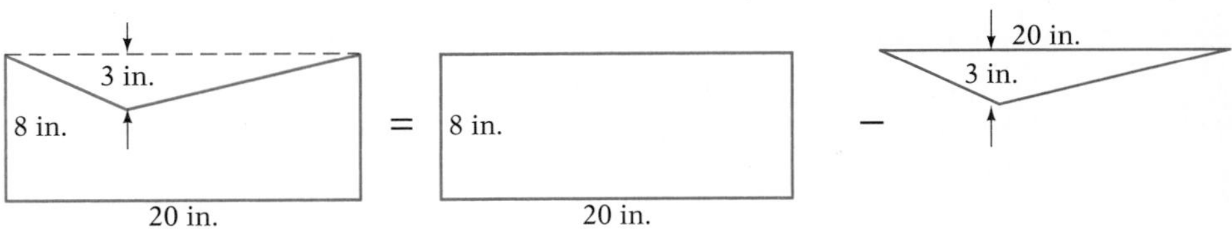

Area = area of the rectangle − area of the triangle

$$= 20 \text{ in.} \cdot 8 \text{ in.} - \frac{1}{2} \cdot 20 \text{ in.} \cdot 3 \text{ in.} = 160 \text{ in.}^2 - 30 \text{ in.}^2 = 130 \text{ in.}^2$$

Example 3

Find the area of the shaded portion of the figure. Use $\pi \approx 3.14$.

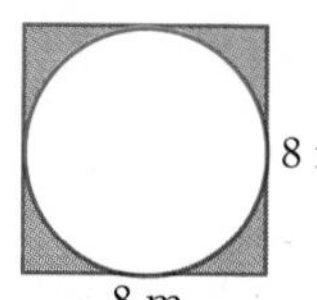

Solution

Area of shaded portion	=	area of square	−	area of circle
	=	$(\text{side})^2$	−	$\pi(\text{radius})^2$
	=	$(8 \text{ m})^2$	−	$\pi(4 \text{ m})^2$
	≈	64 m^2	−	$3.14(16 \text{ m}^2)$
	=	64 m^2	−	50.24 m^2
Area	≈	13.76 m^2		

Example 4

Find the area of the composite figure.

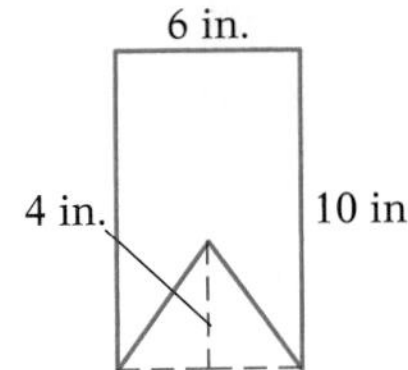

Your solution

Solution on p. A36

Objective C To solve application problems

Example 5

A walkway 2 m wide is built along the front and along both sides of a building, as shown in the figure. Find the area of the walkway.

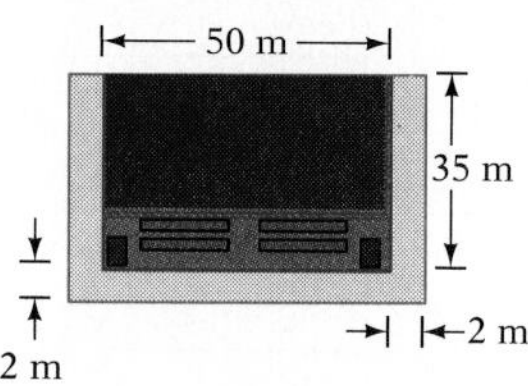

Strategy

To find the area of walkway, add the area of the front section (54 m · 2 m) and the area of the two side sections (each 35 m · 2m).

Solution

$$\text{Area of walkway} = \underbrace{\text{area of front section}} + \underbrace{\text{2(area of one side section)}}$$

$$= (54 \text{ m} \cdot 2 \text{ m}) + 2(35 \text{ m} \cdot 2 \text{ m})$$
$$= 108 \text{ m}^2 + 140 \text{ m}^2$$
$$= 248 \text{ m}^2$$

The area of the walkway is 248 m^2.

Example 6

New carpet is installed in a room measuring 9 ft by 12 ft. Find the area of the room in square yards. ($9 \text{ ft}^2 = 1 \text{ yd}^2$)

Your strategy

Your solution

Solution on p. A36

10.3 EXERCISES

Objective A

In Exercises 1 to 8, find the area of the given figures.

1.

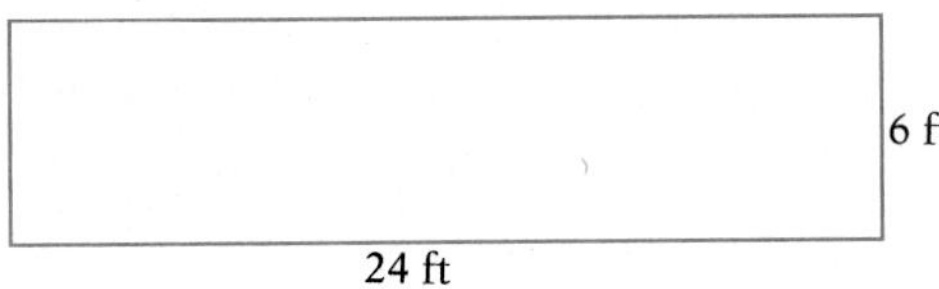

2.

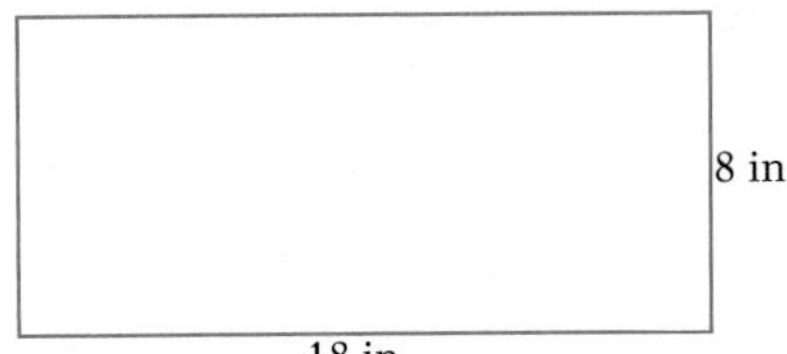

3.

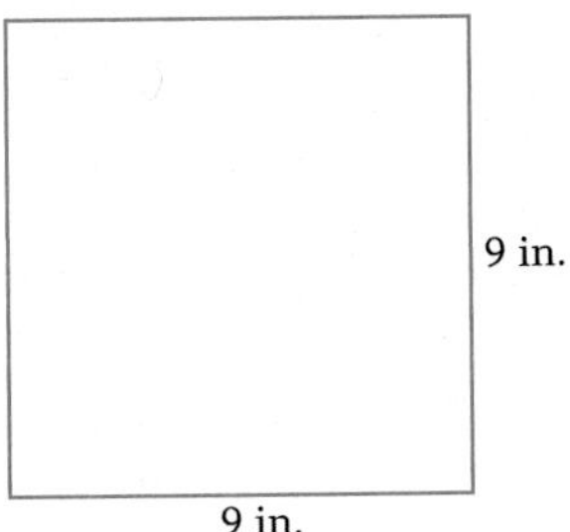

4.

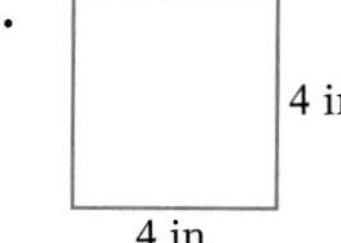

5.

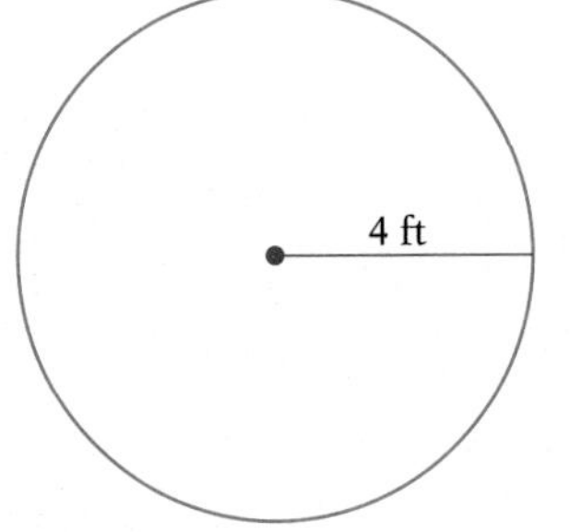

6.

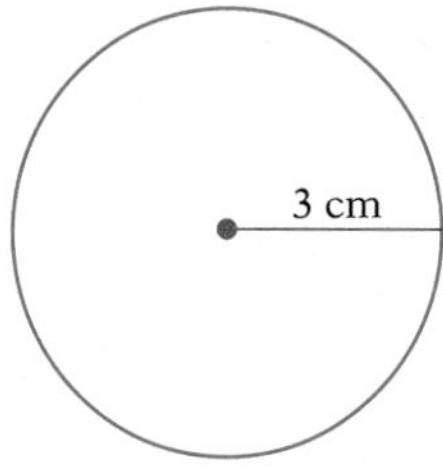

7.

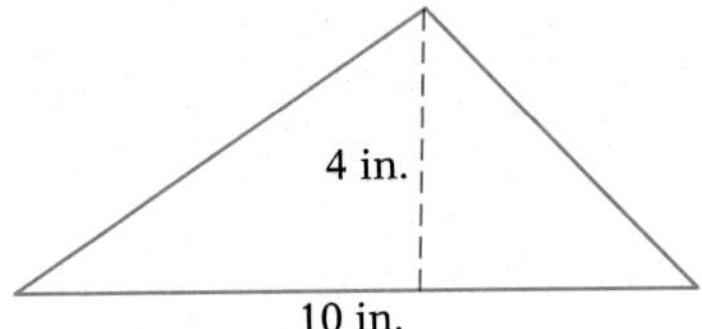

8.

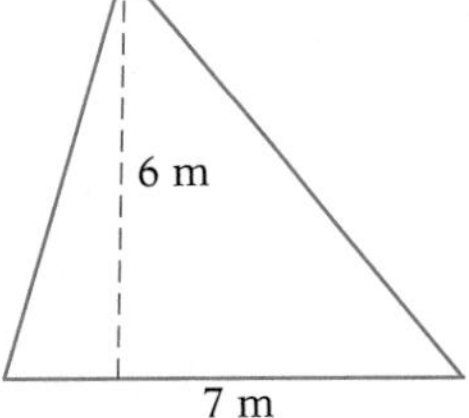

9. Find the area of a right triangle with a base of 3 cm and a height of 1.42 cm.

10. Find the area of a triangle with a base of 3 ft and height of $\frac{2}{3}$ ft.

11. Find the area of a square with a side of 4 ft.

12. Find the area of a square with a side of 10 cm.

13. Find the area of a rectangle with a length of 43 in. and a width of 19 in.

14. Find the area of a rectangle with a length of 82 cm and a width of 20 cm.

15. Find the area of a circle with a radius of 7 in. Use $\pi \approx \frac{22}{7}$.

16. Find the area of a circle with a diameter of 40 cm. Use $\pi \approx 3.14$.

Objective B

In the figures accompanying Exercises 17 to 24, find the area. Use $\pi \approx 3.14$.

17.

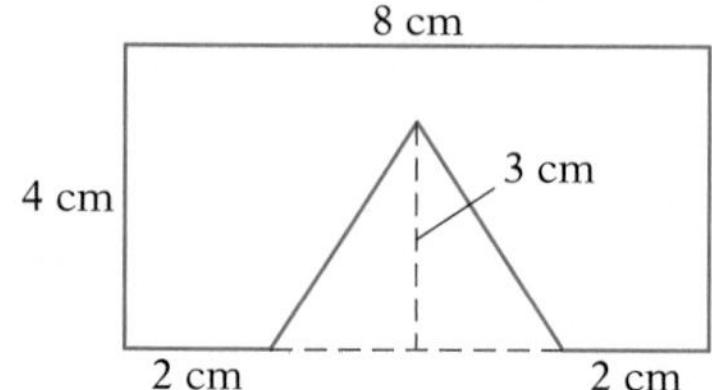

18.

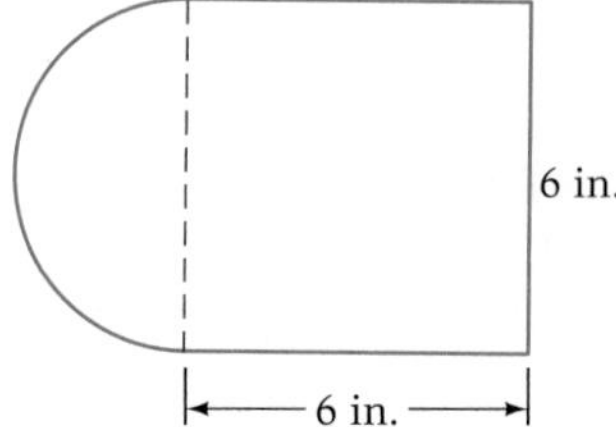

19.

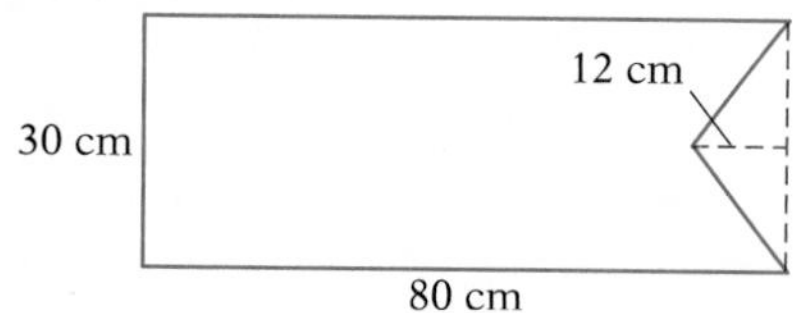

20.

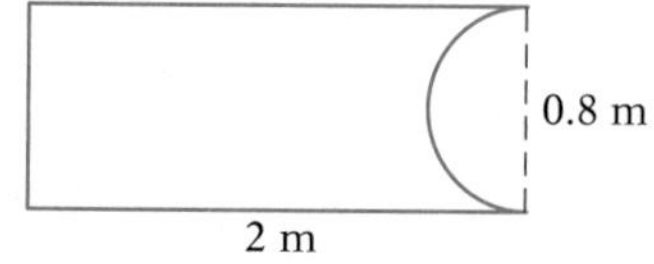

21.

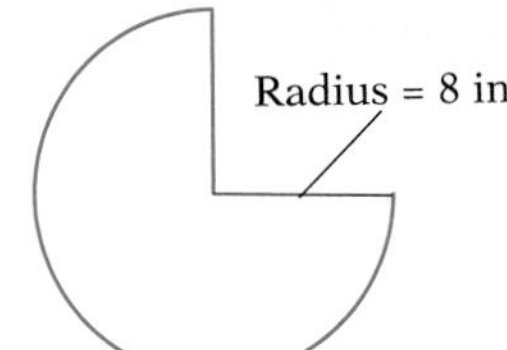

22.

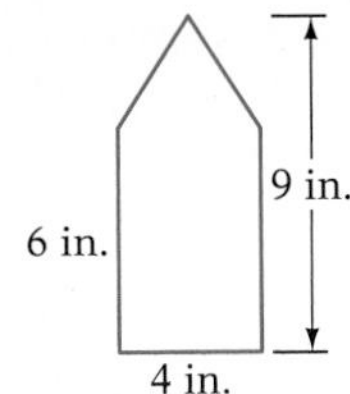

23.

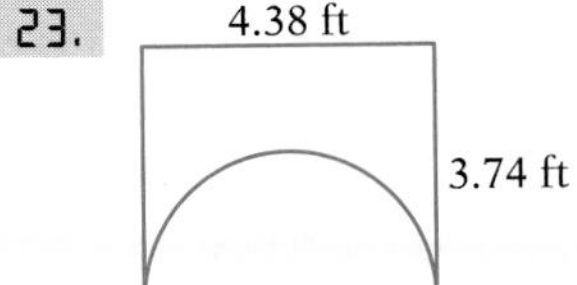

24.

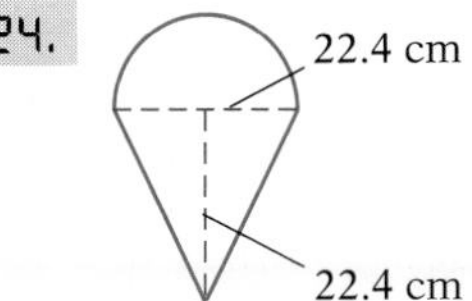

Objective C *Application Problems*

25. Find the area of a rectangular park that has a length of 54 yd and a width of 32 yd.

26. A rectangular garden has a length of 17 ft and a width of 12 ft. Find the area of the garden.

27. The telescope lens located on Mt. Palomar has a diameter of 200 in. Find the area of the lens. Leave the answer in terms of π.

28. An irrigation system waters a circular field that has a 50-ft radius. Find the area watered by the irrigation system. Use $\pi \approx 3.14$.

29. A carpet is to be placed in one room and a hallway as shown in the diagram below. At \$16.50 per square meter, how much will it cost to carpet the area?

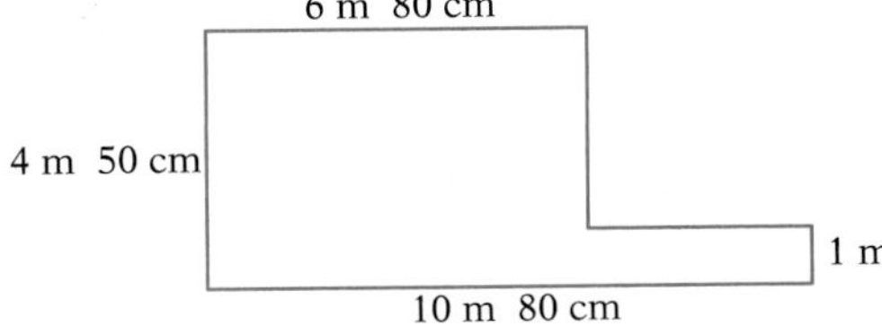

30. Find the area of the concrete driveway with the measurements shown in the figure.

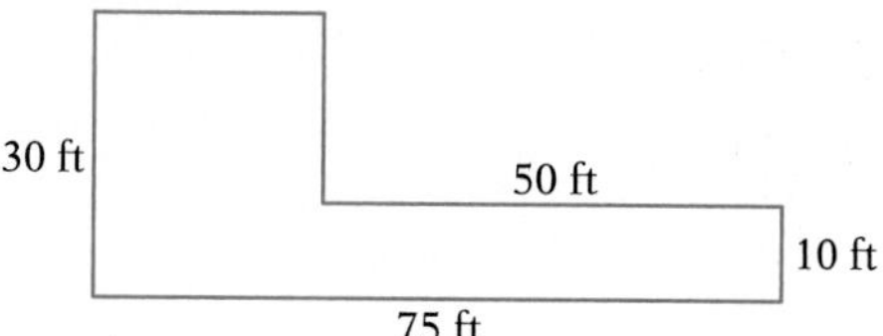

31. Find the area of the 2-m boundary around the swimming pool shown in the figure.

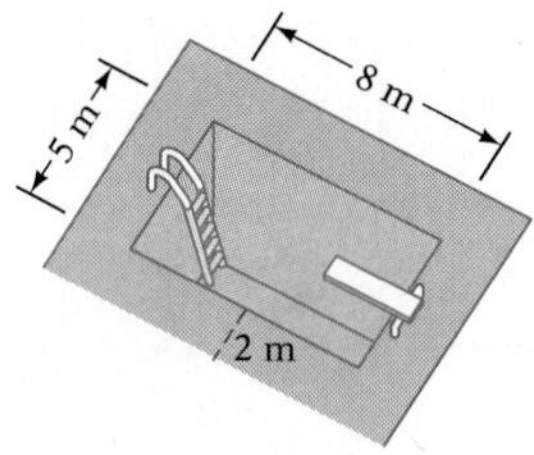

32. How much hardwood floor is needed to cover the roller rink shown in the figure? Use $\pi \approx 3.14$.

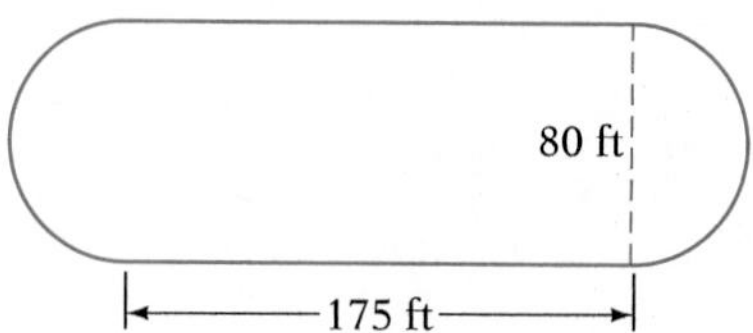

33. A circle has a radius of 8 in. Find the increase in area when the radius is increased by 2 in. Use $\pi \approx 3.14$.

34. A circle has a radius of 5 cm. Find the increase in area when the radius is doubled. Use $\pi \approx 3.14$.

35. Find the total area of the national park with the dimensions shown in the figure.

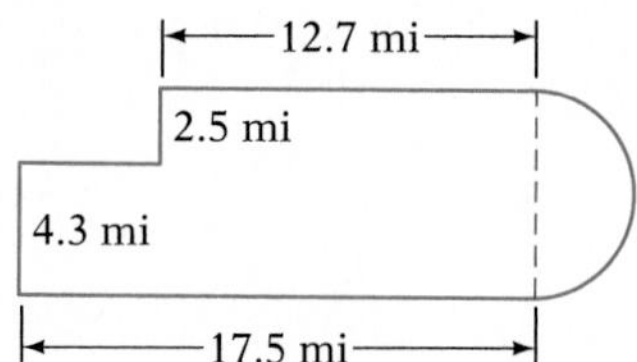

36. Find the cost of plastering the walls of a room 22 ft long, 25 ft 6 in. wide, and 8 ft high. Subtract 120 ft^2 for windows and doors. The cost is \$.75 per square foot.

SECTION 10.4 Volume

Objective A

To find the volume of geometric solids

Volume is a measure of the amount of space inside a closed surface. Volume can be used to describe the amount of heating gas used for cooking, the amount of sand and concrete delivered, and the amount of water in storage for a city's water supply.

Volume is measured in cubic units.

A solid 1 ft long, 1 ft wide, and 1 ft high has a volume of 1 cubic foot, written 1 ft^3.

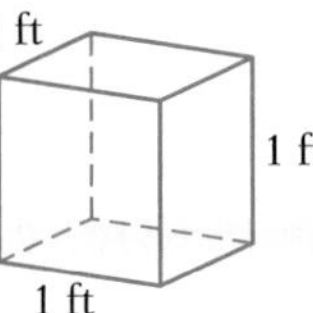

Volumes of common geometric solids are given by the following formulas.

Rectangular solid

Volume = length · width · height
= 4 m · 3 m · 2 m
= 24 m^3

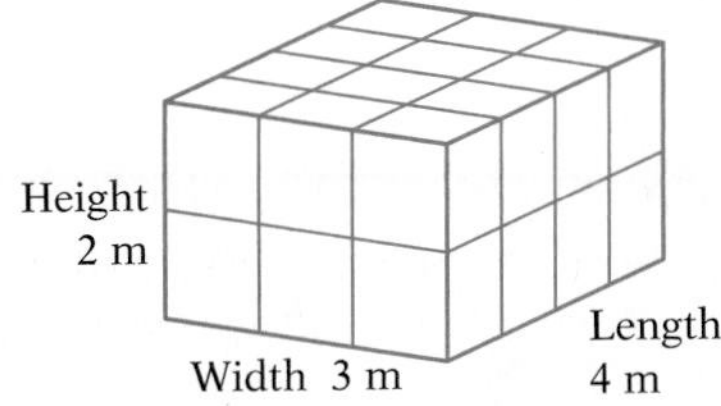

Note that length · width is the area of the base of the solid.

The volume of the rectangular solid shown in the accompanying figure is found by multiplying the length times the width times the height.

Volume = length · width · height
= 9 in. · 3 in. · 4 in.
= 108 $in.^3$

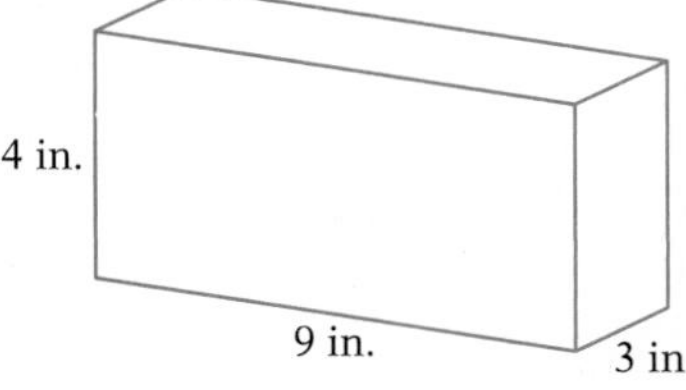

Cube

$$\text{Volume} = \text{side} \cdot \text{side} \cdot \text{side} = (\text{side})^3$$
$$= (4\text{ cm})^3$$
$$= 64\text{ cm}^3$$

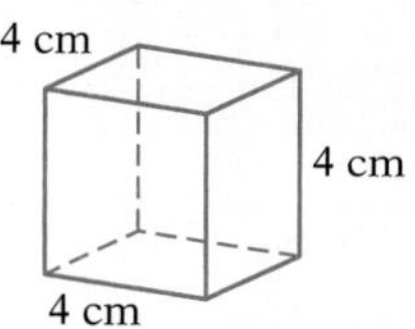

The volume of the cube shown is found by multiplying the side of the cube times itself three times (side cubed).

3 ft
3 ft
3 ft

$$\text{Volume} = (\text{side})^3 = (3\text{ ft})^3 = 27\text{ ft}^3$$

Sphere

$$\text{Volume} = \frac{4}{3}\pi\,(\text{radius})^3$$
$$\approx \frac{4}{3} \cdot 3.14(5\text{ cm})^3$$
$$= \frac{4}{3} \cdot 3.14 \cdot 125\text{ cm}^3$$
$$\text{Volume} \approx 523.33\text{ cm}^3$$

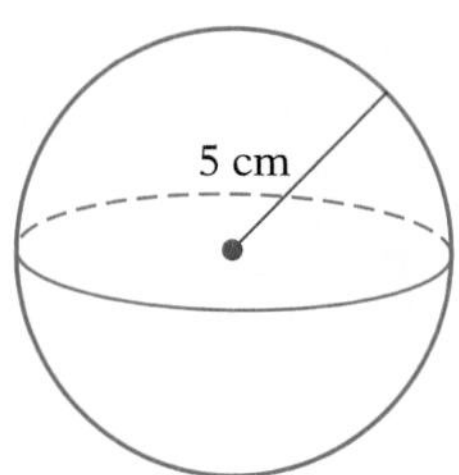

The volume of the sphere shown is found by multiplying four-thirds times pi times the radius cubed.

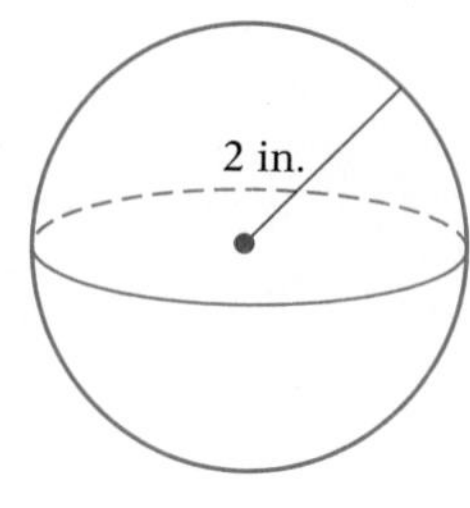

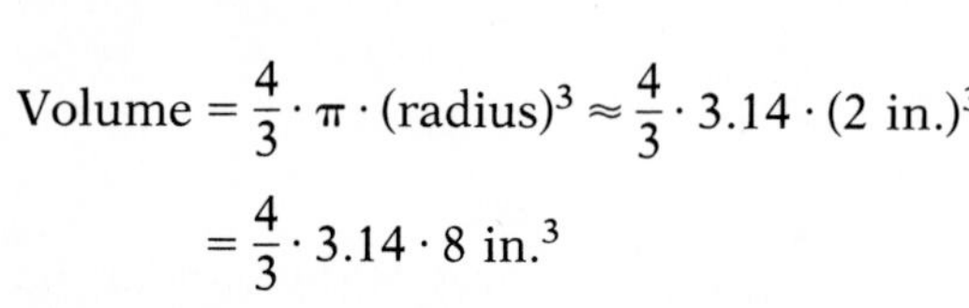

$$\text{Volume} = \frac{4}{3} \cdot \pi \cdot (\text{radius})^3 \approx \frac{4}{3} \cdot 3.14 \cdot (2\text{ in.})^3$$
$$= \frac{4}{3} \cdot 3.14 \cdot 8\text{ in.}^3$$
$$\text{Volume} \approx 33.49\text{ in.}^3$$

Cylinder

$$\text{Volume} = \pi(\text{radius})^2 \cdot \text{height}$$
$$= 3.14(4\text{ in.})^2 \cdot 10\text{ in.}$$
$$= 502.40\text{ in.}^3$$

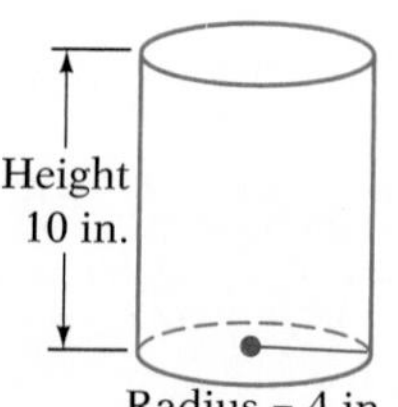

Note that $\pi(\text{radius})^2$ is the area of the base of the cylinder.

The volume of the cylinder in the figure is found by multiplying the base (a circle) of the cylinder times the height.

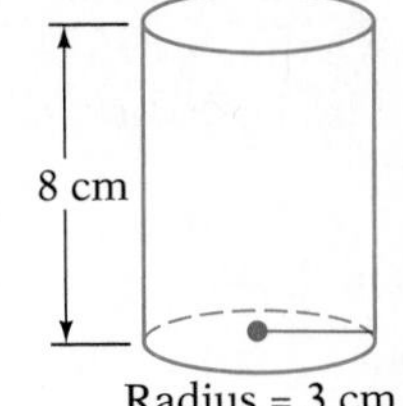

$$\text{Volume} = \pi \cdot (\text{radius})^2 \cdot \text{height} \approx 3.14 \cdot (3 \text{ cm})^2 \cdot 8 \text{ cm}$$
$$= 3.14 \cdot 9 \text{ cm}^2 \cdot 8 \text{ cm}$$
$$\text{Volume} \approx 226.08 \text{ cm}^3$$

Example 1

Find the volume of a rectangular solid with a length of 3 ft, a width of 1.5 ft, and a height of 2 ft.

Solution

$$\text{Volume} = \text{length} \cdot \text{width} \cdot \text{height}$$
$$= 3 \text{ ft} \cdot 1.5 \text{ ft} \cdot 2 \text{ ft}$$
$$= 9 \text{ ft}^3$$

Example 2

Find the volume of a cube with a side of 5 cm.

Your solution

Example 3

Find the volume of a cylinder with a radius of 12 cm and a height of 65 cm. Use $\pi \approx 3.14$.

Solution

$$\text{Volume} = \pi(\text{radius})^2 \cdot \text{height}$$
$$\approx 3.14(12 \text{ cm})^2 \cdot 65 \text{ cm}$$
$$\text{Volume} \approx 29{,}390.4 \text{ cm}^3$$

Example 4

Find the volume of a cylinder with a diameter of 14 in. and a height of 15 in. Use $\pi \approx \frac{22}{7}$.

Your solution

Example 5

Find the volume of a sphere with a diameter of 12 in. Use $\pi \approx 3.14$.

Solution

Radius = 6 in.

$$\text{Volume} = \frac{4}{3}\pi(\text{radius})^3$$
$$\approx \frac{4}{3} \cdot 3.14(6 \text{ in.})^3$$
$$\text{Volume} \approx 904.32 \text{ in}^3$$

Example 6

Find the volume of a sphere with a radius of 3 m. Use $\pi \approx 3.14$.

Your solution

Solutions on p. A37

Objective B

To find the volume of composite geometric solids

Composite geometric solids are solids made from two or more geometric solids. The solid shown is made from a cylinder and one-half of a sphere.

Volume of the composite solid $= \pi(\text{radius})^2 \cdot \text{height} + \frac{1}{2} \cdot \frac{4}{3}\pi(\text{radius})^3$

Find the volume of the composite solid shown above if the radius of the base of the cylinder is 3 in. and the height of the cylinder is 10 in. Use $\pi \approx 3.14$.

The volume equals the volume of a cylinder plus one-half the volume of a sphere. The radius of the sphere equals the radius of the base of the cylinder.

$$\text{Volume} = \pi \cdot (\text{radius})^2 \cdot \text{height} + \frac{1}{2} \cdot \frac{4}{3} \cdot \pi \cdot (\text{radius})^3$$

$$\approx 3.14 \cdot (3\text{ in.})^2 \cdot 10\text{ in.} + \frac{2}{3} \cdot 3.14 \cdot (3\text{ in.})^3$$

$$= 282.6\text{ in.}^3 + 56.52\text{ in.}^3$$

Volume $\approx$ 339.12 in.^{3}

Example 7

Find the volume of the solid in the figure. Use $\pi \approx 3.14$.

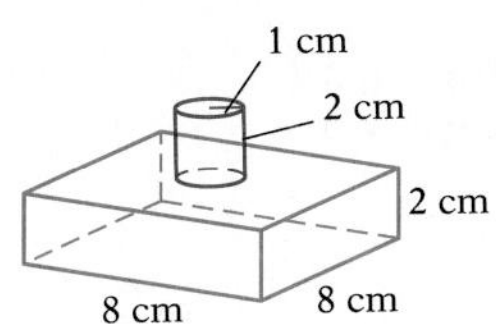

Solution

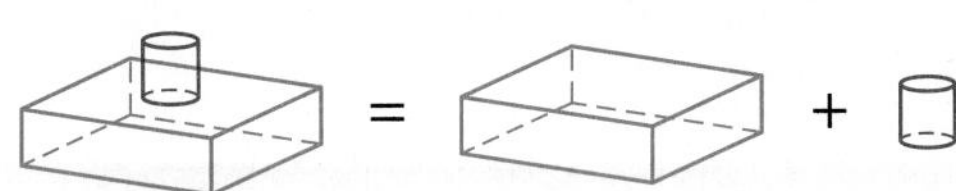

Volume of the solid = Volume of rectangular solid + Volume of cylinder

= (length · width · height) + π(radius)^2 · height

$\approx$ (8 cm · 8 cm · 2 cm) + 3.14(1 cm)^2 · 2 cm

= 128 cm^3 + 6.28 cm^3

Volume $\approx$ 134.28 cm^3

Example 8

Find the volume of the solid in the figure. Use $\pi \approx 3.14$.

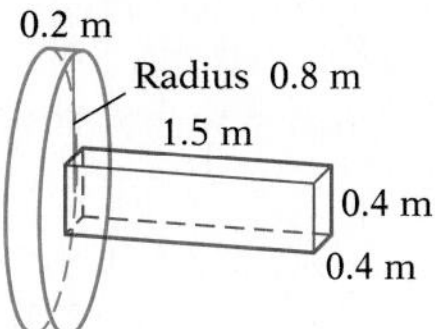

Your solution

Example 9

Find the volume of the solid in the figure. Use $\pi \approx 3.14$.

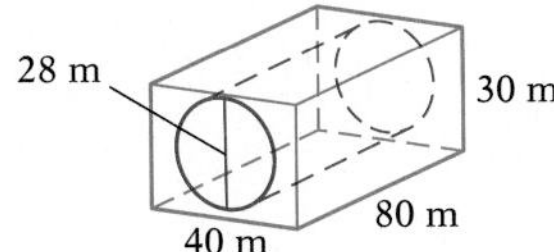

Solution

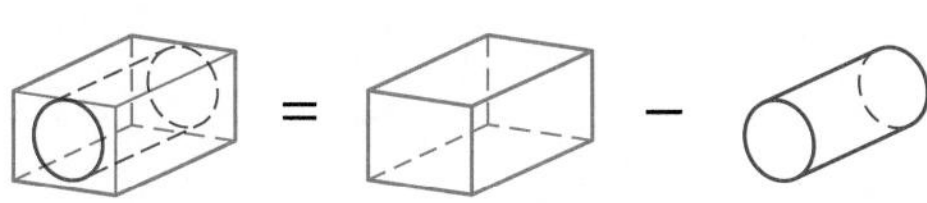

Volume of the solid = Volume of rectangular solid − Volume of cylinder

= (length · width · height) − π(radius)^2 · height

$\approx$ (80 m · 40 m · 30 m) − 3.14 · (14 m)^2 · 80 m

= 96,000 m^3 − 49,235.2 m^3

Volume $\approx$ 46,764.8 m^3

Example 10

Find the volume of the solid in the figure. Use $\pi \approx 3.14$.

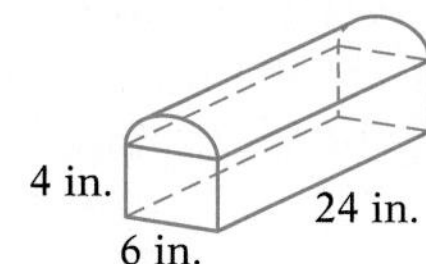

Your solution

Solutions on p. A37

Objective C To solve application problems

Example 11

An aquarium is 28 in. long, 14 in. wide, and 16 in. high. Find the volume of the aquarium.

Strategy

To find the volume of the aquarium, use the formula for the volume of a rectangular solid.

Solution

$$\begin{aligned}\text{Volume} &= \text{length} \cdot \text{width} \cdot \text{height} \\ &= 28 \text{ in.} \cdot 14 \text{ in.} \cdot 16 \text{ in.} \\ &= 6272 \text{ in.}^3\end{aligned}$$

The volume of the aquarium is 6272 in.^{3}

Example 12

Find the volume of a freezer that is 7 ft long, 3 ft high, and 2.5 ft wide.

Your strategy

Your solution

Example 13

Find the volume of the bushing shown in the figure below. Use $\pi \approx 3.14$.

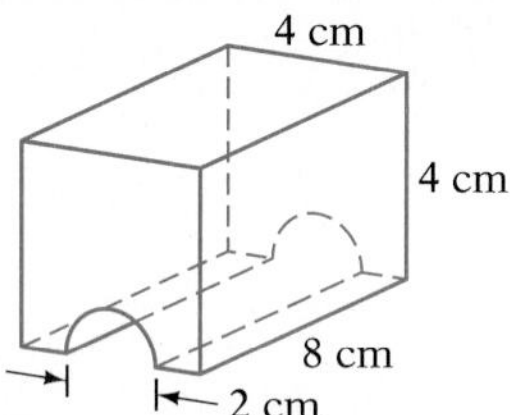

Strategy

To find the volume of the bushing, subtract the volume of the half cylinder from the volume of the rectangular solid.

Solution

$$\text{Volume of the bushing} = \text{volume of rectangular solid} - \frac{1}{2} \text{ of the volume of cylinder}$$

$$\begin{aligned}&= (\text{length} \cdot \text{width} \cdot \text{height}) - \frac{1}{2}\pi(\text{radius})^2 \cdot \text{height} \\ &\approx (4 \text{ cm} \cdot 4 \text{ cm} \cdot 8 \text{ cm}) - \frac{1}{2} \cdot 3.14(1 \text{ cm})^2 \cdot 8 \text{ cm} \\ &= 128 \text{ cm}^3 - 12.56 \text{ cm}^3\end{aligned}$$

$$\text{Volume} \approx 115.44 \text{ cm}^3$$

The volume of the bushing is approximately 115.44 cm^3

Example 14

Find the volume of the channel iron shown in the figure below.

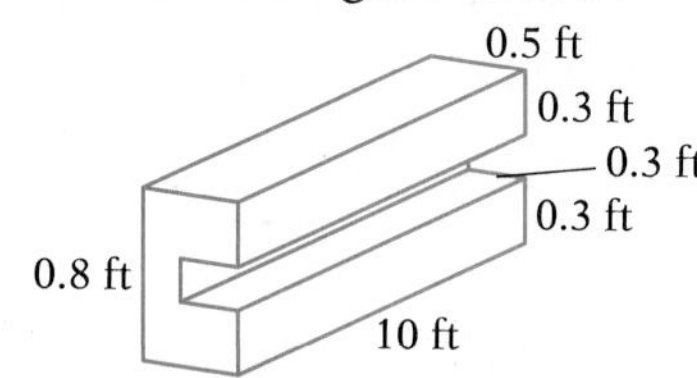

Your strategy

Your solution

Solutions on p. A37

10.4 EXERCISES

Objective A

In Exercises 1 to 8, find the volume of the solids in the accompanying figures. Round to the nearest hundredth. Use $\pi \approx 3.14$.

1.

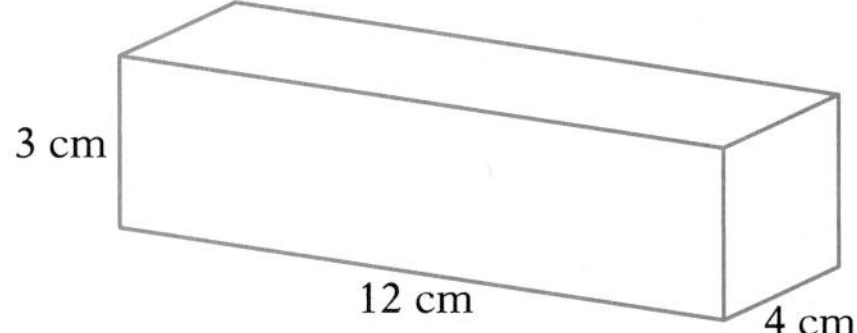

2.

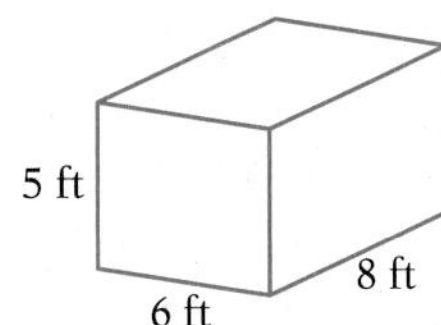

3.

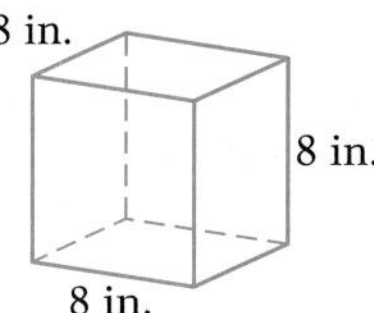

4.

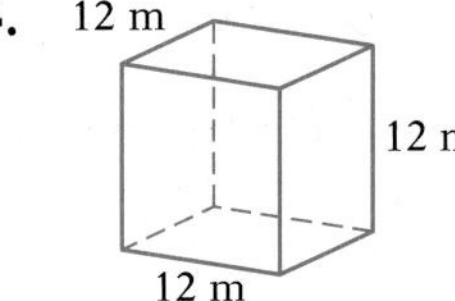

5.

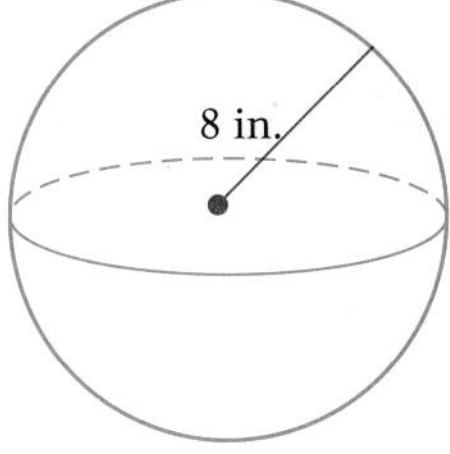

6.

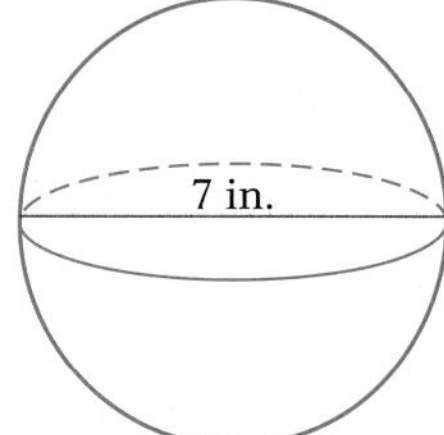

7.

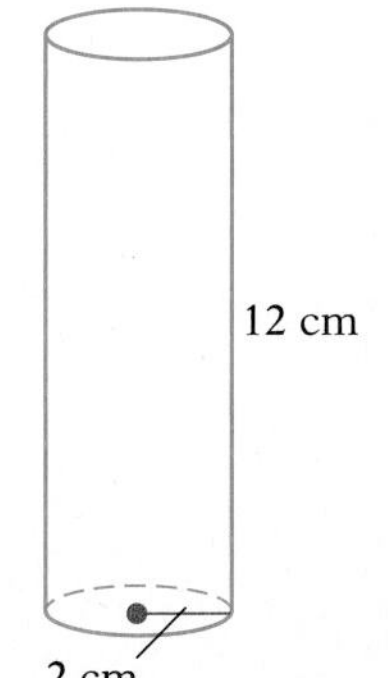

8.

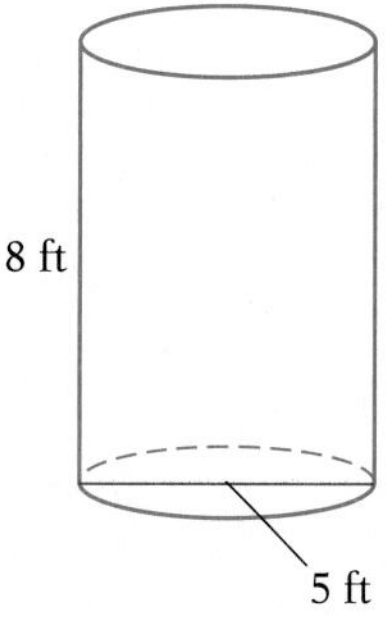

In Exercises 9 to 16, find the volume. Write measurements involving two units in terms of the larger unit before working the problem.

9. Find the volume of a rectangular solid with a length of 2 m, a width of 80 cm, and a height of 4 m.

10. Find the volume of a cylinder with a radius of 7 cm and a height of 14 cm. Use $\pi \approx \frac{22}{7}$.

11. Find the volume of a sphere with a 11-mm radius. Use $\pi \approx 3.14$. Round to the nearest hundredth.

12. Find the volume of a cube with a side of 2 m 14 cm. Round to the nearest tenth.

13. Find the volume of a cylinder with a diameter of 12 ft and a height of 30 ft. Use $\pi \approx 3.14$.

14. Find the volume of a sphere with a 6-ft diameter. Use $\pi \approx 3.14$.

15. Find the volume of a cube with a side of $3\frac{1}{2}$ ft.

16. Find the volume of a rectangular solid with a length of 1 m 15 cm, a width of 60 cm, and a height of 25 cm.

Objective B

In Exercises 17 to 22, find the volume of the accompanying figures. Write measurements involving two units in terms of the larger unit before working the problem. Use $\pi \approx 3.14$.

17.

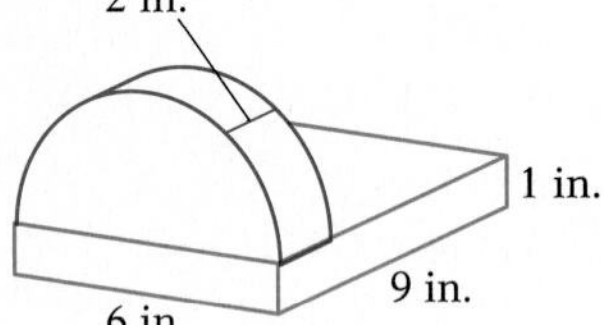

18.

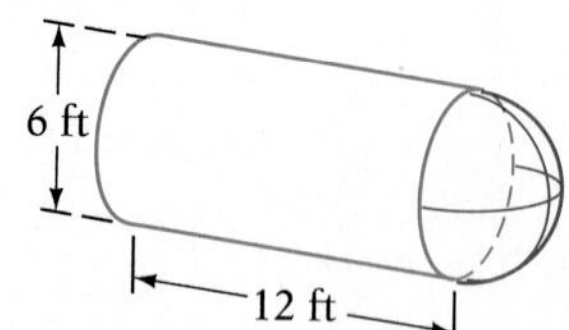

19.

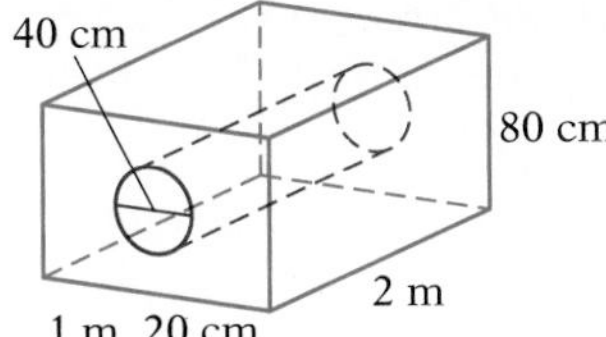

20.

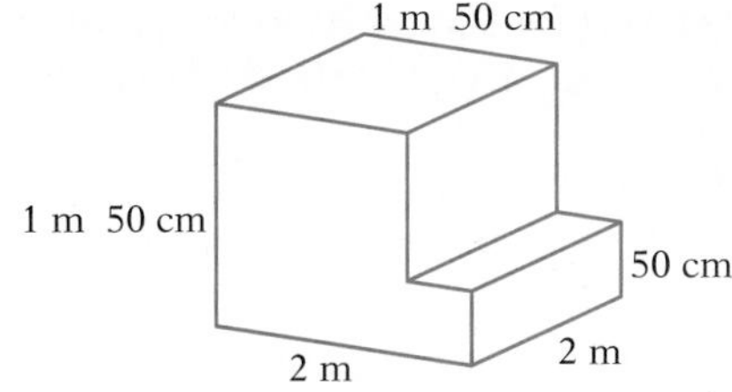

21.

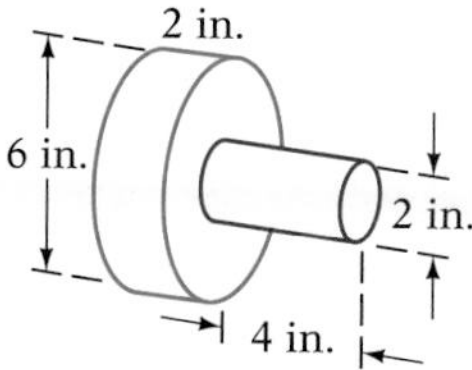

22.

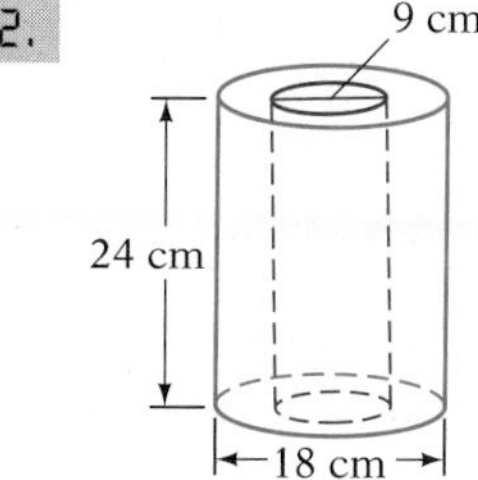

Objective C *Application Problems*

23. A rectangular tank at the fish hatchery is 9 m long, 3 m wide, and 1.5 m deep. Find the volume of the water in the tank when the tank is full.

24. A fuel tank in a booster rocket is a cylinder 10 ft in diameter and 52 ft high. Find the volume of the booster. Use $\pi \approx 3.14$.

25. A hot air balloon is in the shape of a sphere. Find the volume of a hot air balloon that is 32 ft in diameter. Use $\pi \approx 3.14$. (Round to hundredths.)

26. A storage tank for propane is in the shape of a sphere. Find the volume of the storage tank that is 9 m in diameter. Leave the answer in terms of π.

27. A silo, which is in the shape of a cylinder, is 16 ft in diameter and has a height of 36 ft. Find the volume of the silo. Leave the answer in terms of π.

28. An oil storage tank, which is in the shape of a cylinder, is 4 m high and has a 6-m diameter. Find the volume of the oil storage tank. Use $\pi \approx 3.14$.

29. How many gallons of water will fill an aquarium that is 12 in. wide, 18 in. long, and 16 in. high? Round to the nearest tenth. (1 gal = 231 in.^{3})

30. How many gallons of water will fill a fish tank that is 12 in. long, 8 in. wide, and 9 in. high? Round to the nearest tenth. (1 gal = 231 in.^{3})

31. An architect is designing the heating for an auditorium and needs to know the volume of the structure. Find the volume of the auditorium with the measurements shown in the figure below. Use $\pi \approx 3.14$.

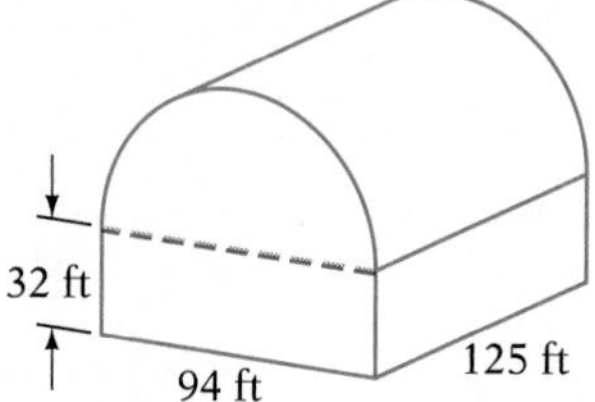

32. Find the volume of the bushing shown in the figure below. Use $\pi \approx 3.14$.

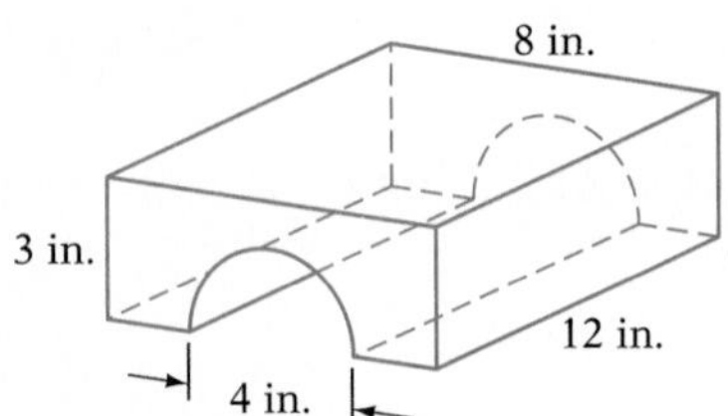

33. A swimming pool contains 65,000 ft^3 of water. Find the total weight of the water in the swimming pool. (1 ft^3 of water weighs 62.4 lb.)

34. The floor of a building is shown in the figure below. The concrete floor is 6 in. thick. Find the cost of the floor at $2.75 per cubic foot. Use $\pi \approx 3.14$. (Round to the nearest cent.)

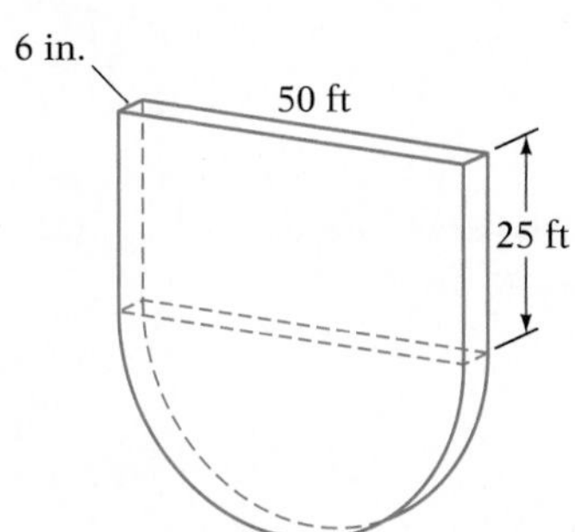

SECTION 10.5 The Pythagorean Theorem

Objective A To use a table to find the square root of a number

The area of the square is 36 in.^{2}. What is the length of one side?

Area of the square = $(\text{side})^2$

$36 = \text{side} \cdot \text{side}$

Area = 36 in.^{2}

What number multiplied times itself equals 36?

$36 = 6 \cdot 6$

The side of the square is 6 in.

The **square root** of a number is one of two identical factors of that number. The square root symbol is $\sqrt{\ }$.

The square root of 36 is 6.

$\sqrt{36} = 6$

A **perfect square** is the product of a whole number times itself.

1, 4, 9, 16, 25, and 36 are perfect squares.

The square root of a perfect square is a whole number.

$1 \cdot 1 = 1$	$\sqrt{1} = 1$
$2 \cdot 2 = 4$	$\sqrt{4} = 2$
$3 \cdot 3 = 9$	$\sqrt{9} = 3$
$4 \cdot 4 = 16$	$\sqrt{16} = 4$
$5 \cdot 5 = 25$	$\sqrt{25} = 5$
$6 \cdot 6 = 36$	$\sqrt{36} = 6$

If a number is not a perfect square, its square root can only be approximated. The approximate square roots of the whole numbers up to 200 are found in the appendix on page A3. The square roots have been rounded to the nearest thousandth.

Number	*Square Root*
33	5.745
34	5.831
35	5.916

$\sqrt{33} \approx 5.745$

$\sqrt{35} \approx 5.916$

Example 1 Find the square roots of the perfect squares 49 and 81.

Solution $\sqrt{49} = 7$ $\sqrt{81} = 9$

Example 2 Find the square roots of the perfect squares 16 and 169.

Your solution

Solution on p. A38

Example 3	Find the square roots of 27 and 108. Use the table on page A3.	**Example 4**	Find the square roots of 32 and 162. Use the table on page A3.
Solution	$\sqrt{27} \approx 5.196$ $\quad \sqrt{108} \approx 10.392$	**Your solution**	

Solution on p. A38

Objective B

To find the unknown side of a right triangle using the Pythagorean Theorem

The Greek mathematician Pythagoras is generally credited with the discovery that the square of the hypotenuse of a right triangle is equal to the sum of the squares of the two legs. This is called the **Pythagorean Theorem.** However, the Babylonians used this theorem more than 1000 years before Pythagoras lived.

Square of the hypotenuse	equals	sum of the squares of the two legs
5^2	=	$3^2 + 4^2$
25	=	9 + 16
25	=	25

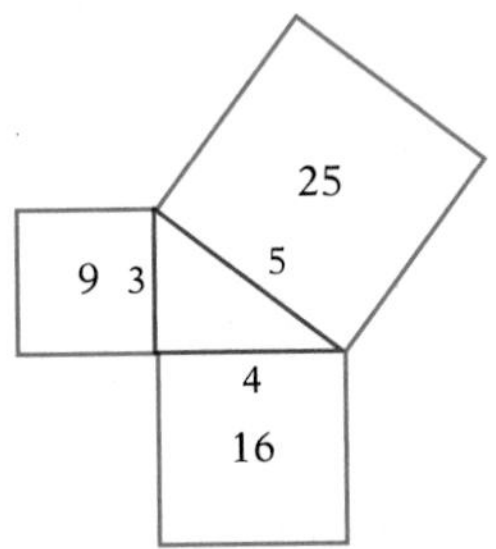

If the length of one side of a right triangle is unknown, one of the following formulas can be used to find its length.

If the hypotenuse is unknown, use

$$\begin{aligned} \text{Hypotenuse} &= \sqrt{(\text{leg})^2 + (\text{leg})^2} \\ &= \sqrt{(3)^2 + (4)^2} \\ &= \sqrt{9 + 16} \\ &= \sqrt{25} \\ &= 5 \end{aligned}$$

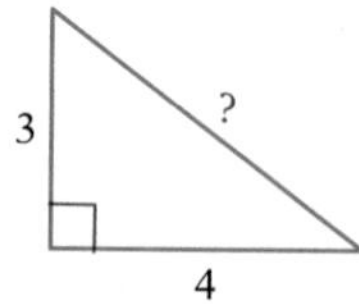

If the length of a leg is unknown, use

$$\begin{aligned} \text{Leg} &= \sqrt{(\text{hypotenuse})^2 - (\text{leg})^2} \\ &= \sqrt{(5)^2 - (4)^2} \\ &= \sqrt{25 - 16} \\ &= \sqrt{9} \\ &= 3 \end{aligned}$$

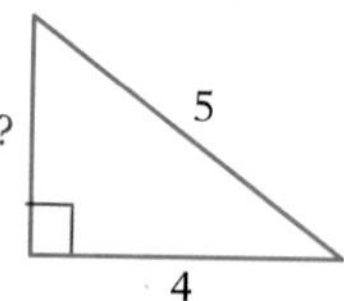

By using the Pythagorean Theorem and several facts from geometry, a relationship among the sides of two special right triangles can be found.

The first special right triangle has two 45° angles and is called a **45°-45°-90° triangle.** The sides opposite the 45° angles are also equal.

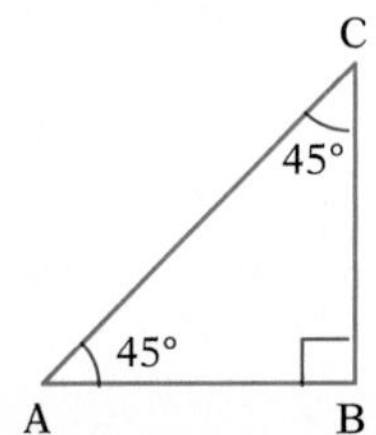

In a 45°-45°-90° triangle, the hypotenuse is equal to $\sqrt{2}$ × (length of the leg).

Find the length of the hypotenuse for a 45°-45°-90° triangle in which the length of one leg is 26 m.

$$\text{Hypotenuse} = \sqrt{2} \times (\text{leg})$$
$$= \sqrt{2} \times 26 \text{ m}$$
$$\approx 1.414 \times 26 \text{ m}$$
$$\text{Hypotenuse} \approx 36.764 \text{ m}$$

The second special right triangle is the **30°-60°-90° triangle.**

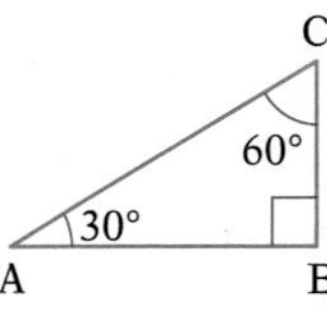

For a 30°-60°-90° triangle, the length of the leg opposite the 30° angle is $\frac{1}{2}$ the length of the hypotenuse.

$$BC = \frac{1}{2} \times AC$$

Find the length of the two legs of a 30°-60°-90° triangle that has a hypotenuse 16 cm in length.

Because the right triangle is a 30° -60° -90° triangle, the leg opposite the 30° angle is $\frac{1}{2}$ the hypotenuse.

$$\text{Leg} = \frac{1}{2} \times 16 \text{ cm}$$
$$\text{Leg} = 8 \text{ cm}$$

Use the Pythagorean Theorem to find the other leg.

$$\text{Leg} = \sqrt{(\text{hypotenuse})^2 - (\text{leg})^2}$$
$$= \sqrt{16^2 - 8^2} = \sqrt{256 - 64}$$
$$= \sqrt{192}$$
$$= 13.856$$
$$\text{Leg} \approx 13.856 \text{ cm}$$

The lengths of the legs are 8 cm and 13.856 cm.

Example 5

Find the hypotenuse of the triangle in the figure.

Solution

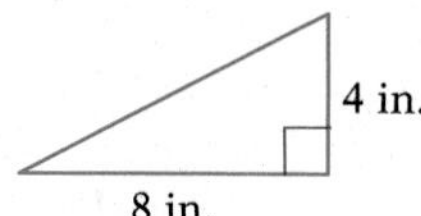

$$\text{Hypotenuse} = \sqrt{(\text{leg})^2 + (\text{leg})^2}$$
$$= \sqrt{(8 \text{ in.})^2 + (4 \text{ in.})^2}$$
$$= \sqrt{64 \text{ in.}^2 + 16 \text{ in.}^2}$$
$$= \sqrt{80 \text{ in.}^2}$$

Use the table on page A3 to find $\sqrt{80}$.

Hypotenuse ≈ 8.944 in.

Example 6

Find the hypotenuse of the triangle in the figure.

Your solution

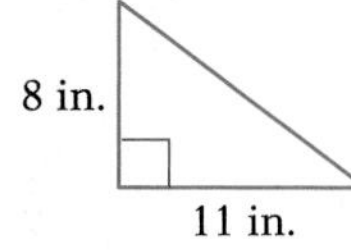

Solution on p. A38

Example 7

Find the length of the leg of the triangle in the figure.

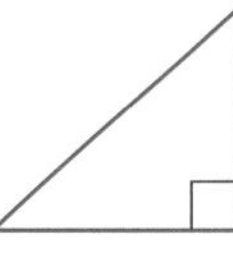

Solution

$$\text{Leg} = \sqrt{(\text{hypotenuse})^2 - (\text{leg})^2}$$
$$= \sqrt{(12\text{ cm})^2 - (9\text{ cm})^2}$$
$$= \sqrt{144\text{ cm}^2 - 81\text{ cm}^2}$$
$$= \sqrt{63\text{ cm}^3}$$

Use the table on page A3 to find $\sqrt{63}$.

Leg $\approx$ 7.937 cm

Example 8

Find the length of the leg of the triangle in the figure.

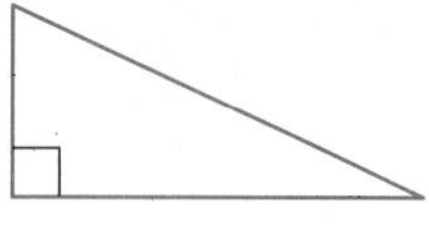

Your solution

Solution on p. A38

Objective C To solve application problems

Example 9

A 25-ft ladder is placed against a building at a point 21 ft from the ground as shown in the figure. Find the distance from the base of the building to the base of the ladder.

Strategy

To find the distance from the base of the building to the base of the ladder, use the Pythagorean Theorem. The hypotenuse is the length of the ladder (25 ft). One leg is the distance along the building from the ground to the top of the ladder. The distance from the base of the building to the base of the ladder is the unknown leg.

Solution

$$\text{Leg} = \sqrt{(\text{hypotenuse})^2 - (\text{leg})^2}$$
$$= \sqrt{(25\text{ ft})^2 - (21\text{ ft})^2}$$
$$= \sqrt{625\text{ ft}^2 - 441\text{ ft}^2}$$
$$= \sqrt{184\text{ ft}^2}$$

Leg $\approx$ 13.565 ft

The distance is 13.565 ft.

Example 10

Find the distance between the centers of the holes in the metal plate in the figure.

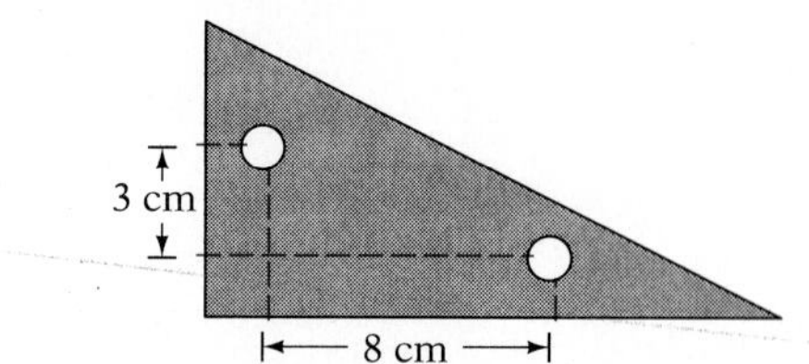

Your strategy

Your solution

Solution on p. A38

10.5 EXERCISES

Objective A

Find the square root. Use the table on page A3.

1. 7
2. 34
3. 42
4. 64
5. 165
6. 144
7. 189
8. 130

Objective B

Find the unknown side of the triangle in the figures accompanying Exercises 9 to 17.

9. 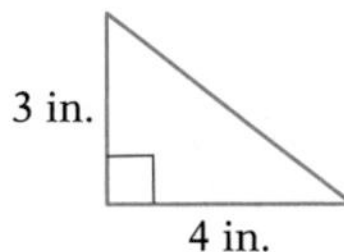

10. 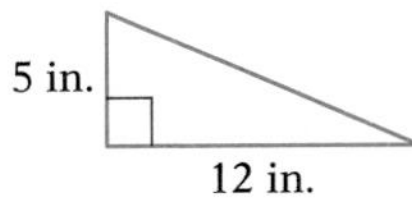

11. 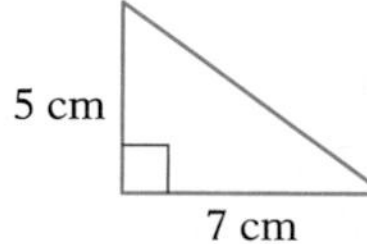

12. 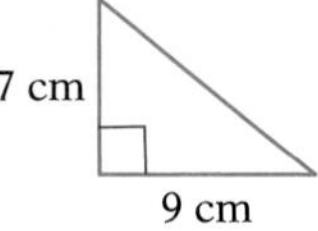

13. 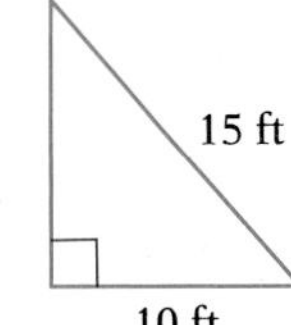

14. 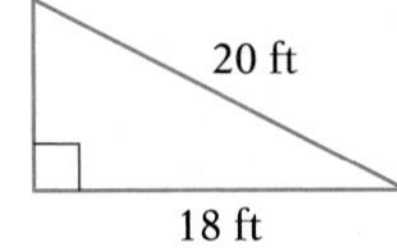

15. 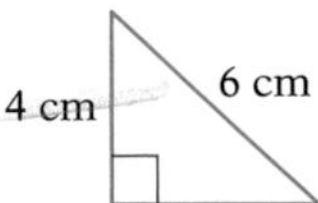

16. 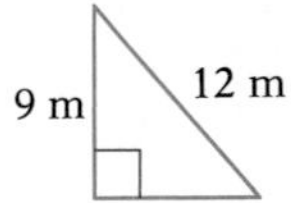

17. 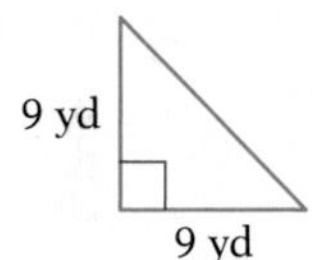

Find the lengths of the two legs in the figures accompanying Exercises 18 to 20.

18. 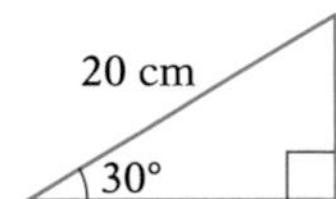

19. 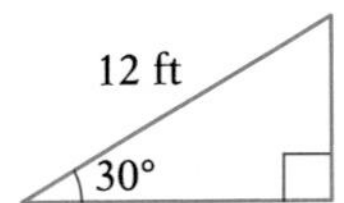

20. 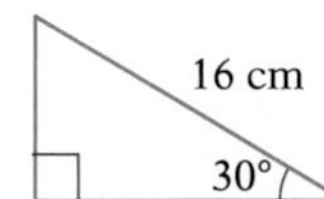

Find the hypotenuse of the triangle in the figures accompanying Exercises 21 to 26.

21. 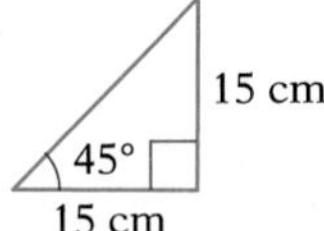

22. 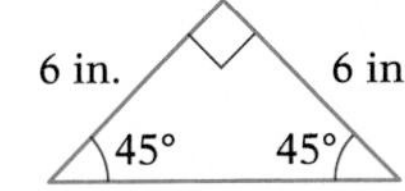

23. 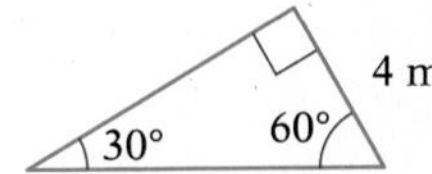

24. 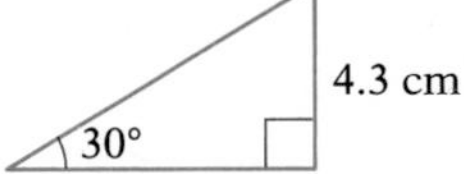

25. 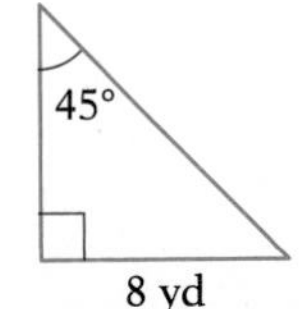

26. 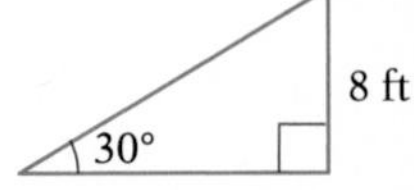

Objective C *Application Problems*

27. A television tower is anchored by a cable from a point 28 m high on the tower to a point 21 m from the base. Find the length of the cable.

28. How high on a building will a 13-ft ladder reach when the bottom of the ladder is 5 ft from the building?

29. A $\frac{1}{4}$-in. steel plate is formed into a brace as shown in the figure. Find the length of the piece of steel before it is bent.

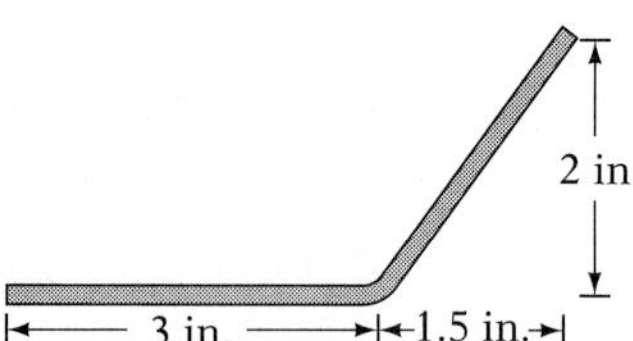

30. A circular brass fitting is to be placed over a 2 in. square shaft as shown in the diagram. Find the diameter of the brass fitting. (Round to the nearest hundredth.) Use $\pi \approx 3.14$.

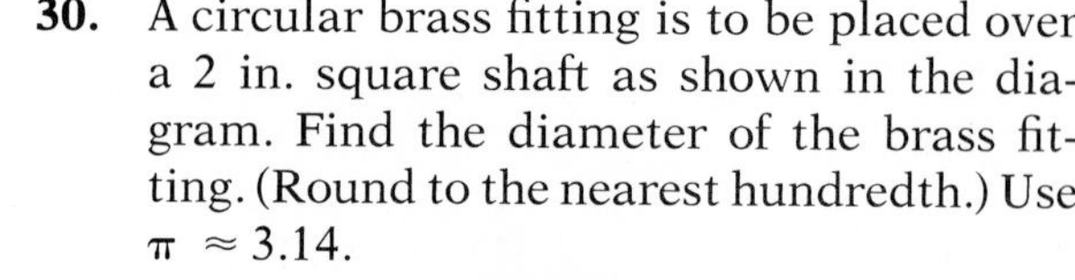

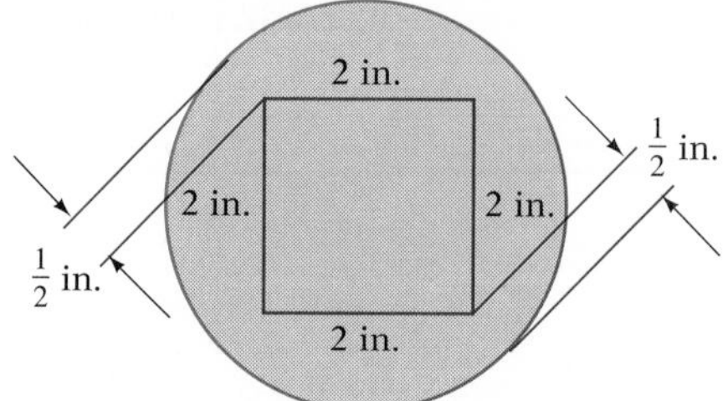

31. Find the distance between the holes in the metal plate in the figure shown. (Round to the nearest hundredth.)

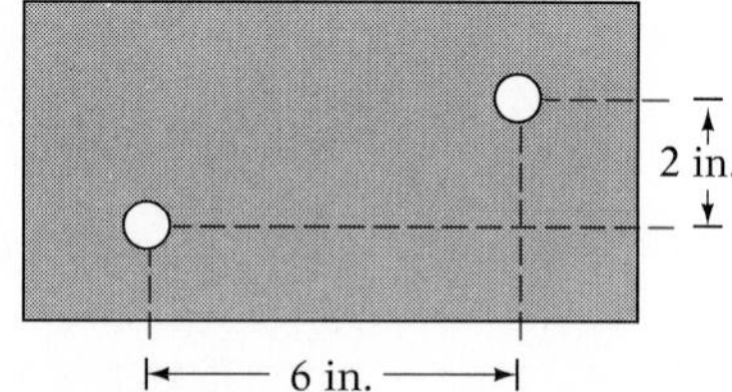

32. Find the length of the ramp used to roll barrels up to the 3.5-ft high loading ramp shown in the figure. (Round to the nearest hundredth.)

3.5 ft

9 ft

33. A fence is built around the plot shown in the figure. At $4.20 per meter, how much did it cost to fence the plot? (Use the Pythagorean Theorem to find the unknown length.)

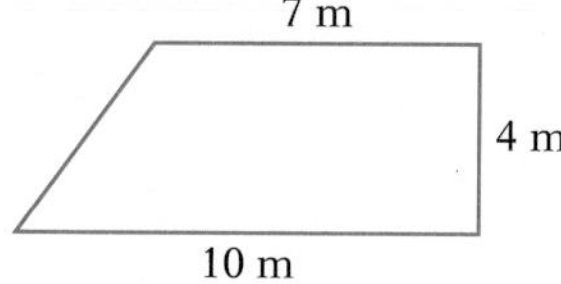

34. Four holes are drilled in the circular plate in the figure shown. The centers of the holes are 3 in. from the center. Find the distance between the centers of adjacent holes.

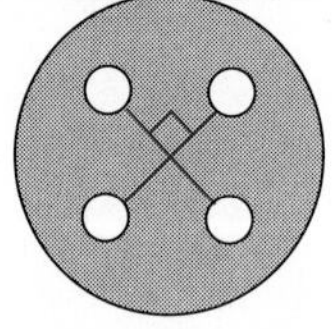

35. A conduit is run in a building as shown in the figure. Find the total length of the conduit.

8 ft

4 ft

8 ft 3 ft 12 ft 6 ft

36. Find the offset distance of the length of pipe as shown in the diagram below. The total length of the pipe is 62 in.

d

$20\frac{3}{4}$ in. 9 in. $31\frac{1}{2}$ in.

SECTION 10.6 Similar Triangles

Objective A To solve similar and congruent triangles

Similar objects have the same shape but not necessarily the same size. A baseball is similar to a basketball. A model airplane is similar to an actual airplane.

Similar objects have corresponding parts; for example, the propellers on the model airplane correspond to the propellers on the actual airplane. The relationship between the sizes of each of the corresponding parts can be written as a ratio, and each ratio will be the same. If the propellers on the model plane are $\frac{1}{50}$ the size of the propellers on the actual plane, then the model wing is $\frac{1}{50}$ the size of the actual wing, the model fuselage is $\frac{1}{50}$ the size of the actual fuselage, and so on.

The two triangles *ABC* and *DEF* shown are similar. The ratios of corresponding sides are equal.

$\frac{AB}{DE} = \frac{2}{6} = \frac{1}{3}, \frac{BC}{EF} = \frac{3}{9} = \frac{1}{3}$, and $\frac{AC}{DF} = \frac{4}{12} = \frac{1}{3}$

The ratio of corresponding sides $= \frac{1}{3}$.

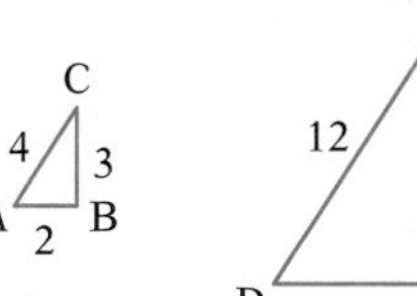

Because the ratios of corresponding sides are equal, three proportions can be formed.

$\frac{AB}{DE} = \frac{BC}{EF}, \frac{AB}{DE} = \frac{AC}{DF}$, and $\frac{BC}{EF} = \frac{AC}{DF}$.

The ratio of corresponding heights equals the ratio of corresponding sides, as shown in the figure.

Ratio of corresponding sides $= \frac{1.5}{6} = \frac{1}{4}$

Ratio of heights $= \frac{2}{8} = \frac{1}{4}$

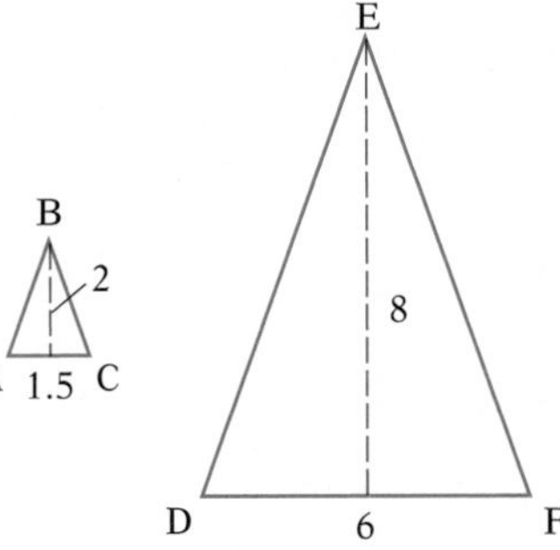

Congruent objects have the same shape *and* the same size.

The two triangles shown are congruent. They have exactly the same size.

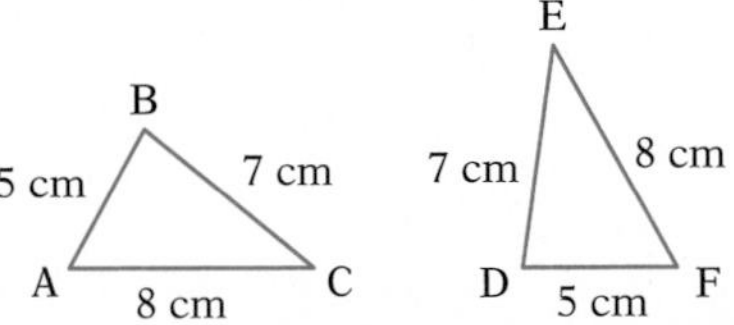

For triangles, congruent means that the corresponding sides *and* angles of the triangle must be equal, unlike similar triangles, which just have corresponding angles equal but not necessarily corresponding sides.

Here are two major rules that can be used to determine if two triangles are congruent.

Side-Side-Side Rule (SSS)

> Two triangles are congruent when three sides of one triangle equal the corresponding sides of the second triangle.

In the two triangles at the right $AB = DE$, $AC = DF$, and $BC = EF$. The corresponding sides of triangles ABC and DEF are equal. The triangles are congruent by the SSS rule.

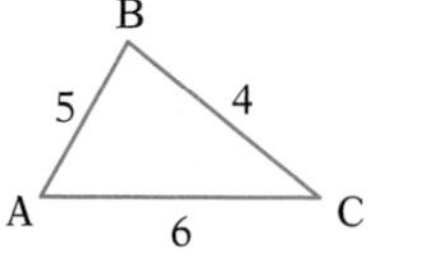

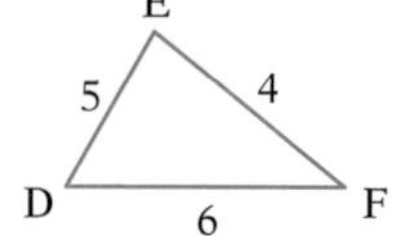

Side-Angle-Side Rule (SAS)

> Two triangles are congruent when two sides and the included angle of one triangle equal the corresponding sides and angle of the second triangle.

In the two triangles at the right $AB = EF$, $AC = DE$, and angle $BAC =$ angle DEF. The triangles are congruent by the SAS rule.

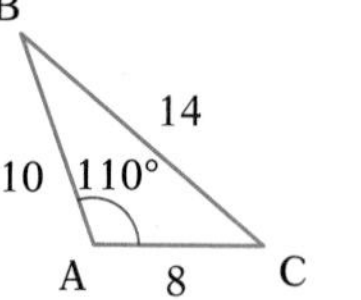

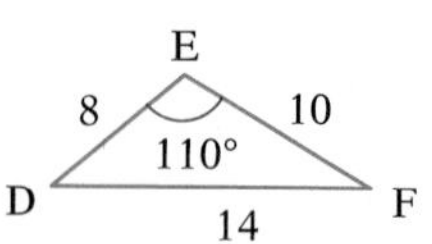

Determine if the two triangles in the adjacent figure are congruent.

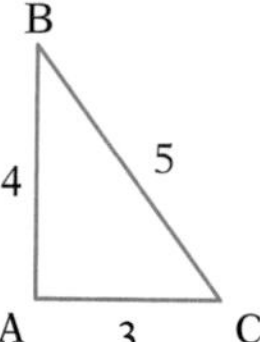

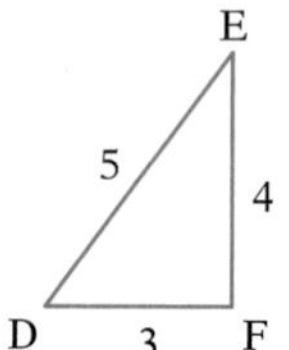

Because $AC = DF$, $AB = FE$, and $BC = DE$ all three sides of one triangle equal the corresponding sides of the second triangle, the triangles are congruent by the SSS Rule.

Example 1 Find the ratio of corresponding sides for the similar triangles ABC and DEF in the figure.

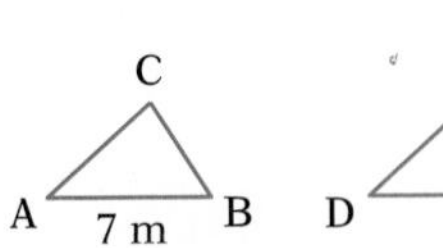

Solution $\frac{7 \cancel{m}}{12 \cancel{m}} = \frac{7}{12}$

Example 2 Find the ratio of corresponding sides for the similar triangles ABC and DEF in the figure.

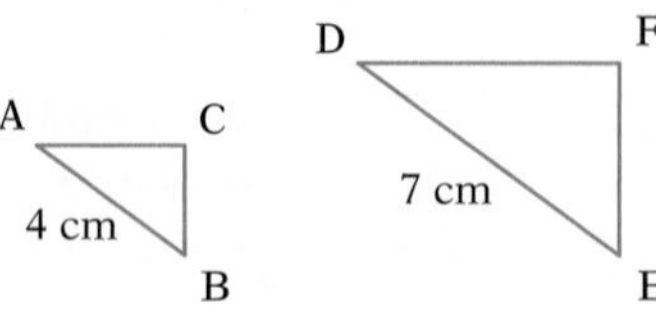

Your solution

Solution on p. A38

Example 3 Triangles *ABC* and *DEF* in the figure are similar. Find side *EF*.

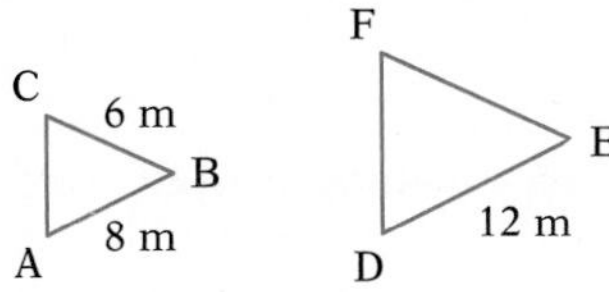

Solution

$$\frac{AB}{DE} = \frac{BC}{EF}$$

$$\frac{8\ \cancel{m}}{12\ \cancel{m}} = \frac{6\text{ m}}{EF}$$

$$8 \times EF = 12 \times 6\text{ m}$$
$$8 \times EF = 72\text{ m}$$
$$EF = 72\text{ m} \div 8$$
$$EF = 9\text{ m}$$

Example 4 Triangles *ABC* and *DEF* in the figure are similar. Find side *DF*.

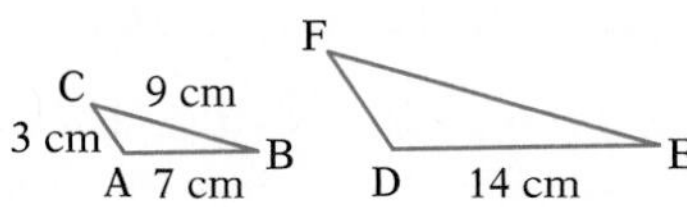

Your solution

Example 5 Determine if triangle *ABC* in the figure is congruent to triangle *DEF*.

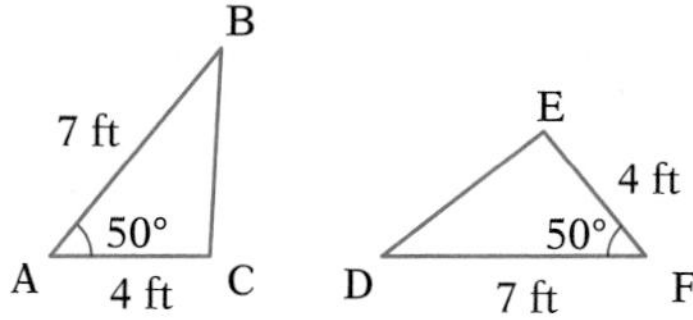

Solution Since $AB = DF$, $AC = EF$, and angle BAC = angle DFE, the triangles are congruent by the SAS rule.

Example 6 Determine if triangle *ABC* in the figure is congruent to triangle *DEF*.

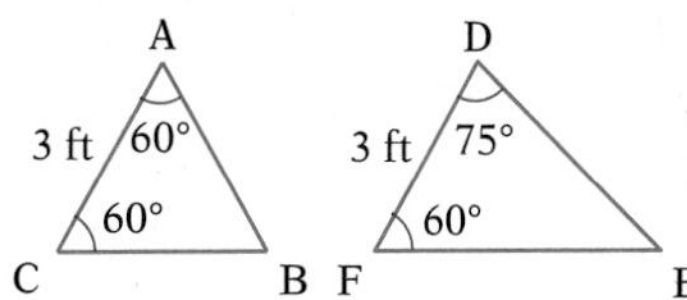

Your solution

Example 7 Triangles *ABC* and *DEF* in the figure are similar. Find the height of *FG*.

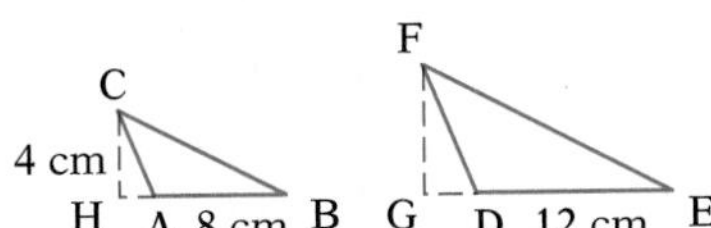

Solution To find the height of *FG*:

$$\frac{AB}{DE} = \frac{\text{height } CH}{\text{height } FG}$$

$$\frac{8\ \cancel{cm}}{12\ \cancel{cm}} = \frac{4\text{ cm}}{\text{height}}$$

$$8 \times \text{height} = 12 \times 4\text{ cm}$$
$$8 \times \text{height} = 48\text{ cm}$$
$$\text{height} = 48\text{ cm} \div 8$$
$$\text{height} = 6\text{ cm}$$

The height *FG* is 6 cm.

Example 8 Triangles *ABC* and *DEF* in the figure are similar. Find the height of *FG*.

Your solution

Solutions on pp. A38–A39

Objective B To solve application problems

Example 9 Triangles ABC and DEF in the figure are similar. Find the area of triangle DEF.

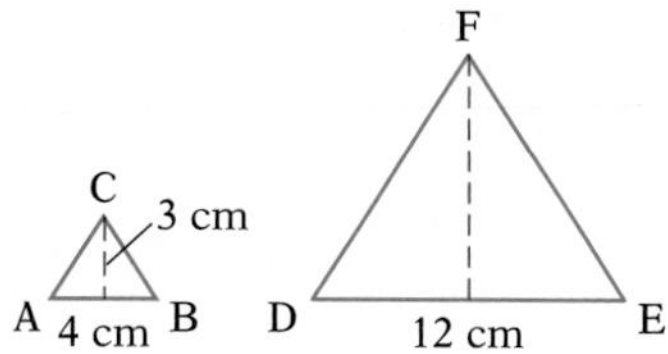

Strategy To find the area of triangle DEF:

- Solve a proportion to find the height of triangle DEF.
- Use the formula Area $= \frac{1}{2} \times$ base $\times$ height.

Solution

$$\frac{AB}{DE} = \frac{\text{height of triangle } ABC}{\text{height of triangle } DEF}$$

$$\frac{4\ \text{cm}}{12\ \text{cm}} = \frac{3\ \text{cm}}{\text{height}}$$

$$4 \times \text{height} = 12 \times 3\ \text{cm}$$
$$4 \times \text{height} = 36\ \text{cm}$$
$$\text{height} = 36\ \text{cm} \div 4$$
$$\text{height} = 9\ \text{cm}$$

$$\text{Area} = \frac{1}{2} \times \text{base} \times \text{height}$$
$$= \frac{1}{2} \times 12\ \text{cm} \times 9\ \text{cm}$$
$$= 54\ \text{cm}^2$$

The area is 54 cm^2.

Example 10 Triangles ABC and DEF in the figure are similar. Find the perimeter of triangle ABC.

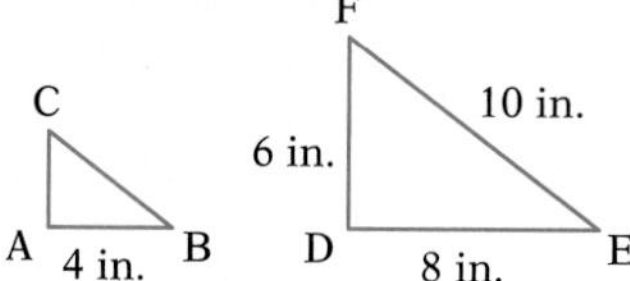

Your strategy

Your solution

Solution on pp. A38–A39

10.6 EXERCISES

Objective A

Find the ratio of corresponding sides for the similar triangles in the figures accompanying Exercises 1 to 4.

1.

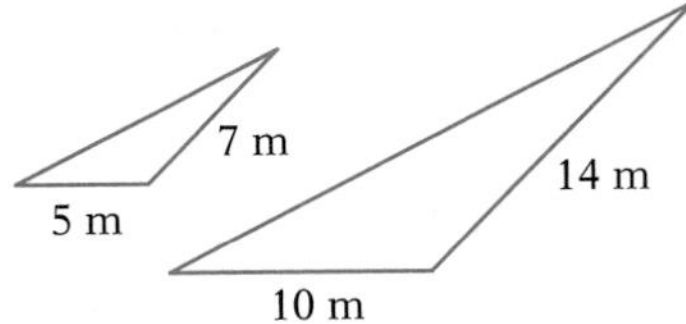

2.

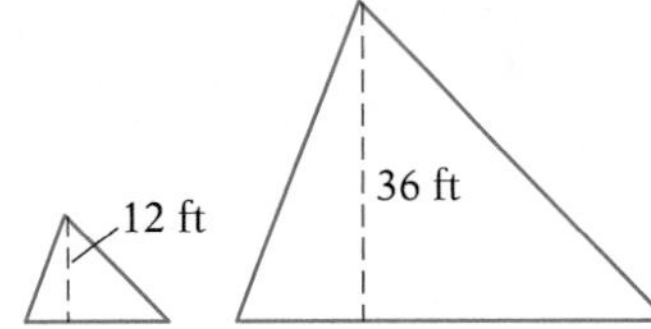

3.

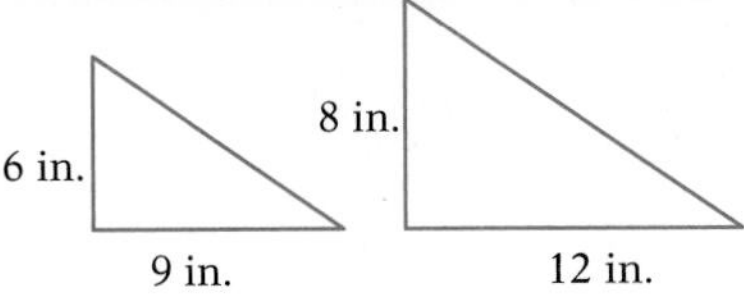

4.

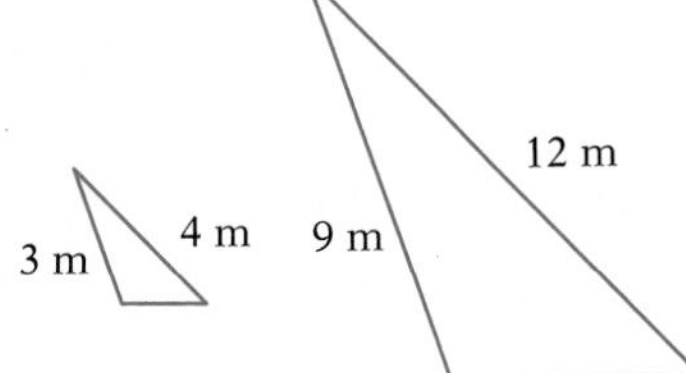

Determine if the two triangles in Exercises 5 to 8 are congruent.

5.

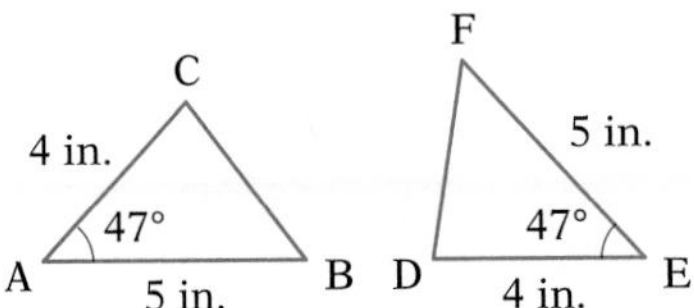

6.

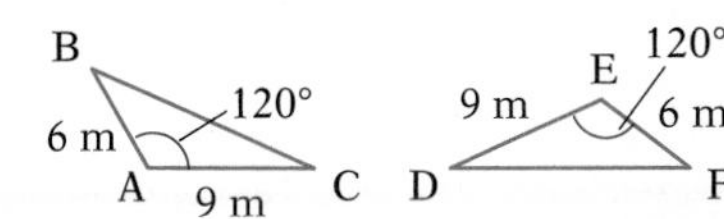

7.

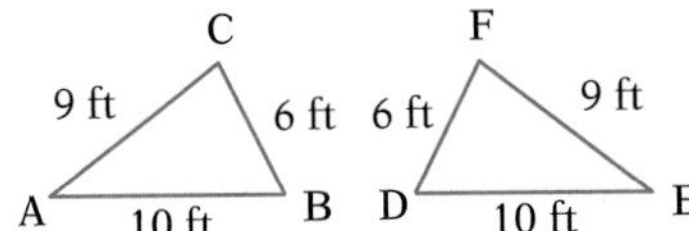

8.

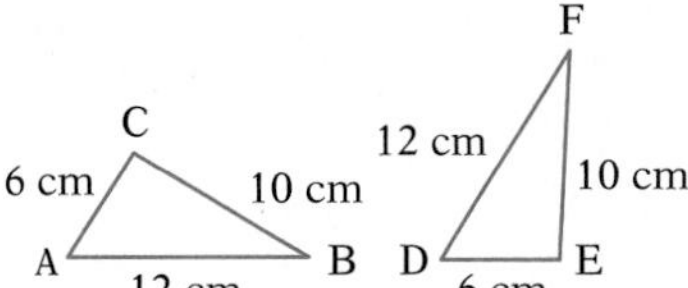

Triangles *ABC* and *DEF* in Exercises 9 to 12 are similar. Find the indicated distance. Round to the nearest tenth.

9. Find side *DE*.

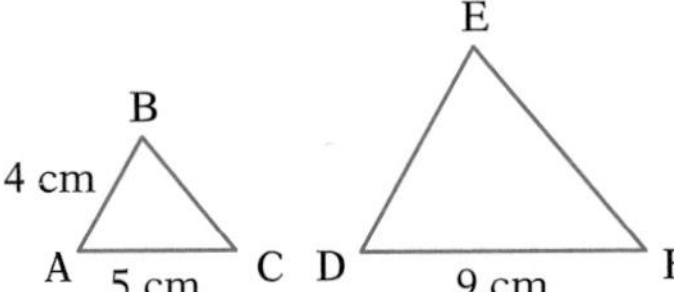

10. Find side *DE*.

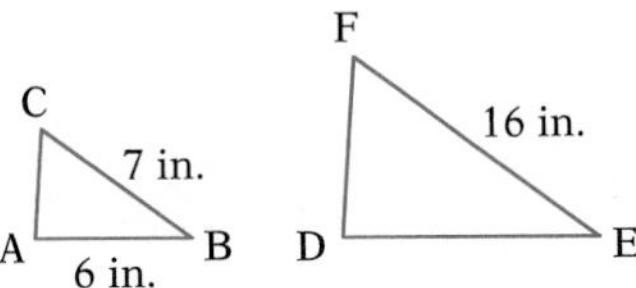

11. Find the height of triangle *DEF*.

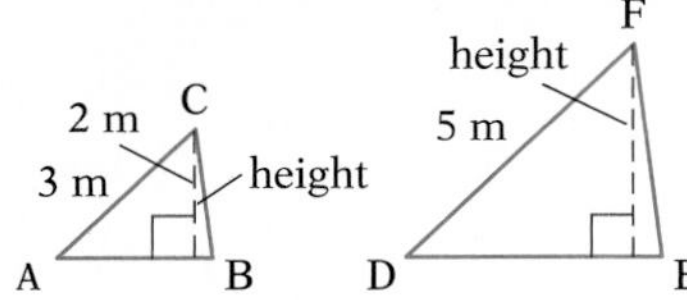

12. Find the height of triangle *ABC*.

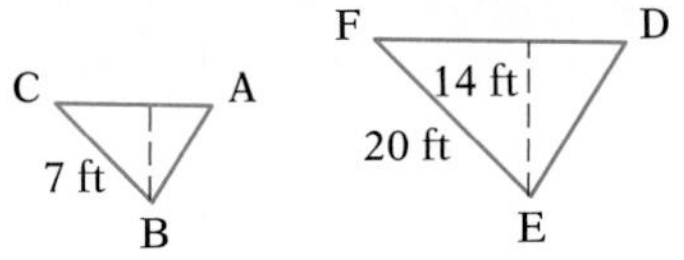

Objective B *Application Problems*

The sun's rays, objects on Earth, and the shadows cast by them form similar triangles.

13. Find the height of the flagpole shown.

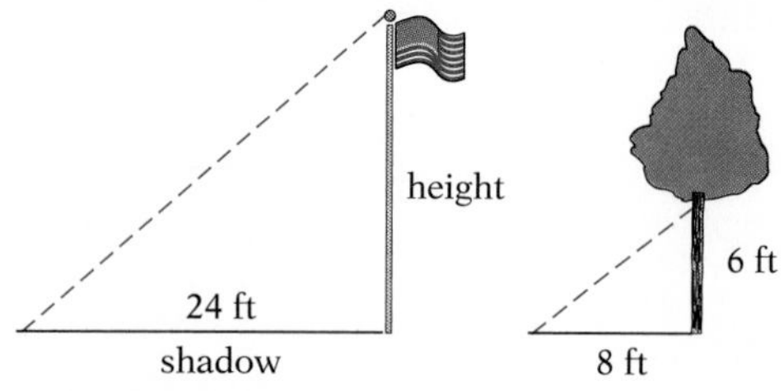

14. Find the height of the flagpole shown.

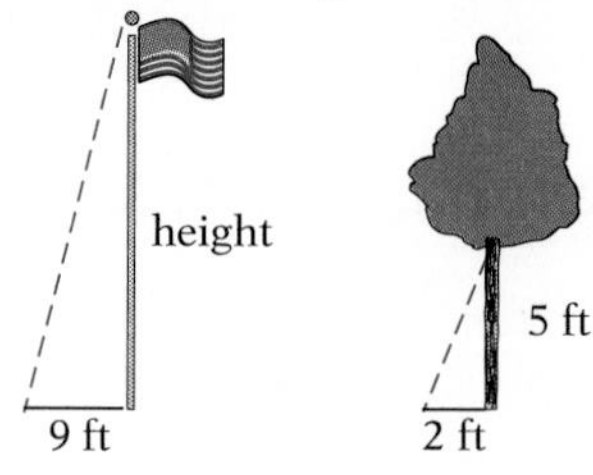

15. Find the height of the building shown.

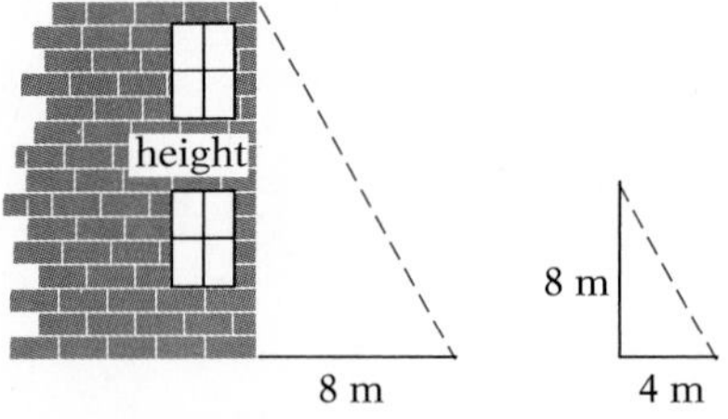

16. Find the height of the building in the figure below.

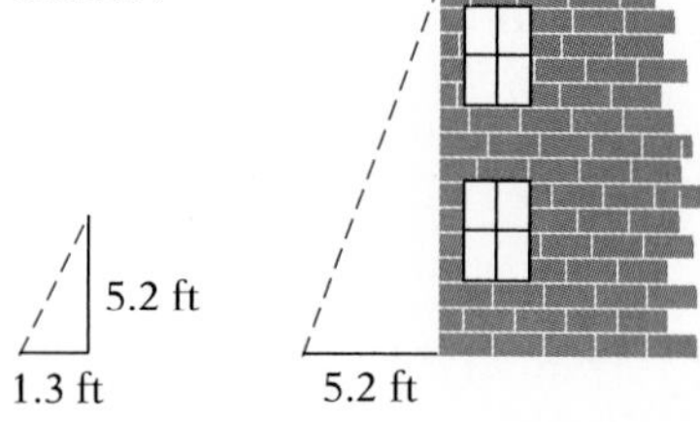

In Exercises 5 to 10, triangles *ABC* and *DEF* are similar.

17. Find the perimeter of triangle *ABC*.

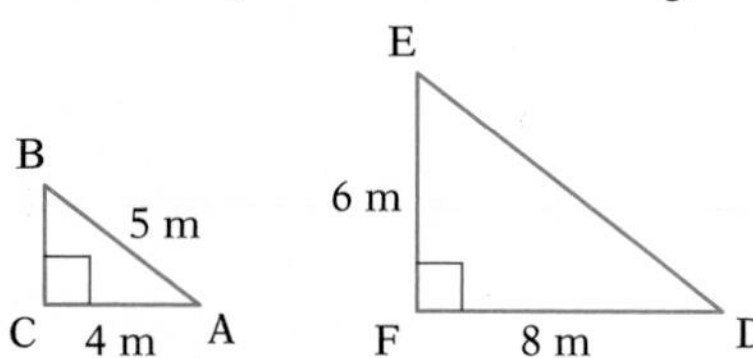

18. Find the perimeter of triangle *DEF*.

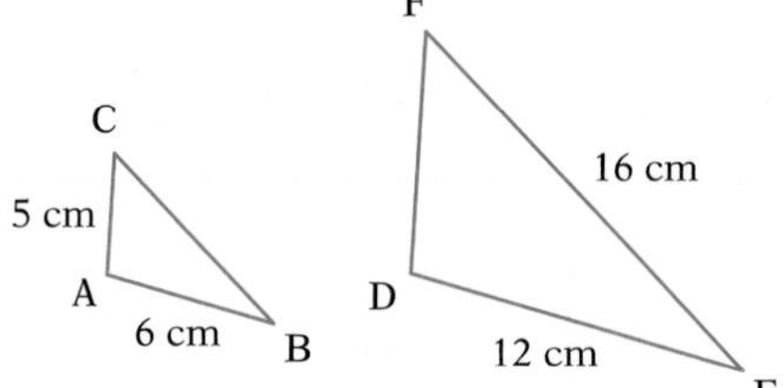

19. Find the area of triangle *ABC*.

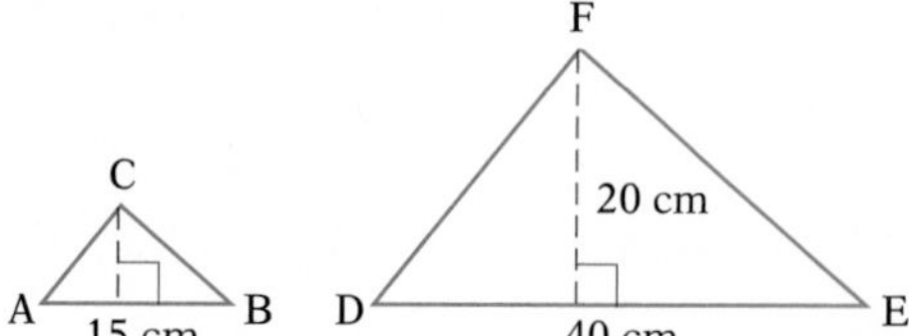

20. Find the area of triangle *DEF*.

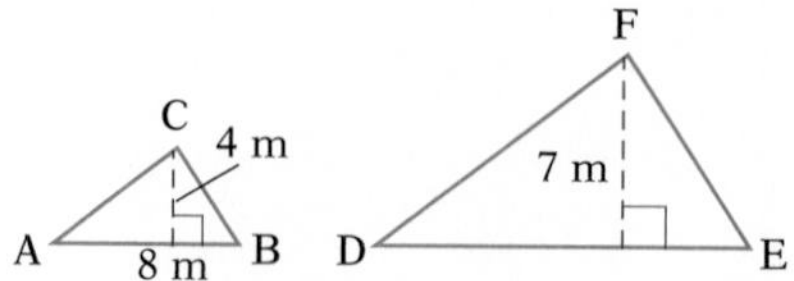

Calculators and Computers

Buffon Needle Problem

A computer simulation is a program that attempts to approximate an actual situation. For example, a pilot uses a flight simulator to practice flight maneuvers, such as engine failure, which would be hazardous in actual flight. The flight simulator is a computer program that simulates the actual conditions, thereby giving the pilot practice.

The Buffon Needle Problem is another example of a simulation. A computer program simulates dropping a needle onto a piece of paper on which are drawn a series of parallel lines. The distance between each line is the same as the length of the needle.

The program BUFFON NEEDLE PROBLEM on the Math ACE disk will allow you to simulate dropping a needle 100, 500, or more times without you actually having to do it. An interesting result of the program is that the ratio of the number of times the needle crosses a line to the total number of times the needle is dropped can be used to approximate a value for pi. The relationship is $\pi \approx \frac{2N}{C}$, where N is the number of times the needle is dropped, and C is the number of times the needle crosses a line.

Chapter Summary

Key Words

An *angle* is formed when two rays start from the same point. An angle is measured in *degrees.*

A 90° angle is called a *right angle.*

Perpendicular lines are intersecting lines that form right angles.

Complementary angles are two angles whose sum is 90°.

Supplementary angles are two angles whose sum is 180°.

A *right triangle* contains one right angle. The side opposite the right angle in a right triangle is called the *hypotenuse.* The other two sides are called *legs.*

A *45°-45°-90° triangle* is a special right triangle in which the sides opposite the 45° angles are also equal.

A *30°-60°-90° triangle* is a special right triangle in which the length of the leg opposite the 30° angle is $\frac{1}{2}$ the length of the hypotenuse.

A *quadrilateral* is a four-sided plane figure.

A *rectangle* is a parallelogram that has four right angles.

A *square* is a rectangle that has four equal sides.

A *circle* is a plane figure in which all points are the same distance from the center of the circle. The *diameter* is a line segment across a circle going through the center. The *radius* is equal to one-half the diameter.

A *rectangular solid* is a solid in which all six faces are rectangles.

A *cube* is a rectangular solid in which all six faces are squares.

A *sphere* is a solid in which all points are the same distance from the center of the sphere.

Perimeter is the distance around a plane figure.

Area is a measure of the amount of surface in a region.

Volume is a measure of the amount of space inside a closed surface.

The *square root* of a number is one of two identical factors of that number.

Similar objects have the same shape but not necessarily the same size.

Congruent objects have the same shape and the same size.

Composite geometric solids are solids made from two or more geometric solids.

Essential Rules

Perimeter Equations

Triangle:	Perimeter = side 1 + side 2 + side 3
Square:	Perimeter = 4 × side
Rectangle:	Perimeter = 2 × length + 2 × width
Circle:	Circumference = 2 × π × radius or = π × diameter

Area Equations

Triangle:	Area = $\frac{1}{2}$ (base × height)
Square:	Area = side × side = $(\text{side})^2$
Rectangle:	Area = length × width
Circle:	Area = π × $(\text{radius})^2$

Volume Equations

Rectangular solid:	Volume = length × width × height
Cube:	Volume = $(\text{side})^3$
Sphere:	Volume = $\frac{4}{3}$ × π × $(\text{radius})^3$
Cylinder:	Volume = π × $(\text{radius})^2$ × height

Pythagorean Theorem

Hypotenuse = $\sqrt{\text{leg}^2 + \text{leg}^2}$

Side-Side-Side (SSS)

Two triangles are congruent when three sides of one triangle equal the corresponding sides of the second triangle.

Side-Angle-Side (SAS)

Two triangles are congruent when two sides and the included angle of one triangle equal the corresponding sides and included angle of the second triangle.

Chapter Review

SECTION 10.1

1. Given $AB = 15$, $CD = 6$, and $AD = 24$, find the length of BC.

A B C D ℓ

2. Find the supplement of a 105° angle.

3. The diameter of a sphere is 1.5 m. Find the radius of the sphere.

4. A right triangle has a 32° angle. Find the measures of the other two angles.

5. Given that $\ell_1 \parallel \ell_2$, find the measures of angles a and b.

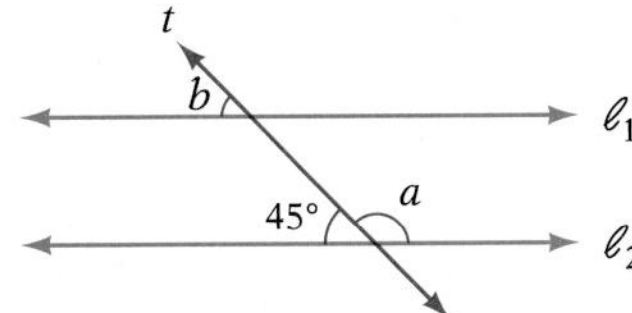

6. Given that $\ell_1 \parallel \ell_2$, find the measures of angles a and b.

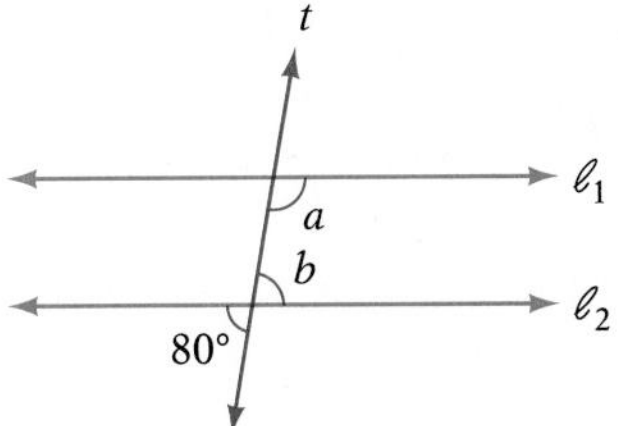

SECTION 10.2

7. Find the perimeter of the rectangle in the figure below.

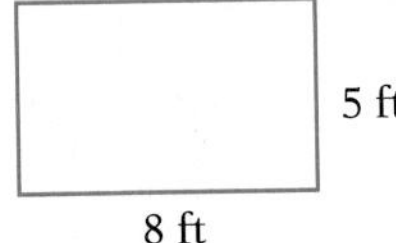

8. Find the circumference of the circle in the figure below.

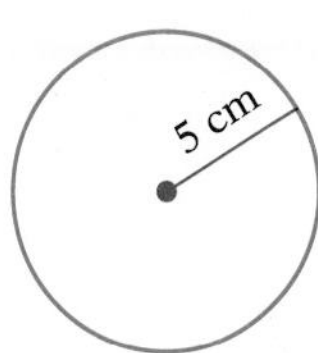

9. Find the perimeter of the composite figure shown. Use $\pi \approx 3.14$.

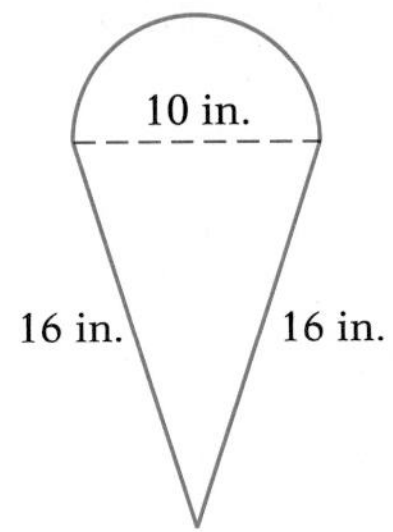

10. A bicycle tire has a diameter of 28 in. How many feet does the bicycle travel if the wheel makes 10 revolutions? Use $\pi \approx 3.14$. Round to the nearest tenth of a foot.

SECTION 10.3

11. Find the area of the rectangle shown in the figure.

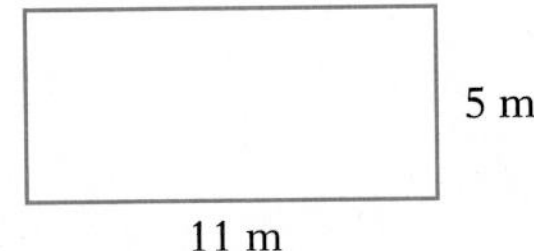

12. Find the area of the circle shown in the figure. Use $\pi \approx 3.14$.

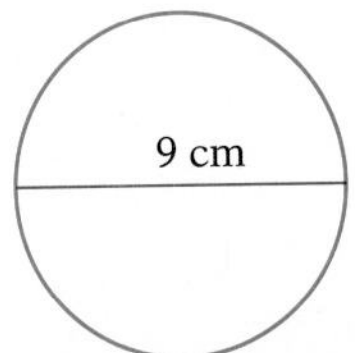

13. Find the area of the composite figure shown. Use $\pi \approx 3.14$.

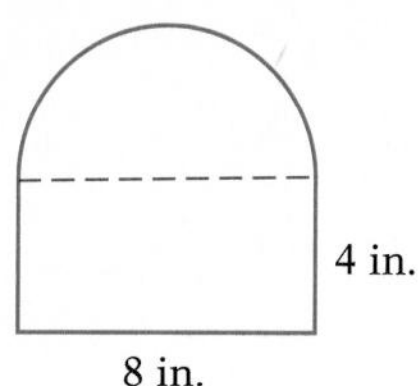

14. New carpet is installed in a room measuring 18 ft by 14 ft. Find the area of the room in square yards. ($9 \text{ ft}^2 = 1 \text{ yd}^2$)

SECTION 10.4

15. Find the volume of the rectangular solid shown in the figure.

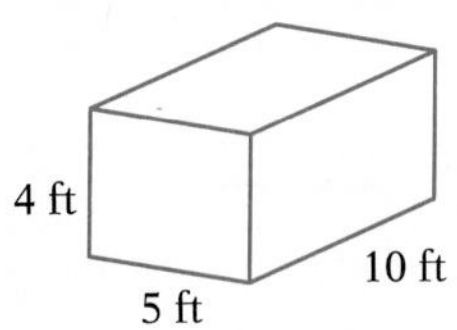

16. Find the volume of a sphere with a diameter of 8 ft. Use $\pi \approx 3.14$. Round to the nearest tenth.

17. Find the volume of the composite figure shown below.

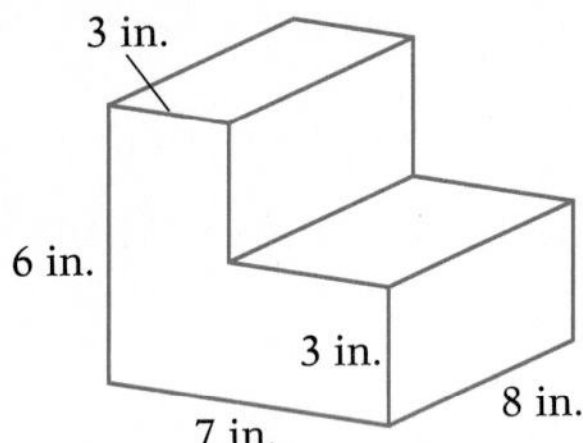

18. A silo, which is in the shape of a cylinder, is 9 ft in diameter and has a height of 18 ft. Find the volume of the silo. Use $\pi \approx 3.14$.

SECTION 10.5

19. Find the square root of 15. Use the table on page A3.

20. Find the unknown side of the triangle in the figure below.

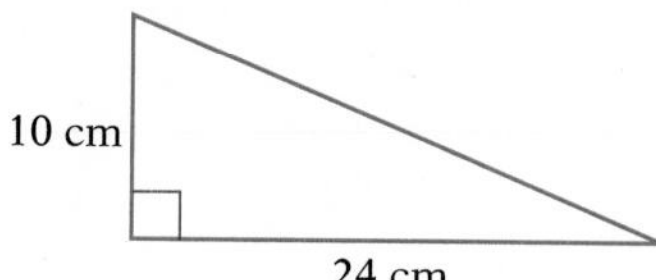

21. How high on a building will a 17-ft ladder reach when the bottom of the ladder is 8 ft from the building?

SECTION 10.6

22. Triangles *ABC* and *DEF* are similar. Find the height of triangle *DEF*.

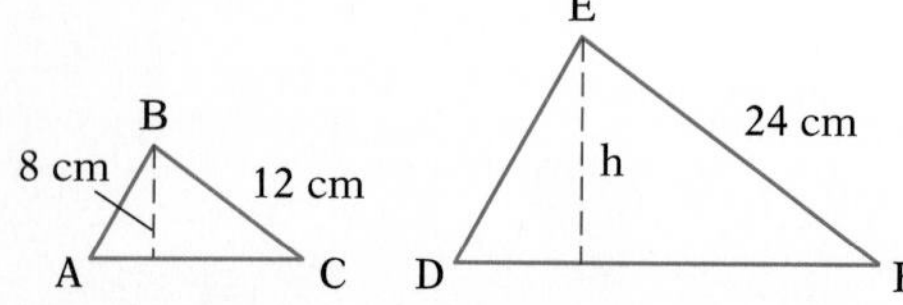

23. Triangles *ABC* and *DEF* are similar. Find the area of triangle *DEF*.

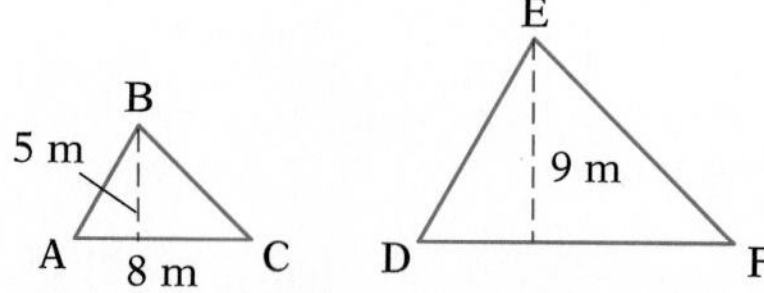

Chapter Test

1. Find the complement of a 32° angle.

2. A right triangle has a 40° angle. Find the measure of the other two angles.

3. In the accompanying figure, lines ℓ_1 and ℓ_2 are parallel. $\angle x = 30°$. Find $\angle y$.

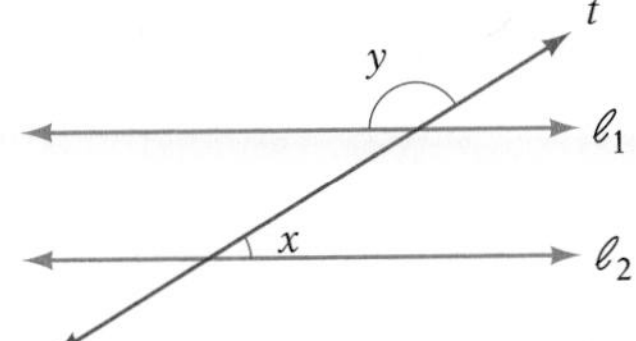

4. In the accompanying figure, lines ℓ_1 and ℓ_2 are parallel. $\angle x = 45°$. Find $\angle a + \angle b$.

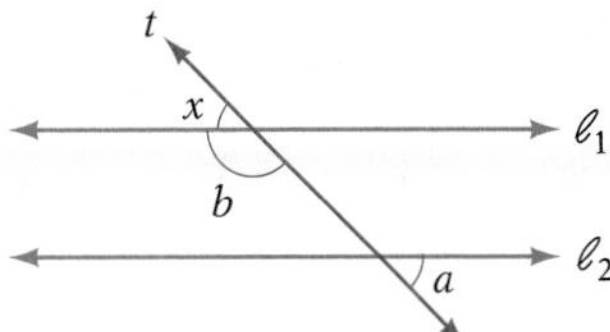

5. Find the perimeter of a rectangle with a length of 2 m and a width of 1.4 m.

6. Find the perimeter of the composite figure. Use $\pi \approx 3.14$.

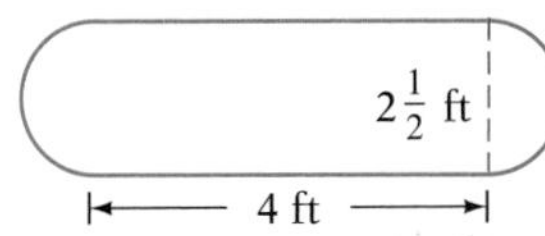

7. A carpet is to be placed as shown in the diagram below. At $13.40 per square yard, how much will it cost to carpet the area? (Round to the nearest cent.)

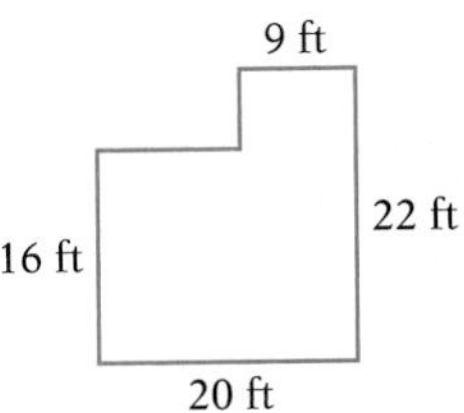

8. Find the area of a circle with a diameter of 2 m. Use $\frac{22}{7}$ for π.

9. Find the area of the composite figure.

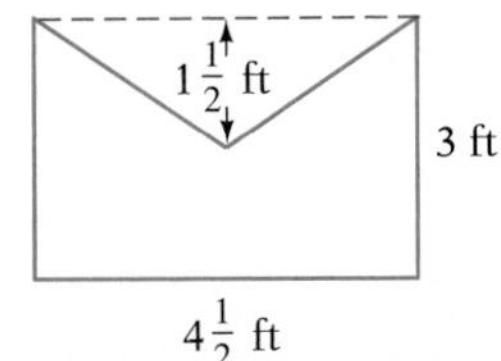

10. Find the cross-sectional area of a redwood tree that is 11 ft 6 in. in diameter. Use 3.14 for π. (Round to hundredths.)

11. How much more pizza is contained in a pizza with radius 10 in. than in one with radius 8 in.? Use $\pi \approx 3.14$.

12. Find the volume of a cylinder with a height of 6 m and a radius of 3 m. Use 3.14 for π.

13. Find the volume of the composite figure.

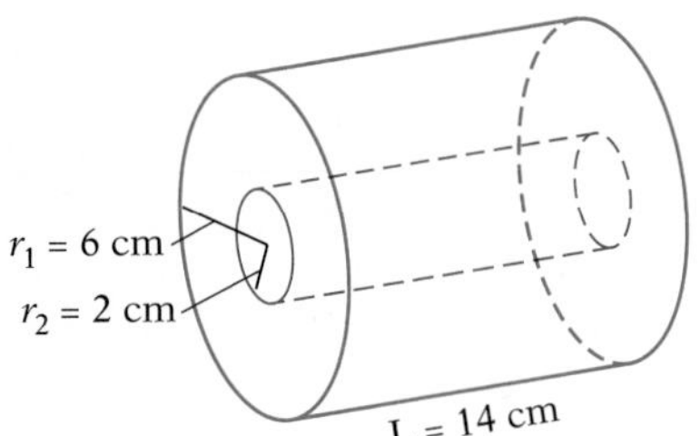

14. A toolbox is 1 ft 2 in. long, 9 in. wide, and 8 in. high. The sides and bottom of the toolbox are $\frac{1}{2}$ in. thick. Find the volume of the interior of the toolbox in cubic inches.

15. Find the square root of 189. Use the table on page A3.

16. Find the unknown side of the triangle shown in the figure.

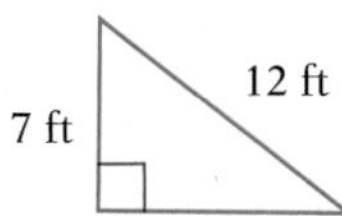

17. Find the length of the rafter for the roof shown in the figure.

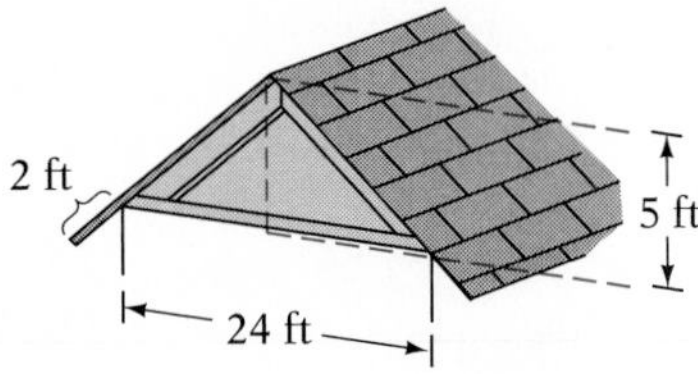

18. Triangles *ABC* and *DEF* below are similar. Find side *BC*.

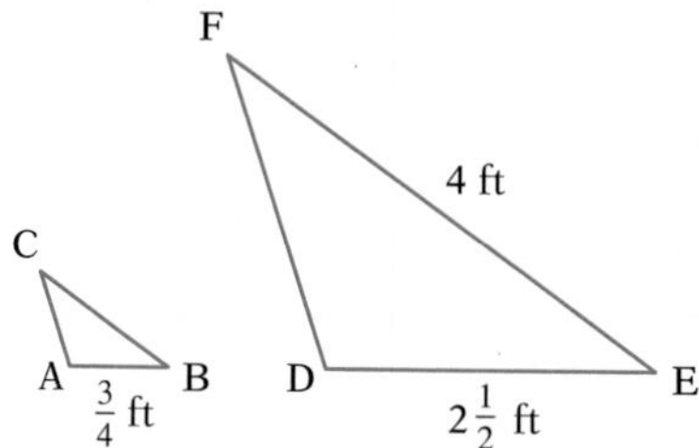

19. Find the width of the canal shown in the figure below.

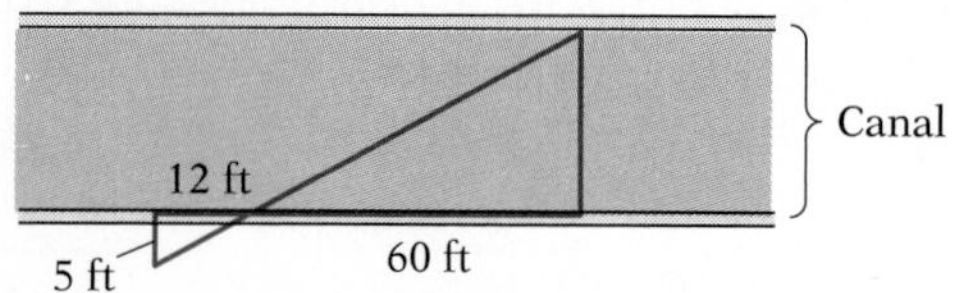

20. In the accompanying figure, triangles *ABC* and *DEF* are congruent right triangles. Find the length of *FE*.

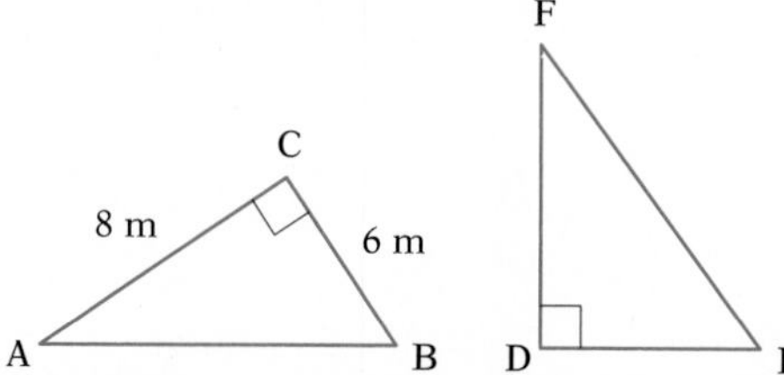

Cumulative Review

1. Find the GCF of 96 and 144.

2. Add: $3\frac{5}{12} + 2\frac{9}{16} + 1\frac{7}{8}$

3. Divide: $4\frac{1}{3} \div 6\frac{2}{9}$

4. Simplify: $\left(\frac{2}{3}\right)^2 \div \left(\frac{1}{3} + \frac{1}{2}\right) - \frac{2}{5}$

5. Divide: $0.027\overline{)1.0395}$

6. Write "\$348.80 earned in 40 hours" as a unit rate.

7. Solve the proportion $\frac{3}{8} = \frac{n}{100}$.

8. Write $37\frac{1}{2}\%$ as a fraction.

9. 6.4 is what percent of 80?

10. 30.94 is 36.4% of what number?

11. Convert 18 in. to feet.

12. Convert $\frac{2}{5}$ ton to pounds.

13. Convert 32.5 km to meters.

14. Subtract: 32 m − 42 cm

15. Convert 1270 mg to grams.

16. Add: 6 L 32 ml + 2 L 695 ml

17. You bought a car for $7516 and made a down payment of $1000. You paid the balance in 36 equal monthly installments. Find the monthly payment.

18. The sales tax on a suit costing $175 is $6.75. At the same rate, find the sales tax on a stereo system costing $1220.

19. A heavy equipment operator receives an hourly wage of \$16.06 an hour after receiving a 10% wage increase. Find the operator's hourly wage before the increase.

20. An after-Christmas sale has a markdown rate of 55%. Find the sale price of a dress that had a regular price of \$120.

21. An IRA pays 10% annual interest compounded daily. What would the value on an investment of \$25,000 after 20 years be? Use the table on page A4.

22. A 4×4-inch tile weighs 6 oz. Find the weight of a package of 144 tiles in pounds.

23. Twenty-five rivets are used to fasten two steel plates together. The plates are 5 m 40 cm long, and the rivets are equally spaced with a rivet at each end. Find the distance in centimeters between the rivets.

24. Find the complement of a 76° angle.

25. Given that the lines ℓ_1 and ℓ_2 in the figure below are parallel. Find the angles a and b.

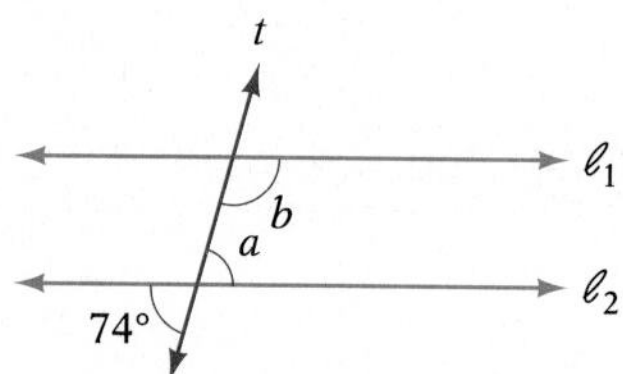

26. Find the perimeter of the composite figure. Use 3.14 for π.

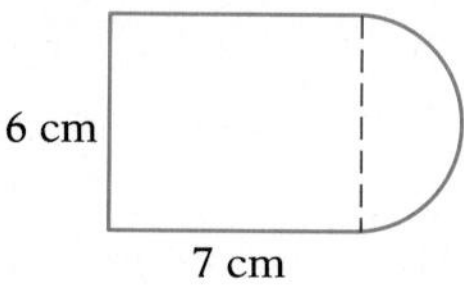

27. Find the area of the composite figure.

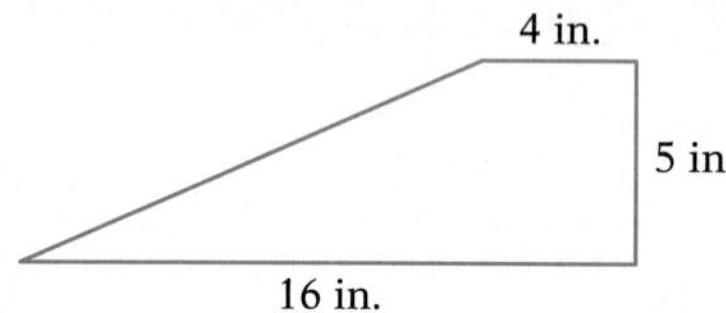

28. Find the volume of the composite figure. Use 3.14 for π.

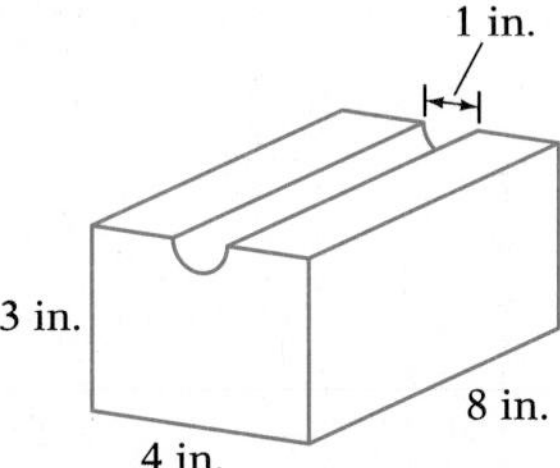

29. Find the unknown side of the triangle shown in the figure below. Round to the nearest hundredth.

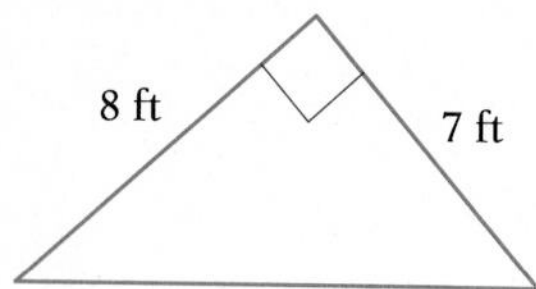

30. Triangles ABC and DEF below are similar. Find the perimeter of DEF.

11 Rational Numbers

OBJECTIVES

- To identify the order relation between two integers
- To evaluate expressions containing the absolute value symbol
- To add integers
- To subtract integers
- To multiply integers
- To divide integers
- To solve application problems
- To add or subtract rational numbers
- To multiply or divide rational numbers
- To use the Order of Operations Agreement to simplify expressions

History of Negative Numbers

When a temperature is "below zero," that temperature is represented by placing the symbol − before the number. Thus −4 degrees means that the temperature is 4 degrees below zero. The number −4 is read "negative four" and is an example of a negative number.

The earliest evidence for the use of negative numbers dates from China, about 300 B.C. The Chinese used two sets of calculating rods, one red for positive numbers and one black for negative numbers. The idea seems to have remained with the Chinese and did not spread quickly to the Middle East or to Europe.

Between the years 300 B.C. and 1500 A.D., there were a few further attempts to deal with negative numbers. Then, in 1500, negative numbers began appearing more and more frequently. In most cases, however, the numbers were considered "fictitious" numbers or "false" numbers. The idea of using a negative number to refer to a quantity below zero was still not an accepted idea.

By the 18th century, negative numbers were discussed in most mathematics textbooks. However, operations such as multiplying two negative numbers were not included in some books.

As the 18th century came to a close, some 2000 years after the first evidence of negative numbers from the Chinese, negative numbers were finally accepted. This concept, as with other mathematical ideas, took a long time to completely develop. Do you suppose that today there is some mathematical idea that will take 2000 years before being accepted?

SECTION 11.1 Introduction to Integers

Objective A To identify the order relation between two integers

Thus far, we have encountered only zero and the numbers greater than zero in the text. The numbers greater than zero are called **positive numbers.** However, the phrases "12 degree below zero," "\$25 in debt," and "15 feet below sea level" refer to numbers less than zero. These numbers are called **negative numbers.**

The integers are . . . $-4, -3, -2, -1, 0, 1, 2, 3, 4, \ldots$

Each integer can be shown on a number line. The integers to the left of zero on the number line are called **negative integers** and are represented by a negative sign ($-$) placed in front of the number. The integers to the right of zero are called **positive integers.** The positive integers are also called natural numbers. Zero is neither a positive nor negative integer.

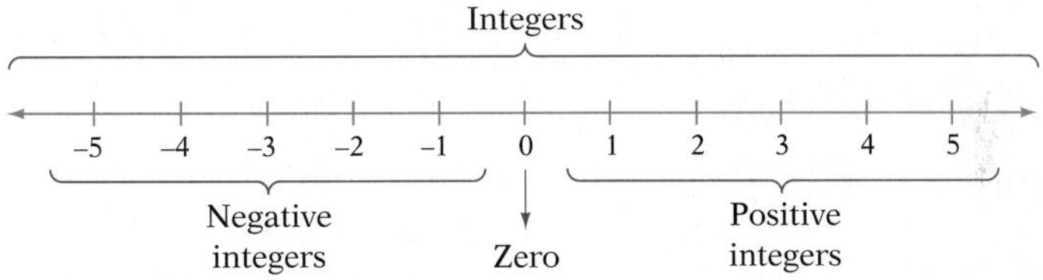

A number line can be used to visualize the order relation of two integers. A number that appears to the left of a given number is less than ($<$) the given number. A number that appears to the right of a given number is greater than ($>$) the given number.

2 is greater than negative 4.
$2 > -4$

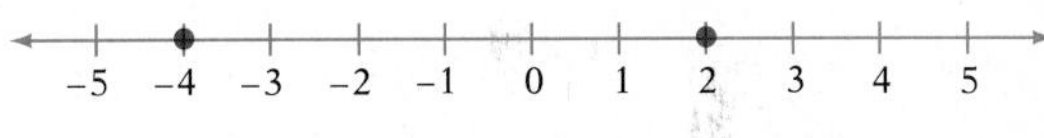

Negative 5 is less than negative 3.
$-5 < -3$

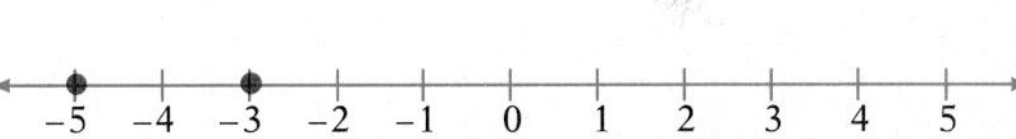

Example 1 The temperature at the North pole was recorded as 87 degrees below zero. Represent this temperature as an integer.

Solution -87

Example 2 The surface of the Salton Sea is 232 ft below sea level. Represent this depth by a signed number.

Your solution

Example 3 Graph -2 on the number line.

Solution

-4 -3 -2 -1 0 1 2 3 4

Example 4 Graph -4 on the number line.

Your solution

-4 -3 -2 -1 0 1 2 3 4

Example 5 Place the correct symbol, $<$ or $>$, between the two numbers.

a. $-5 \quad -7$ **b.** $\frac{1}{2} \quad -2$

Solution **a.** $-5 > -7$ **b.** $\frac{1}{2} > -2$

Example 6 Place the correct symbol, $<$ or $>$, between the two numbers.

a. $-12 \quad -8$ **b.** $-5 \quad 0$

Your solution

Solutions on p. A39

Objective B To evaluate expressions containing the absolute value symbol

11

Two numbers that are the same distance from zero on the number line but on opposite sides of zero are called **opposites.**

−4 is the opposite of 4

and

4 is the opposite of −4.

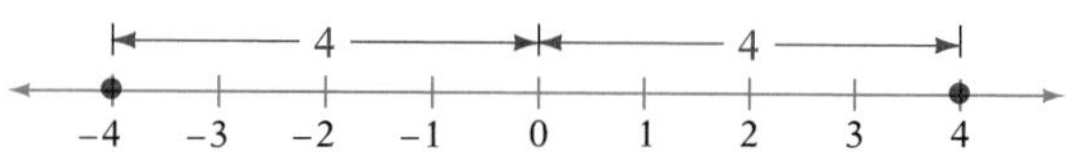

Note that a negative sign can be read as "the opposite of."

$-(4) = -4$ The opposite of positive 4 is negative 4.

$-(-4) = 4$ The opposite of negative 4 is positive 4.

The **absolute value** of a number is the distance between zero and the number on the number line. Therefore the absolute value of a number is a positive number or zero. The symbol for absolute value is "| |".

The distance from 0 to 4 is 4. Thus $|4| = 4$ (the absolute value of 4 is 4).

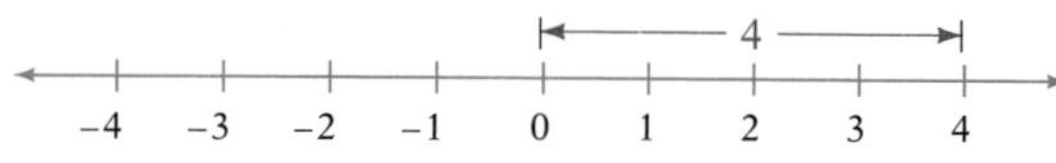

The distance from 0 to −4 is 4. Thus $|-4| = 4$ (the absolute value of −4 is 4).

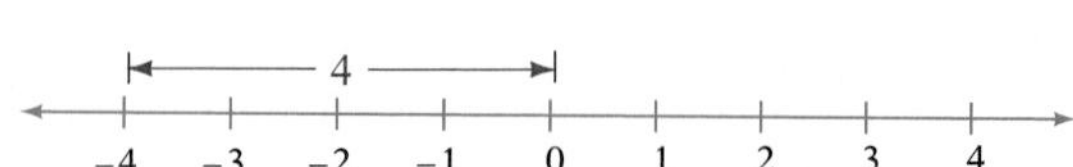

The absolute value of a positive number is the number itself. The absolute value of a negative number is the opposite of the negative number. The absolute value of zero is zero.

Example 7 Find the absolute value of 2 and −3.

Solution $|2| = 2$
$|-3| = 3$

Example 8 Find the absolute value of −7 and 21.

Your solution

Example 9 Evaluate $|-34|$ and $|0|$.

Solution $|-34| = 34$
$|0| = 0$

Example 10 Evaluate $|2|$ and $|-9|$.

Your solution

Example 11 Evaluate $-|-4|$.

Solution $-|-4| = -4$
The minus sign *in front of* the absolute value sign is not affected by the absolute value sign.

Example 12 Evaluate $-|-12|$.

Your solution

Solutions on p. A39

11.1 EXERCISES

Objective A

Place the correct symbol, < or >, between the two numbers.

1. 3 5

2. 7 4

3. −2 −5

4. −6 −1

5. −16 1

6. −2 13

7. 3 −7

8. 5 −6

9. −11 −8

10. −4 −10

11. −1 −6

12. −9 −4

13. 0 −3

14. 8 0

15. 6 −8

16. 8 −6

17. −14 16

18. −12 1

19. 35 28

20. 42 19

21. −42 27

22. −36 49

23. 21 −34

24. 53 −46

25. −27 −39

26. −51 −20

27. −87 63

28. −75 92

29. 68 −79

30. 95 −71

31. −62 −84

32. −91 −70

33. 94 83

34. 76 81

35. 59 −67

36. 48 −66

37. −93 −55

38. −64 −86

39. −88 57

40. −58 82

41. 0 129

42. −136 0

43. −131 101

44. 127 −150

45. −194 −180

Objective B

Find the opposite number.

46. 4 **47.** 16 **48.** -2 **49.** -3

50. 22 **51.** 45 **52.** -31 **53.** -88

Evaluate.

54. $|2|$ **55.** $|-2|$ **56.** $|-6|$ **57.** $|6|$ **58.** $|8|$

59. $|5|$ **60.** $|-9|$ **61.** $|-1|$ **62.** $-|-1|$ **63.** $-|-8|$

64. $-|-5|$ **65.** $-|0|$ **66.** $|16|$ **67.** $|19|$ **68.** $|-12|$

69. $|-22|$ **70.** $-|29|$ **71.** $-|20|$ **72.** $-|-14|$ **73.** $-|-18|$

74. $|-15|$ **75.** $|-23|$ **76.** $-|33|$ **77.** $-|27|$ **78.** $-|-36|$

79. $-|-41|$ **80.** $|32|$ **81.** $|25|$ **82.** $|-38|$ **83.** $|-30|$

84. $-|37|$ **85.** $-|34|$ **86.** $-|-42|$ **87.** $-|-45|$ **88.** $|44|$

89. $|36|$ **90.** $|-74|$ **91.** $|-61|$ **92.** $-|88|$ **93.** $-|52|$

94. $-|-81|$ **95.** $-|-93|$ **96.** $|-107|$ **97.** $|-119|$

SECTION 11.2 Addition and Subtraction of Integers

Objective A To add integers

This section begins with translating the sum of two integers into words.

$8 + 5$	positive 8 plus positive 5
$(-8) + 5$	negative 8 plus positive 5
$8 + (-5)$	positive 8 plus negative 5
$(-8) + (-5)$	negative 8 plus negative 5

As shown in the examples above, parentheses are frequently used when writing a negative integer.

Besides an integer being graphed as a dot on a number line, an integer can be represented anywhere along a number line by an arrow. A positive number is represented by an arrow pointing to the right. A negative number is represented by an arrow pointing to the left. The absolute value of the number is represented by the length of the arrow. The integers 5 and -4 are shown on the number line in the figure below.

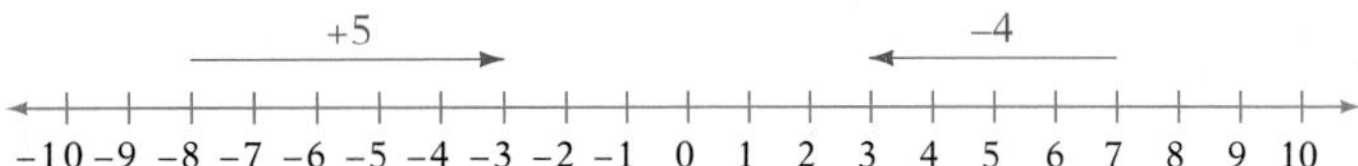

The sum of two integers can be shown on a number line. To add two integers, find the point on the number line corresponding to the first addend. At that point draw an arrow representing the second addend. The sum is the number directly below the tip of the arrow.

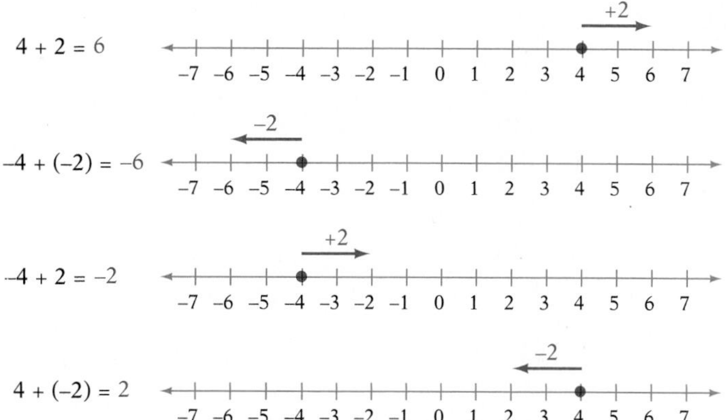

The sums of the integers shown above can be placed in two groups:

Group 1. Addends have the same sign

$4 + 2$	positive 4 plus positive 2
$(-4) + (-2)$	negative 4 plus negative 2

Group 2. Addends have different signs

$(-4) + 2$	negative 4 plus positive 2
$4 + (-2)$	positive 4 plus negative 2

The rule used for adding two integers depends on whether the signs of the addends are the same or different.

Rule for Adding Two Integers

Same signs: To add integers with the same signs, add the absolute values of the numbers. Then attach the sign of the addends.

Different signs: To add integers with different signs, subtract the number with smaller absolute value from the number with larger absolute value. Then attach the sign of the addend with the larger absolute value.

Add: $(-4) + (-9)$

Because the signs of the addends are the same, add the absolute values of the numbers.

$$|-4| = 4, \ |-9| = 9$$
$$4 + 9 = 13$$

Then attach the sign of the addends.

$$(-4) + (-9) = -13$$

Add: $6 + (-13)$

Because the signs of the addends are different, subtract the number of smaller absolute value from the number of larger absolute value.

$$|6| = 6, \ |-13| = 13$$
$$13 - 6 = 7$$

Then attach the sign of the number with the larger absolute value. Because $|-13| > |6|$, attach the negative sign.

$$6 + (-13) = -7$$

Add: $162 + (-247)$

Because the signs are different, find the difference between the absolute values of the numbers and attach the sign of the number with the greater absolute value.

$$162 + (-247) = -85$$

Add: $-14 + (-47)$

Because the signs are the same, add the absolute values of the numbers and attach the sign of the addends.

$$-14 + (-47) = -61$$

When adding more than two integers, start from the left and add the first two numbers. Then add the sum to the third number. Continue the process until all the numbers have been added.

Add: $(-4) + (-6) + (-8) + 9$

$$\underbrace{(-4) + (-6)} + (-8) + 9$$

Add the first two numbers.

$$\underbrace{(-10) + (-8)} + 9$$

Add the sum to the third number.

$$\underbrace{(-18) + 9}$$

Continue adding until all numbers have been added.

$$-9$$

Example 1 Add: $-162 + 98$

Solution $-162 + 98 = -64$

Example 2 Add: $-154 + (-37)$

Your solution

Solution on p. A39

Example 3	Add: $42 + (-12) + (-30)$	**Example 4**	Add: $-36 + 17 + (-21)$
Solution	$42 + (-12) + (-30)$ $30 + (-30)$ 0	**Your solution**	
Example 5	Add: $-2 + (-7) + 4 + (-6)$	**Example 6**	Add: $-5 + (-2) + 9 + (-3)$
Solution	$-2 + (-7) + 4 + (-6)$ $-9 + 4 + (-6)$ $-5 + (-6)$ -11	**Your solution**	

Solutions on p. A39

Objective B

To subtract integers

Before the rules for subtracting two integers are explained, first look at the translation into words of an expression that is the difference of two integers:

$9 - 3$	positive 9 minus positive 3
$(-9) - 3$	negative 9 minus positive 3
$9 - (-3)$	positive 9 minus negative 3
$(-9) - (-3)$	negative 9 minus negative 3

Notice from above that the sign − is used two different ways. One way is as a negative sign, as in (-9), *negative* 9. The second way is to indicate the operation of subtraction, as in $9 - 3$, 9 *minus* 3.

Look at the next four subtraction expressions and decide whether the second number in each expression is a positive number or a negative number:

1. $(-10) - 8$ **3.** $10 - (-8)$

2. $(-10) - (-8)$ **4.** $(-10) - 8$

In 1 and 4, the second number is a positive 8. In 2 and 3, the second number is a negative 8.

Rule for Subtracting Two Integers

> To subtract two integers, add the opposite of the second integer to the first integer.

This rule states that to subtract two integers, rewrite the subtraction expression as the sum of the first number and the opposite of the second number.

For example, rewrite $(-9) - (-15)$ (negative 9) − (negative 15)

as $(-9) + 15$ (negative 9) + (the opposite of negative 15)

Therefore $(-9) - (-15) = (-9) + 15 = 6$

Here are more examples:

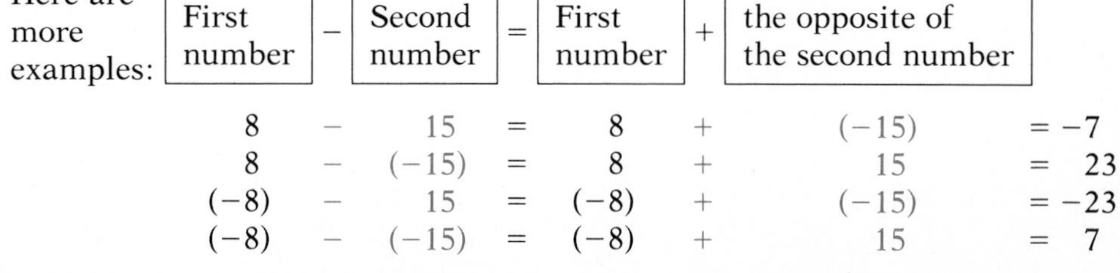

First number	−	Second number	=	First number	+	the opposite of the second number	
8	−	15	=	8	+	(−15)	= −7
8	−	(−15)	=	8	+	15	= 23
(−8)	−	15	=	(−8)	+	(−15)	= −23
(−8)	−	(−15)	=	(−8)	+	15	= 7

Subtract: $(-15) - 75$

Rewrite the subtraction operation as the sum of the first number and the opposite of the second number. Then add.

$(-15) - 75$
$(-15) + (-75)$
-90

Subtract: $27 - (-32)$

Rewrite the subtraction operation as the sum of the first number and the opposite of the second number. Then add.

$27 - (-32)$
$27 + 32$
59

When subtraction occurs several times in an expression, rewrite each subtraction as addition of the opposite and then add.

Subtract: $-13 - 5 - (-8)$

Rewrite each subtraction as the addition of the opposite and then add.

$-13 - 5 - (-8)$
$-13 + (-5) + 8$
$-18 + 8$
-10

Example 7 Subtract: $-12 - 8$

Solution
$-12 - 8$
$-12 + (-8)$
-20

Example 8 Subtract: $-8 - 14$

Your solution

Example 9 Subtract: $8 - 6 - (-20)$

Solution
$8 - 6 - (-20)$
$8 + (-6) + 20$
$2 + 20$
22

Example 10 Subtract: $3 - (-4) - 15$

Your solution

Example 11 Subtract: $-8 - 30 - (-12) - 7 - (-14)$

Solution
$-8 - 30 - (-12) - 7 - (-14)$
$-8 + (-30) + 12 + (-7) + 14$
$-38 + 12 + (-7) + 14$
$-26 + (-7) + 14$
$-33 + 14$
-19

Example 12 Subtract: $4 - (-3) - 12 - (-7) - 20$

Your solution

Example 13 Subtract: $12 - 12 - (-3) - 5 - 7$

Solution
$12 - 12 - (-3) - 5 - 7$
$12 + (-12) + 3 + (-5) + (-7)$
$0 + 3 + (-5) + (-7)$
$3 + (-5) + (-7)$
$-2 + (-7)$
-9

Example 14 Subtract: $17 - 10 - 2 - (-6) - 9$

Your solution

Solutions on pp. A39–A40

11.2 EXERCISES

Objective A

Add.

1. $3 + (-5)$
2. $-4 + 2$
3. $8 + 12$
4. $16 + 23$
5. $-3 + (-8)$
6. $-12 + (-1)$
7. $-4 + (-5)$
8. $-12 + (-12)$
9. $6 + (-9)$
10. $4 + (-9)$
11. $-6 + 7$
12. $-12 + 6$
13. $2 + (-3) + (-4)$
14. $7 + (-2) + (-8)$
15. $-3 + (-12) + (-15)$
16. $9 + (-6) + (-16)$
17. $-17 + (-3) + 29$
18. $13 + 62 + (-38)$
19. $-3 + (-8) + 12$
20. $-27 + (-42) + (-18)$
21. $13 + (-22) + 4 + (-5)$
22. $-14 + (-3) + 7 + (-6)$
23. $-22 + 10 + 2 + (-18)$
24. $-6 + (-8) + 13 + (-4)$
25. $-16 + (-17) + (-18) + 10$
26. $-25 + (-31) + 24 + 19$
27. $-126 + (-247) + (-358) + 339$
28. $-651 + (-239) + 524 + 487$

Objective B

Subtract.

29. $16 - 8$

30. $12 - 3$

31. $7 - 14$

32. $6 - 9$

33. $-7 - 2$

34. $-9 - 4$

35. $7 - (-2)$

36. $3 - (-4)$

37. $-6 - (-3)$

38. $-4 - (-2)$

39. $6 - (-12)$

40. $-12 - 16$

41. $-4 - 3 - 2$

42. $4 - 5 - 12$

43. $12 - (-7) - 8$

44. $-12 - (-3) - (-15)$

45. $4 - 12 - (-8)$

46. $13 - 7 - 15$

47. $-6 - (-8) - (-9)$

48. $7 - 8 - (-1)$

49. $-30 - (-65) - 29 - 4$

50. $42 - (-82) - 65 - 7$

51. $-16 - 47 - 63 - 12$

52. $42 - (-30) - 65 - (-11)$

53. $-47 - (-67) - 13 - 15$

54. $-18 - 49 - (-84) - 27$

55. $167 - 432 - (-287) - 359$

56. $-521 - (-350) - 164 - (-299)$

SECTION 11.3 Multiplication and Division of Integers

Objective A To multiply integers

Multiplication is the repeated addition of the same number.

Several different symbols are used to indicate multiplication:

$3 \times 2 = 6$ $\quad 3 \cdot 2 = 6$ $\quad (3)(2) = 6$

When 5 is multiplied by a sequence of decreasing integers, each product decreases by 5.

$$5 \times 3 = 15$$
$$5 \times 2 = 10$$
$$5 \times 1 = 5$$
$$5 \times 0 = 0$$

The pattern developed can be continued so that 5 is multiplied by a sequence of negative numbers. The resulting products must be negative in order to maintain the pattern of decreasing by 5.

$$5 \times (-1) = -5$$
$$5 \times (-2) = -10$$
$$5 \times (-3) = -15$$
$$5 \times (-4) = -20$$

This example illustrates that the product of a positive number and a negative number is negative.

When -5 is multiplied by a sequence of decreasing integers, each product increases by 5.

$$-5 \times 3 = -15$$
$$-5 \times 2 = -10$$
$$-5 \times 1 = -5$$
$$-5 \times 0 = 0$$

The pattern developed can be continued so that -5 is multiplied by a sequence of negative numbers. The resulting products must be positive in order to maintain the pattern of increasing by 5.

$$-5 \times (-1) = 5$$
$$-5 \times (-2) = 10$$
$$-5 \times (-3) = 15$$
$$-5 \times (-4) = 20$$

This example illustrates that the product of two negative numbers is positive.

The pattern for multiplication shown above is summarized in the following rules for multiplying integers.

Same Signs

> To multiply numbers with the same signs, multiply the absolute values of the factors. The product is positive.

$$4 \cdot 8 = 32$$
$$(-4)(-8) = 32$$

Different Signs

> To multiply numbers with different signs, multiply the absolute values of the factors. The product is negative.

$$-4 \cdot 8 = -32$$
$$(4)(-8) = -32$$

Multiply: $2(-3)(-5)(-7)$

To multiply more than two numbers, multiply the first two numbers. Then multiply the product by the third number. Continue until all the numbers have been multiplied.

$2(-3)(-5)(-7)$

$-6 \cdot (-5)(-7)$

$30 \cdot (-7)$

-210

Example 1 Multiply: $(-2) \cdot 3 \cdot 6$

Solution $(-2) \cdot 3 \cdot 6$
$-6 \cdot 6$
-36

Example 2 Multiply: $(-3) \cdot 4 \cdot (-5)$

Your solution

Example 3 Multiply: $-42 \cdot 62$

Solution $-42 \cdot 62$
-2604

Example 4 Multiply: $-38 \cdot 51$

Your solution

Example 5 Multiply: $-2 \cdot (-10) \cdot 7 \cdot 12$

Solution $-2 \cdot (-10) \cdot 7 \cdot 12$
$20 \cdot 7 \cdot 12$
$140 \cdot 12$
1680

Example 6 Multiply: $-6 \cdot 8 \cdot (-11) \cdot 3$

Your solution

Example 7 Multiply: $-5(-4)(6)(-3)$

Solution $-5(-4)(6)(-3)$
$20(6)(-3)$
$120(-3)$
-360

Example 8 Multiply: $-7(-8)(9)(-2)$

Your solution

Solutions on p. A40

Objective B To divide integers

11

For every division problem there is a related multiplication problem.

Division: $\frac{8}{2} = 4$ Related multiplication $4 \cdot 2 = 8$

This fact can be used to illustrate the rules for dividing signed numbers.

Same Signs

> The quotient of two numbers with the same signs is positive.

$\frac{12}{3} = 4$ because $4 \cdot 3 = 12$

$\frac{-12}{-3} = 4$ because $4(-3) = -12$

Different Signs

> The quotient of two numbers with different signs is negative.

$\frac{12}{-3} = -4$ because $-4(-3) = 12$

$\frac{-12}{3} = -4$ because $-4 \cdot 3 = -12$

Note that $\frac{12}{-3}$, $\frac{-12}{3}$, and $-\frac{12}{3}$ are all equal to -4.

If a and b are two integers, then $\frac{a}{-b} = \frac{-a}{b} = -\frac{a}{b}$.

Properties of Zero and One in Division

> Zero divided by any number other than zero is zero.

$\frac{0}{a} = 0$ because $0 \cdot a = 0$

> Division by zero is not defined.

$\frac{4}{0} = ?$ $? \times 0 = 4$ There is no number whose product with zero is 4.

> Any number other than zero divided by itself is 1.

$\frac{a}{a} = 1$ because $1 \cdot a = a$

Example 9 Divide: $(-120) \div (-8)$

Solution $(-120) \div (-8)$

15

Example 10 Divide: $(-135) \div (-9)$

Your solution

Example 11 Divide: $-81 \div 3$

Solution $-81 \div 3$

-27

Example 12 Divide: $-72 \div 4$

Your solution

Example 13 Divide: $95 \div (-5)$

Solution $95 \div (-5)$

-19

Example 14 Divide: $84 \div (-6)$

Your solution

Solutions on p. A40

Objective C To solve application problems

Example 15

The average temperature on the sunlit side of the moon is approximately 215°F. On the dark side, it is approximately −250°F. Find the difference between these average temperatures.

Strategy

To find the difference, subtract the average temperature on the dark side of the moon (−250°) from the average temperature on the sunlit side (215°).

Solution

$215 - (-250)$
$215 + 250$
465

The difference is 465°F.

Example 16

The average temperature throughout the earth's stratosphere is −70°F. The average temperature on the earth's surface is 57°F. Find the difference between these average temperatures.

Your strategy

Your solution

Example 17

The daily high temperatures during one week were recorded as follows: −9°, 3°, 0°, −8°, 2°, 1°, 4°. Find the average daily high temperature for the week.

Strategy

To find the average daily high temperature:

- Add the seven temperature readings.
- Divide by 7.

Solution

$-9 + 3 + 0 + (-8) + 2 + 1 + 4$
$-6 + 0 + (-8) + 2 + 1 + 4$
$-6 + (-8) + 2 + 1 + 4$
$-14 + 2 + 1 + 4$
$-12 + 1 + 4$
$-11 + 4$
-7
$-7 \div 7 = -1$

The average daily high temperature was −1°.

Example 18

The daily low temperatures during one week were recorded as follows: −6°, −7°, 1°, 0°, −5°, −10°, −1°. Find the average daily low temperature for the week.

Your strategy

Your solution

Solutions on p. A40

11.3 Exercises

Objective A

Multiply.

1. 14×3
2. 62×9
3. $-4 \cdot 6$
4. $-7 \cdot 3$
5. $-2 \cdot (-3)$
6. $-5 \cdot (-1)$
7. $(9)(2)$
8. $(3)(8)$
9. $5(-4)$
10. $4(-7)$
11. $-8(2)$
12. $-9(3)$
13. $(-5)(-5)$
14. $(-3)(-6)$
15. $(-7)(0)$
16. -32×4
17. -24×3
18. $19 \cdot (-7)$
19. $6(-17)$
20. $-8(-26)$
21. $-4(-35)$
22. $-5 \cdot (23)$
23. $-6 \cdot (38)$
24. $9(-27)$
25. $8(-40)$
26. $-7(-34)$
27. $-4(39)$
28. $4 \cdot (-8) \cdot 3$
29. $5 \times 7 \times (-2)$
30. $8 \cdot (-6) \cdot (-1)$

Multiply.

31. $(-9)(-9)(2)$
32. $-8(-7)(-4)$
33. $-5(8)(-3)$
34. $(-6)(5)(7)$
35. $-1(4)(-9)$
36. $6(-3)(-2)$
37. $4(-4) \cdot 6(-2)$
38. $-5 \cdot 9(-7) \cdot 3$
39. $-9(4) \cdot 3(1)$
40. $8(8)(-5)(-4)$
41. $(-6) \cdot 7 \cdot (-10)(-5)$
42. $-9(-6)(11)(-2)$
43. $-6(-5)(12)(0)$
44. $7(9) \cdot 10 \cdot (-1)$
45. $-19(28)(-43)(-11)$
46. $-65(13)(-47)(-92)$

Objective B

Divide.

47. $12 \div (-6)$
48. $18 \div (-3)$
49. $(-72) \div (-9)$
50. $(-64) \div (-8)$
51. $0 \div (-6)$
52. $-49 \div 7$
53. $45 \div (-5)$
54. $-24 \div 4$
55. $-36 \div 4$
56. $-56 \div 7$
57. $-81 \div (-9)$
58. $-40 \div (-5)$
59. $72 \div (-3)$
60. $44 \div (-4)$
61. $-60 \div 5$
62. $-66 \div 6$
63. $-93 \div (-3)$
64. $-98 \div (-7)$
65. $(-85) \div (-5)$
66. $(-60) \div (-4)$
67. $120 \div 8$

Divide.

68. $144 \div 9$

69. $78 \div (-6)$

70. $84 \div (-7)$

71. $-72 \div 4$

72. $-80 \div 5$

73. $-114 \div (-6)$

74. $-91 \div (-7)$

75. $-104 \div (-8)$

76. $-126 \div (-9)$

77. $57 \div (-3)$

78. $162 \div (-9)$

79. $-136 \div (-8)$

80. $-128 \div 4$

81. $-130 \div (-5)$

82. $(-280) \div 8$

83. $(-92) \div (-4)$

84. $-196 \div (-7)$

85. $-150 \div (-6)$

86. $(-261) \div 9$

87. $204 \div (-6)$

88. $165 \div (-5)$

89. $-132 \div (-12)$

90. $-156 \div (-13)$

91. $-182 \div 14$

92. $-144 \div 12$

93. $143 \div 11$

94. $168 \div 14$

95. $-180 \div (-15)$

96. $-169 \div (-13)$

97. $154 \div (-11)$

98. $274{,}883 \div 367$

99. $398{,}750 \div 1375$

100. $841{,}662 \div 2461$

Objective C *Application Problems*

101. Find the temperature after a rise of 7°C from −8°C.

102. Find the temperature after a rise of 5°C from −19°C.

103. During a card game of Hearts, you had a score of 11 points before your opponent "shot the moon," subtracting a score of 26 from your total. What was your score after your opponent shot the moon?

104. In a card game of Hearts, you had a score of −19 before you shot the moon, entitling you to add 26 points to your score. What was your score after you shot the moon?

The elevation, or height, of places on earth is measured in relation to sea level, or the average level of the ocean's surface. The table below shows height above sea level as a positive number, depth below sea level as a negative number.

Place	*Elevation (in feet)*
Mt. Everest	29,028
Mt. Aconcagua	23,035
Mt. McKinley	20,320
Mt. Kilimanjaro	19,340
Salinas Grandes	−131
Death Valley	−282
Qattara Depression	−436
Dead Sea	−1286

105. Use the table to find the difference in elevation between Mt. McKinley and Death Valley.

106. Use the table to find the difference in elevation between Mt. Kilimanjaro and the Qattara Depression (the highest and lowest points in Africa).

107. Use the table to find the difference in elevation between Mt. Everest and the Dead Sea (the highest and lowest points in Asia).

108. Use the table to find the difference in elevation between Mt. Aconcagua and Salinas Grandes (the highest and lowest points in South America).

109. The daily low temperatures during one week were recorded as follows: 5°, −4°, 9°, 0°, −11°, −13°, −7°. Find the mean daily low temperature for the week.

110. The daily high temperatures during one week were recorded as follows: −7°, −10°, 4°, 6°, −2°, −8°, −4°. Find the mean daily high temperature for the week.

SECTION 11.4 Operations with Rational Numbers

Objective A To add or subtract rational numbers

In Chapter 2, operations with fractions were discussed. In this section, operations with *rational numbers* will be discussed. The rational numbers include all the positive and negative fractions.

Rational Numbers

The rational numbers are the numbers that can be written as the ratio of two integers.

Examples of rational numbers are $\frac{-4}{7}$ and $\frac{9}{2}$. Because $4 = \frac{4}{1}$ (the ratio of two integers), 4 is a rational number. The rational numbers include all the integers.

To add or subtract rational numbers in fractional form, first find the least common multiple (LCM) of the denominators.

Simplify: $-\frac{7}{8} + \frac{5}{6}$

Find the LCM of the denominators. $8 = 2 \cdot 2 \cdot 2; \quad 6 = 2 \cdot 3$
$LCM = 2 \cdot 2 \cdot 2 \cdot 3 = 24$

Rewrite each fraction, using the LCM of the denominators as the common denominator. Add the numerators.

$$-\frac{7}{8} + \frac{5}{6} = -\frac{21}{24} + \frac{20}{24} = \frac{-21 + 20}{24} = \frac{-1}{24} = -\frac{1}{24}$$

Simplify: $-\frac{7}{9} - \frac{5}{12}$

$9 = 3 \cdot 3; \quad 12 = 2 \cdot 2 \cdot 3; \quad LCM = 2 \cdot 2 \cdot 3 \cdot 3 = 36$

$$-\frac{7}{9} - \frac{5}{12} = -\frac{28}{36} - \frac{15}{36} = \frac{(-28)}{36} + \frac{(-15)}{36} = \frac{(-28) + (-15)}{36} = \frac{-43}{36} = -\frac{43}{36} = -1\frac{7}{36}$$

Because numbers written as terminating decimals can be expressed as the ratio of two integers, terminating decimals are also rational numbers. For example,

$$0.75 = \frac{75}{100} = \frac{3}{4} \quad \text{and} \quad 6.3 = 6\frac{3}{10} = \frac{63}{10}$$

To add or subtract decimals, write the numbers so that the decimal points are in a vertical line. Then use the rules for adding integers. Write the decimal point in the answer directly below the decimal points in the problem.

Simplify: $47.034 + (-56.91)$

Because the signs are different, find the difference between the absolute values of the numbers.

$$\begin{array}{r} 56.91 \\ -47.034 \\ \hline 9.876 \end{array}$$

Attach the sign of the number with the greater absolute value. $47.034 - 56.91 = -9.876$

Simplify: $-39.09 - 102.98$

Recall that to subtract two numbers, add the opposite of the second number. $(-39.09) + (-102.98)$

Because the signs are the same, find the sum of the absolute values of the numbers.

$$\begin{array}{r} 39.09 \\ +102.98 \\ \hline 142.07 \end{array}$$

Attach the common sign. $-39.09 - 102.98 = -142.07$

Example 1 Simplify: $\frac{5}{16} - \frac{7}{40}$

Solution The LCM of 16 and 40 is 80.

$$\frac{5}{16} - \frac{7}{40} = \frac{25}{80} - \frac{14}{80}$$
$$= \frac{25}{80} + \frac{-14}{80}$$
$$= \frac{25 + (-14)}{80} = \frac{11}{80}$$

Example 2 Simplify: $\frac{5}{9} - \frac{11}{12}$

Your solution

Example 3 Simplify: $-\frac{3}{4} + \frac{1}{6} - \frac{5}{8}$

Solution The LCM of 4, 6, and 8 is 24.

$$-\frac{3}{4} + \frac{1}{6} - \frac{5}{8} = -\frac{18}{24} + \frac{4}{24} - \frac{15}{24}$$
$$= \frac{-18}{24} + \frac{4}{24} + \frac{-15}{24}$$
$$= \frac{-18 + 4 + (-15)}{24} = \frac{-29}{24} = -1\frac{5}{24}$$

Example 4 Simplify: $-\frac{7}{8} - \frac{5}{6} + \frac{2}{3}$

Your solution

Example 5 Simplify:
$4.027 + 19.66 + 3.09$

Solution

$$\begin{array}{r} 4.027 \\ 19.66 \\ +\ 3.09 \\ \hline 26.777 \end{array}$$

Example 6 Simplify:
$3.907 + 4.9 + 6.63$

Your solution

Example 7 Simplify: $42.987 - 98.61$

Solution

$$\begin{array}{r} 98.51 \\ -42.987 \\ \hline 55.623 \end{array}$$

$42.987 - 98.61 = -55.623$

Example 8 Simplify: $16.127 - 67.91$

Your solution

Example 9 Simplify:
$1.02 + (-3.6) + 9.24$

Solution $1.02 + (-3.6) + 9.24$
$-2.58 + 9.24$
6.66

Example 10 Simplify:
$2.7 + (-9.44) + 6.2$

Your solution

Solutions on pp. A40–A41

Objective B To multiply and divide rational numbers

The product of two rational numbers written as fractions is the product of the numerators over the product of the denominators. Use the sign rules for multiplying integers.

Simplify: $-\frac{3}{8} \times \frac{12}{17}$

Because the signs are different, the product is negative.

$$-\frac{3}{8} \times \frac{12}{17} = -\frac{3 \times 12}{8 \times 17} = -\frac{36}{136} = -\frac{9}{34}$$

To divide rational numbers written as fractions, invert the divisor and then multiply.

Simplify: $-\frac{3}{10} \div \left(-\frac{18}{25}\right)$

Because the signs are the same, the quotient is positive.

$$-\frac{3}{10} \div \left(-\frac{18}{25}\right) = \frac{3}{10} \times \frac{25}{18} = \frac{3 \times 25}{10 \times 18} = \frac{75}{180} = \frac{5}{12}$$

To multiply or divide rational numbers written in decimal form, use the sign rules for integers and the methods shown in Chapter 3.

Simplify: $(-6.89) \times (-0.00035)$

The signs are the same. Multiply the absolute values.

$$\begin{array}{rl} 6.89 & \text{2 decimal places} \\ \times\ 0.00035 & \text{5 decimal places} \\ \hline 3445 & \\ 2067 & \\ \hline 0.0024115 & \text{7 decimal places} \end{array}$$

The product is positive. $(-6.89) \times (-0.00035) = 0.0024115$

Simplify $1.32 \div (-0.27)$. Round to the nearest tenth.

Divide the absolute values. Move the decimal point two places in the divisor and then in the dividend. Place the decimal point in the quotient.

$$\begin{array}{r} 4.88 \approx 4.9 \\ .27.\overline{)1.32.00} \\ -1\ 08 \\ \hline 24\ 0 \\ -21\ 6 \\ \hline 2\ 40 \\ -2\ 16 \\ \hline 24 \end{array}$$

The signs are different. The quotient is negative.

$1.32 \div (-0.27) \approx -4.9$

Example 11 Simplify: $-\frac{7}{12} \times \frac{9}{14}$

Solution The product is negative.

$$-\frac{7}{12} \times \frac{9}{14} = -\left(\frac{7 \cdot 9}{12 \cdot 14}\right)$$
$$= -\frac{63}{168} = -\frac{3}{8}$$

Example 12 Simplify: $\left(-\frac{2}{3}\right)\left(-\frac{9}{10}\right)$

Your solution

Example 13 Simplify: $-\frac{3}{8} \div -\frac{5}{12}$

Solution The quotient is positive.

$$-\frac{3}{8} \div \left(-\frac{5}{12}\right) = \frac{3}{8} \cdot \frac{12}{5}$$
$$= \frac{3 \cdot 12}{8 \cdot 5}$$
$$= \frac{36}{40} = \frac{9}{10}$$

Example 14 Simplify: $-\frac{5}{8} \div \frac{5}{40}$

Your solution

Example 15 Simplify: -4.29×8.2

Solution The product is negative.

$$\begin{array}{r} 4.29 \\ \times\ 8.2 \\ \hline 858 \\ 3432 \\ \hline 35.178 \end{array}$$

$-4.29 \times 8.2 = -35.178$

Example 16 Simplify: -5.44×3.8

Your solution

Example 17 Simplify: $-3.2 \times (-0.4) \times 6.9$

Solution $-3.2 \times (-0.4) \times 6.9$
1.28×6.9
8.832

Example 18 Simplify: $3.44 \times (-1.7) \times 0.6$

Your solution

Example 19 Simplify: $-0.0792 \div (-0.42)$
Round to the nearest hundredth.

Solution

$$\begin{array}{r} 0.188 \approx 0.19 \\ 0.42.\overline{)0.07.920} \\ -4\,2 \\ \hline 3\,72 \\ -3\,36 \\ \hline 360 \\ -336 \\ \hline 24 \end{array}$$

$-0.0792 \div (-0.42) \approx 0.19$

Example 20 Simplify: $-0.394 \div 1.7$
Round to the nearest hundredth.

Your solution

Solutions on p. A41

11.4 EXERCISES

Objective A

Simplify.

1. $\frac{2}{3} + \frac{5}{12}$

2. $\frac{1}{2} + \frac{3}{8}$

3. $\frac{5}{8} - \frac{5}{6}$

4. $\frac{1}{9} - \frac{5}{27}$

5. $-\frac{5}{12} - \frac{3}{8}$

6. $-\frac{5}{6} - \frac{5}{9}$

7. $-\frac{6}{13} + \frac{17}{26}$

8. $-\frac{7}{12} + \frac{5}{8}$

9. $-\frac{5}{8} - \left(-\frac{11}{12}\right)$

10. $-\frac{3}{4} - \frac{5}{8}$

11. $-\frac{7}{12} - \left(-\frac{7}{8}\right)$

12. $\frac{5}{12} - \frac{11}{15}$

13. $\frac{2}{5} - \frac{14}{15}$

14. $-\frac{2}{3} - \frac{5}{8}$

15. $-\frac{3}{4} - \left(-\frac{2}{3}\right)$

16. $\frac{3}{4} - \frac{3}{7}$

17. $-\frac{5}{2} - \left(-\frac{13}{4}\right)$

18. $-\frac{7}{3} - \left(-\frac{3}{2}\right)$

19. $\frac{1}{3} + \frac{5}{6} - \frac{2}{9}$

20. $\frac{1}{2} - \frac{2}{3} + \frac{1}{6}$

21. $-\frac{3}{8} - \frac{5}{12} - \frac{3}{16}$

22. $-\frac{5}{16} + \frac{3}{4} - \frac{7}{8}$

23. $\frac{1}{2} - \frac{3}{8} - \left(-\frac{1}{4}\right)$

24. $\frac{3}{4} - \left(-\frac{7}{12}\right) - \frac{7}{8}$

25. $\frac{1}{3} - \frac{1}{4} - \frac{1}{5}$

26. $\frac{2}{3} - \frac{1}{2} + \frac{5}{6}$

27. $\frac{5}{16} + \frac{1}{8} - \frac{1}{2}$

Simplify.

28. $\frac{5}{8} - \left(-\frac{5}{12}\right) + \frac{1}{3}$

29. $\frac{1}{8} - \frac{11}{12} + \frac{1}{2}$

30. $-\frac{7}{9} + \frac{14}{15} + \frac{8}{21}$

31. $\frac{1}{2} + \left(-\frac{3}{8}\right) + \frac{5}{12}$

32. $-\frac{3}{8} + \frac{3}{4} - \left(-\frac{3}{16}\right)$

33. $\frac{1}{3} - \frac{3}{5} - \frac{5}{6}$

34. $3.4 + (-6.8)$

35. $-4.9 + 3.27$

36. $-8.32 + (-0.57)$

37. $-3.5 + 7$

38. $-4.8 + (-3.2)$

39. $6.2 + (-4.29)$

40. $-4.6 + 3.92$

41. $7.2 + (-8.42)$

42. $3.09 + 6.025$

43. $-45.71 + (-135.8)$

44. $124.09 + (-67.5)$

45. $-35.274 + 12.47$

46. $4.2 + (-6.8) + 5.3$

47. $6.7 + 3.2 + (-10.5)$

48. $-4.5 + 3.2 + (-19.4)$

49. $2.09 - 6.72 - 5.4$

50. $16.4 + 3.09 - 7.93$

51. $-18.39 + 4.9 - 23.7$

52. $19 - (-3.72) - 82.75$

53. $-3.07 - (-2.97) - 17.4$

54. $-3.09 - 4.6 - 27.3$

55. $-3.89 + (-2.9) + 4.723 + 0.2$

56. $4.207 + 6.91 + (-3.825) + (-10.04)$

57. $-3.005 + (-3.925) + (-6.002)$

58. $-4.02 + 6.809 - (-3.57) - (-0.419)$

59. $0.0153 + (-1.0294) + (-1.0726)$

60. $0.27 + (-3.5) - (-0.27) + (-5.44)$

Objective B

Simplify.

61. $\frac{1}{2} \times \left(-\frac{3}{4}\right)$

62. $-\frac{2}{9} \times \left(-\frac{3}{14}\right)$

63. $\left(-\frac{3}{8}\right)\left(-\frac{4}{15}\right)$

64. $\left(-\frac{3}{4}\right)\left(-\frac{8}{27}\right)$

65. $-\frac{1}{2} \times \frac{8}{9}$

66. $\frac{5}{12} \times \left(-\frac{8}{15}\right)$

67. $\frac{5}{8} \times \left(-\frac{7}{12}\right) \times \frac{16}{25}$

68. $\left(\frac{1}{2}\right)\left(-\frac{3}{4}\right)\left(-\frac{5}{8}\right)$

69. $\left(\frac{5}{12}\right)\left(-\frac{8}{15}\right)\left(-\frac{1}{3}\right)$

70. $\left(-\frac{5}{12}\right)\left(\frac{42}{65}\right)$

71. $\left(\frac{3}{8}\right)\left(-\frac{15}{41}\right)$

72. $\left(\frac{2}{9}\right)\left(-\frac{15}{17}\right)$

73. $\left(-\frac{15}{8}\right)\left(-\frac{16}{3}\right)$

74. $\left(-\frac{5}{7}\right)\left(-\frac{14}{15}\right)$

75. $\left(\frac{3}{8}\right)\left(\frac{15}{41}\right)$

76. $\frac{1}{3} \div \left(-\frac{1}{2}\right)$

77. $-\frac{3}{8} \div \frac{7}{8}$

78. $\left(-\frac{3}{4}\right) \div \left(-\frac{7}{40}\right)$

79. $\frac{3}{8} \div \frac{1}{4}$

80. $\frac{5}{6} \div \left(-\frac{3}{4}\right)$

81. $-\frac{5}{12} \div \frac{15}{32}$

82. $-\frac{7}{8} \div \frac{4}{21}$

83. $\frac{7}{10} \div \frac{2}{5}$

84. $-\frac{15}{64} \div \left(-\frac{3}{40}\right)$

85. $-\frac{5}{16} \div \left(-\frac{3}{8}\right)$

86. $\frac{5}{9} \div \left(-\frac{15}{32}\right)$

87. $\left(-\frac{8}{19}\right) \div \frac{7}{38}$

88. $\left(-\frac{2}{3}\right) \div 4$

89. $(-6) \div \frac{4}{9}$

90. $\left(-\frac{3}{8}\right) \div \left(-\frac{5}{12}\right)$

91. 4.2×0.9

92. $-6.7 \times (-4.2)$

93. $-8.9 \times (-3.5)$

94. -1.6×4.9

95. 7.8×9.6

96. -14.3×7.9

97. $(-0.78)(-0.15)$

98. $(0.49)(-3.9)$

99. $(-1.21)(-0.03)$

Simplify.

100. $(-6.5)(8)$

101. $(-7)(-0.42)$

102. $(12.1)(-0.5)$

103. $(8.4)(-9.5)$

104. $(-3.9)(-1.9)$

105. $(-0.84)(-0.6)$

106. $(-27.08) \div (0.4)$

107. $(-0.396) \div (3.6)$

108. $(-2.501) \div (0.41)$

109. $(-8.919) \div (-0.9)$

110. $-77.6 \div (-0.8)$

111. $59.01 \div (-0.7)$

112. $(-7.04) \div (-3.2)$

113. $(-84.66) \div 1.7$

114. $-3.312 \div (0.8)$

115. $-0.1314 \div (-0.018)$

116. $1.003 \div (-0.59)$

117. $26.22 \div (-6.9)$

Divide. Round to the nearest hundredth.

118. $(-19.08) \div 0.45$

119. $21.792 \div (-0.96)$

120. $(-38.665) \div (-9.5)$

121. $(-3.171) \div (-45.3)$

122. $27.738 \div (-60.3)$

123. $(-13.97) \div (-25.4)$

Use your calculator to simplify.

124. $3.17 \times (-16) \times 3.027 \times (-5)$

125. $(-3.2) \times (-6.9) \times 3.25 \times (-7.25)$

126. $426 \times (-3.98) \times 6.92 \times (-7.55)$

127. $0.024 \times (-3.29) \times 162.5 \times 0.075$

128. $4256 \div (-0.0235)$

129. $-0.00723 \div (-261)$

130. $-1.23759 \div 6.2333$

131. $6.5432 \div (-666.72)$

SECTION 11.5 The Order of Operations Agreement

Objective A To use the Order of Operations Agreement to simplify expressions

The Order of Operations Agreement has been used throughout this textbook. In simplifying expressions with rational numbers, the same Order of Operations Agreement is used. This agreement is restated here.

The Order of Operations Agreement

Step 1 Do all operations inside parentheses.
Step 2 Simplify any expressions containing exponents.
Step 3 Do multiplication and division as they occur from left to right.
Step 4 Rewrite subtraction as the addition of the opposite. Then do additions as they occur from left to right.

Exponents may be confusing in expressions with signed numbers.

$(-3)^2 = (-3) \times (-3) = 9$
$-3^2 = -(3)^2 = -(3 \times 3) = -9$

Note that -3 is squared only when the negative sign is *inside* parentheses.

Simplify: $(-3)^2 - 2 \times (8 - 3) + (-5)$

$(-3)^2 - 2 \times \underbrace{(8 - 3)} + (-5)$ — 1. Perform operations inside parentheses.

$\underbrace{(-3)^2} - 2 \times 5 + (-5)$ — 2. Simplify expressions with exponents.

$9 - \underbrace{2 \times 5} + (-5)$ — 3. Do multiplications and divisions as they occur from left to right.

$9 - 10 + (-5)$

$\underbrace{9 + (-10)} + (-5)$ — 4. Rewrite subtraction as the addition of the opposite. Then add from left to right.

$\underbrace{(-1) + (-5)}$

-6

Simplify: $\left(\frac{1}{4} - \frac{1}{2}\right)^2 \div \frac{3}{8}$

$\underbrace{\left(\frac{1}{4} - \frac{1}{2}\right)}^2 \div \frac{3}{8}$ — 1. Perform operations inside parentheses.

$\underbrace{\left(-\frac{1}{4}\right)^2} \div \frac{3}{8}$ — 2. Simplify expressions with exponents.

$\frac{1}{16} \div \frac{3}{8}$ — 3. Do multiplication and division as they occur from left to right.

$\underbrace{\frac{1}{16} \times \frac{8}{3}}$

$\frac{1}{6}$

Example 1 Simplify: $8 - 4 \div (-2)$

Solution
$8 - 4 \div (-2)$
$8 - (-2)$
$8 + 2$
10

Example 2 Simplify: $9 - 9 \div (-3)$

Your solution

Example 3 Simplify: $12 \div (-2)^2 + 5$

Solution
$12 \div (-2)^2 + 5$
$12 \div 4 + 5$
$3 + 5$
8

Example 4 Simplify: $8 \div 4 \cdot 4 - (-2)^2$

Your solution

Example 5 Simplify: $12 - (-10) \div (8 - 3)$

Solution
$12 - (-10) \div (8 - 3)$
$12 - (-10) \div 5$
$12 - (-2)$
$12 + 2$
14

Example 6 Simplify: $8 - (-15) \div (2 - 7)$

Your solution

Example 7 Simplify:
$(-3)^2 \times (5 - 7)^2 - (-9) \div 3$

Solution
$(-3)^2 \times (5 - 7)^2 - (-9) \div 3$
$(-3)^2 \times (-2)^2 - (-9) \div 3$
$9 \times 4 - (-9) \div 3$
$36 - (-9) \div 3$
$36 - (-3)$
$36 + 3$
39

Example 8 Simplify:
$(-2)^2 \times (3 - 7)^2 -$
$(-16) \div (-4)$

Your solution

Example 9 Simplify: $3 \div \left(\frac{1}{2} - \frac{1}{4}\right) - 3$

Solution
$3 \div \left(\frac{1}{2} - \frac{1}{4}\right) - 3$
$3 \div \frac{1}{4} - 3$
$\frac{3}{1} \times \frac{4}{1} - 3$
$12 - 3$
$12 + (-3)$
9

Example 10 Simplify: $7 \div \left(\frac{1}{7} - \frac{3}{14}\right) - 9$

Your solution

Solutions on p. A41

11.5 EXERCISES

Objective A

Simplify.

1. $8 \div 4 + 2$
2. $3 - 12 \div 2$
3. $4 + (-7) + 3$
4. $-16 \div 2 + 8$
5. $4^2 - 4$
6. $6 - 2^2$
7. $2 \times (3 - 5) - 2$
8. $2 - (8 - 10) \div 2$
9. $4 - (-3)^2$
10. $(-2)^2 - 6$
11. $4 - (-3) - 5$
12. $6 + (-8) - (-3)$
13. $16 \div 8 \times 4$
14. $4 \times 8 \div 16$
15. $3^2 + 2^2 + 5$
16. $(-3)^2 - (-2)^2 - 5$
17. $4 \times (2 - 4) - 4$
18. $6 - 2 \times (1 - 3)$
19. $4 - (-2)^2 + (-3)$
20. $-3 + (-6)^2 - 1$
21. $3^2 - 4 \times 2$
22. $9 \div 3 - (-3)^2$
23. $3 \times (6 - 2) \div 6$
24. $4 \times (2 - 7) \div 5$
25. $2^2 - (-3)^2 + 2$
26. $3 \times (8 - 5) + 4$
27. $6 - 2 \times (1 - 5)$
28. $4 \times 2 \times (3 - 6)$
29. $(-2)^2 - (-3)^2 + 1$
30. $4^2 - 3^2 - 4$
31. $6 - (-3) \times (-3)^2$
32. $4 - (-5) \times (-2)^2$
33. $4 \times 2 - 3 \times 7$
34. $16 \div 2 - 9 \div 3$
35. $(-2)^2 - 5 \times 3 - 1$
36. $4 - 2 \times 7 - 3^2$

Simplify.

37. $7 \times 6 - 5 \times 6 + 3 \times 2 - 2 + 1$

38. $3 \times 2^2 + 5 \times (3 + 2) - 17$

39. $-4 \times 3 \times (-2) + 12 \times (3 - 4) + (-12)$

40. $3 \times 4^2 - 16 - 4 + 3 - (1 - 2)^2$

41. $-12 \times (6 - 8) + 1^2 \times 3^2 \times 2 - 6 \times 2$

42. $-3 \times (-2)^2 \times 4 \div 8 - (-12)$

43. $10 \times 9 - (8 + 7) \div 5 + 6 - 7 + 8$

44. $-27 - (-3)^2 - 2 - 7 + 6 \times 3$

45. $3^2 \times (4 - 7) \div 9 + 6 - 3 - 4 \times 2$

46. $16 - 4 \times 8 + 4^2 - (-18) \div (-9)$

47. $(-3)^2 \times (5 - 7)^2 - (-9) \div 3$

48. $-2 \times 4^2 - 3 \times (2 - 8) - 3$

49. $(1.2)^2 - 4.1 \times 0.3$

50. $2.4 \times (-3) - 2.5$

51. $1.6 - (-1.6)^2$

52. $4.1 \times 8 \div (-4.1)$

53. $(4.1 - 3.9) - 0.7^2$

54. $1.8 \times (-2.3) - 2$

55. $(-0.4)^2 \times 1.5 - 2$

56. $(6.2 - 1.3) \times (-3)$

57. $4.2 - (-3.9) - 6$

58. $-\frac{1}{2} + \frac{3}{8} \div \left(-\frac{3}{4}\right)$

59. $\left(\frac{3}{4}\right)^2 - \frac{3}{8}$

60. $\left(\frac{1}{2}\right)^2 - \left(-\frac{1}{2}\right)^2$

61. $\frac{5}{16} - \frac{3}{8} + \frac{1}{2}$

62. $\frac{2}{7} \div \frac{5}{7} - \frac{3}{14}$

63. $\frac{1}{2} \times \frac{1}{4} \times \frac{1}{2} - \frac{3}{8}$

64. $\frac{2}{3} \times \frac{5}{8} \div \frac{2}{7}$

65. $\frac{1}{2} - \left(\frac{3}{4} - \frac{3}{8}\right) \div \frac{1}{3}$

66. $\frac{3}{8} \div \left(-\frac{1}{2}\right)^2 + 2$

Calculators and Computers

The [+/−] Key on a Calculator

The [+/−] key on a calculator is called the change sign key. It is used to change the sign of the number that is currently in the display. Turn on your calculator and enter 4. Now press [+/−]. Notice that −4 is now displayed. Press the [+/−] key again; the −4 is changed to 4.

The [+/−] key is used to calculate with negative numbers. Here are some examples:

Example 1: Find −34.982 − (−64.72).

Enter	*Display*	*Comment*
34.982	34.982	
[+/−]	−34.982	
[−]	−34.982	
64.72	64.72	
[+/−]	−64.72	
[=]	29.738	The answer is 29.738

Example 2: Find 48.93 ÷ (−21.3).

Enter	*Display*	*Comment*
48.93	48.93	
[÷]	48.93	
21.3	21.3	
[+/−]	−21.3	
[=]	−2.297183	The answer is −2.297183.

Chapter Summary

Key Words

Positive numbers are numbers greater than zero.

Negative numbers are numbers less than zero.

The *integers* are . . . −4, −3, −2, −1, 0, 1, 2, 3, 4, . . .

Negative integers are integers to the left of zero on the number line.

Positive integers are integers to the right of zero on the number line.

The *absolute value* of a number is its distance from zero on a number line.

A *rational number* is a number that can be written as the ratio of two numbers.

Essential Rules

Addition of Integers with the Same Sign To add numbers with the same signs, add the absolute values of the numbers. Then attach the sign of the addends.

Addition of Integers with Different Signs To add numbers with different signs, find the difference between the absolute values of the numbers. Then attach the sign of the number with the greater absolute value.

Subtraction of Integers To subtract one integer from another, add the opposite of the second integer to the first integer.

Multiplication of Integers with the Same Sign To multiply numbers with the same signs, multiply the absolute values of the numbers. The product is positive.

Multiplication of Integers with Different Signs To multiply numbers with different signs, multiply the absolute values of the numbers. The product is negative.

Division of Integers with the Same Sign The quotient of two numbers with the same signs is positive.

Division of Integers with Different Signs The quotient of two numbers with different signs is negative.

Order of Operations Agreement

Step 1 Perform operations inside parentheses.
Step 2 Simplify exponential expressions.
Step 3 Do multiplication and division as they occur from left to right.
Step 4 Rewrite subtraction as the addition of the opposite. Then do additions as they occur from left to right.

Chapter Review

SECTION 11.1

1. Place the correct symbol, < or >, between the two numbers.
   $-2 \quad -40$

2. Place the correct symbol, < or >, between the two numbers.
   $0 \quad -3$

3. Find the opposite of -4.

4. Find the opposite of 22.

5. Evaluate $|-5|$.

6. Evaluate $-|-6|$.

SECTION 11.2

7. Add: $-22 + 14 + (-18)$

8. Subtract: $-8 - (-2) - (-10) - 3$

SECTION 11.3

9. Multiply: $2 \times (-13)$

10. $-18 \div (-3)$

11. Find the temperature after a rise of 18° from $-22°$.

SECTION 11.4

12. Simplify: $\frac{5}{8} - \frac{5}{6}$

13. Simplify: $\frac{5}{12} + \left(-\frac{2}{3}\right)$

14. Simplify: $-\frac{3}{8} + \frac{5}{12} + \frac{2}{3}$

15. Simplify: $-\frac{5}{12} + \frac{7}{9} - \frac{1}{3}$

16. Simplify: $-0.33 + 1.98 - 1.44$

17. Simplify: $-33.4 + 9.8 - (-16.2)$

18. Simplify: $\frac{1}{3} \times \left(-\frac{3}{4}\right)$

19. Simplify: $\frac{6}{34} \times \frac{17}{40}$

20. Simplify: $\left(-\frac{2}{3}\right)\left(\frac{6}{11}\right)\left(-\frac{22}{25}\right)$

21. Simplify: $\left(-\frac{3}{8}\right) \div \left(-\frac{4}{5}\right)$

22. Simplify: $-\frac{7}{12} \div \left(-\frac{14}{39}\right)$

23. Simplify: $1.2 \times (-0.035)$

24. Simplify: -0.08×16

25. Simplify: $-1.464 \div 18.3$

SECTION 11.5

26. Simplify: $12 - 6 \div 3$

27. Simplify: $16 \div 8 \times 4$

28. Simplify: $16 + 4 \div 2$

29. Simplify: $3^2 - 9 + 2$

30. Simplify: $16 \div 4(8 - 2)$

31. Simplify: $-0.4 \times 5 - (-3.33)$

32. Simplify: $\left(\frac{2}{3}\right)^2 - \frac{5}{6}$

33. Simplify: $-\frac{1}{2} + \frac{3}{8} \div \frac{9}{20}$

Chapter Test

1. Place the correct symbol, < or >, between the two numbers.
   $-8 \quad -10$

2. Place the correct symbol, < or >, between the two numbers.
   $0 \quad -4$

3. Evaluate $-|-2|$.

4. Add: $-2 + 3 + (-8)$

5. Add: $16 + (-10) + (-20)$

6. Subtract: $-5 - (-8)$

7. Subtract: $16 - 4 - (-5) - 7$

8. Multiply: -4×12

9. Multiply: $-5 \times (-6) \times 3$

10. Divide: $-72 \div 8$

11. Find the temperature after a rise of 11°C from

12. The daily low temperature readings for a 3-day period were as follows: $-7°$, $9°$, $-8°$. Find the average low temperature for the 3-day period.

13. Add: $-\frac{2}{5} + \frac{7}{15}$

14. Add: $-\frac{1}{2} + \frac{1}{3} + \frac{1}{4}$

15. Subtract: $-\frac{3}{8} + \frac{2}{3}$

16. Subtract: $-\frac{2}{5} - \left(-\frac{7}{10}\right)$

17. Multiply: $\frac{3}{8} \times \left(-\frac{5}{6}\right) \times \left(-\frac{4}{15}\right)$

18. Divide: $-\frac{2}{3} \div \frac{5}{6}$

19. Add: $1.22 + (-3.1)$

20. Subtract: $-1.004 - 3.01$

21. Subtract: $2.113 - (-1.1)$

22. Multiply: $0.032 \times (-1.9)$

23. Divide: $-15.64 \div (-4.6)$

24. Simplify: $4 \times (4 - 7) \div (-2) - 4 \times 8$

25. Simplify: $(-2)^2 - (-3)^2 \div (1 - 4)^2 \times 2 - 6$

Cumulative Review

1. Simplify: $16 - 4 \cdot (3 - 2)^2 \cdot 4$

2. Subtract: $8\frac{1}{2} - 3\frac{4}{7}$

3. Divide: $3\frac{7}{8} \div 1\frac{1}{2}$

4. Simplify: $\frac{3}{8} \div \left(\frac{3}{8} - \frac{1}{4}\right) \div \frac{7}{3}$

5. Subtract: $2.907 - 1.09761$

6. Solve the proportion $\frac{7}{12} = \frac{n}{32}$. Round to the nearest hundredth.

7. 22 is 160% of what number?

8. Convert: 7 qt = ____ gal ____ qt

9. Convert: 6 L 692 ml = ____ L

10. Convert 4.2 ft to meters. Round to the nearest hundredth. (1 m = 3.28 ft)

11. Find the circumference of a circle with a radius of 21 cm. Use $\pi \approx \frac{22}{7}$.

12. Find the volume of a sphere with a diameter of 6 in. Use $\pi \approx 3.14$.

13. Add: $-8 + 5$

14. Add: $3\frac{1}{4} + \left(-6\frac{5}{8}\right)$

15. Subtract: $-6\frac{1}{8} - 4\frac{5}{12}$

16. Subtract: $-12 - (-7) - 3(-8)$

17. Multiply: $-3.2 \times (-1.09)$

18. Multiply: $-6 \times 7 \times \left(-\frac{3}{4}\right)$

19. Divide: $42 \div (-6)$

20. Divide: $-2\frac{1}{7} \div \left(-3\frac{3}{5}\right)$

21. Simplify: $3 \times (3 - 7) \div 6 - 2$

22. Simplify:
$4 - (-2)^2 \div (1 - 2)^2 \times 3 + 4$

23. A board $5\frac{2}{3}$ ft long is cut from a board 8 ft long. What is the length of the board remaining?

24. You had a balance of \$763.56 in your checkbook before writing checks of \$135.88 and \$47.81 and making a deposit of \$223.44. Find your new checkbook balance.

25. A suit that regularly sells for \$165 is on sale for \$120. Find the percent decrease in price. Round to the nearest tenth of a percent.

26. A reception is planned for 80 guests. How many gallons of coffee should be prepared to provide 2 c of coffee for each guest?

27. Find the perimeter of a rectangle with a length of 14 m and a width of 7.2 m.

28. Find the area of the composite figure. Use $\pi \approx 3.14$.

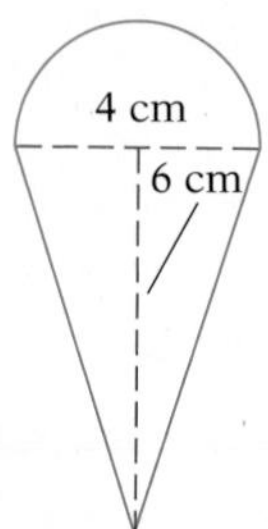

29. Find the height of the platform.

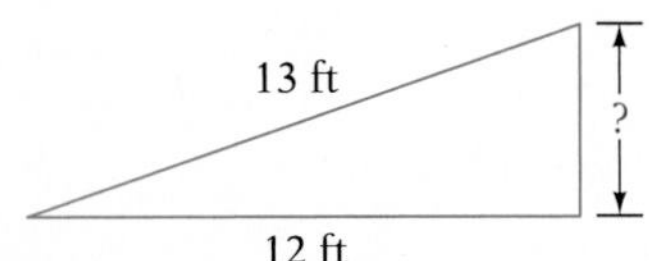

30. The daily high temperature readings for a 4-day period were recorded as follows: $-19°$, $-7°$, $1°$, and $9°$. Find the average high temperature for the 4-day period.

12 Introduction to Algebra

OBJECTIVES

- To evaluate variable expressions
- To simplify variable expressions containing no parentheses
- To simplify variable expressions containing parentheses
- To determine if a given value is a solution of an equation
- To solve an equation of the form $x + a = b$
- To solve an equation of the form $ax = b$
- To solve application problems
- To solve an equation of the form $ax + b = c$
- To solve application problems
- To solve an equation of the form $ax + b = cx + d$
- To solve an equation containing parentheses
- To translate a verbal expression into a mathematical expression given the variable
- To translate a verbal expression into a mathematical expression by assigning the variable
- To translate a sentence into an equation and solve
- To solve application problems

History of Equal Signs

A portion of a page of the first book that used an equals sign, =, is shown below. This book was written in 1557 by Robert Recorde and was titled *The Whetstone of Witte.*

Notice in the illustration above the words "bicause noe 2 thynges can be moare equalle." Recorde decided that two things could not be more equal than two parallel lines of the same length. Therefore it made sense to use this symbol to show equality.

This page also illustrates the use of the plus sign, +, and the minus sign, −. These symbols had been widely used for only about 100 years when this book was written.

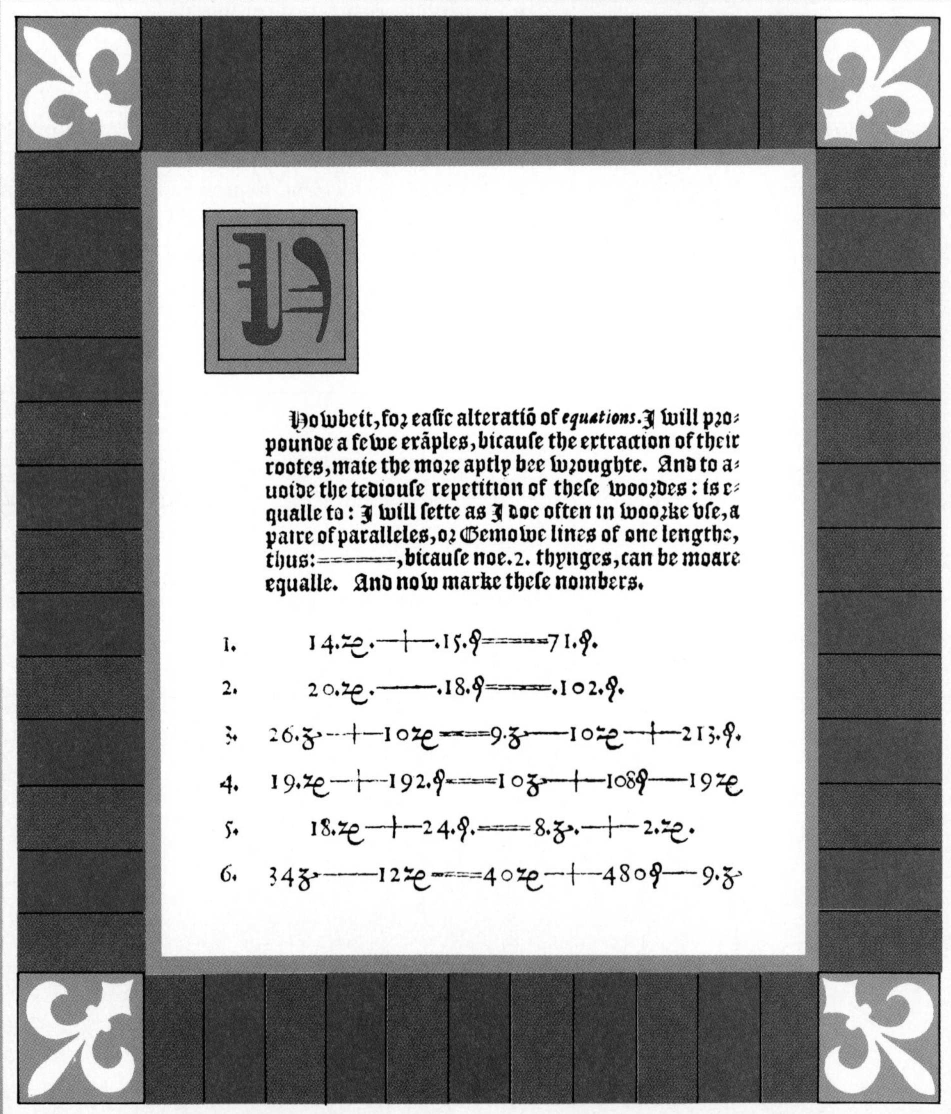

Howbeit, foꝛ eaſie alteratiō of *equations*. I will pꝛo-
pounde a fewe exāples, bicauſe the extraction of their
rootes, maie the moꝛe aptly bee wꝛoughte. And to a-
uoide the tediouſe repetition of theſe wooꝛdes : is e-
qualle to : I will ſette as I doe often in wooꝛke vſe, a
paire of paralleles, oꝛ Gemowe lines of one lengthe,
thus: ═════, bicauſe noe. 2. thynges, can be moare
equalle. And now marke theſe nombers.

1. 14.ze. — + — .15.ϱ ═════ 71.ϱ.
2. 20.ze. ——— .18.ϱ ═════ .102.ϱ.
3. 26.ʒ — + — 10ze ═════ 9.ʒ —— 10ze — + — 213.ϱ.
4. 19.ze — + — 192.ϱ ═════ 10ʒ — + — 108ϱ —— 19ze
5. 18.ze — + — 24.ϱ. ═════ 8.ʒ. — + — 2.ze.
6. 34ʒ ——— 12ze ═════ 40ze — + — 480ϱ —— 9.ʒ

SECTION 12.1 Variable Expressions

Objective A To evaluate variable expressions

Often we discuss a quantity without knowing its exact value, for example, next year's inflation rate, the price of gasoline next summer, or the interest rate on a new-car loan next fall. In mathematics, a letter of the alphabet is used to stand for a quantity that is unknown or that can change or *vary.* The letter is called a **variable.** An expression that contains one or more variables is called a **variable expression.**

A company's business manager has determined that the company will make a \$2 profit on each radio it sells. The manager wants to describe the company's total profit from the sale of radios. Because the number of radios that the company will sell is unknown, the manager lets the variable n stand for that number. Then the variable expression $2 \cdot n$, or simply $2n$, describes the company's profit for selling n radios.

The company's profit for selling n radios is $2 \cdot n = 2n$.

If the company sells 12 radios, its profit is $2 \cdot 12 = \$24$.

If the company sells 75 radios, its profit is $2 \cdot 75 = \$150$.

Replacing the variable or variables in a variable expression and then simplifying the resulting numerical expression is called **evaluating the variable expression.**

Evaluate $3x^2 + xy - z$ when $x = -2$, $y = 3$, and $z = -4$.

$3x^2 + xy - z$

Replace each variable in the expression with the number it stands for.

$3(-2)^2 + (-2)(3) - (-4)$

Use the Order of Operations Agreement to simplify the resulting numerical expression.

$3 \cdot 4 + (-2)(3) - (-4)$

$12 + (-6) - (-4)$

$12 + (-6) + 4$

$6 + 4$

10

The value of the variable expression $3x^2 + xy - z$ when $x = -2$, $y = 3$, and $z = -4$ is 10.

Example 1 Evaluate $3x - 4y$ when $x = -2$ and $y = 3$.

Solution
$3x - 4y$
$3(-2) - 4(3)$
$-6 - 12$
$-6 + (-12)$
-18

Example 2 Evaluate $6a - 5b$ when $a = -3$ and $b = 4$.

Your solution

Example 3 Evaluate $-x^2 - 6 \div y$ when $x = -3$ and $y = 2$.

Solution
$-x^2 - 6 \div y$
$-(-3)^2 - 6 \div 2$
$-9 - 6 \div 2$
$-9 - 3$
$-9 + (-3)$
-12

Example 4 Evaluate $-3s^2 - 12 \div t$ when $s = -2$ and $t = 4$.

Your solution

Example 5 Evaluate $-\frac{1}{2}y^2 - \frac{3}{4}z$ when $y = 2$ and $z = -4$.

Solution
$-\frac{1}{2}y^2 - \frac{3}{4}z$
$-\frac{1}{2}(2)^2 - \frac{3}{4}(-4)$
$-\frac{1}{2} \cdot 4 - \frac{3}{4}(-4)$
$-2 - (-3)$
$-2 + (3)$
1

Example 6 Evaluate $-\frac{2}{3}m + \frac{3}{4}n^3$ when $m = 6$ and $n = 2$.

Your solution

Example 7 Evaluate $-2ab + b^2 + a^2$ when $a = -\frac{3}{5}$ and $b = \frac{2}{5}$.

Solution
$-2ab + b^2 + a^2$
$-2\left(-\frac{3}{5}\right)\left(\frac{2}{5}\right) + \left(\frac{2}{5}\right)^2 + \left(-\frac{3}{5}\right)^2$
$-2\left(-\frac{3}{5}\right)\left(\frac{2}{5}\right) + \left(\frac{4}{25}\right) + \left(\frac{9}{25}\right)$
$\frac{12}{25} + \frac{4}{25} + \frac{9}{25}$
$\frac{25}{25} = 1$

Example 8 Evaluate $-3yz - z^2 + y^2$ when $y = -\frac{2}{3}$ and $z = \frac{1}{3}$.

Your solution

Solutions on p. A42

Objective B — To simplify variable expressions containing no parentheses

12

The **terms** of a variable expression are the addends of the expression. The variable expression at the right has four terms.

4 terms

$7x^2 + (-6xy) + x + (-8)$

variable terms: $7x^2$, $-6xy$, x; constant term: -8

Three of the terms are **variable terms:** $7x^2$, $(-6xy)$, and x.

One of the terms is a **constant term:** (-8). A constant term has no variables.

Each variable term is composed of a **numerical coefficient** and a **variable part** (the variable or variables and their exponents).

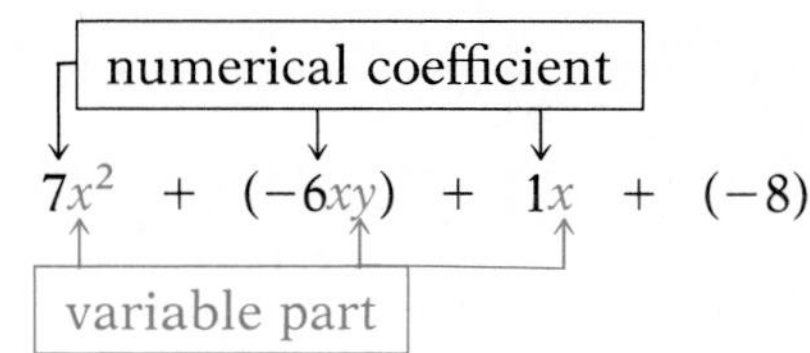

When the numerical coefficient is 1, the 1 is usually not written. ($1x = x$)

Like terms of a variable expression are the terms with the same variable part. (Since $y^2 = y \cdot y$, y^2 and y are not like terms.)

In variable expressions that contain constant terms, the constant terms are like terms.

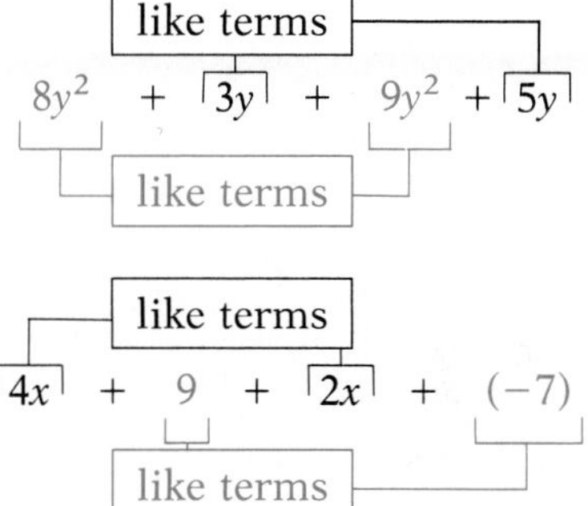

The Commutative and Associative Properties of Addition are used to simplify variable expressions. These properties can be stated in general form using variables.

Commutative Property of Addition

If a and b are two numbers, then $a + b = b + a$.

Associative Property of Addition

If a, b, and c are three numbers, then $a + (b + c) = (a + b) + c$.

To **simplify a variable expression.** *combine* like terms by adding their numerical coefficients. For example, to simplify $2y + 3y$, think of this as

$$2y + 3y = (y + y) + (y + y + y) = 5y$$

Simplify: $8z - 5 + 2z$

Use the Commutative and Associative Properties of Addition to group like terms. Combine the like terms $8z + 2z = 10z$.

$$\begin{aligned} 8z - 5 + 2z &= 8z + 2z - 5 \\ &= 10z - 5 \end{aligned}$$

Simplify: $12a - 4b - 8a + 2b$

Use the Commutative and Associative Properties of Addition to group like terms. Combine like terms.

$$12a - 4b - 8a + 2b = 12a - 8a - 4b + 2b$$
$$= 4a - 2b$$

Simplify: $6z^2 + 3 - z^2 - 7$

Use the Commutative and Associative Properties to group like terms. Combine like terms.

$$6z^2 + 3 - z^2 - 7 = 6z^2 - z^2 + 3 - 7$$
$$= 5z^2 - 4$$

Example 9 Simplify:
$6xy - 8x + 5x - 9xy$

Solution
$6xy - 8x + 5x - 9xy$
$6xy + (-8)x + 5x + (-9)xy$
$6xy + (-9)xy + (-8)x + 5x$
$-3xy + (-3)x$
$-3xy - 3x$

Example 10 Simplify:
$5a^2 - 6b^2 + 7a^2 - 9b^2$

Your solution

Example 11 Simplify:
$-4z^2 + 8 + 5z^2 - 3$

Solution
$-4z^2 + 8 + 5z^2 - 3$
$-4z^2 + 8 + 5z^2 + (-3)$
$-4z^2 + 5z^2 + 8 + (-3)$
$z^2 + 5$

Example 12 Simplify:
$-6x + 7 + 9x - 10$

Your solution

Example 13 Simplify:
$\frac{1}{4}m^2 - \frac{1}{2}n^2 + \frac{1}{2}m^2$

Solution
$\frac{1}{4}m^2 - \frac{1}{2}n^2 + \frac{1}{2}m^2$
$= \frac{1}{4}m^2 + \frac{1}{2}m^2 - \frac{1}{2}n^2$
$= \frac{3}{4}m^2 - \frac{1}{2}n^2$

Example 14 Simplify:
$\frac{3}{8}w + \frac{1}{2} - \frac{1}{4}w - \frac{2}{3}$

Your solution

Solutions on p. A42

Objective C

To simplify variable expressions containing parentheses

The Commutative and Associative Properties of Multiplication and the Distributive Property are used to simplify variable expressions that contain parentheses. These properties can be stated in general form using variables.

Commutative Property of Multiplication

If a and b are two numbers, then $a \cdot b = b \cdot a$.

Associative Property of Multiplication

If a, b, and c are three numbers, then $a \cdot (b \cdot c) = (a \cdot b) \cdot c$.

The Associative and Commutative Properties of Multiplication are used to simplify variable expressions such as the following.

Simplify: $-5(4x)$

Use the Associative Property of Multiplication.

$$-5(4x) = (-5 \cdot 4)x = -20x$$

Simplify: $(6y) \cdot 5$

Use the Commutative Property of Multiplication.
Use the Associative Property of Multiplication.

$$(6y) \cdot 5 = 5 \cdot (6y) = (5 \cdot 6)y = 30y$$

The Distributive Property is used to remove parentheses from variable expressions that contain both multiplication and addition.

Distributive Property

If a, b, and c are three numbers, then $a(b + c) = ab + ac$.

Simplify $4(z + 5)$.

The Distributive Property is used to rewrite the variable expression without parentheses.

$$4(z + 5) = 4z + 4(5) = 4z + 20$$

Simplify: $-3(2x + 7)$

Use the Distributive Property.

$$-3(2x + 7) = -3(2x) + (-3)(7) = -6x + (-21)$$

Recall that $a + (-b) = a - b$.

$$= -6x - 21$$

The Distributive Property can also be stated in terms of subtraction.

$$a(b - c) = ab - ac$$

Simplify: $8(2r - 3s)$

Use the Distributive Property. $8(2r - 3s) = 8(2r) - 8(3s)$
$= 16r - 24s$

Simplify: $-5(2x - 4y)$

Use the Distributive Property. $-5(2x - 4y) = (-5) \cdot (2x) - (-5) \cdot (4y)$
$= -10x - (-20y)$

Recall that $a - (-b) = a + b$. $= -10x + 20y$

Simplify $12 - 5(m + 2) + 2m$. $12 - 5(m + 2) + 2m$

Use the Distributive Property to simplify the expression $-5(m + 2)$. $12 - 5m + (-5)(2) + 2m$
$12 - 5m + (-10) + 2m$

Use the Commutative and Associative Properties to group like terms. $12 + (-10) - 5m + 2m$

Combine like terms by adding their numerical coefficients. Add constant terms. $2 - 3m$

Example 15 Simplify: $4(x - 3)$

Solution $4(x - 3) = 4x - 4(3)$
$= 4x - 12$

Example 16 Simplify: $5(a - 2)$

Your solution

Example 17 Simplify: $5n - 3(2n - 4)$

Solution $5n - 3(2n - 4)$
$= 5n - 3(2n) - (-3)(4)$
$= 5n - 3(2n) + 3(4)$
$= 5n - 6n + 12$
$= -n + 12$

Example 18 Simplify: $8s - 2(3s - 5)$

Your solution

Example 19 Simplify:
$3(c - 2) + 2(c + 6)$

Solution $3(c - 2) + 2(c + 6)$
$= 3c - 3(2) + 2c + 2(6)$
$= 3c + 2c - 6 + 12$
$= 5c + 6$

Example 20 Simplify:
$4(x - 3) - 2(x + 1)$

Your solution

Solution on p. A42

12.1 EXERCISES

Objective A

Evaluate the variable expression when $a = -3$, $b = 6$, and $c = -2$.

1. $5a - 3b$
2. $4c - 2b$
3. $2a + 3c$
4. $2c + 4a$
5. $a^2 + b$
6. $c^2 + 3b$
7. $c - ab$
8. $a - bc$
9. $a^2 - b$
10. $c^2 - b$
11. $2c^2 - a$
12. $3a^2 - c$
13. $-c^2$
14. $-a^2$
15. $b - a^2$
16. $b - c^2$
17. $ab - c^2$
18. $bc - a^2$
19. $2ab - c^2$
20. $3bc - a^2$
21. $a - (b \div a)$
22. $c - (b \div c)$
23. $2ac - (b \div a)$
24. $4ac \div (b \div a)$
25. $b^2 - c^2$
26. $b^2 - a^2$
27. $b^2 \div (ac)$
28. $3c^2 \div (ab)$
29. $c^2 - (b \div c)$
30. $a^2 - (b \div a)$
31. $a^2 + b^2 + c^2$
32. $a^2 - b^2 - c^2$
33. $ac + bc + ab$
34. $ac - bc - cb$
35. $a^2 + b^2 - ab$
36. $b^2 + c^2 - bc$
37. $2b - (3c + a^2)$
38. $\frac{2}{3}b + \left(\frac{1}{2}c - a\right)$
39. $\frac{1}{3}a + \left(\frac{1}{2}b - \frac{2}{3}a\right)$
40. $-\frac{2}{3}b - \left(\frac{1}{2}c + a\right)$
41. $\frac{1}{6}b + \frac{1}{3}(c + a)$
42. $\frac{1}{2}c + \left(\frac{1}{3}b - a\right)$

Evaluate the variable expression when $a = -\frac{1}{2}$, $b = \frac{3}{4}$, and $c = \frac{1}{4}$.

43. $a + (b - c)$

44. $c + (b - 2c)$

45. $4a + (3b - c)$

46. $2b + (c - 3a)$

47. $2a - b^2 \div c$

48. $b \div (-c) + 2a$

Evaluate the variable expression when $a = 3.72$, $b = -2.31$, and $c = -1.74$.

49. $a^2 - b^2$

50. $a^2 - b \cdot c$

51. $3ac - (c \div a)$

52. $3b - (3b - a^2)$

53. $2c + (b^2 - c)$

54. $abc - 2(b \div c^2)$

Objective B

Simplify.

55. $7z + 9z$

56. $6x + 5x$

57. $12m - 3m$

58. $5y - 12y$

59. $5at + 7at$

60. $12mn + 11mn$

61. $-4yt + 7yt$

62. $-12yt + 5yt$

63. $-3x - 12y$

64. $-12y - 7y$

65. $3t^2 - 5t^2$

66. $7t^2 + 8t^2$

67. $6c - 5 + 7c$

68. $7x - 5 + 3x$

69. $2t + 3t - 7t$

70. $-5y + 7y - 3y$

71. $4xy - 6xy + 3xy$

72. $10xy - 12xy + 3xy$

73. $5uv - 12uv + uv$

74. $t^2 - 4tv - 5t^2$

75. $x^2 - xy + 3x^2$

Simplify.

76. $9x^2 - 5 - 3x^2$

77. $7y^2 - 2 - 4y^2$

78. $3w - 7u + 4w$

79. $6w - 8u + 8w$

80. $4 - 6xy - 7xy$

81. $10 - 11xy - 12xy$

82. $7t^2 - 5t^2 - 4t^2$

83. $3v^2 - 6v^2 - 8v^2$

84. $5ab - 7a - 10ab$

85. $2bc - 7c - 10bc$

86. $-7xy - 3xy - 2xy$

87. $-ab - 3ab - 7ab$

88. $3x^2 - 4y^2 - 7x^2$

89. $7y^2 - x^2 - 10y^2$

90. $-4xy - 3y + 2xy$

91. $-10ab - 3a + 2ab$

92. $-4x^2 - x + 2x^2$

93. $-3y^2 - y + 7y^2$

94. $4x^2 - 8y - x^2 + y$

95. $2a - 3b^2 - 5a + b^2$

96. $8y - 4z - y + 2z$

97. $3x^2 - 7x + 4x^2 - x$

98. $5y^2 - y + 6y^2 - 5y$

99. $6s - t - 9s + 7t$

100. $5w - 2v - 9w + 5v$

101. $4m + 8n - 7m + 2n$

102. $z + 9y - 4z + 3y$

103. $-5ab + 7ac + 10ab - 3ac$

104. $-2x^2 - 3x - 11x^2 + 14x$

105. $\frac{4}{9}a^2 - \frac{1}{5}b^2 + \frac{2}{9}a^2 + \frac{4}{5}b^2$

106. $\frac{6}{7}x^2 + \frac{2}{5}x - \frac{3}{7}x^2 - \frac{4}{5}x$

107. $4.235x - 0.297x + 3.056x$

108. $8.092y - 3.0793y + 0.063y$

109. $7.81m + 3.42n - 6.25m - 7.19n$

110. $8.34y^2 - 4.21y - 6.07y^2 - 5.39y$

Objective C

Simplify.

111. $5(x + 4)$

112. $3(m + 6)$

113. $(y - 3)4$

114. $(z - 3)7$

115. $-2(a + 4)$

116. $-5(b + 3)$

117. $9(s - t)$

118. $8(x - y)$

119. $6(2a + 4)$

120. $3(5x + 10)$

121. $2(4m - 7)$

122. $5(3c - 5)$

123. $-4(w - 3)$

124. $-3(v - 6)$

125. $3m + 4(m + z)$

126. $5x + 2(x + 7)$

127. $6z - 3(z + 4)$

128. $8y - 4(y + 2)$

129. $7w - 2(w - 3)$

130. $9x - 4(x - 6)$

131. $-5m + 3(m + 4)$

132. $-2y + 3(y - 2)$

133. $5m + 3(m + 4) - 6$

134. $4n + 2(n + 1) - 5$

135. $8z - 2(z - 3) + 8$

136. $9y - 3(y - 4) + 8$

137. $6 - 4(a + 4) + 6a$

138. $3x + 2(x + 2) + 5x$

139. $7x + 4(x + 1) + 3x$

140. $-7t + 2(t - 3) - t$

141. $-3y + 2(y - 4) - y$

142. $z - 2(1 - z) - 2z$

143. $2y - 3(2 - y) + 4y$

144. $3(y - 2) - 2(y - 6)$

145. $7(x + 2) + 3(x - 4)$

146. $2(t - 3) + 7(t + 3)$

147. $3(y - 4) - 2(y - 3)$

148. $3t - 6(t - 4) + 8t$

149. $5x + 3(x - 7) - 9t$

SECTION 12.2 Introduction to Equations

Objective A To determine if a given value is a solution of an equation

An **equation** expresses the equality of two mathematical expressions. The expressions can be either numerical or variable expressions.

$$\left.\begin{array}{l} 5 + 4 = 9 \\ 3x + 13 = x - 8 \\ y^2 + 4 = 6y + 1 \\ x = -3 \end{array}\right\} \text{Equations}$$

In the equation at the right, if the variable is replaced by 4, the equation is true.

$x + 3 = 7$

$4 + 3 = 7$ A true equation

If the variable is replaced by 6, the equation is false.

$6 + 3 = 7$ A false equation

A **solution** of an equation is a value of the variable that results in a true equation. 4 is a solution of the equation $x + 3 = 7$. 6 is not a solution of the equation $x + 3 = 7$.

Is -2 a solution of the equation $x^2 + 1 = 2x + 9$?

$$\begin{array}{r|l} \multicolumn{2}{c}{x^2 + 1 = 2x + 9} \\ \hline (-2)^2 + 1 & 2(-2) + 9 \\ 4 + 1 & -4 + 9 \\ \multicolumn{2}{c}{5 = 5} \end{array}$$

Replace the variable by the given value.

Evaluate the numerical expressions.

Compare the results. If the results are equal, the given value is a solution. If the results are not equal, the given value is not a solution.

Yes, -2 is a solution of the equation $x^2 + 1 = 2x + 9$.

Example 1 Is $\frac{1}{2}$ a solution of $2x(x + 2) = 3x + 1$?

Solution

$$\begin{array}{r|l} \multicolumn{2}{c}{2x(x + 2) = 3x + 1} \\ \hline 2\left(\frac{1}{2}\right)\left(\frac{1}{2} + 2\right) & 3\left(\frac{1}{2}\right) + 1 \\ 2\left(\frac{1}{2}\right)\left(\frac{5}{2}\right) & 3\left(\frac{1}{2}\right) + 1 \\ \multicolumn{2}{c}{\frac{5}{2} = \frac{5}{2}} \end{array}$$

Yes, $\frac{1}{2}$ is a solution.

Example 2 Is -2 a solution of $x(x + 3) = 4x + 6$?

Your solution

Solution on p. A43

Example 3 Is 5 a solution of $(x-2)^2 = x^2 - 4x + 2$?

Solution

$$\begin{array}{r|l} \multicolumn{2}{c}{(x-2)^2 = x^2 - 4x + 2} \\ \hline (5-2^2) & 5^2 - 4(5) + 2 \\ 3^2 & 25 - 4(5) + 2 \\ 9 & 25 - 20 + 2 \\ & 25 + (-20) + 2 \end{array}$$

$9 \neq 7$ ($\neq$ means is not equal to)

No. 5 is not a solution.

Example 4 Is -3 a solution of $x^2 - x = 3x + 7$?

Your solution

Solution on p. A43

Objective B To solve an equation of the form $x + a = b$

A solution of an equation is a value of the variable that, when substituted in the equation, results in a true equation. To **solve** an equation means to find a solution of the equation.

The simplest equation to solve is an equation of the form **variable = constant.** The constant is the solution of the equation.

If $x = 7$, then 7 is the solution of the equation because $7 = 7$ is a true equation.

In solving an equation of the form $x + a = b$, the goal is to simplify the given equation to one of the form *variable = constant.* The Addition Properties that follow are used to simplify equations to this form.

Addition Property of Equations

If a, b, and c are algebraic expressions, then the equations $a = b$ and $a + c = b + c$ have the same solutions.

The Addition Property of Equations states that the same quantity can be added to each side of an equation without changing the solution of the equation.

Addition Property of Zero

The sum of a term and zero is the term.

$$a + 0 = a \qquad 0 + a = a$$

In solving an equation, the goal is to rewrite the given equation in the form *variable = constant*. The Addition Property of Equations is used to remove a term from one side of the equation by adding the opposite of that term to each side of the equation.

Solve: $x - 7 = -2$

The goal is to simplify the equation to one of the form *variable = constant.*

$$x - 7 = -2$$

Add the opposite of the constant term -7 to each side of the equation. After simplifying and using the addition Property of Zero, the equation will be in the form *variable = constant.*

$$x - 7 + 7 = -2 + 7$$
$$x + 0 = 5$$
$$x = 5$$

variable = constant

Check:
$$x - 7 = -2$$
$$5 - 7 = -2$$
$$-2 = -2$$ A true equation

The solution is 5.

Because subtraction is defined in terms of addition, the Addition Property of Equations allows the same number to be subtracted from each side of an equation.

Solve: $x + 8 = 5$

The goal is to simplify the equation to one of the form *variable = constant.*

$$x + 8 = 5$$

Add the opposite of the constant term 8 to each side of the equation. This procedure is equivalent to subtracting 8 from each side of the equation.

$$x + 8 - 8 = 5 - 8$$
$$x + 0 = -3$$
$$x = -3$$

You should check this solution.

The solution is -3.

Example 5 Solve: $4 + m = -2$

Solution

$$4 + m = -2$$
$$4 - 4 + m = -2 - 4$$
$$0 + m = -6$$
$$m = -6$$

The solution is -6.

Example 6 Solve: $-2 + y = -5$

Your solution

Example 7 Solve: $3 = y - 2$

Solution

$$3 = y - 2$$
$$3 + 2 = y - 2 + 2$$
$$5 = y + 0$$
$$5 = y$$

The solution is 5.

Example 8 Solve: $7 = y + 8$

Your solution

Example 9 Solve: $\frac{2}{7} = \frac{5}{7} + t$

Solution

$$\frac{2}{7} = \frac{5}{7} + t$$
$$\frac{2}{7} - \frac{5}{7} = \frac{5}{7} - \frac{5}{7} + t$$
$$-\frac{3}{7} = 0 + t$$
$$-\frac{3}{7} = t$$

The solution is $-\frac{3}{7}$.

Example 10 Solve: $\frac{1}{5} = z + \frac{4}{5}$

Your solution

Solution on p. A43

Objective C

To solve an equation of the form $ax = b$

In solving an equation of the form $ax = b$, the goal is to simplify the given equation to one of the form *variable* = *constant*. The Multiplication Properties that follow are used to simplify equations to this form.

Multiplication Property of Equations

> If a, b, and c are algebraic expressions and $c \neq 0$, then the equation $a = b$ has the same solutions as the equation $ac = bc$.

This property states that each side of an equation can be multiplied by the same nonzero number without changing the solutions of the equation.

Multiplication Property of Reciprocals

> The product of a nonzero term and its reciprocals equals 1.

$$a\left(\frac{1}{a}\right) = 1 \qquad \frac{1}{a}(a) = 1$$

or

$$\left(\frac{a}{b}\right)\left(\frac{b}{a}\right) = 1 \qquad \left(\frac{b}{a}\right)\left(\frac{a}{b}\right) = 1$$

Multiplication Property of One

> The product of a term and 1 is the term.
>
> $a \cdot 1 = a \qquad 1 \cdot a = a$

Recall that the goal of solving an equation is to rewrite the equation in the form *variable* = *constant*. The Multiplication Property of Equations is used to rewrite an equation in this form by multiplying each side of the equation by the reciprocal of the coefficient.

Solve: $\frac{2}{3}x = 8$

Multiply each side of the equation by $\frac{3}{2}$, the reciprocal of $\frac{2}{3}$. After simplifying and using the Multiplication Property of 1, the equation will be in the form *variable* = *constant*.

$$\frac{2}{3}x = 8$$

$$\left(\frac{3}{2}\right)\left(\frac{2}{3}\right)x = \left(\frac{3}{2}\right)8$$

$$1 \cdot x = 12$$

$$x = 12$$

Check:

$$\frac{2}{3}x = 8$$

$$\left(\frac{2}{3}\right)12 = 8$$

$$8 = 8$$

The solution is 12.

Because division is defined in terms of multiplication, the Multiplication Property of Equations allows each side of an equation to be divided by the same nonzero quantity.

Solve: $-4x = 24$

The goal is to rewrite the equation in the form *variable = constant.*

$$-4x = 24$$

Multiply each side of the equation by the reciprocal of -4. This procedure is equivalent to dividing each side of the equation by -4. Simplify by using the Multiplication Property of 1.

$$\frac{-4x}{-4} = \frac{24}{-4}$$

$$1x = -6$$

$$x = -6$$

The solution is -6.

You should check this solution.

When using the Multiplication Property of Equations, it is usually easier to multiply each side of the equation by the reciprocal of the coefficient when the coefficient is a fraction. Divide each side of the equation by the coefficient when the coefficient is an integer or decimal.

Example 11 Solve: $-2x = 6$

Solution

$$-2x = 6$$

$$\frac{-2x}{-2} = \frac{6}{-2}$$

$$1x = -3$$

$$x = -3$$

The solution is -3.

Example 12 Solve: $4z = -20$

Your solution

Example 13 Solve: $-9 = \frac{3}{4}y$

Solution

$$-9 = \frac{3}{4}y$$

$$\left(\frac{4}{3}\right)(-9) = \left(\frac{4}{3}\right)\left(\frac{3}{4}y\right)$$

$$-12 = 1y$$

$$-12 = y$$

The solution is -12.

Example 14 Solve: $8 = \frac{2}{5}n$

Your solution

Example 15 Solve: $6z - 8z = -5$

Solution

$$6z - 8z = -5 \quad \text{Combine like terms.}$$

$$-2z = -5$$

$$\frac{-2z}{-2} = \frac{-5}{-2}$$

$$1z = \frac{5}{2}$$

$$z = \frac{5}{2}$$

The solution is $2\frac{1}{2}$.

Example 16 Solve: $\frac{2}{3}t - \frac{1}{3}t = -2$

Your solution

Solutions on p. A43

Objective D To solve application problems

Example 17

An accountant for an auto parts store found that the weekly profit for the store was $850 and that the total amount spent during the week was $1200. Use the formula $P = R - C$, where P is the profit, R is the revenue, and C is the amount spent, to find the revenue for the week.

Strategy

To find the revenue for the week, replace the variables P and C in the formula by the given values and solve for R.

Solution

$$P = R - C$$
$$850 = R - 1200$$
$$850 + 1200 = R - 1200 + 1200$$
$$2050 = R + 0$$
$$2050 = R$$

The revenue for the week was $2050.

Example 18

A clothing store's regular price for a pair of slacks is $30. During a storewide sale, the slacks were priced at $22. Use the formula $S = R - D$, where S is the sale price, R is the regular price, and D is the discount, to find the discount.

Your strategy

Your solution

Example 19

A store manager uses the formula $S = R - D \cdot R$, where S is the sale price, R is the regular price, and D is the discount rate. During a clearance sale, all items are discounted 20%. Find the regular price of a jacket that is on sale for $40.

Strategy

To find the regular price of the jacket, replace the variables S and D in the formula by the given values and solve for R.

Solution

$$S = R - D \cdot R$$
$$40 = 1 \cdot R - 0.20R$$
$$40 = 0.8R$$
$$\frac{40}{0.8} = \frac{0.8R}{0.8}$$
$$50 = R$$

The regular price of the jacket is $50.

Example 20

An investor uses the formula $A = P + I \cdot P$, where A is the value of the investment in 1 year, P is the original investment, and I is the interest rate for the investment. Find the interest rate for an original investment of $1000 that had a value of $1120 after 1 year.

Your strategy

Your solution

Solutions on p. A43

12.2 EXERCISES

Objective A

1. Is 2 a solution of $3x = 15$?

2. Is 4 a solution of $4x = 18$?

3. Is 2 a solution of $3x + 5 = 11$?

4. Is -3 a solution of $2x + 9 = 3$?

5. Is -2 a solution of $5x + 7 = 12$?

6. Is 2 a solution of $4 - 2x = 8$?

7. Is -2 a solution of $4x + 1 = 3x$?

8. Is 4 a solution of $5 - 2x = 4x$?

9. Is 3 a solution of $3x - 2 = x + 4$?

10. Is 2 a solution of $4x + 8 = 4 - 2x$?

11. Is 3 a solution of $3x - 7 = 5 - x$?

12. Is 3 a solution of $4 - 2x = 1 - x$?

13. Is 3 a solution of $x^2 - 5x + 1 = 10 - 5x$?

14. Is -5 a solution of $x^2 - 3x - 1 = 9 - 6x$?

15. Is -1 a solution of $x^2 - 4x + 4 = (x - 2)^2$?

16. Is 7 a solution of $x^2 + 2x + 1 = (x + 1)^2$?

17. Is -1 a solution of $2x(x - 1) = 3 - x$?

18. Is 2 a solution of $3x(x - 3) = x - 8$?

19. Is 2 a solution of $x(x - 2) = x^2 - 4$?

20. Is -4 a solution of $x(x + 4) = x^2 + 16$?

21. Is $-\frac{2}{3}$ a solution of $3x + 6 = 4$?

22. Is $\frac{1}{2}$ a solution of $2x - 7 = -3$?

23. Is $\frac{1}{4}$ a solution of $2x - 3 = 1 - 14x$?

24. Is $-\frac{1}{3}$ a solution of $5x - 2 = 1 - 2x$?

25. Is $-\frac{1}{4}$ a solution of $2 - 3x = 1 - 7x$?

26. Is $\frac{2}{3}$ a solution of $4 - 2x = 6 - 5x$?

27. Is $\frac{3}{4}$ a solution of $3x(x - 2) = x - 4$?

28. Is $\frac{2}{5}$ a solution of $5x(x + 1) = x + 3$?

29. Is $\frac{1}{2}$ a solution of $2x(x - 2) = x + 3$?

30. Is $-\frac{1}{3}$ a solution of $3x(2x + 4) = 3x - 7$?

31. Is 1.32 a solution of $x^2 - 3x = -0.8776 - x$?

32. Is -1.9 a solution of $x^2 - 3x = x + 3.8$?

33. Is 1.05 a solution of $x^2 + 3x = x(x + 3)$?

Objective B

Solve.

34. $x + 3 = 9$

35. $x + 7 = 5$

36. $y - 6 = 16$

37. $z - 4 = 10$

38. $3 + n = 4$

39. $6 + x = 8$

40. $z + 7 = 2$

41. $w + 9 = 5$

42. $x - 3 = -7$

43. $m - 4 = -9$

44. $y + 6 = 6$

45. $t - 3 = -3$

46. $v - 7 = -4$

47. $x - 3 = -1$

48. $1 + x = 0$

49. $3 + y = 0$

50. $x - 10 = 5$

51. $y - 7 = 3$

52. $x + 4 = -7$

53. $t - 3 = -8$

54. $w + 5 = -5$

55. $z + 6 = -6$

56. $x + 7 = -8$

57. $x + 2 = -5$

58. $x + \frac{1}{2} = -\frac{1}{2}$

59. $x - \frac{5}{6} = -\frac{1}{6}$

60. $y + \frac{7}{11} = -\frac{3}{11}$

61. $\frac{2}{5} + x = -\frac{3}{5}$

62. $\frac{7}{8} + y = -\frac{1}{8}$

63. $\frac{1}{3} + x = \frac{2}{3}$

64. $x + \frac{1}{2} = -\frac{1}{3}$

65. $y + \frac{3}{8} = \frac{1}{4}$

66. $y + \frac{2}{3} = -\frac{3}{8}$

67. $t + \frac{1}{4} = -\frac{1}{2}$

68. $x + \frac{1}{3} = \frac{5}{12}$

69. $y + \frac{2}{3} = -\frac{5}{12}$

70. $x - 1.0792 = 5.4092$

71. $w + 3.057 = 0.07293$

72. $v - 27.092 = -11.08$

73. $4.039 + y = -4.3092$

74. $-0.139 + t = -1.392$

75. $x + 1.2903 = -0.976$

Objective C

Solve.

76. $3y = 12$

77. $5x = 30$

78. $5z = -20$

79. $3z = -27$

80. $-2x = 6$

81. $-4t = 20$

82. $-5x = -40$

83. $-2y = -28$

84. $40 = 8x$

85. $24 = 3y$

86. $-24 = 4x$

87. $-21 = 7y$

88. $\frac{x}{3} = 5$

89. $\frac{y}{2} = 10$

90. $\frac{n}{4} = -2$

91. $\frac{y}{7} = -3$

92. $-\frac{x}{4} = 1$

93. $\frac{-y}{3} = 5$

94. $\frac{2}{3}w = 4$

95. $\frac{5}{8}x = 10$

96. $\frac{3}{4}v = -3$

97. $\frac{2}{7}x = -12$

98. $-\frac{1}{3}x = -2$

99. $-\frac{1}{5}y = -3$

100. $\frac{3}{8}x = -24$

101. $\frac{5}{12}y = -16$

102. $-4 = -\frac{2}{3}z$

103. $-8 = -\frac{5}{6}x$

104. $-12 = -\frac{3}{8}y$

105. $-9 = \frac{5}{6}t$

106. $\frac{2}{3}x = -\frac{2}{7}$

107. $\frac{3}{7}y = \frac{5}{6}$

108. $4x - 2x = 7$

109. $3a - 6a = 8$

110. $\frac{4}{5}m - \frac{1}{5}m = 9$

111. $\frac{1}{3}b - \frac{2}{3}b = -1$

112. $4.203x = 1.7092$

113. $-1.723x = 14.809$

114. $\frac{t}{1.723} = -4.19$

115. $\frac{y}{17.94} = -0.0235$

116. $23.025x = -106.98$

117. $\frac{x}{-1.985} = -3.689$

Objective B *Application Problems*

In Exercises 118 and 119, use the formula $A = P + I$, where A is the value of the investment after 1 year, P is the original investment, and I is the increase in value of the investment.

118. The value of an investment in a high tech company after 1 year was \$17,700. The increase in value during the year was \$2700. Find the amount of the original investment.

119. The original investment in a mutual fund was \$8000. The value of the mutual fund after 1 year was \$11,420. Find the increase in the value of the investment.

In Exercises 120 and 121, use the formula $V = N - D$, where V is the value of the car now, N is the original value of the car, and D is the depreciation.

120. The depreciation of a car that originally cost \$16,500 is \$8000. Find the value of the car now.

121. The value of a camper that originally cost \$22,400 is now \$16,000. Find the depreciation.

In Exercises 122 and 123, use the formula $D = M \cdot G$, where D is distance, M is miles per gallon, and G is the number of gallons.

122. A sales executive averaged 27 mi/gal on a 621-mi trip. Find the number of gallons of gasoline used.

123. The manufacturer of a subcompact car estimates that the car can travel 570 mi on a 15-gal tank of gas. Find the miles per gallon.

In Exercises 124 and 125, use the formula $S = \frac{D}{t}$, where S is the speed in miles per hour, D is the distance, and t is the time.

124. A student drove 288 mi in 6 h. Find the average speed for the trip.

125. A truck driver drove 12 h at an average speed of 54 mi/h. Find the distance the truck driver drove in the 12 h.

In Exercises 126 and 127, use the formula $S = C + M$, where S is the selling price, C is the cost, and M is the markup.

126. A computer store sells a computer for \$2240. The computer has a markup of \$420. Find the cost of the computer.

127. A department store buys shirts for \$13.50 and sells the shirts for \$19.80. Find the markup on each shirt.

In Exercises 128 and 129, use the formula $S = C + R \cdot C$, where S is the selling price, C is cost, and R is the markup rate.

128. A store manager uses a markup of 24% on all appliances. Find the cost of a blender that sells for \$52.70.

129. A record store uses a markup rate of 30%. Find the cost of a compact disc that sells for \$13.39.

SECTION 12.3 General Equations: Part I

Objective A To solve an equation of the form $ax + b = c$

To solve an equation of the form $ax + b = c$, it is necessary to use both the Addition and Multiplication Properties to simplify the equation to one of the form *variable* = *constant.*

Solve: $\frac{x}{4} - 1 = 3$

The goal is to simplify the equation to one of the form *variable* = *constant.*

$$\frac{x}{4} - 1 = 3$$

Add the opposite of the constant term -1 to each side of the equation. Then simplify (Addition Properties).

$$\frac{x}{4} - 1 + 1 = 3 + 1$$
$$\frac{x}{4} + 0 = 4$$
$$\frac{x}{4} = 4$$

Multiply each side of the equation by the reciprocal of the numerical coefficient of the variable term. Then simplify (Multiplication Properties).

$$4 \cdot \frac{x}{4} = 4 \cdot 4$$
$$1x = 16$$
$$x = 16$$

Write the solution.

The solution is 16.

Example 1 Solve: $3x + 7 = 2$

Solution

$$3x + 7 = 2$$
$$3x + 7 - 7 = 2 - 7$$
$$3x = -5$$
$$\frac{3x}{3} = \frac{-5}{3}$$
$$x = -\frac{5}{3}$$

The solution is $-1\frac{2}{3}$.

Example 2 Solve: $5x + 8 = 6$

Your solution

Example 3 Solve: $5 - x = 6$

Solution

$$5 - x = 6$$
$$5 - 5 - x = 6 - 5$$
$$-x = 1$$
$$(-1)(-x) = (-1) \cdot 1$$
$$x = -1$$

The solution is -1.

Example 4 Solve: $7 - x = 3$

Your solution

Solutions on p. A44

Objective B To solve application problems

Example 5

Find the Celsius temperature when the Fahrenheit temperature is 212°. Use the formula $F = \frac{9}{5}C + 32$, where F is the Fahrenheit temperature and C is the Celsius temperature.

Strategy

To find the Celsius temperature, replace the variable F in the formula by the given value and solve for C.

Solution

$$F = \frac{9}{5}C + 32$$
$$212 = \frac{9}{5}C + 32$$
$$212 - 32 = \frac{9}{5}C + 32 - 32$$
$$180 = \frac{9}{5}C$$
$$\frac{5}{9} \cdot 180 = \frac{5}{9} \cdot \frac{9}{5}C$$
$$100 = C$$

The Celsius temperature is 100°.

Example 6

Find the Celsius temperature when the Fahrenheit temperature is −22°. Use the formula $F = \frac{9}{5}C + 32$, where F is the Fahrenheit temperature and C is the Celsius temperature.

Your strategy

Your solution

Example 7

To find the total cost of production, an economist uses the formula $T = U \cdot N + F$, where T is the total cost, U is the cost per unit, N is the number of units made, and F is the fixed cost. Find the number of units made during a week when the total cost was $8000, the cost per unit was $16, and the fixed costs were $2000.

Strategy

To find the number of units made, replace the variables T, U, and F in the formula by the given values and solve for N.

Solution

$$T = U \cdot N + F$$
$$8000 = 16 \cdot N + 2000$$
$$8000 - 2000 = 16 \cdot N + 2000 - 2000$$
$$6000 = 16 \cdot N$$
$$\frac{6000}{16} = \frac{16 \cdot N}{16}$$
$$375 = N$$

The number of units made was 375.

Example 8

Find the cost per unit during a week when the total cost was $4500, the number of units produced was 250, and the fixed costs were $1500. Use the formula $T = U \cdot N + F$, where T is the total cost, U is the cost per unit, N is the number of units made, and F is the fixed cost.

Your strategy

Your solution

Solutions on p. A44

12.3 EXERCISES

Objective A

Solve.

1. $3x + 5 = 14$
2. $5z + 6 = 31$
3. $2n - 3 = 7$
4. $4y - 4 = 20$
5. $5w + 8 = 3$
6. $3x + 10 = 1$
7. $3z - 4 = -16$
8. $6x - 1 = -13$
9. $5 + 2x = 7$
10. $12 + 7x = 33$
11. $6 - x = 3$
12. $4 - x = -2$
13. $3 - 4x = 11$
14. $2 - 3x = 11$
15. $5 - 4x = 17$
16. $8 - 6x = 14$
17. $3x + 6 = 0$
18. $5x - 20 = 0$
19. $-3x - 4 = -1$
20. $-7x - 22 = -1$
21. $12x - 30 = 6$
22. $9x - 7 = 2$
23. $3x + 7 = 4$
24. $8x + 13 = 5$
25. $-2x + 11 = -3$
26. $-4x + 15 = -1$
27. $14 - 5x = 4$
28. $7 - 3x = 4$
29. $-8x + 7 = -9$
30. $-7x + 13 = -8$
31. $9x + 13 = 13$
32. $-2x + 7 = 7$
33. $7x - 14 = 0$
34. $5x + 10 = 0$
35. $4x - 4 = -4$
36. $-13x - 1 = -1$
37. $3x + 5 = 7$
38. $4x + 6 = 9$
39. $2x - 3 = 15$
40. $6x - 1 = 16$

Solve.

41. $8x + 9 = 4$
42. $7x + 8 = 3$
43. $3x - 5 = -2$
44. $4x - 7 = -4$

45. $4 + 6x = 7$
46. $8 + 9x = 14$
47. $3 - 2x = 8$
48. $5 - 3x = 4$

49. $4 - 7x = -12$
50. $1 - 5x = -8$
51. $12x - 5 = -7$
52. $8x - 1 = -5$

53. $3x - 4 = 17$
54. $12x - 3 = 7$
55. $2x - 3 = -8$
56. $5x - 3 = -12$

57. $-6x + 2 = -7$
58. $-3x + 9 = -1$
59. $-2x - 3 = -7$
60. $-5x - 7 = -4$

61. $3x + 8 = 2$
62. $2x - 9 = 8$
63. $3x - 7 = 0$
64. $7x - 2 = 0$

65. $-5x + 2 = 7$
66. $-2x + 9 = 12$
67. $-7x + 3 = 1$
68. $-2x + 9 = 6$

69. $12x - 12 = 1$
70. $7x - 7 = 1$
71. $4x - 7 = 13$
72. $7x - 9 = 21$

73. $\frac{1}{2}x - 2 = 3$
74. $\frac{1}{3}x + 1 = 4$
75. $\frac{3}{5}w - 1 = 2$
76. $\frac{2}{5}w + 5 = 6$

77. $\frac{2}{9}t - 3 = 5$
78. $\frac{5}{9}t - 3 = 2$
79. $\frac{y}{3} - 6 = -8$
80. $\frac{y}{2} - 2 = 3$

81. $\frac{x}{3} - 2 = -5$
82. $\frac{x}{4} - 3 = 5$
83. $\frac{5}{8}v + 6 = 3$
84. $\frac{2}{3}v - 4 = 3$

85. $\frac{4}{7}z + 10 = 5$
86. $\frac{3}{8}v - 3 = 4$
87. $\frac{2}{9}x - 3 = 5$
88. $\frac{1}{2}x + 3 = -8$

Solve.

89. $\frac{3}{4}x - 5 = -4$

90. $\frac{2}{3}x - 5 = -8$

91. $1.5x - 0.5 = 2.5$

92. $2.5w - 1.3 = 3.7$

93. $0.8t + 1.1 = 4.3$

94. $0.3v + 2.4 = 1.5$

95. $0.4x - 2.3 = 1.3$

96. $1.2t + 6.5 = 2.9$

97. $3.5y - 3.5 = 10.5$

98. $1.9x - 1.9 = -1.9$

99. $0.32x + 4.2 = 3.2$

100. $2.2y - 12 = 3.4$

101. $1.4t - 8.4 = -14$

102. $0.25x + 3 = -4.5$

103. $5x - 3x + 2 = 8$

104. $6m + 2m - 3 = 5$

105. $4a - 7a - 8 = 4$

106. $3y - 8y - 9 = 6$

107. $x - 4x + 5 = 11$

108. $-2y + y - 3 = 6$

109. $1.94x - 3.925 = 16.94$

110. $0.032x - 0.0194 = 0.139$

111. $423.8x - 49.6 = 182.75$

112. $-3.256x + 42.38 = -16.9$

113. $6.09x + 17.33 = 16.805$

114. $1.925x + 32.87 = -16.994$

Objective B *Application Problems*

In Exercises 115 and 116, use the relationship between Fahrenheit temperature and Celsius temperature, which is given by the formula $F = \frac{9}{5}C + 32$, where F is the Fahrenheit temperature and C is the Celsius temperature.

115. Find the Celsius temperature when the Fahrenheit temperature is $-40°$.

116. Find the Celsius temperature when the Fahrenheit temperature is 72°. Round to the nearest tenth of a degree.

In Exercises 117 and 118, use the formula $V = V_0 + 32t$, where V is the final velocity of a falling obect, V_0 is the starting velocity of a falling object, and t is the time for the object to fall.

117. Find the time required for an object to reach a velocity of 480 ft/s when dropped from the top of a large building. (The starting velocity is 0 ft/s.)

118. Find the time required for a falling object to increase in velocity from 16 ft/s to 128 ft/s.

In Exercises 119 and 120, use the formula $T = U \cdot N + F$, where T is the total cost, U is the cost per unit, N is the number of units made, and F is the fixed cost.

119. Find the cost per unit during a week when the total cost was $80,000, the total number of units produced was 500, and the fixed costs were $15,000.

120. Find the number of units made during a week when the total cost was $25,000, the cost per unit was $8, and the fixed costs were $5000.

In Exercises 121 and 122, use the formula $T = I \cdot R + B$, where T is the monthly tax, I is the monthly income, R is the income tax rate, and B is the base monthly tax.

121. The monthly tax a mechanic pays is $476. The mechanic's monthly tax rate is 22%, and the base monthly tax is $80. Find the mechanic's monthly salary.

122. The monthly tax a teacher pays is $770. The teacher's monthly income is $3100, and the base monthly tax is $150. Find the teacher's income tax rate.

In Exercises 123 to 126, use the formula $M = S \cdot R + B$, where M is the monthly earnings, S is total sales, R is the commission rate, and B is the base monthly salary.

123. A book representative earns a base monthly salary of $600 plus a 9% commission on total sales. Find the total sales during a month the representative earned $3480.

124. A sales executive earns a base monthly salary of $1000 plus a 5% commission on total sales. Find the total sales during a month the executive earned $2800.

125. A sales executive earns a base monthly salary of $750. Find the executive's commission rate during a month when total sales were $42,000 and the executive earned $2640.

126. A sales executive earns a base monthly salary of $500. Find the executive's commission rate during a month when total sales were $42,500 and the executive earned $3560.

SECTION 12.4 General Equations: Part II

Objective A To solve an equation of the form $ax + b = cx + d$

When a variable occurs on each side of an equation, the Addition Properties are used to rewrite the equation so that variable terms are on one side of the equation and constant terms are on the other side of the equation. Then the Multiplication Properties are used to simplify the equation to one of the form *variable = constant.*

Solve: $4x - 6 = 8 - 3x$

The goal is to write the equation in the form *variable = constant.*

$$4x - 6 = 8 - 3x$$

Add $3x$ to each side of the equation. Then simplify (Addition Properties). Now only one variable term occurs in the equation.

$$4x + 3x - 6 = 8 - 3x + 3x$$
$$7x - 6 = 8 + 0$$
$$7x - 6 = 8$$

Add 6 to each side of the equation. Then simplify (Addition Properties). Now only one constant term occurs in the equation.

$$7x - 6 + 6 = 8 + 6$$
$$7x + 0 = 14$$
$$7x = 14$$

Divide each side of the equation by the numerical coefficient of the variable term. Then simplify (Multiplication Properties).

$$\frac{7x}{7} = \frac{14}{7}$$
$$1x = 2$$
$$x = 2$$

Write the solution.

The solution is 2.

Example 1

Solve: $\frac{2}{9}x - 3 = \frac{7}{9}x + 2$

Solution

$$\frac{2}{9}x - 3 = \frac{7}{9}x + 2$$
$$-\frac{7}{9}x + \frac{2}{9}x - 3 = -\frac{7}{9}x + \frac{7}{9}x + 2$$
$$-\frac{5}{9}x - 3 = 2$$
$$-\frac{5}{9}x - 3 + 3 = 2 + 3$$
$$-\frac{5}{9}x = 5$$
$$\left(-\frac{9}{5}\right)\left(-\frac{5}{9}\right)x = \left(-\frac{9}{5}\right)5$$
$$x = -9$$

The solution is -9.

Example 2

Solve: $\frac{1}{5}x - 2 = \frac{2}{5}x + 4$

Your solution

Solution on p. A44

Objective B To solve an equation containing parentheses

When an equation contains parentheses, one of the steps in solving the equation requires use of the Distributive Property.

$$a(b + c) = ab + ac$$

The Distributive Property is used to rewrite a variable expression without parentheses.

Solve: $4(3 + x) - 2 = 2(x - 4)$

The goal is write the equation in the form *variable = constant.*

$$4(3 + x) - 2 = 2(x - 4)$$

Use the Distributive Property to simplify terms containing parentheses.

$$12 + 4x - 2 = 2x - 8$$
$$10 + 4x = 2x - 8$$

Subtract $2x$ from each side of the equation.

$$10 + 4x - 2x = 2x - 2x - 8$$
$$10 + 2x = -8$$

Subtract 10 from each side of the equation.

$$10 - 10 + 2x = -8 - 10$$
$$2x = -18$$

Divide each side of the equation by the numerical coefficient of the variable term.

$$\frac{2x}{2} = -\frac{18}{2}$$
$$x = -9$$

Write the solution.

The solution is -9.

Example 3 Solve: $3(x + 2) - x = 11$

Solution

$$3(x + 2) - x = 11$$
$$3x + 6 - x = 11$$
$$2x + 6 = 11$$
$$2x + 6 - 6 = 11 - 6$$
$$2x = 5$$
$$\frac{2x}{2} = \frac{5}{2}$$
$$x = \frac{5}{2}$$
$$x = 2\frac{1}{2}$$

The solution is $2\frac{1}{2}$.

Example 4 Solve: $4(x - 1) - x = 5$

Your solution

Solution on p. A44

12.4 EXERCISES

Objective A

Solve.

1. $6x + 3 = 2x + 5$
2. $7x + 1 = x + 19$
3. $3x + 3 = 2x + 2$
4. $6x + 3 = 3x + 6$
5. $5x + 4 = x - 12$
6. $3x - 12 = x - 8$
7. $7x - 2 = 3x - 6$
8. $2x - 9 = x - 8$
9. $9x - 4 = 5x - 20$
10. $8x - 7 = 5x + 8$
11. $2x + 1 = 16 - 3x$
12. $3x + 2 = -23 - 2x$
13. $5x - 2 = -10 - 3x$
14. $4x - 3 = 7 - x$
15. $2x + 7 = 4x + 3$
16. $7x - 6 = 10x - 15$
17. $x + 4 = 6x - 11$
18. $x - 6 = 4x - 21$
19. $3x - 7 = x - 7$
20. $2x + 6 = 7x + 6$
21. $3 - 4x = 5 - 3x$
22. $6 - 2x = 9 - x$
23. $7 + 3x = 9 + 5x$
24. $12 + 5x = 9 - 3x$
25. $5 + 2x = 7 + 5x$
26. $9 + x = 2 + 3x$
27. $8 - 5x = 4 - 6x$
28. $9 - 4x = 11 - 5x$
29. $6x + 1 = 3x + 2$
30. $7x + 5 = 4x + 7$
31. $5x + 8 = x + 5$
32. $9x + 1 = 3x - 4$
33. $2x - 3 = 6x - 4$

Solve.

34. $x - 5 = 7x - 8$

35. $2x - 9 = 4x + 5$

36. $3x + 7 = 5x + 10$

37. $2x - 3 = 4x + 3$

38. $6x - 7 = 4x + 7$

39. $4 - x = 7 - 2x$

40. $7 - 2x = 5 - x$

41. $6 - 4x = -7 + 2x$

42. $12 - 3x = -4 + 3x$

43. $4 - 3x = 4 - 5x$

44. $6 - 3x = 6 - 5x$

45. $2x + 7 = 4x - 3$

46. $6x - 2 = 2x - 9$

47. $4x - 7 = -3x + 2$

48. $6x - 3 = -5x + 8$

49. $7x - 5 = 3x + 9$

50. $-6x - 2 = -8x - 4$

51. $-7x + 2 = 3x - 8$

52. $-3 - 4x = 7 - 2x$

53. $-8 + 5x = 8 + 6x$

54. $3 - 7x = -2 + 5x$

55. $3x - 2 = 7 - 5x$

56. $5x + 8 = 4 - 2x$

57. $4 - 3x = 6x - 8$

58. $12x - 9 = 3x + 12$

59. $4x + 13 = -6x + 9$

60. $\frac{5}{7}x - 3 = \frac{2}{7}x + 6$

61. $\frac{4}{5}x - 1 = \frac{1}{5}x + 5$

62. $\frac{3}{7}x + 5 = \frac{5}{7}x - 1$

63. $\frac{3}{4}x + 2 = \frac{1}{4}x - 9$

64. $4.92x - 1.987 = 3x + 4.9283$

65. $102.9x + 30.4 = 1.987 - 4.7x$

Objective B

Solve.

66. $6x + 2(x - 1) = 14$

67. $3x + 2(x + 4) = 13$

68. $-3 + 4(x + 3) = 5$

69. $8x - 3(x - 5) = 30$

70. $6 - 2(x + 4) = 6$

71. $5 - 3(x + 2) = 8$

72. $5 + 7(x + 3) = 20$

73. $6 - 3(x - 4) = 12$

74. $2x + 3(x - 5) = 10$

75. $3x - 4(x + 3) = 9$

76. $3(x - 4) + 2x = 3$

77. $4 + 3(x - 9) = -12$

78. $2x - 3(x - 4) = 12$

79. $4x - 2(x - 5) = 10$

80. $2x + 3(x + 4) = 7$

81. $3(x + 2) + 7 = 12$

82. $3(x - 2) + 5 = 5$

83. $4(x - 5) + 7 = 7$

84. $3x + 7(x - 2) = 5$

85. $-3x - 3(x - 3) = 3$

86. $4x - 2(x + 9) = 8$

87. $3x - 6(x - 3) = 9$

88. $3x + 5(x - 2) = 10$

89. $3x - 5(x - 1) = -5$

90. $3x + 4(x + 2) = 2(x + 9)$

91. $5x + 3(x + 4) = 4(x + 2)$

Solve.

92. $2x - 3(x - 4) = 2(x + 6)$

93. $3x - 4(x - 1) = 3(x - 2)$

94. $7 - 2(x - 3) = 3(x - 1)$

95. $4 - 3(x + 2) = 2(x - 4)$

96. $6x - 2(x - 3) = 11(x - 2)$

97. $9x - 5(x - 3) = 5(x + 4)$

98. $6x - 3(x + 1) = 5(x + 2)$

99. $2x - 7(x - 2) = 3(x - 4)$

100. $7 - (x + 1) = 3(x + 3)$

101. $12 + 2(x - 9) = 3(x - 12)$

102. $2x - 3(x + 4) = 2(x - 5)$

103. $3x + 2(x - 7) = 7(x - 1)$

104. $x + 5(x - 4) = 3(x - 8) - 5$

105. $2x - 2(x - 1) = 3(x - 2) + 7$

106. $9x - 3(x - 4) = 13 + 2(x - 3)$

107. $3x - 4(x - 2) = 15 - 3(x - 2)$

108. $3(x - 4) + 3x = 7 - 2(x - 1)$

109. $2(x - 6) + 7x = 5 - 3(x - 2)$

110. $3.67x - 5.3(x - 1.932) = 6.99$

111. $4.06x + 4.7(x + 3.22) = 1.774$

112. $8.45(x - 10) = 3(x - 3.854)$

113. $4(x - 1.99) - 3.92 = 3(x - 1.77)$

SECTION 12.5 Translating Verbal Expressions into Mathematical Expressions

Objective A

To translate a verbal expression into a mathematical expression given the variable

One of the major skills required in applied mathematics is to translate a verbal expression into a mathematical expression. Doing so requires recognizing the verbal phrases that translate into mathematical operations. Following is a partial list of the verbal phrases used to indicate the different mathematical operations:

Addition	more than	5 more than x	$x + 5$
	the sum of	the sum of w and 3	$w + 3$
	the total of	the total of 6 and z	$6 + z$
	increased by	x increased by 7	$x + 7$
Subtraction	less than	5 less than y	$y - 5$
	the difference between	the difference between w and 3	$w - 3$
	decreased by	8 decreased by x	$8 - x$
Multiplication	times	3 times x	$3x$
	the product of	the product of 4 and t	$4t$
	of	two thirds of v	$\frac{2}{3}v$
Division	divided by	n divided by 3	$\frac{n}{3}$
	the quotient of	the quotient of z and 4	$\frac{z}{4}$
	the ratio of	the ratio of s to 6	$\frac{s}{6}$

Translating phrases that contain the words *sum, difference, product,* and *quotient* can sometimes cause a problem. In the examples at the right, note where the operation symbol is placed.

the *sum* of x *and* y (+)	$x + y$
the *difference* of x *and* y (−)	$x - y$
the *product* of x *and* y (·)	$x \cdot y$
the *quotient* of x *and* y (÷)	$\frac{x}{y}$

Note the placement of the fraction bar when translating the word *ratio.*

the *ratio* of x *to* y (÷)	$\frac{x}{y}$

Translate "the quotient of n and the sum of n and 6" into a mathematical expression.

the *quotient* of n *and* the *sum* of n *and* 6 (÷, +) $\qquad \frac{n}{n + 6}$

Example 1 Translate "the sum of n and the product of 4 and n" into a mathematical expression.

Solution $n + 4n$

Example 2 Translate "the difference between t and twice t" into a mathematical expression.

Your solution

Solution on p. A45

Example 3 Translate "the product of 3 and the difference between z and 4" into a mathematical expression.

Solution $3(z - 4)$

Example 4 Translate "the quotient of 5 and the product of 7 and x" into a mathematical expression.

Your solution

Solution on p. A45

Objective B

To translate a verbal expression into a mathematical expression by assigning the variable

In most applications that involve translating phrases into mathematical expressions, the variable to be used is not given. To translate these phrases, a variable must be assigned to the unknown quantity before the mathematical expression can be written.

Translate "the difference between a number and twice the number" into a mathematical expression.

Identify the phrases that indicate the mathematical operations.	The difference between a number and twice the number
Assign a variable to one of the unknown quantities.	The unknown number: n
Use the assigned variable to write an expression for any other unknown quantity.	Twice the number: $2n$
Use the identified operations to write the mathematical expression.	$n - 2n$

Example 5 Translate "four less than some number" into a mathematical expression.

Solution The unknown number: x

$x - 4$

Example 6 Translate "twelve decreased by some number" into a mathematical expression.

Your solution

Example 7 Translate "the total of a number and the square of the number" into a mathematical expression.

Solution The unknown number: c
The square of the number: c^2

$c + c^2$

Example 8 Translate "the product of a number and one-half of the number" into a mathematical expression.

Your solution

Solutions on p. A45

12.5 EXERCISES

Objective A

Translate into a mathematical expression.

1. 9 less than y

2. w divided by 7

3. z increased by 3

4. the product of -2 and x

5. the sum of two thirds of n and n

6. the difference between the square of r and r

7. the quotient of m and the difference between m and 3

8. v increased by twice v

9. the product of 9 and the sum of 4 more than x

10. the total of a and the quotient of a and 7

11. the difference between n and the product of -5 and n

12. x decreased by the quotient of x and 2

13. the product of c and one fourth of c

14. the quotient of 3 less than z and z

15. the total of the square of m and twice the square of m

16. the product of y and the sum of y and 4

17. 2 times the sum of t and 6

18. the quotient of r and the difference between 8 and r

19. x divided by the total of 9 and x

20. the sum of z and the product of 6 and z

21. three times the sum of b and 6

22. the ratio of w to the sum of w and 8

Objective B

Translate into a mathematical expression.

23. the square of a number

24. five less than some number

25. a number divided by twenty

26. the difference between a number and twelve

27. four times some number

28. the quotient of five and a number

29. three fourths of a number

30. the sum of a number and seven

31. four increased by some number

32. the ratio of a number and nine

33. the difference between five times a number and the number

34. six less than the total of three and a number

35. the product of a number and two more than the number

36. the quotient of six and the sum of nine and a number

37. seven times the total of a number and eight

38. the difference between ten and the quotient of a number and two

39. the square of a number plus the product of three and the number

40. a number decreased by the product of five and the number

41. the sum of three more than a number and one half of the number

42. eight more than twice the sum of a number and seven

43. the quotient of three times a number and the number

44. the square of a number divided by the sum of the number and twelve

SECTION 12.6 Translating Sentences into Equations and Solving

Objective A

To translate a sentence into an equation and solve

An equation states that two mathematical expressions are equal. Therefore, to translate a sentence into an equation requires recognition of the words or phrases that mean "equals." Some of these phrases are listed below:

equals
is
is equal to
amounts to
represents } translate to =

Once the sentence is translated into an equation, the equation can be simplified to one of the form *variable* = *constant* and the solution found.

Translate "three more than twice a number is seventeen" into an equation and solve.

Assign a variable to the unknown quantity. — The unknown number: n

Find two verbal expressions for the same value. — | Three more than twice a number | is | seventeen |

Write a mathematical expression for each verbal expression. Write the equals sign.

Solve the resulting equation.

$$2n + 3 = 17$$
$$2n + 3 - 3 = 17 - 3$$
$$2n = 14$$
$$\frac{2n}{2} = \frac{14}{2}$$
$$n = 7$$

The number is 7.

Example 1 Translate "a number decreased by six equals fifteen" into an equation and solve.

Solution The unknown number: x

| A number decreased by six | equals | fifteen |

$$x - 6 = 15$$
$$x - 6 + 6 = 15 + 6$$
$$x = 21$$

The number is 21.

Example 2 Translate "a number increased by four equals twelve" into an equation and solve.

Your solution

Solution on p. A45

Example 3 The quotient of a number and six is five. Find the number.

Solution The unknown number: z

The quotient of a number and six	is	five

$$\frac{z}{6} = 5$$

$$6 \cdot \frac{z}{6} = 6 \cdot 5$$

$$z = 30$$

The number is 30.

Example 4 The product of two and a number is ten. Find the number.

Your solution

Example 5 Eight decreased by twice a number is four. Find the number.

Solution The unknown number: t

Eight decreased by twice a number	is	four

$$8 - 2t = 4$$

$$8 - 8 - 2t = 4 - 8$$

$$-2t = -4$$

$$\frac{-2t}{-2} = \frac{-4}{-2}$$

$$t = 2$$

The number is 2.

Example 6 The sum of three times a number and six equals four. Find the number.

Your solution

Example 7 Three less than a number divided by seven is one. Find the number.

Solution The unknown number: x

Three less than a number divided by seven	is	one

$$\frac{x}{7} - 3 = 1$$

$$\frac{x}{7} - 3 + 3 = 1 + 3$$

$$\frac{x}{7} = 4$$

$$7 \cdot \frac{x}{7} = 7 \cdot 4$$

$$x = 28$$

The number is 28.

Example 8 Three more than four times a number is ten. Find the number.

Your solution

Solutions on p. A45

Objective B To solve application problems

Example 9

The cost of a television with remote control is \$649. This amount is \$125 more than the cost without remote control. Find the cost of the television without remote control.

Strategy

To find the cost of the television without remote control, write and solve an equation using C to represent the cost of the television without remote control.

Solution

\$649	is	\$125 more than the television without remote control

$$649 = C + 125$$
$$649 - 125 = C + 125 - 125$$
$$524 = C$$

The cost of the television without remote control is \$524.

Example 10

The sale price of a pair of slacks is \$18.95. This amount is \$6 less than the regular price. Find the regular price.

Your strategy

Your solution

Example 11

By purchasing a fleet of cars, a company receives a discount of \$972 on each car purchased. This amount is 18% off the regular price. Find the regular price.

Strategy

To find the regular price, write and solve an equation using P to represent the regular price of the car.

Solution

\$972	is	18% off the regular price

$$972 = 0.18 \cdot P$$
$$\frac{972}{0.18} = \frac{0.18P}{0.18}$$
$$5400 = P$$

The regular price is \$5400.

Example 12

At a certain speed, the engine rpm (revolutions per minute) of a car in fourth gear is 2500. This is two-thirds of the rpm of the engine in third gear. Find the rpm of the engine when in third gear.

Your strategy

Your solution

Solutions on p. A45

Example 13

A plumber charged \$815 for work done in an office building. This charge included \$90 for parts and \$25 per hour for labor. Find the number of hours the plumber worked in the office building.

Strategy

To find the number of hours worked, write and solve an equation using N to represent the number of hours worked.

Solution

\$815	included	\$90 for parts and \$25 per hour for labor

$$\begin{aligned} 815 &= 90 + 25N \\ 815 - 90 &= 90 - 90 + 25N \\ 725 &= 25N \\ \frac{725}{25} &= \frac{25N}{25} \\ 29 &= N \end{aligned}$$

The plumber worked 29 h.

Example 14

A sales accountant earned \$2500 last month. This amount was the sum of a base monthly salary of \$800 and an 8% commission on total sales. Find the total sales for the month.

Your strategy

Your solution

Example 15

The state income tax for an employee last month was \$128. This amount is \$5 more than 8% of the employee's monthly salary. Find the employee's monthly salary.

Strategy

To find the employee's monthly salary, write and solve an equation using S to represent the employee's monthly salary.

Solution

\$128	is	\$5 more than 8% of the monthly salary

$$\begin{aligned} 128 &= 0.08 \cdot S + 5 \\ 128 - 5 &= 0.08 \cdot S + 5 - 5 \\ 123 &= 0.08 \cdot S \\ \frac{123}{0.08} &= \frac{0.08 \cdot S}{0.08} \\ 1537.50 &= S \end{aligned}$$

The employee's monthly salary is \$1537.50.

Example 16

The total cost to make a model ZY television is \$300. The cost includes \$100 for materials plus \$12.50 per hour for labor. How many hours of labor are required to make a model ZY television?

Your strategy

Your solution

Solutions on p. A46

12.6 EXERCISES

Objective A

Translate into an equation and solve.

1. The sum of a number and seven is twelve. Find the number.

2. A number decreased by seven is five. Find the number.

3. The product of three and a number is eighteen. Find the number.

4. The quotient of a number and three is one. Find the number.

5. Five more than a number is three. Find the number.

6. A number divided by four is six. Find the number.

7. Six times a number is fourteen. Find the number.

8. Seven less than a number is three. Find the number.

9. Five sixths of a number is fifteen. Find the number.

10. The total of twenty and a number is five. Find the number.

11. The sum of three times a number and four is eight. Find the number.

12. The sum of one third of a number and seven is twelve. Find the number.

13. Seven less than four times a number is nine. Find the number.

14. The total of a number divided by four and nine is two. Find the number.

15. The ratio of a number to nine is fourteen. Find the number.

16. Five increased by five times a number is equal to 30. Find the number.

Translate into an equation and solve.

17. Six less than the quotient of a number and four is equal to negative two. Find the number.

18. The product of a number plus three and two is eight. Find the number.

19. The difference between seven and twice a number is thirteen. Find the number.

20. Five more than the product of three and a number is eight. Find the number.

21. Nine decreased by the quotient of a number and two is five. Find the number.

22. The total of ten times a number and seven is twenty-seven. Find the number.

23. The sum of three fifths of a number and eight is two. Find the number.

24. Five less than two thirds of a number is three. Find the number.

25. The difference between a number divided by 4.186 and 7.92 is 12.529. Find the number.

26. The total of 5.68 times a number and 132.7 is the number minus 29.265. Find the number.

27. 37.35 decreased by a number is equal to 3.5 times the number. Find the number.

28. 24.325 less than the product of 1.25 and a number is -13.58. Find the number.

Objective B *Application Problems*

Write an equation and solve.

29. A shoe store sells a shoe for $72.50. This amount is $4.25 less than that at a competitor's store. Find the price at the competitor's store.

30. A restaurant manager is paid a salary of $832 a week, which is $58 more a week than the salary paid last year. Find the weekly salary paid to the manager last year.

31. The value of a camper this year is $15,000, which is four-fifths of what its value was last year. Find the value of the camper last year.

32. The value of a house this year is $125,000. This amount is twice the value of the house 4 years ago. Find its value 4 years ago.

33. A department uses a markup rate of 40%. Find the selling price of a camcorder that cost the department store $750.

34. A nurse is now making a salary of $3200 per month, which is four times the salary the nurse was making 10 years ago. Find the nurse's salary 10 years ago.

35. A family spends $680 on the house payment and utilities, which amounts to one fourth of the family's monthly income. Find the family's monthly income.

36. The cost of a scientific calculator is now one third of what it cost 5 years ago. The value of a scientific calculator is now $24. Find the cost of the calculator 5 years ago.

37. A computer manufacturing company has increased its monthly output by 40 computers. This amount represents an 8% increase over last year's production. Find the monthly output a year ago.

38. The population of a midwestern city has decreased by 12,000 in the last 5 years. This amount represents a 3% decrease. Find the city's population 5 years ago.

39. During a sale, a camera is discounted $40, which is 20% off the regular price. Find the regular price.

40. The cost of a pair of skis is $240. This price includes the store's cost for the skis plus a markup rate of 25%. Find the store's cost for the skis.

41. A plumber charged \$385 for a water softener and installation. The charge included \$310 for the water softener and \$25 per hour for labor. How long did it take the plumber to install the water softener?

42. The monthly salary for a sales representative was \$2580. This amount included the sales representative's base monthly salary of \$600 plus a 3% commission on total sales. Find the total sales for the month.

43. A pair of hockey skates that cost \$108 are sold for \$151.20. Find the markup rate on the pair of skates.

44. A car is purchased for \$5392. A down payment of \$1000 is made. The remainder is paid in 36 equal monthly installments. Find the monthly payment.

45. There is a total of 600 ft^2 of window space in a house. This amount is 300 ft^2 less than 10% of the house's total wall space. Find the house's total wall space.

46. A contractor charges \$75 plus \$24 for each yard of cement. How many yards of cement can be purchased for \$363?

47. A water flow restrictor has reduced the flow of water to 2 gal/min. This amount is 1 gal/min less than three-fifths the original water flow rate. Find the original rate.

48. The Fahrenheit temperature equals the sum of 32 and nine-fifths of the Celsius temperature. Find the Celsius temperature when the Fahrenheit temperature is 104°.

49. A sales executive receives a base monthly salary of \$600 plus an 8.25% commission on total sales per month. During 1 month the sales executive received \$4109.55. Find the total sales for the month.

50. A farmer harvested 28,336 bushels of corn. This amount was a 12% increase over last year's crop. How many bushels of corn did the farmer harvest last year?

51. A mechanic charges \$932.65 for repairing an automobile. This cost includes \$460.15 for materials and \$22.50 per hour for labor. How many hours of labor are required to repair the automobile?

52. A gift store paid \$27,500 in state income tax. This amount was \$5000 more than 9% of the store's total income. Find the store's total income.

Calculators and Computers

The y^x Key and Compound Interest

The y^x key is used to calculate powers of a number. For example, to evaluate 9^5, use the following keystrokes.

Enter 9. Press y^x. Enter 5. Press =.

The display should be 59049.

The y^x key is useful in calculating compound interest. Recall studying compound interest in Chapter 6. In that chapter, tables were used to find the value of a compound interest investment. Those tables were calculated by using the compound interest formula.

Compound Interest Formula $P = A(1 + i)^n$

In this formula, A is the amount invested, i is the interest rate per period, n is the number of periods the amount is invested, and P is the principal or value of the investment at the end of n periods.

Example 1 An investment of $350 is placed in an account that earns 10% annual interest compounded quarterly. Find the principal after 5 years.

Because the interest is compounded quarterly (4 times a year), $i = \frac{0.10}{4} = 0.025$. In 5 years there are 20 quarters (4×5). $A = 350$. Enter the following on your calculator:

350 × (1 + .025) y^x 20 =

The display should be 573.5157.
The principal is $573.52, rounded to the nearest cent.

Example 2 An investment of $175 is placed in an account that earns 7% annual interest compounded daily. Find the principal after 3 years.

$i = \frac{0.07}{365} = 0.000192$; $n = 3 \times 365 = 1095$; $A = 175$

Enter the following on your calculator:

175 × (1 + 0.000192) y^x 1095 =

The display should be 215.9411.
The principal is $215.94, rounded to the nearest cent.

Chapter Summary

Key Words A *variable* is a letter that is used to stand for a quantity that is unknown.

A *variable expression* is an expression that contains one or more variables.

The *terms* of a variable expression are the addends of the expression.

A *variable term* is composed of a numerical coefficient and a variable part.

Like terms of a variable expression are the terms with the same variable part.

An *equation* expresses the equality of two mathematical expressions.

A *solution* of an equation is a value of the variable that results in a true equation.

Essential Rules

The Commutative Property of Addition	If a and b are two numbers, then $\mathbf{a + b = b + a}$.
The Associative Property of Addition	If a, b, and c are three numbers, then $\mathbf{a + (b + c) = (a + b) + c}$.
The Commutative Property of Multiplication	If a and b are two numbers, then $\mathbf{a \cdot b = b \cdot a}$.
The Associative Property of Multiplication	If a, b, and c are three numbers, then $\mathbf{a(b \cdot c) = (a \cdot b)c}$.
The Distributive Property	If a, b, and c are three numbers, then $\mathbf{a(b + c) = ab + bc}$.
Addition Property of Equations	If a, b, and c are algebraic expressions, then the equation $a = b$ has the same solution as the equation $a + c = b + c$.
Multiplication Property of Equations	If a, b, and c are algebraic expressions and $c \neq 0$, then the equation $a = b$ has the same solution as the equation $ac = bc$.

Chapter Review

SECTION 12.1

1. Evaluate $a^2 - 3b$ when $a = 2$ and $b = -3$.

2. Evaluate $a^2 - (b \div c)$ when $a = -2, b = 8$, and $c = -4$.

3. Simplify: $6bc - 7bc + 2bc - 5bc$

4. Simplify: $\frac{1}{2}x^2 - \frac{1}{3}x^2 + \frac{1}{5}x^2 + 2x^2$

5. Simplify: $-2(a - b)$

6. Simplify: $3x - 2(3x - 2)$

SECTION 12.2

7. Is -2 a solution of $3x - 2 = -8$?

8. Is 5 a solution of $3x - 5 = -10$?

9. Solve: $x - 3 = -7$

10. Solve: $x + 3 = -2$

11. Solve: $-3x = 27$

12. Solve: $-\frac{3}{8}x = -\frac{15}{32}$

13. A tourist drove a rental car 621 mi on 27 gal of gas. Find the number of miles per gallon of gas. Use the formula $D = M \cdot G$, where D is distance, M is miles per gallon, and G is the number of gallons.

SECTION 12.3

14. Solve: $-2x + 5 = -9$

15. Solve: $35 - 3x = 5$

16. Solve: $\frac{2}{3}x + 3 = -9$

17. Solve: $\frac{5}{6}x - 4 = 5$

18. Find the Celsius temperature when the Fahrenheit temperature is 100°. Use the formula $F = \frac{9}{5}C + 32$, where F is the Fahrenheit temperature and C is the Celsius temperature. Round to the nearest tenth.

SECTION 12.4

19. Solve:
$7 - 3x = 2 - 5x$

20. Solve:
$6x - 9 = -3x + 36$

21. Solve:
$3(x - 2) + 2 = 11$

22. Solve:
$5x - 3(1 - 2x) = 4(2x - 1)$

SECTION 12.5

23. Translate "the total of n and the quotient of n and 5" into a mathematical expression.

24. Translate "the sum of five more than a number and one third of the number" into a mathematical expression.

SECTION 12.6

25. The difference between nine and twice a number is five. Find the number.

26. The product of five and a number is fifty. Find the number.

27. A compact disc player is now on sale for 40% off the regular price of \$380. Find the sale price.

Chapter Test

1. Evaluate $c^2 - (2a + b^2)$ when $a = 3$, $b = -6$, and $c = -2$.

2. Evaluate $\frac{x^2}{y} - \frac{y^2}{x}$ for $x = 3$ and $y = -2$.

3. Simplify: $3y - 2x - 7y - 9x$

4. Simplify: $3y + 5(y - 3) + 8$

5. Is 3 a solution of $x^2 + 3x - 7 = 3x - 2$?

6. Solve: $x - 12 = 14$

7. Solve: $-5x = 14$

8. Solve: $\frac{5}{8}x = -10$

9. A loan of \$6600 is to be paid in 48 equal monthly installments. Find the monthly payment. Use the formula $L = P \cdot N$, where L is the loan amount, P is the monthly payment, and N is the number of months.

10. Solve: $3x - 12 = -18$

11. Solve: $\frac{x}{5} - 12 = 7$

12. Solve: $5 = 3 - 4x$

13. A clock manufacturer's fixed costs per month are \$5000. The unit cost for each clock is \$15. Find the number of clocks made during a month in which the total cost was \$65,000. Use the formula $T = U \cdot N + F$, where T is the total cost, U is the cost per unit, N is the number of units made, and F is the fixed cost.

14. Solve: $8 - 3x = 2x - 8$

15. Solve: $3x - 4(x - 2) = 8$

16. Translate "the sum of x and one third of x" into a mathematical expression.

17. Translate "five times the sum of a number and three" into a mathematical expression.

18. Translate "three less than two times a number is seven" into an equation and solve.

19. The total of five and three times a number is the number minus two. Find the number.

20. A sales executive earned \$3600 last month. This salary is the sum of the base monthly salary of \$1200 and a 6% commission on total sales. Find the executive's total sales for the month.

Cumulative Review

1. Simplify:
$6^2 - (18 - 6) \div 4 + 8$

2. Subtract: $3\frac{1}{6} - 1\frac{7}{15}$

3. Simplify:
$\left(\frac{3}{8} - \frac{1}{4}\right) \div \frac{3}{4} + \frac{4}{9}$

4. Multiply: 9.67×0.0049

5. Write "\$84 earned in 20 hours" as a unit rate.

6. Solve the proportion $\frac{2}{3} = \frac{n}{40}$. Round to the nearest hundredth.

7. Write $5\frac{1}{3}\%$ as a fraction.

8. What percent of 30 is 42?

9. 8 is 125% of what number?

10. Multiply: 3 ft 9 in. × 5

11. Convert $1\frac{3}{8}$ lb to ounces.

12. Convert 2 g 82 mg to grams.

13. Add: $-2 + 5 + (-8) + 4$

14. Subtract: $13 - (-6)$

15. Simplify: $(-2)^2 - (-8) \div (3 - 5)^2$

16. Evaluate $3ab - 2ac$ when $a = -2$, $b = 6$, and $c = -3$.

17. Simplify: $3z - 2x + 5z - 8x$

18. Simplify: $6y - 3(y - 5) + 8$

19. Solve: $2x - 5 = -7$

20. Solve: $7x - 3(x - 5) = -10$

21. Solve: $-\frac{2}{3}x = 5$

22. Solve: $\frac{x}{3} - 5 = -12$

23. In a mathematics class of 34 students, 6 received an A grade. Find the percent of the students in the mathematics class who received an A grade. Round to the nearest tenth of a percent.

24. The manager of a pottery store used a markup rate of 40%. Find the price of a piece of pottery that cost the store \$28.50.

25. Find the perimeter of the composite figure. Use $\pi \approx 3.14$. Round to the nearest hundredth.

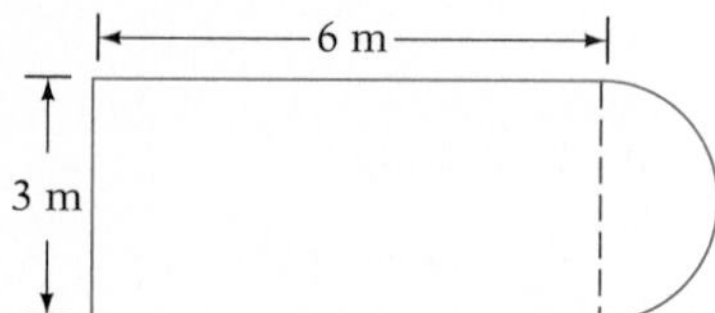

26. A toy store borrowed \$80,000 at a simple interest rate of 11% for 4 months. What is the simple interest due on the loan? Round to the nearest cent.

27. Translate "The sum of three times a number and 4" into a mathematical expression.

28. A car travels 318 mi in 6 h. Find the average speed in miles per hour. Use the formula $S = \frac{D}{t}$, where S is the average speed, D is the distance traveled, and t is the time of travel.

29. A sales executive receives a base salary of \$800 plus an 8% commission on total sales. Find the total sales during a month in which the sales executive earned \$3400. Use the formula $M = S \cdot R + B$, where M is the monthly earnings, S is the total sales, R is the commission rate, and B is the base monthly salary.

30. Three less than eight times a number is three more than five times the number. Find the number.

Final Examination

1. Subtract:
$100{,}914 - 97{,}655$

2. Divide: $67\overline{)34{,}773}$

3. Simplify:
$3^2 \cdot (5 - 3)^2 \div 3 + 4$

4. Find the LCM of 9, 12, and 16.

5. Add: $\frac{3}{8} + \frac{5}{6} + \frac{1}{5}$

6. Subtract: $7\frac{5}{12} - 3\frac{13}{16}$

7. Multiply: $3\frac{5}{8} \times 1\frac{5}{7}$

8. Divide: $1\frac{2}{3} \div 3\frac{3}{4}$

9. Simplify: $\left(\frac{2}{3}\right)^3 \cdot \left(\frac{3}{4}\right)^2$

10. Simplify: $\left(\frac{2}{3}\right)^2 \div \left(\frac{3}{4} + \frac{1}{3}\right) - \frac{1}{3}$

11. Add:

$$\begin{array}{r} 4.972 \\ 28.6 \\ 1.88 \\ +128.725 \\ \hline \end{array}$$

12. Subtract:

$$\begin{array}{r} 90.001 \\ -29.796 \\ \hline \end{array}$$

13. Multiply:

$$\begin{array}{r} 2.97 \\ \times 0.0094 \\ \hline \end{array}$$

14. Divide: $0.062\overline{)0.0426}$
Round to the nearest hundredth.

15. Convert 0.45 to a fraction in simplest form.

16. Write "323.4 miles on 13.2 gallons of gas" as a unit rate.

17. Solve the proportion $\frac{12}{35} = \frac{n}{160}$.
Round to the nearest tenth.

18. Write $22\frac{1}{2}\%$ as a fraction.

19. Write 1.35 as a percent.

20. Write $\frac{5}{4}$ as a percent.

21. Find 120% of 30.

22. 12 is what percent of 9?

23. 42 is 60% of what number?

24. Convert $1\frac{2}{3}$ ft to inches.

25. Subtract:
3 ft 2 in. − 1 ft 10 in.

26. Convert 40 oz to pounds.

27. Add:
$$\begin{array}{r} 3 \text{ lb } 12 \text{ oz} \\ +2 \text{ lb } 10 \text{ oz} \\ \hline \end{array}$$

28. Convert 18 pt to gallons.

29. Divide: $3\overline{)5 \text{ gal } 1 \text{ qt}}$

30. Convert 2.48 m to cm.

31. Multiply:
$$\begin{array}{r} 4 \text{ m } 62 \text{ cm} \\ \times \quad 5 \\ \hline \end{array}$$

32. Convert 1 kg 614 g to kg.

33. Subtract: 2 kg − 742 g

34. Convert 2 L 67 ml to milliliters.

35. Divide: 3 L 642 ml ÷ 4

36. Convert 55 mi to kilometers. Round to the nearest hundredth. (1.61 km = 1 mi)

37. Find the perimeter of a rectangle with a length of 1.2 m and a width of 0.75 m.

38. Find the area of a rectangle with a length of 9 in. and a width of 5 in.

39. Find the volume of a box with a length of 20 cm, a width of 12 cm, and a height of 5 cm.

40. Add: $-2 + 8 + (-10)$

41. Subtract: $-30 - (-15)$

42. Multiply: $2\frac{1}{2} \times \left(-\frac{1}{5}\right)$

43. Divide: $-1\frac{3}{8} \div 5\frac{1}{2}$

44. Simplify:
$(-4)^2 \div (1 - 3)^2 - (-2)$

45. Simplify:
$2x - 3(x - 4) + 5$

46. Solve: $\frac{2}{3}x = -12$

47. Solve: $3x - 5 = 10$

48. Solve: $8 - 3x = x + 4$

49. You have $872.48 in your checking account. You write checks of $321.88 and $34.23 and then make a deposit of $443.56. Find your new checking account balance.

50. In a pre-election survey, it is estimated that 5 out of 8 eligible voters will vote in an election. How many people will vote in an election with 102,000 eligible voters?

51. A new computer has a sales price of $1800. The price of the computer is 40% of what the price was 4 years ago. What was the price of the computer 4 years ago?

52. A sales executive received commissions of $4320, $3572, $2864, and $4420 during a 4-month period. Find the average income for the 4 months.

53. A contractor borrows $120,000 for 9 months at an annual interest rate of 10%. What is the simple interest due on the loan?

54. The circle graph shows the population of the five most populous countries. Find the percent of the population of China to the total population of the top five countries. Round to the nearest tenth of a percent.

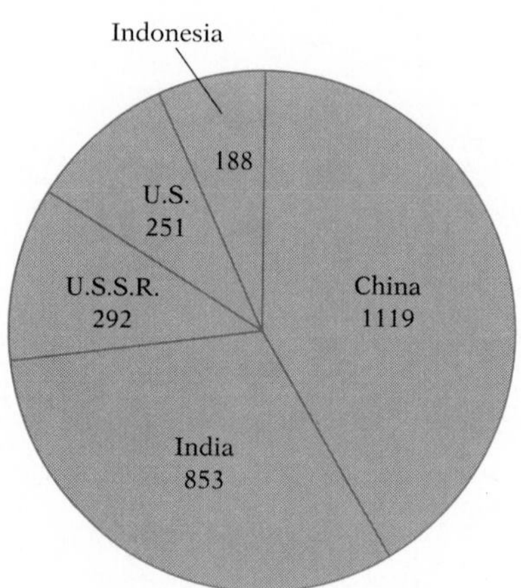

Population in Millions of People.

55. A compact disc player that regularly sells for $314.00 is on sale for $226.08. What is the discount rate?

56. An 8 × 8-in. tile weighs 9 oz. Find the weight of a box containing 144 tiles in pounds.

57. Find the perimeter of the composite figure. Use $\pi \approx 3.14$.

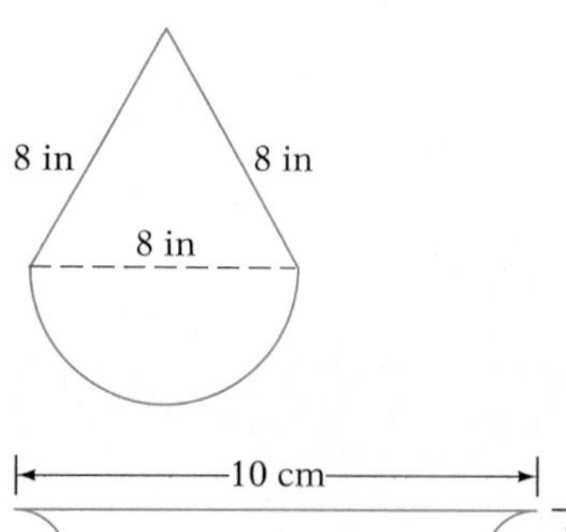

58. Find the area of the composite figure. Use $\pi \approx 3.14$.

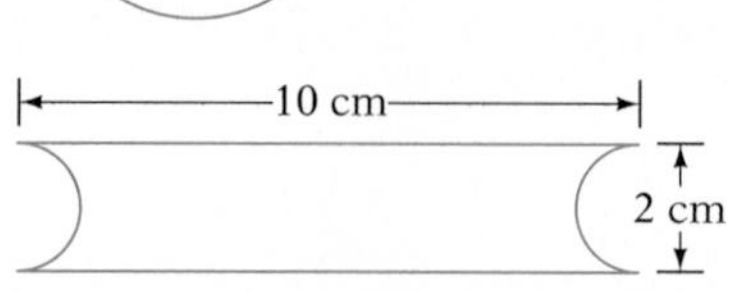

59. Five less than the quotient of a number and two is equal to three. Find the number.

APPENDIX

Addition Table

+	0	1	2	3	4	5	6	7	8	9	10	11	12
0	0	1	2	3	4	5	6	7	8	9	10	11	12
1	1	2	3	4	5	6	7	8	9	10	11	12	13
2	2	3	4	5	6	7	8	9	10	11	12	13	14
3	3	4	5	6	7	8	9	10	11	12	13	14	15
4	4	5	6	7	8	9	10	11	12	13	14	15	16
5	5	6	7	8	9	10	11	12	13	14	15	16	17
6	6	7	8	9	10	11	12	13	14	15	16	17	18
7	7	8	9	10	11	12	13	14	15	16	17	18	19
8	8	9	10	11	12	13	14	15	16	17	18	19	20
9	9	10	11	12	13	14	15	16	17	18	19	20	21
10	10	11	12	13	14	15	16	17	18	19	20	21	22
11	11	12	13	14	15	16	17	18	19	20	21	22	23
12	12	13	14	15	16	17	18	19	20	21	22	23	24

Multiplication Table

×	0	1	2	3	4	5	6	7	8	9	10	11	12
0	0	0	0	0	0	0	0	0	0	0	0	0	0
1	0	1	2	3	4	5	6	7	8	9	10	11	12
2	0	2	4	6	8	10	12	14	16	18	20	22	24
3	0	3	6	9	12	15	18	21	24	27	30	33	36
4	0	4	8	12	16	20	24	28	32	36	40	44	48
5	0	5	10	15	20	25	30	35	40	45	50	55	60
6	0	6	12	18	24	30	36	42	48	54	60	66	72
7	0	7	14	21	28	35	42	49	56	63	70	77	84
8	0	8	16	24	32	40	48	56	64	72	80	88	96
9	0	9	18	27	36	45	54	63	72	81	90	99	108
10	0	10	20	30	40	50	60	70	80	90	100	110	120
11	0	11	22	33	44	55	66	77	88	99	110	121	132
12	0	12	24	36	48	60	72	84	96	108	120	132	144

Table of Square Roots

Decimal approximations have been rounded to the nearest thousandth.

Number	Square Root	Number	Square Root	Number	Square Root	Number	Square Root
1	1	51	7.141	101	10.050	151	12.288
2	1.414	52	7.211	102	10.100	152	12.329
3	1.732	53	7.280	103	10.149	153	12.369
4	2	54	7.348	104	10.198	154	12.410
5	2.236	55	7.416	105	10.247	155	12.450
6	2.449	56	7.483	106	10.296	156	12.490
7	2.646	57	7.550	107	10.344	157	12.530
8	2.828	58	7.616	108	10.392	158	12.570
9	3	59	7.681	109	10.440	159	12.610
10	3.162	60	7.746	110	10.488	160	12.649
11	3.317	61	7.810	111	10.536	161	12.689
12	3.464	62	7.874	112	10.583	162	12.728
13	3.606	63	7.937	113	10.630	163	12.767
14	3.742	64	8	114	10.677	164	12.806
15	3.873	65	8.062	115	10.724	165	12.845
16	4	66	8.124	116	10.770	166	12.884
17	4.123	67	8.185	117	10.817	167	12.923
18	4.243	68	8.246	118	10.863	168	12.961
19	4.359	69	8.307	119	10.909	169	13
20	4.472	70	8.367	120	10.954	170	13.038
21	4.583	71	8.426	121	11	171	13.077
22	4.690	72	8.485	122	11.045	172	13.115
23	4.796	73	8.544	123	11.091	173	13.153
24	4.899	74	8.602	124	11.136	174	13.191
25	5	75	8.660	125	11.180	175	13.229
26	5.099	76	8.718	126	11.225	176	13.267
27	5.196	77	8.775	127	11.269	177	13.304
28	5.292	78	8.832	128	11.314	178	13.342
29	5.385	79	8.888	129	11.358	179	13.379
30	5.477	80	8.944	130	11.402	180	13.416
31	5.568	81	9	131	11.446	181	13.454
32	5.657	82	9.055	132	11.489	182	13.491
33	5.745	83	9.110	133	11.533	183	13.528
34	5.831	84	9.165	134	11.576	184	13.565
35	5.916	85	9.220	135	11.619	185	13.601
36	6	86	9.274	136	11.662	186	13.638
37	6.083	87	9.327	137	11.705	187	13.675
38	6.164	88	9.381	138	11.747	188	13.711
39	6.245	89	9.434	139	11.790	189	13.748
40	6.325	90	9.487	140	11.832	190	13.784
41	6.403	91	9.539	141	11.874	191	13.820
42	6.481	92	9.592	142	11.916	192	13.856
43	6.557	93	9.644	143	11.958	193	13.892
44	6.633	94	9.695	144	12	194	13.928
45	6.708	95	9.747	145	12.042	195	13.964
46	6.782	96	9.798	146	12.083	196	14
47	6.856	97	9.849	147	12.124	197	14.036
48	6.928	98	9.899	148	12.166	198	14.071
49	7	99	9.950	149	12.207	199	14.107
50	7.071	100	10	150	12.247	200	14.142

Compound Interest Table

Compounded Annually

	5%	6%	7%	8%	9%	10%	11%
1 year	1.05000	1.06000	1.07000	1.08000	1.09000	1.10000	1.11000
5 years	1.27628	1.33823	1.40255	1.46933	1.53862	1.61051	1.68506
10 years	1.62890	1.79085	1.96715	2.15893	2.36736	2.59374	2.83942
15 years	2.07893	2.39656	2.75903	3.17217	3.64248	4.17725	4.78459
20 years	2.65330	3.20714	3.86968	4.66095	5.60441	6.72750	8.06239

Compounded Semiannually

	5%	6%	7%	8%	9%	10%	11%
1 year	1.05062	1.06090	1.07123	1.08160	1.09203	1.10250	1.11303
5 years	1.28008	1.34392	1.41060	1.48024	1.55297	1.62890	1.70814
10 years	1.63862	1.80611	1.98979	2.19112	2.41171	2.65330	2.91776
15 years	2.09757	2.42726	2.80679	3.24340	3.74531	4.32194	4.98395
20 years	2.68506	3.26204	3.95926	4.80102	5.81634	7.03999	8.51331

Compounded Quarterly

	5%	6%	7%	8%	9%	10%	11%
1 year	1.05094	1.06136	1.07186	1.08243	1.09308	1.10381	1.11462
5 years	1.28204	1.34686	1.41478	1.48595	1.56051	1.63862	1.72043
10 years	1.64362	1.81402	2.00160	2.20804	2.43519	2.68506	2.95987
15 years	2.10718	2.44322	2.83182	3.28103	3.80013	4.39979	5.09225
20 years	2.70148	3.29066	4.00639	4.87544	5.93015	7.20957	8.76085

Compounded Daily

	5%	6%	7%	8%	9%	10%	11%
1 year	1.05127	1.06183	1.07250	1.08328	1.09416	1.10516	1.11626
5 years	1.28400	1.34983	1.41902	1.49176	1.56823	1.64861	1.73311
10 years	1.64866	1.82203	2.01362	2.22535	2.45933	2.71791	3.00367
15 years	2.11689	2.45942	2.85736	3.31968	3.85678	4.48077	5.20569
20 years	2.71810	3.31979	4.05466	4.95217	6.04830	7.38703	9.02203

To use this table:

1. Locate the section which gives the desired compounding period.
2. Locate the interest rate in the top row of that section.
3. Locate the number of years in the left-hand column of that section.
4. Locate the number where the interest-rate column and the number-of-years row meet. This is the compound interest factor.

Example An investment yields an annual interest rate of 10% compounded quarterly for 5 years.
The compounding period is "compounded quarterly."
The interest rate is 10%.
The number of years is 5.
The number where the row and column meet is 1.63862. This is the compound interest factor.

Compound Interest Table

Compounded Annually

	12%	13%	14%	15%	16%	17%	18%
1 year	1.12000	1.13000	1.14000	1.15000	1.16000	1.17000	1.18000
5 years	1.76234	1.84244	1.92542	2.01136	2.10034	2.19245	2.28776
10 years	3.10585	3.39457	3.70722	4.04556	4.41144	4.80683	5.23384
15 years	5.47357	6.25427	7.13794	8.13706	9.26552	10.53872	11.97375
20 years	9.64629	11.52309	13.74349	16.36654	19.46076	23.10560	27.39303

Compounded Semiannually

	12%	13%	14%	15%	16%	17%	18%
1 year	1.12360	1.13423	1.14490	1.15563	1.16640	1.17723	1.18810
5 years	1.79085	1.87714	1.96715	2.06103	2.15893	2.26098	2.36736
10 years	3.20714	3.52365	3.86968	4.24785	4.66096	5.11205	5.60441
15 years	5.74349	6.61437	7.61226	8.75496	10.06266	11.55825	13.26768
20 years	10.28572	12.41607	14.97446	18.04424	21.72452	26.13302	31.40942

Compounded Quarterly

	12%	13%	14%	15%	16%	17%	18%
1 year	1.12551	1.13648	1.14752	1.15865	1.16986	1.18115	1.19252
5 years	1.80611	1.89584	1.98979	2.08815	2.19112	2.29891	2.41171
10 years	3.26204	3.59420	3.95926	4.36038	4.80102	5.28497	5.81636
15 years	5.89160	6.81402	7.87809	9.10513	10.51963	12.14965	14.02741
20 years	10.64089	12.91828	15.67574	19.01290	23.04980	27.93091	33.83010

Compounded Daily

	12%	13%	14%	15%	16%	17%	18%
1 year	1.12747	1.13880	1.15024	1.16180	1.17347	1.18526	1.19716
5 years	1.82194	1.91532	2.01348	2.11667	2.22515	2.33918	2.45906
10 years	3.31946	3.66845	4.05411	4.48031	4.95130	5.47178	6.04696
15 years	6.04786	7.02625	8.16288	9.48335	11.01738	12.79950	14.86983
20 years	11.01883	13.45751	16.43582	20.07316	24.51534	29.94039	36.56578

Monthly Payment Table

	8%	9%	10%	11%	12%	13%
1 year	0.0869884	0.0874515	0.0879159	0.0883817	0.0888488	0.0893173
2 years	0.0452273	0.0456847	0.0461449	0.0466078	0.0470735	0.0475418
3 years	0.0313364	0.0317997	0.0322672	0.0327387	0.0332143	0.0336940
4 years	0.0244129	0.0248850	0.0253626	0.0258455	0.0263338	0.0268275
5 years	0.0202764	0.0207584	0.0212470	0.0217424	0.0222445	0.0227531
20 years	0.0083644	0.0089973	0.0096502	0.0103219	0.0110109	0.0117158
25 years	0.0077182	0.0083920	0.0090870	0.0098011	0.0105322	0.0112784
30 years	0.0073376	0.0080462	0.0087757	0.0095232	0.0102861	0.0110620

To use this table:

1. Locate the desired interest rate in the top row.
2. Locate the number of years in the left-hand column.
3. Locate the number where the interest-rate column and the number-of-years row meet. This is the monthly payment factor.

Example: A home has a 30-year mortgage at an annual interest rate of 12%.
The interest rate is 12%.
The number of years is 30.
The number where the row and column meet is 0.0102861. This is the monthly payment factor.

SOLUTIONS to Chapter 1 Examples

SECTION 1.1 *pages 3–6*

Example 2 Number line from 0 to 12 with a dot at 9:
0 1 2 3 4 5 6 7 8 9 10 11 12

Example 4 **a.** $45 > 29$ **b.** $27 > 0$

Example 6 Thirty-six million four hundred sixty-two thousand seventy-five

Example 8 452,007

Example 10 60,000 + 8000 + 200 + 80 + 1

Example 12 100,000 + 9000 + 200 + 7

Example 14 370,000

Example 16 4000

SECTION 1.2 *pages 9–12*

Example 2

$$\begin{array}{r} 3{,}508 \\ +92{,}170 \\ \hline 95{,}678 \end{array}$$

Example 4

$$\begin{array}{r} 102 \\ 7351 \\ 1024 \\ +\ 410 \\ \hline 8887 \end{array}$$

Example 6

$$\begin{array}{r} {\scriptstyle 2} \\ 95 \\ 88 \\ +67 \\ \hline 250 \end{array}$$

Example 8

$$\begin{array}{r} {\scriptstyle 1\,1\,2\,1} \\ 392 \\ 4{,}079 \\ 89{,}035 \\ +\ 4{,}992 \\ \hline 98{,}498 \end{array}$$

Example 10

Strategy To find the odometer reading after the trip, add the mileage before the trip (23,972) to the miles traveled during the trip (297).

Solution

$$\begin{array}{r} 23{,}972 \\ +\quad 297 \\ \hline 24{,}269 \end{array}$$

The odometer reading after the trip was 24,269 miles.

Example 12

Strategy To find the total amount budgeted for the three items each month, add the three amounts ($275, $75, and $64).

Solution

$$\begin{array}{r} \$275 \\ 75 \\ +\ 64 \\ \hline \$414 \end{array}$$

The total amount budgeted for the three items is $414.

SECTION 1.3 *pages 19–22*

Example 2

```
 8925
-6413
 2512
```

Example 4

```
 17,504
- 9,302
  8,202
```

Example 6

```
  2 14 7 11
  3 4 8 1
-   8 6 5
  2 6 1 6
```

Example 8

```
         15
       4 5 12
  5 4, 5 6 2
 -1 4, 4 8 5
  4 0, 0 7 7
```

Example 10

```
     13  9  9
  5  3 10 10 13
  6  4, 0  0  3
 -5  4, 9  3  6
     9, 0  6  7
```

Example 12

Strategy To find the amount that remains to be paid, subtract the down payment ($675) from the cost ($3250).

Solution

```
    11 14
   2 1  4 10
 $3 2  5  0
 -   6  7  5
 $2  5  7  5
```

$2575 remains to be paid.

Example 14

Strategy To find your take-home pay:

- Add to find the total of the deductions ($127 + $18 + $35).
- Subtract the total of the deductions from your total salary ($638).

Solution

```
 $127              $638
   18             -180
 + 35              $458
 $180 deductions
```

Your take-home pay is $458.

SECTION 1.4 *pages 27–30*

Example 2

```
  4
  78
 × 6
 468
```

Example 4

```
  3 5
  648
 ×  7
 4536
```

Example 6

```
     756
    ×305
    3780
  22680
 230,580
```

Example 8

Strategy To find the number of cars the dealer will receive in 12 months, multiply the number of months (12) by the number of cars received each month (37).

Solution

```
  37
 ×12
  74
 37
 444
```

The dealer will receive 444 cars in 12 months.

Example 10

Strategy To find the total cost of the order:
- Find the cost of the sports jackets by multiplying the number of jackets (25) by the cost for each jacket ($23).
- Add the product to the cost for the suits ($4800).

Solution

```
 $23            $4800
 ×25          +  575
 115           $5375
 46
$575  cost for jackets
```

The total cost of the order is $5375.

SECTION 1.5 *pages 35–42*

Example 2

```
   7
9)63
```

Check: 7 × 9 = 63

Example 4

```
     453
9)  4077
   -36
     47
    -45
     27
    -27
      0
```

Check: 453 × 9 = 4077

Example 6

```
     705
9)  6345
   -63
     04
    - 0
     45
    -45
      0
```

Check: 705 × 9 = 6345

Example 8

```
     870 r5
6)  5225
   -48
     42
    -42
     05
    - 0
      5
```

Check: (870 × 6) + 5 =
5220 + 5 = 5225

Example 10

```
    3,058 r3
7)  21,409
   -21
    0 4
   - 0
     40
    -35
     59
    -56
      3
```

Check: (3058 × 7) + 3 =
21,406 + 3 = 21,409

Example 12

```
      109
42)  4578
    -42
      37
     - 0
     378
    -378
       0
```

Check: 109 × 42 = 4578

Example 14

```
        470 r29
 39) 18,359
    -15 6
      2 75
     -2 73
        29
       - 0
        29
```

Check: (470 × 39) + 29 =
18,330 + 29 = 18,359

Example 16

```
          62 r111
 534) 33,219
     -32 04
       1 179
      -1 068
         111
```

Check: (62 × 534) + 111 =
33,108 + 111 = 33,219

Example 18

```
          421 r33
 515) 216,848
     -206 0
       10 84
      -10 30
          548
         -515
           33
```

Check: (421 × 515) + 33 =
216,815 + 33 = 216,848

Example 20

Strategy To find the number of tires to be stored on each shelf, divide the number of tires (270) by the number of shelves (15).

Solution

```
      18
 15) 270
    -15
     120
    -120
       0
```

Each shelf can store 18 tires.

Example 22

Strategy To find the number of cases produced in 8 hours:

- Find the number of cases produced in one hour by dividing the number of cans produced (12,600) by the number of cans to a case (24).
- Multiply the number of cases produced in one hour by 8.

Solution

```
        525 cases produced in one hour
 24) 12,600
    -12 0
        60
       -48
       120
      -120
         0
```

```
  525
×   8
 4200
```

In 8 hours, 4200 cases are produced.

SECTION 1.6 *pages 47–48*

Example 2 $2^4 \cdot 3^3$

Example 4 10^7

Example 6 $(2 \cdot 2 \cdot 2) \cdot (5 \cdot 5) = 8 \cdot 25 = 200$

Example 8
$5 \cdot 4 \div 4 - 2$
$20 \div 4 - 2$
$5 - 2$
3

SECTION 1.7 *pages 51–52*

Example 2 1, 2, 4, 5, 8, 10, 20 and 40 are factors of 40.

Example 4 $44 = 2 \cdot 2 \cdot 11$

Example 6 $177 = 3 \cdot 59$

SOLUTIONS to Chapter 2 Examples

SECTION 2.1 *pages 63–64*

Example 2

	2	3	5	7
50 =	2		5 · 5	
84 =	2 · 2	3		7
135 =		3 · 3 · 3	5	

The LCM $= 2 \cdot 2 \cdot 3 \cdot 3 \cdot 3 \cdot 5 \cdot 5 \cdot 7$
$= 18{,}900$

Example 4

	2	3	5
36 =	2 · 2	3 · 3	
60 =	2 · 2	3	5
72 =	2 · 2 · 2	3 · 3	

The GCF $= 2 \cdot 2 \cdot 3 = 12$

Example 6

	2	3	5	7	11
11 =					11
24 =	2 · 2 · 2	3			
30 =	2	3	5		

Since no numbers are circled, the GCF = 1.

SECTION 2.2 *pages 67–68*

Example 2 $4\frac{1}{4}$

Example 4 $\frac{17}{4}$

Example 6 $4\frac{2}{5}$

Example 8 4

Example 10 $\frac{117}{8}$

SECTION 2.3 *pages 71–72*

Example 2 $45 \div 5 = 9$ $\qquad \frac{3}{5} = \frac{3 \cdot 9}{5 \cdot 9} = \frac{27}{45}$

$\frac{27}{45}$ is equivalent to $\frac{3}{5}$.

Example 4 Write 6 as $\frac{6}{1}$.

$18 \div 1 = 18$ $\qquad 6 = \frac{6 \cdot 18}{1 \cdot 18} = \frac{108}{18}$

$\frac{108}{18}$ is equivalent to 6.

Example 6 $\frac{16}{24} = \frac{\overset{1}{\cancel{2}} \cdot \overset{1}{\cancel{2}} \cdot \overset{1}{\cancel{2}} \cdot 2}{\underset{1}{\cancel{2}} \cdot \underset{1}{\cancel{2}} \cdot \underset{1}{\cancel{2}} \cdot 3} = \frac{2}{3}$

Example 8 $\frac{8}{56} = \frac{\overset{1}{\cancel{2}} \cdot \overset{1}{\cancel{2}} \cdot \overset{1}{\cancel{2}}}{\underset{1}{\cancel{2}} \cdot \underset{1}{\cancel{2}} \cdot \underset{1}{\cancel{2}} \cdot 7} = \frac{1}{7}$

Example 10 $\frac{15}{32} = \frac{3 \cdot 5}{2 \cdot 2 \cdot 2 \cdot 2 \cdot 2} = \frac{15}{32}$

Example 12 $\frac{48}{36} = \frac{\overset{1}{\cancel{2}} \cdot \overset{1}{\cancel{2}} \cdot 2 \cdot 2 \cdot \overset{1}{\cancel{3}}}{\underset{1}{\cancel{2}} \cdot \underset{1}{\cancel{2}} \cdot \underset{1}{\cancel{3}} \cdot 3} = \frac{4}{3} = 1\frac{1}{3}$

SECTION 2.4 *pages 75–78*

Example 2

$$\begin{array}{r} \frac{3}{8} \\ +\frac{7}{8} \\ \hline \frac{10}{8} = \frac{5}{4} = 1\frac{1}{4} \end{array}$$

Example 4

$$\begin{array}{r} \frac{5}{12} = \frac{20}{48} \\ +\frac{9}{16} = \frac{27}{48} \\ \hline \frac{47}{48} \end{array}$$

Example 6

$$\begin{array}{r} \frac{7}{8} = \frac{105}{120} \\ +\frac{11}{15} = \frac{88}{120} \\ \hline \frac{193}{120} = 1\frac{73}{120} \end{array}$$

Example 8

$$\begin{array}{r} \frac{3}{4} = \frac{30}{40} \\ \frac{4}{5} = \frac{32}{40} \\ +\frac{5}{8} = \frac{25}{40} \\ \hline \frac{87}{40} = 2\frac{7}{40} \end{array}$$

Example 10 $7 + \frac{6}{11} = 7\frac{6}{11}$

Example 12

$$\begin{array}{r} 29 \\ +17\frac{5}{12} \\ \hline 46\frac{5}{12} \end{array}$$

Example 14

$$\begin{array}{r} 7\frac{4}{5} = 7\frac{24}{30} \\ 6\frac{7}{10} = 6\frac{21}{30} \\ +13\frac{11}{15} = 13\frac{22}{30} \\ \hline 26\frac{67}{30} = 28\frac{7}{30} \end{array}$$

Example 16

$$\begin{array}{r} 9\frac{3}{8} = 9\frac{45}{120} \\ 17\frac{7}{12} = 17\frac{70}{120} \\ +10\frac{14}{15} = 10\frac{112}{120} \\ \hline 36\frac{227}{120} = 37\frac{107}{120} \end{array}$$

Example 18

Strategy To find the total time spent on the activities, add the three times $\left(4\frac{1}{2}, 3\frac{3}{4}, 1\frac{1}{3}\right)$.

Solution

$$\begin{array}{r} 4\frac{1}{2} = 4\frac{6}{12} \\ 3\frac{3}{4} = 3\frac{9}{12} \\ +1\frac{1}{3} = 1\frac{4}{12} \\ \hline 8\frac{19}{12} = 9\frac{7}{12} \end{array}$$

The total time spent on the three activities was $9\frac{7}{12}$ hours.

Example 20

Strategy To find the overtime pay:

- Find the total number of overtime hours $\left(1\frac{2}{3} + 3\frac{1}{3} + 2\right)$.
- Multiply the total number of hours by the overtime hourly wage ($24).

Solution

$$\begin{array}{r} 1\frac{2}{3} \\ 3\frac{1}{3} \\ +2 \\ \hline 6\frac{3}{3} = 7 \text{ hours} \end{array} \qquad \begin{array}{r} \$24 \\ \times\ 7 \\ \hline \$168 \end{array}$$

The carpenter earned $168 in overtime pay.

SECTION 2.5 *pages 83–86*

Example 2

$$\begin{array}{r} \frac{16}{27} \\ -\frac{7}{27} \\ \hline \frac{9}{27} = \frac{1}{3} \end{array}$$

Example 4

$$\begin{array}{r} \frac{5}{6} = \frac{20}{24} \\ -\frac{3}{8} = \frac{9}{24} \\ \hline \frac{11}{24} \end{array}$$

Example 6

$$\begin{array}{r} \frac{13}{18} = \frac{52}{72} \\ -\frac{7}{24} = \frac{21}{72} \\ \hline \frac{31}{72} \end{array}$$

Example 8

$$\begin{array}{r} 17\frac{5}{9} = 17\frac{20}{36} \\ -11\frac{5}{12} = 11\frac{15}{36} \\ \hline 6\frac{5}{36} \end{array}$$

Example 10

$$\begin{array}{r} 8 = 7\frac{13}{13} \\ -2\frac{4}{13} = 2\frac{4}{13} \\ \hline 5\frac{9}{13} \end{array}$$

Example 12

$$\begin{array}{r} 21\frac{7}{9} = 21\frac{28}{36} = 20\frac{64}{36} \\ -\ 7\frac{11}{12} = 7\frac{33}{36} = 7\frac{33}{36} \\ \hline 13\frac{31}{36} \end{array}$$

Example 14

$$\begin{array}{r} 81\frac{17}{80} = 80\frac{97}{80} \\ -17\frac{29}{80} = 17\frac{29}{80} \\ \hline 63\frac{68}{80} = 63\frac{17}{20} \end{array}$$

Example 16

Strategy To find the time remaining before the plane lands, subtract the number of hours already in the air $\left(2\frac{3}{4}\right)$ from the total time of the trip $\left(5\frac{1}{2}\right)$.

Example 18

Strategy To find the amount of weight to be lost during the third month:

- Find the total weight loss during the first two months $\left(7\frac{1}{2} + 5\frac{3}{4}\right)$.
- Subtract the total weight loss from the goal (24 pounds).

Solution

$$5\frac{1}{2} = 5\frac{2}{4} = 4\frac{6}{4}$$
$$-2\frac{3}{4} = 2\frac{3}{4} = 2\frac{3}{4}$$
$$2\frac{3}{4} \text{ hours}$$

The plane will land in $2\frac{3}{4}$ hours.

Solution

$$7\frac{1}{2} = 7\frac{2}{4}$$
$$+5\frac{3}{4} = 5\frac{3}{4}$$
$$12\frac{5}{4} = 13\frac{1}{4} \text{ pounds lost}$$

$$24 = 23\frac{4}{4}$$
$$-13\frac{1}{4} = 13\frac{1}{4}$$
$$10\frac{3}{4} \text{ pounds}$$

The patient must lose $10\frac{3}{4}$ pounds to achieve the goal.

SECTION 2.6 *pages 91–94*

Example 2 $\frac{12}{25} \times \frac{5}{16} = \frac{12 \cdot 5}{25 \cdot 16} = \frac{\cancel{2} \cdot \cancel{2} \cdot 3 \cdot \cancel{5}}{5 \cdot \cancel{5} \cdot \cancel{2} \cdot \cancel{2} \cdot 2 \cdot 2} = \frac{3}{20}$

Example 4 $\frac{4}{21} \times \frac{7}{44} = \frac{4 \cdot 7}{21 \cdot 44} = \frac{\cancel{2} \cdot \cancel{2} \cdot \cancel{7}}{3 \cdot \cancel{7} \cdot \cancel{2} \cdot \cancel{2} \cdot 11} = \frac{1}{33}$

Example 6 $\frac{2}{21} \times \frac{10}{33} = \frac{2 \cdot 10}{21 \cdot 33} = \frac{2 \cdot 2 \cdot 5}{3 \cdot 7 \cdot 3 \cdot 11} = \frac{20}{693}$

Example 8 $\frac{16}{5} \times \frac{15}{24} = \frac{16 \cdot 15}{5 \cdot 24} = \frac{\cancel{2} \cdot \cancel{2} \cdot \cancel{2} \cdot 2 \cdot \cancel{3} \cdot \cancel{5}}{\cancel{5} \cdot \cancel{2} \cdot \cancel{2} \cdot \cancel{2} \cdot \cancel{3}} = 2$

Example 10 $5\frac{2}{5} \times \frac{5}{9} = \frac{27}{5} \times \frac{5}{9} = \frac{27 \cdot 5}{5 \cdot 9} = \frac{\cancel{3} \cdot \cancel{3} \cdot 3 \cdot \cancel{5}}{\cancel{5} \cdot \cancel{3} \cdot \cancel{3}} = 3$

Example 12 $9 \times \frac{7}{12} = \frac{9}{1} \times \frac{7}{12} = \frac{9 \cdot 7}{1 \cdot 12} = \frac{\cancel{3} \cdot 3 \cdot 7}{1 \cdot 2 \cdot 2 \cdot \cancel{3}} = \frac{21}{4} = 5\frac{1}{4}$

Example 14 $3\frac{2}{5} \times 6\frac{1}{4} = \frac{17}{5} \times \frac{25}{4} = \frac{17 \cdot 25}{5 \cdot 4} = \frac{17 \cdot \cancel{5} \cdot 5}{\cancel{5} \cdot 2 \cdot 2} = \frac{85}{4} = 21\frac{1}{4}$

Example 16 $3\frac{2}{7} \times 6 = \frac{23}{7} \times \frac{6}{1} = \frac{23 \cdot 6}{7 \cdot 1} = \frac{23 \cdot 3 \cdot 2}{7 \cdot 1} = \frac{138}{7} = 19\frac{5}{7}$

Example 18

Strategy To find the value of the house today, multiply the old value of the house ($30,000) by $3\frac{1}{2}$.

Solution $\frac{30{,}000}{1} \times \frac{7}{2} = \frac{30{,}000 \cdot 7}{1 \cdot 2} = 105{,}000$

The value of the house today is $105,000.

Example 20

Strategy To find the cost of the air compressor:

- Multiply to find the value of the drying chamber $\left(\frac{4}{5} \times \$60{,}000\right)$.
- Subtract the value of the drying chamber from the total value of the two items ($60,000).

Solution $\frac{4}{5} \times \frac{\$60{,}000}{1} = \frac{\$240{,}000}{5} = \$48{,}000$

value of the drying chamber

$$\begin{array}{r} \$60{,}000 \\ -48{,}000 \\ \hline \$12{,}000 \end{array}$$

The cost of the air compressor is $12,000.

SECTION 2.7 *pages 99–102*

Example 2

$$\frac{3}{7} \div \frac{2}{3} = \frac{3}{7} \times \frac{3}{2} = \frac{3 \cdot 3}{7 \cdot 2} = \frac{9}{14}$$

Example 4

$$\frac{3}{4} \div \frac{9}{10} = \frac{3}{4} \times \frac{10}{9} = \frac{3 \cdot 10}{4 \cdot 9} =$$

$$\frac{\overset{1}{\cancel{3}} \cdot \overset{1}{\cancel{2}} \cdot 5}{\underset{1}{\cancel{2}} \cdot 2 \cdot \underset{1}{\cancel{3}} \cdot 3} = \frac{5}{6}$$

Example 6

$$8 \div \frac{2}{3} = \frac{8}{1} \div \frac{2}{3} = \frac{8}{1} \times \frac{3}{2} =$$

$$\frac{\overset{1}{\cancel{2}} \cdot 2 \cdot 2 \cdot 3}{\underset{1}{\cancel{2}}} = 12$$

Example 8

$$\frac{5}{7} \div 6 = \frac{5}{7} \div \frac{6}{1} =$$

$$\frac{5}{7} \times \frac{1}{6} = \frac{5}{7 \cdot 2 \cdot 3} = \frac{5}{42}$$

Example 10

$$12\frac{3}{5} \div 7 = \frac{63}{5} \div \frac{7}{1} = \frac{63}{5} \times \frac{1}{7} =$$

$$\frac{63 \cdot 1}{5 \cdot 7} = \frac{\cdot 3 \cdot 3 \cdot \overset{1}{\cancel{7}}}{5 \cdot \underset{1}{\cancel{7}}} = \frac{9}{5} = 1\frac{4}{5}$$

Example 12

$$3\frac{2}{3} \div 2\frac{2}{5} = \frac{11}{3} \div \frac{12}{5} =$$

$$\frac{11}{3} \times \frac{5}{12} = \frac{11 \cdot 5}{3 \cdot 12} =$$

$$\frac{11 \cdot 5}{3 \cdot 2 \cdot 2 \cdot 3} = \frac{55}{36} = 1\frac{19}{36}$$

Example 14

$$2\frac{5}{6} \div 8\frac{1}{2} = \frac{17}{6} \div \frac{17}{2} = \frac{17}{6} \times \frac{2}{17} =$$

$$\frac{17 \cdot 2}{6 \cdot 17} = \frac{\overset{1}{\cancel{17}} \cdot \overset{1}{\cancel{2}}}{\underset{1}{\cancel{2}} \cdot 3 \cdot \underset{1}{\cancel{17}}} = \frac{1}{3}$$

Example 16

$$6\frac{2}{5} \div 4 = \frac{32}{5} \div \frac{4}{1} = \frac{32}{5} \times \frac{1}{4} =$$

$$\frac{32 \cdot 1}{5 \cdot 4} = \frac{2 \cdot 2 \cdot 2 \cdot \overset{1}{\cancel{2}} \cdot \overset{1}{\cancel{2}}}{5 \cdot \underset{1}{\cancel{2}} \cdot \underset{1}{\cancel{2}}} = \frac{8}{5} = 1\frac{3}{5}$$

Example 18

Strategy To find the price of one ounce of gold, divide the total price of the coin ($975) by the number of ounces $\left(2\frac{1}{2}\right)$.

Solution

$$\$975 \div 2\frac{1}{2} = \frac{\$975}{1} \div \frac{5}{2} =$$

$$\frac{\$975}{1} \times \frac{2}{5} = \frac{\$975 \cdot 2}{1 \cdot 5} = \$390$$

The price of one ounce of gold is $390.

Example 20

Strategy To find the length of the remaining piece:

- Divide the total length of the board (16 feet) by the length of each shelf ($3\frac{1}{3}$ feet).
- Multiply the fractional part of the result by the length of one shelf to determine the length of the remaining piece.

Solution

$$16 \div 3\frac{1}{3} = 16 \div \frac{10}{3} =$$

$$\frac{16}{1} \times \frac{3}{10} = \frac{16 \cdot 3}{1 \cdot 10} = \frac{\cancel{2} \cdot 2 \cdot 2 \cdot 2 \cdot 3}{\cancel{2} \cdot 5} =$$

$$\frac{24}{5} = 4\frac{4}{5}$$

$$\frac{4}{5} \times \frac{10}{3} = \frac{40}{15} = 2\frac{2}{3}$$

The length of the piece remaining is $2\frac{2}{3}$ feet.

SECTION 2.8 *pages 107–108*

Example 2 $\frac{9}{14} > \frac{13}{21}$

Example 4 $\left(\frac{7}{11}\right)^2 \cdot \left(\frac{2}{7}\right) = \left(\frac{7}{11} \cdot \frac{7}{11}\right) \cdot \left(\frac{2}{7}\right) =$

$\frac{\not{7} \cdot 7 \cdot 2}{11 \cdot 11 \cdot \not{7}} = \frac{14}{121}$

Example 6 $\left(\frac{1}{13}\right)^2 \cdot \left(\frac{1}{4} + \frac{1}{6}\right) \div \frac{5}{13}$

$\left(\frac{1}{13}\right)^2 \cdot \left(\frac{5}{12}\right) \div \frac{5}{13}$

$\left(\frac{1}{169}\right) \cdot \left(\frac{5}{12}\right) \div \frac{5}{13}$

$\left(\frac{1 \cdot 5}{13 \cdot 13 \cdot 12}\right) \div \frac{5}{13}$

$\frac{1 \cdot \overset{1}{\not{5}} \cdot \overset{1}{\not{13}}}{\underset{1}{\not{13}} \cdot 13 \cdot 12 \cdot \underset{1}{\not{5}}}$

$\frac{1}{156}$

SOLUTIONS to Chapter 3 Examples

SECTION 3.1 *pages 121–122*

Example 2 Two hundred nine and five thousand eight hundred thirty-eight hundred-thousandths

Example 4 42,000.000207

Example 6 4.35

Example 8 3.29053

SECTION 3.2 *pages 125–126*

Example 2

```
 12
  4.62
 27.9
+ 0.62054
 33.14054
```

Example 4

```
  1
  6.05
 12.
+ 0.374
 18.424
```

Example 6

Strategy To find the total cost of the four items, add the four costs ($850.65, $282.97, $650.00, and $835.00).

Solution

```
 $850.65
  282.97
  650.00
 +835.00
$2618.62
```

The total cost of the four items is $2618.62.

Example 8

Strategy To find the total income, add the four commissions ($485.60, $599.46, $326.75, and $725.42) to the salary ($425.00).

Solution

```
 $485.60
  599.46
  326.75
  725.42
 +425.00
$2562.23
```

The executive's total income was $2562.23.

SECTION 3.3 *pages 129–130*

Example 2

$$\begin{array}{r} 72.039 \\ -\ 8.47 \\ \hline 63.569 \end{array}$$

Check:

$$\begin{array}{r} 8.47 \\ +63.569 \\ \hline 72.039 \end{array}$$

Example 4

$$\begin{array}{r} 35.00 \\ -\ 9.67 \\ \hline 25.33 \end{array}$$

Check:

$$\begin{array}{r} 9.67 \\ +25.33 \\ \hline 35.00 \end{array}$$

Example 6

$$\begin{array}{r} 3.7000 \\ -1.9715 \\ \hline 1.7285 \end{array}$$

Check:

$$\begin{array}{r} 1.9715 \\ +1.7285 \\ \hline 3.7000 \end{array}$$

Example 8

Strategy To find the amount of change, subtract the amount paid ($3.85) from $5.00.

Solution

$$\begin{array}{r} \$5.00 \\ -3.85 \\ \hline \$1.15 \end{array}$$

Your change is $1.15.

Example 10

Strategy To find the new balance.

- Add to find the total of the three checks ($1025.60 + $79.85 + $162.47).
- Subtract the total from the previous balance ($2472.69).

Solution

$$\begin{array}{r} \$1025.60 \\ 79.85 \\ +\ 162.47 \\ \hline \$1267.92 \end{array} \qquad \begin{array}{r} \$2472.69 \\ -1267.92 \\ \hline \$1204.77 \end{array}$$

The new balance is $1204.77.

SECTION 3.4 *pages 133–136*

Example 2

$$\begin{array}{r} 870 \\ \times\ 4.6 \\ \hline 5220 \\ 3480 \\ \hline 4002.0 \end{array}$$

Example 4

$$\begin{array}{r} 0.28 \\ \times\ 0.7 \\ \hline 0.196 \end{array}$$

Example 6

$$\begin{array}{r} 0.000086 \\ \times\ \ 0.057 \\ \hline 602 \\ 430 \\ \hline 0.000004902 \end{array}$$

Example 8

$$\begin{array}{r} 79.063 \\ \times\ 0.47 \\ \hline 553441 \\ 316252 \\ \hline 37.15961 \end{array}$$

Example 10

$$\begin{array}{r} 4.68 \\ \times 6.03 \\ \hline 1404 \\ 28\ 080 \\ \hline 28.2204 \end{array}$$

Example 12 $6.9 \times 1000 = 6900$

Example 14 $4.0273 \times 10^2 = 402.73$

Example 16

Strategy To find the cost of running the freezer for 210 hours, multiply the hourly cost ($.035) by the number of hours (210).

Solution

$$\begin{array}{r} \$.035 \\ \times 210 \\ \hline \$7.35 \end{array}$$

The cost of running the freezer for 210 hours is $7.35.

Example 18

Strategy To find the total cost of the stereo:

- Multiply the monthly payment ($37.18) by the number of months (18).
- Add the total to the down payment ($175.00).

Solution

$$\begin{array}{r} \$37.18 \\ \times \quad 18 \\ \hline \$669.24 \end{array} \qquad \begin{array}{r} \$699.24 \\ +\ 175.00 \\ \hline \$844.24 \end{array}$$

SECTION 3.5 *pages 141–144*

Example 2

$$\begin{array}{r} 2.7 \\ 0.052.\overline{)0.140.4} \\ -104 \\ \hline 36\ 4 \\ -36\ 4 \\ \hline 0 \end{array}$$

Example 4

$$\begin{array}{r} 0.4873 \approx 0.487 \\ 76\overline{)37.0420} \\ -30\ 4 \\ \hline 6\ 64 \\ -6\ 08 \\ \hline 562 \\ -532 \\ \hline 300 \\ -228 \\ \hline \end{array}$$

Example 6

$$\begin{array}{r} 72.73 \approx 72.7 \\ 5.09.\overline{)370.20.00} \\ -356\ 3 \\ \hline 13\ 90 \\ -10\ 18 \\ \hline 3\ 720 \\ -3\ 563 \\ \hline 1570 \\ -1527 \\ \hline \end{array}$$

Example 8 $309.21 \div 10{,}000 = 0.030921$

Example 10 $42.93 \div 10^4 = 0.004293$

Example 12

Strategy To find the trucker's monthly income, divide the annual income ($23,741.40) by the number of months (12).

Example 14

Strategy To find the store's total income:

- Divide the total amount paid in taxes ($138.60) by the tax on each lamp ($11.55).
- Multiply the quotient by the price of one lamp ($95.60).

Solution

$$\begin{array}{r} \$1{,}978.45 \\ 12\overline{)\ \$23{,}741.40} \\ \underline{-12} \\ 11\,7 \\ \underline{-10\,8} \\ 94 \\ \underline{-84} \\ 101 \\ \underline{-\ 96} \\ 5\,4 \\ \underline{-4\,8} \\ 60 \\ \underline{-60} \\ 0 \end{array}$$

The trucker's monthly income is \$1978.45.

Solution

$$\begin{array}{r} 12. \\ \$11.55.\overline{)\$138.60.} \\ \underline{-115\,5} \\ 23\,10 \\ \underline{-23\,10} \\ 0 \end{array}$$

$$\begin{array}{r} \$95.60 \\ \underline{\times\quad 12} \\ \$1147.20 \end{array}$$

The store's total income was \$1147.20.

SECTION 3.6 *pages 149–150*

Example 2 $\begin{array}{r} 0.56 \approx 0.6 \\ 16\overline{)9.00} \end{array}$

Example 4 $4\frac{1}{6} = \frac{25}{6}$ $\begin{array}{r} 4.166 \approx 4.17 \\ 6\overline{)25.00} \end{array}$

Example 6 $0.56 = \frac{56}{100} = \frac{14}{25}$

$5.35 = 5\frac{35}{100} = 5\frac{7}{20}$

Example 8 $0.12\frac{7}{8} = \frac{12\frac{7}{8}}{100} = 12\frac{7}{8} \div 100 =$

$\frac{103}{8} \times \frac{1}{100} = \frac{103}{800}$

Example 10 $64.009 < 64.99$

Example 12 $0.63 > \frac{5}{8}$

SOLUTIONS to Chapter 4 Examples

SECTION 4.1 *pages 163–164*

Example 2 $\frac{20 \text{ pounds}}{24 \text{ pounds}} = \frac{20}{24} = \frac{5}{6}$

20 pounds:24 pounds = 20:24 = 5:6

20 pounds TO 24 pounds = 20 TO 24 = 5 TO 6

Example 4 $\frac{64 \text{ miles}}{8 \text{ miles}} = \frac{64}{8} = \frac{8}{1}$

64 miles:8 miles = 64:8 = 8:1

64 miles TO 8 miles = 64 TO 8 = 8 TO 1

Example 6

Strategy To find the ratio, write the ratio of board feet of cedar (12,000) to board feet of ash (18,000) in simplest form.

Solution $\frac{12{,}000}{18{,}000} = \frac{2}{3}$

The ratio is $\frac{2}{3}$.

Example 8

Strategy To find the ratio, write the ratio of the amount spent on radio advertising (\$15,000) to the amount spent on radio and television advertising (\$15,000 + \$20,000).

Solution $\frac{\$15{,}000}{\$15{,}000 + \$20{,}000} = \frac{\$15{,}000}{\$35{,}000} = \frac{3}{7}$

The ratio is $\frac{3}{7}$.

SECTION 4.2 *pages 167–168*

Example 2 $\frac{15 \text{ pounds}}{12 \text{ trees}} = \frac{5 \text{ pounds}}{4 \text{ trees}}$

Example 4 $\frac{260 \text{ miles}}{8 \text{ hours}} = \frac{32.5 \text{ miles}}{1 \text{ hour}}$

$= 32.5$ miles/hour

Example 6

Strategy To find the miles per hour rate, divide the number of miles (47) by the number of hours (3).

Solution $3\overline{)47.00}$ = 15.66

The rate is 15.7 miles/hour.

Example 8

Strategy To find the jeweler's profit per ounce:

- Find the total profit by subtracting the cost ($1980) from the selling price ($2075).
- Find the profit per ounce by dividing the total profit by the number of ounces (5).

Solution
$2075
−1980
$ 110 total profit

$5\overline{)\$95}$ = $19

The profit per ounce is $19.

SECTION 4.3 *pages 171–174*

Example 2 $\frac{6}{10}$ and $\frac{9}{15}$ → $10 \times 9 = 90$, → $6 \times 15 = 90$

The proportion is true.

Example 4 $\frac{32}{6}$ and $\frac{90}{8}$ → $6 \times 90 = 540$, → $32 \times 8 = 256$

The proportion is not true.

Example 6

$n \times 7 = 14 \times 3$
$n \times 7 = 42$

$n = 42 \div 7$
$n = 6$

Check: $\frac{6}{14}$ and $\frac{3}{7}$ → $14 \times 3 = 42$, → $6 \times 7 = 42$

Example 8

$5 \times 20 = 8 \times n$
$100 = 8 \times n$

$100 \div 8 = n$
$12.5 = n$

Example 10

$15 \times n = 20 \times 12$
$15 \times n = 240$

$n = 240 \div 15$
$n = 16$

Check: $\frac{15}{20}$ and $\frac{12}{16}$ → $20 \times 12 = 240$, → $15 \times 16 = 240$

Example 12

$12 \times 4 = 7 \times n$
$48 = 7 \times n$

$48 \div 7 = n$
$6.86 \approx n$

Example 14

$n \times 1 = 12 \times 4$
$n \times 1 = 48$

$n = 48 \div 1$
$n = 48$

Check: $\frac{48}{12}$ and $\frac{4}{1}$ → $12 \times 4 = 48$, → $48 \times 1 = 48$

Example 16

$3 \times n = 12 \times 8$
$3 \times n = 96$

$n = 96 \div 3$
$n = 32$

Check: $\frac{3}{8}$ and $\frac{12}{32}$ → $8 \times 12 = 96$, → $3 \times 32 = 96$

Example 18

Strategy To find the number of jars that can be packed in 15 boxes, write and solve a proportion using n to represent the number of jars.

Example 20

Strategy To find the number of tablespoons of fertilizer needed, write and solve a proportion using n to represent the number of tablespoons of fertilizer.

Solution $\frac{24 \text{ jars}}{6 \text{ boxes}} = \frac{n \text{ jars}}{15 \text{ boxes}}$

$24 \times 15 = 6 \times n$
$360 = 6 \times n$
$360 \div 6 = n$
$60 = n$

60 jars can be packed in 15 boxes.

Solution $\frac{3 \text{ tablespoons}}{4 \text{ gallons}} = \frac{n \text{ tablespoons}}{10 \text{ gallons}}$

$3 \times 10 = 4 \times n$
$30 = 4 \times n$
$30 \div 4 = n$
$7\frac{1}{2} = n$

For 10 gallons of water, $7\frac{1}{2}$ tablespoons of fertilizer are required.

SOLUTIONS to Chapter 5 Examples

SECTION 5.1 *pages 189–190*

Example 2 $125\% = 125 \times \frac{1}{100} = \frac{125}{100} = 1\frac{1}{4}$

$125\% = 125 \times 0.01 = 1.25$

Example 4 $33\frac{1}{3}\% = 33\frac{1}{3} \times \frac{1}{100} = \frac{100}{3} \times \frac{1}{100}$

$= \frac{100}{300} = \frac{1}{3}$

Example 6 $0.25\% = 0.25 \times 0.01 = 0.0025$

Example 8 $0.048 = 0.048 \times 100\% = 4.8\%$

Example 10 $3.67 = 3.67 \times 100\% = 367\%$

Example 12 $0.62\frac{1}{2} = 0.62\frac{1}{2} \times 100\% = 62\frac{1}{2}\%$

Example 14 $\frac{5}{6} = \frac{5}{6} \times 100\% = \frac{500\%}{6} = 83\frac{1}{3}\%$

Example 16 $1\frac{4}{9} = \frac{13}{9} = \frac{13}{9} \times 100\%$

$= \frac{1300\%}{9} \approx 144.4\%$

SECTION 5.2 *pages 193–194*

Example 2 $n = 0.063 \times 150$

$n = 9.45$

Example 4 $n = \frac{1}{6} \times 66$

$n = 11$

Example 6

Strategy To find the sales tax, write and solve a basic percent equation using n to represent the sales tax (amount). The percent is 6% and the base is $450.

Solution $6\% \times \$450 = n$
$.06 \times \$450 = n$
$\$27 = n$

The sales tax was $27.

Example 8

Strategy To find the new hourly wage:

- Find the amount of the raise. Write and solve a basic percent equation using n to represent the amount of the raise (amount). The percent is 8%. The base is $13.50.
- Add the amount of the raise to the old wage.

Solution $8\% \times \$13.50 = n$
$0.08 \times \$13.50 = n$
$\$1.08 = n$

$$\begin{array}{r} \$13.50 \\ +\ 1.08 \\ \hline \$14.58 \end{array}$$

The new hourly wage is $14.58.

SECTION 5.3 *pages 197–198*

Example 2

$$n \times 32 = 16$$
$$n = 16 \div 32$$
$$n = 0.50$$
$$n = 50\%$$

Example 4

$$n \times 15 = 48$$
$$n = 48 \div 15$$
$$n = 3.20$$
$$n = 320\%$$

Example 6

$$30 = n \times 45$$
$$30 \div 45 = n$$
$$0.66\tfrac{2}{3} = n$$
$$66\tfrac{2}{3}\% = n$$

Example 8

Strategy To find what percent of the income the income tax is, write and solve a basic percent equation using n to represent the percent. The base is \$20,000 and the amount is \$3000.

Solution

$$n \times \$20{,}000 = \$3000$$
$$n = \$3000 \div \$20{,}000$$
$$n = 0.15 = 15\%$$

The income tax is 15% of the income.

Example 10

Strategy To find the percent that did not favor the candidate:

- Subtract to find the number of people that did not favor the candidate (1000 − 667).
- Write and solve a basic percent equation using n to represent the percent. The base is 1000 and the amount is the number of people who did not favor the candidate.

Solution $1000 - 667 = 333$ number who did not favor the candidate

$$n \times 1000 = 333$$
$$n = 333 \div 1000$$
$$n = 0.333 = 33.3\%$$

33.3% of the people did not favor the candidate.

SECTION 5.4 *pages 201–202*

Example 2

$$0.86 \times n = 215$$
$$n = 215 \div 0.86$$
$$n = 250$$

Example 4

$$0.025 \times n = 15$$
$$n = 15 \div 0.025$$
$$n = 600$$

Example 6

$$\frac{1}{6} \times n = 5$$
$$n = 5 \div \frac{1}{6}$$
$$n = 30$$

Example 8

Strategy To find the original value of the car, write and solve a basic percent equation using n to represent the original value (base). The percent is 51%. The amount is \$3876.

Example 10

Strategy To find the difference between the original price and the selling price:

- Find the original price. Write and solve a basic percent equation using n to represent the original price (base). The percent is 80%. The amount is \$44.80.
- Subtract the sale price (\$44.80) from the original price.

Solution $51\% \times n = \$3876$

$0.51 \times n = \$3876$

$n = \$3876 \div 51$

$n = \$7600$

The original value of the car was $7600.

Solution $80\% \times n = \$44.80$

$0.80 \times n = \$44.80$

$n = \$44.80 \div 0.80$

$n = \$56.00$ original price

$\$56.00 - \$44.80 = \$11.20$

The difference between the original price and the sale price is $11.20.

SECTION 5.5 *pages 205–206*

Example 2 $\frac{26}{100} = \frac{22}{n}$

$26 \times n = 100 \times 22$

$26 \times n = 2200$

$n = 2200 \div 26$

$n \approx 84.62$

Example 4 $\frac{16}{100} = \frac{n}{132}$

$16 \times 132 = 100 \times n$

$2112 = 100 \times n$

$2112 \div 100 = n$

$21.12 = n$

Example 6

Strategy To find the number of days it snowed, write and solve a proportion using n to represent the number of days (amount). The percent is 64%. The base is 150.

Solution $\frac{64}{100} = \frac{n}{150}$

$64 \times 150 = 100 \times n$

$9600 = 100 \times n$

$9600 \div 100 = n$

$96 = n$

It snowed 96 days.

Example 8

Strategy To find the percent of pens which were not defective:

- Subtract to find the number of pens which were not defective (200 − 5).
- Write and solve a proportion using n to represent the percent of pens which were not defective. The base is 200 and the amount is the number of pens not defective.

Solution $200 - 5 = 195$ number of pens not defective

$\frac{n}{100} = \frac{195}{200}$

$200 \times n = 195 \times 100$

$200 \times n = 19500$

$n = 19500 \div 200$

$n = 97.5\%$

97.5% of the pens were not defective.

SOLUTIONS to Chapter 6 Examples

SECTION 6.1 *pages 219–220*

Example 2 **a.** $6.25 \div 5 = 1.25$
Unit cost: \$1.25 per quart
b. $85 \div 4 = 21.25$
Unit cost: 21.3¢ per ear

Example 4

Strategy To find the more economical purchase, compare unit costs.

Solution $2.52 \div 6 = 0.42$
$1.66 \div 4 = 0.415$
$\$.415 < \$.42$

The more economical purchase is 4 cans for \$1.66.

Example 6

Strategy To find the total cost, multiply the unit cost (\$4.96) by the number of units (7).

Solution $4.96 \times 7 = 34.72$

The total cost is \$34.72.

SECTION 6.2 *pages 223–228*

Example 2

Strategy To find the percent increase:
- Find the amount of increase.
- Solve the basic percent equation for *percent.*

Solution

$$\begin{array}{r} 2250 \\ -2000 \\ \hline 250 \end{array}$$

$$\begin{aligned} n \times 2000 &= 250 \\ n &= 250 \div 2000 \\ n &= 0.125 = 12.5\% \end{aligned}$$

The percent increase was 12.5%.

Example 4

Strategy To find the new hourly wage:
- Solve the basic percent equation for *amount.*
- Add the amount of increase to the original wage.

Solution

$$\begin{aligned} 0.14 \times 6.50 &= n \\ 0.91 &= n \end{aligned}$$

$6.50 + 0.91 = 7.41$

The new hourly wage is \$7.41.

Example 6

Strategy To find the markup, solve the basic percent equation for *amount.*

Solution

$$\begin{aligned} 0.20 \times 8 &= n \\ 1.60 &= n \end{aligned}$$

The markup is \$1.60.

Example 8

Strategy To find the selling price:
- Find the markup by solving the basic percent equation for *amount.*
- Add the markup to the cost.

Solution

$$\begin{aligned} 0.55 \times 72 &= n \\ 39.60 &= n \end{aligned}$$

$72 + 39.60 = 111.60$

The selling price is \$111.60.

Example 10

Strategy To find the percent decrease:
- Find the amount of decrease.
- Solve the basic percent equation for *percent.*

Example 12

Strategy To find the visibility:
- Find the amount of decrease by solving the basic percent equation for *amount.*
- Subtract the amount of decrease from the original visibility.

Solution $150 - 120 = 30$
$n \times 150 = 30$
$n = 30 \div 150$
$n = 0.2 = 20\%$

The percent decrease was 20%.

Example 14

Strategy To find the discount rate, solve the basic percent equation for *percent.*

Solution $n \times 125 = 25$
$n = 0.20 = 20\%$

The discount rate is 20%.

Solution $0.40 \times 5 = n$
$2 = n$
$5 - 2 = 3$

The visibility was 3 miles.

Example 16

Strategy To find the sale price:
- Find the discount by solving the basic percent equation for *amount.*
- Subtract to find the sale price.

Solution $0.20 \times 10.25 = n$
$2.05 = n$
$10.25 - 2.05 = 8.20$

The sale price is \$8.20.

SECTION 6.3 *pages 233–234*

Example 2

Strategy To find the simple interest, multiply the principal by the annual interest rate by the time (in years).

Solution $15{,}000 \times 0.12 \times 1.5 = 2700$

The interest due is \$2700.

Example 4

Strategy To find the interest due, multiply the principal by the monthly interest rate by the time (in months).

Solution $400 \times 0.02 \times 2 = 16$

The interest charge is \$16.

Example 6

Strategy To find the interest earned:
- Find the new principal by multiplying the original principal by the factor found in the compound interest table.
- Subtract the original principal from the new principal.

Solution $1000 \times 7.20957 = 7209.57$
$7209.57 - 1000 = 6209.57$

The interest earned is \$6209.57.

SECTION 6.4 *pages 237–240*

Example 2

Strategy To find the mortgage:
- Find the down payment by solving the basic percent equation for *amount*.
- Subtract the down payment from the purchase price.

Solution $0.25 \times 216{,}000 = n$
$54{,}000 = n$

$216{,}000 - 54{,}000 = 162{,}000$

The mortgage is \$162,000.

Example 4

Strategy To find the loan origination fee, solve the basic percent equation for *amount*.

Solution $0.045 \times 80{,}000 = n$
$3600 = n$

The loan origination fee is \$3600.

Example 6

Strategy To find the monthly mortgage payment:
- Subtract the down payment from the purchase price to find the mortgage.
- Multiply the mortgage by the factor found in the monthly payment table.

Solution $75{,}000 - 15{,}000 = 60{,}000$
$60{,}000 \times 0.0110109 = 660.654$

The monthly mortgage payment is \$660.65.

Example 8

Strategy To find the interest:
- Multiply the mortgage by the factor found in the monthly payment table on page A6 to find the monthly mortgage payment.
- Subtract the principal from the monthly mortgage payment.

Solution $125{,}000 \times 0.0083920 = 1049$
$1049 - 492.65 = 556.35$

The interest on the mortgage is \$556.35.

Example 10

Strategy To find the monthly payment:
- Divide the annual property tax by 12 to find the monthly property tax.
- Add the monthly property tax to the monthly mortgage payment.

Solution $744 \div 12 = 62$
$415.20 + 62 = 477.20$

The total monthly payment is \$477.20.

SECTION 6.5 *pages 243–244*

Example 2

Strategy To find the amount financed:
- Find the down payment by solving the basic percent equation for *amount*.
- Subtract the down payment from the purchase price.

Example 4

Strategy To find the license fee, solve the basic percent equation for *amount*.

Solution $0.20 \times 9200 = n$
$1840 = n$
$9200 - 1840 = 7360$

The amount financed is $7360.

Example 6

Strategy To find the cost of operating the car, multiply the cost per mile by the number of miles driven.

Solution $23{,}000 \times 0.22 = 5060$

The cost of operating the car is $5060.

Solution $0.015 \times 7350 = n$
$110.25 = n$

The license fee is $110.25.

Example 8

Strategy To find the cost per mile for the car insurance, divide the cost for insurance by the number of miles driven.

Solution $360 \div 15{,}000 = 0.024$

The cost per mile for insurance is $0.024.

Example 10

Strategy To find the monthly payment:
- Subtract the down payment from the purchase price to find the amount financed.
- Multiply the amount financed by the factor found in the monthly payment table on page A6.

Solution $15{,}900 - 3975 = 11{,}925$
$11{,}925 \times 0.0258455 = 308.2075$

The monthly payment is $308.21.

SECTION 6.6 *pages 247–248*

Example 2

Strategy To find the worker's earnings:
- Find the worker's overtime wage by multiplying the hourly wage by 2.
- Multiply the number of hours worked by the overtime hours.

Solution $8.50 \times 2 = 17$
$17 \times 8 = 136$

The construction worker earns $136.

Example 6

Strategy To find the total earnings:
- Find the commission by multiplying the commission rate by the sales over $50,000.
- Add the commission to the annual salary.

Solution $175{,}000 - 50{,}000 = 125{,}000$
$125{,}000 \times 0.095 = 11{,}875$
$12{,}000 + 11{,}875 = 23{,}875$

The insurance agent earned $23,875.

Example 4

Strategy To find the salary per month, divide the annual salary by the number of months in a year (12).

Solution $28{,}224 \div 12 = 2352$

The contractor's monthly salary is $2352.

SECTION 6.7 *pages 251–256*

Example 2

Strategy To find the current balance:

- Subtract the amount of the check from the old balance.
- Add the amount of each deposit.

Solution

302.46	
− 20.59	check
281.87	
176.86	first deposit
+ 94.73	second deposit
553.46	

The current checking account balance is $553.46.

Example 4

Solution

Current checkbook balance:	653.41
Check: 237	+ 48.73
	702.14
Interest:	+ 2.11
	704.25
Deposit:	−523.84
	180.41

Closing bank balance from bank statement: $180.41.

Checkbook balance: $180.41.

The bank statement and checkbook balance.

SOLUTIONS to Chapter 7 Examples

SECTION 7.1 *pages 271–274*

Example 2

Strategy To find what percent (N) of the total number of housing starts (30) the number of July housing starts (12) is, use the basic percent equation to solve for N.

Solution *Percent* × *base* = *amount*

$$N \times 30 = 12$$

$$N = \frac{12}{30} = \frac{4}{10}$$

$$N = 40\%$$

The number of July housing starts is 40% of the total number of housing starts.

Example 4

Strategy To find the ratio:

- Locate the annual cost of insurance and the annual cost of maintenance in the circle graph.
- Write the ratio of insurance cost to maintenance cost in simplest form.

Example 6

Strategy To find the Federal income tax:

- Locate the percent of the distribution which is Federal income tax in the circle graph.
- Solve the basic percent for amount.

Solution Annual insurance cost: \$400
Annual maintenance cost: \$500

$\frac{\$400}{\$500} = \frac{4}{5}$

The ratio is $\frac{4}{5}$.

Solution Federal income tax: 22%

Percent × base = amount

$0.22 \times 2000 = \$440$

The Federal income tax is \$440.

SECTION 7.2 *pages 279–280*

Example 2

Strategy To find the difference between the company's first quarter profits for 1991 and 1990:

- Read the bar graph to find the first quarter profit for each year.
- Subtract to find the difference between the profits.

Solution 1991 first quarter profit: \$15,000
1990 first quarter profit: \$13,000

\$15,000 − \$13,000 = \$2000.

The difference is \$2000.

Example 4

Strategy To find the difference between the number of commercial plane landings and the number of private plane landings in 1991.

- Read the line graph to find the number of landings for each type of plane in 1991.
- Subtract to find the difference.

Solution Commercial plane landings: 90,000
Private plane landings: 50,000

90,000 − 50,000 = 40,000

The difference is 40,000.

SECTION 7.3 *pages 283–284*

Example 2

Strategy To find the number of employees:

- Read the histogram to find the number of employees whose hourly wage is between \$8 and \$10 and the number whose hourly wage is between \$10 and \$12.
- Add to find the number of employees whose wage is between \$8 and \$12.

Solution Number whose wage is between
\$8 and \$10: 15
\$10 and \$12: 17

15 + 17 = 32

32 employees earn between \$8 and \$12.

Example 4

Strategy To find the number of people who scored between 50 and 70 on the exam:

- Read the frequency polygon to find the number of people who scored between 50 and 60 and the number who scored between 60 and 70.
- Add to find the number of people who scored between 50 and 70.

Solution The number who scored between
50 and 60: 5
60 and 70: 6

5 + 6 = 11

11 people scored between 50 and 70.

SECTION 7.4 *pages 287–288*

Example 2

Strategy To find the monthly mean cost for gasoline:

- Find the sum of the costs.
- Divide the sum by the number of months (6).

Solution

$$\begin{array}{r} \$118 \\ 130 \\ 109 \\ 141 \\ 134 \\ +136 \\ \hline \$768 \end{array} \qquad \begin{array}{r} \$128 \\ 6\overline{)\$768} \end{array}$$

The monthly mean cost for gasoline is $128.

Example 4

Strategy To find the median number of books loaned, arrange the number of books from smallest to largest. The median is the average of the two middle numbers.

Solution

$$\left.\begin{array}{r} 375 \\ 420 \end{array}\right\} \text{2 numbers}$$
$$\left.\begin{array}{r} 450 \\ 480 \end{array}\right\} \text{middle numbers}$$
$$\left.\begin{array}{r} 490 \\ 500 \end{array}\right\} \text{2 numbers}$$

$$\frac{450 + 480}{2} = 465$$

The median number of books loaned was 465.

SOLUTIONS to Chapter 8 Examples

SECTION 8.1 *pages 301–304*

Example 2 $60 \text{ in.} = 60 \cancel{\text{in.}} \times \frac{1 \text{ ft}}{12 \cancel{\text{in.}}} = 5 \text{ ft}$

Example 4 $14 \text{ ft} = 14 \cancel{\text{ft}} \times \frac{1 \text{ yd}}{3 \cancel{\text{ft}}} = 4\frac{2}{3} \text{ yd}$

Example 6

$$9800 \text{ ft} = 9800 \cancel{\text{ft}} \times \frac{1 \text{ mi}}{5280 \cancel{\text{ft}}}$$
$$= 1\frac{113}{132} \text{ mi}$$

Example 8

$$\begin{array}{r} 3 \text{ ft } 6 \text{ in.} \\ 12\overline{)\ 42} \\ -36 \\ \hline 6 \end{array} \qquad 42 \text{ in.} = 3 \text{ ft } 6 \text{ in.}$$

Example 10

$$\begin{array}{r} 4 \text{ yd } 2 \text{ ft} \\ 3\overline{)\ 14} \\ -12 \\ \hline 2 \end{array}$$

14 ft = 4 yd 2 ft

Example 12

$$\begin{array}{r} 3 \text{ ft} \quad 5 \text{ in.} \\ +4 \text{ ft} \quad 9 \text{ in.} \\ \hline 7 \text{ ft } 14 \text{ in.} \end{array} = 8 \text{ ft } 2 \text{ in.}$$

Example 14

$$\begin{array}{r} 3 \text{ ft } 14 \text{ in.} \\ \cancel{4 \text{ ft}} \ \cancel{2 \text{ in.}} \\ -1 \text{ ft} \quad 8 \text{ in.} \\ \hline 2 \text{ ft} \quad 6 \text{ in.} \end{array}$$

Example 16

$$\begin{array}{r} 4 \text{ yd } 1 \text{ ft} \\ \times \qquad 8 \\ \hline 32 \text{ yd } 8 \text{ ft} \end{array} = 34 \text{ yd } 2 \text{ ft}$$

Example 18

$$\begin{array}{r} 3 \text{ yd} \quad 2 \text{ ft} \\ 2\overline{)\ 7 \text{ yd} \quad 1 \text{ ft}} \\ -6 \text{ yd} \qquad\quad \\ \hline 1 \text{ yd} = 3 \text{ ft} \\ 4 \text{ ft} \\ -4 \text{ ft} \\ \hline 0 \end{array}$$

Example 20

$$\begin{array}{r} 6\frac{1}{4} \text{ ft} = 6\frac{3}{12} \text{ ft} = 5\frac{15}{12} \text{ ft} \\ -3\frac{2}{3} \text{ ft} = 3\frac{8}{12} \text{ ft} = 3\frac{8}{12} \text{ ft} \\ \hline 2\frac{7}{12} \text{ ft} \end{array}$$

Example 22

Strategy To find the width of the storage room:
- Multiply the number of tiles (8) by the width of each tile (9 in.).
- Divide the result by the number of inches in one foot (12) to find the width in feet.

Solution 9 in. × 8 = 72 in.

72 ÷ 12 = 6

The width is 6 ft.

Example 24

Strategy To find the length of each piece, divide the total length (9 ft 8 in.) by the number of pieces (4).

Solution

$$\begin{array}{r} 2\text{ ft}\quad 5\text{ in.} \\ 4\overline{)\;9\text{ ft}\quad 8\text{ in.}} \\ \underline{-8\text{ ft}}\qquad\qquad \\ 1\text{ ft} = \underline{12\text{ in.}} \\ 20\text{ in.} \\ \underline{-20\text{ in.}} \\ 0 \end{array}$$

Each piece is 2 ft 5 in. long.

SECTION 8.2 *pages 307–308*

Example 2 $3\text{ lb} = 3\,\not{\text{lb}} \times \dfrac{16\text{ oz}}{1\,\not{\text{lb}}} = 48\text{ oz}$

Example 4 $4200\text{ lb} = 4200\,\not{\text{lb}} \times \dfrac{1\text{ ton}}{2000\,\not{\text{lb}}} = 2\frac{1}{10}\text{ tons}$

Example 6

$$\begin{array}{r} 6\text{ lb }17\text{ oz} \\ \not{7}\text{ lb }\not{1}\text{ oz} \\ \underline{-3\text{ lb }\;4\text{ oz}} \\ 3\text{ lb }13\text{ oz} \end{array}$$

Example 8

$$\begin{array}{r} 3\text{ lb }6\text{ oz} \\ \underline{\times\qquad 4\quad} \\ 12\text{ lb }24\text{ oz} = 13\text{ lb }8\text{ oz} \end{array}$$

Example 10

Strategy To find the weight of each bar of soap:
- Multiply the number of bars (12) by the weight of each bar (9 oz).
- Convert the number of ounces to pounds.

Solution

$$\begin{array}{r} 12 \\ \underline{\times\ 9\text{ oz}} \\ 108\text{ oz} \end{array}$$

$108\,\not{\text{oz}} \times \dfrac{1\text{ lb}}{16\,\not{\text{oz}}} = 6\frac{3}{4}\text{ lb}$

Each bar weighs $6\frac{3}{4}$ lb.

SECTION 8.3 *pages 311–312*

Example 2

$$18 \text{ pt} = 18 \cancel{\text{pt}} \times \frac{1 \cancel{\text{qt}}}{2 \cancel{\text{pt}}} \times \frac{1 \text{ gal}}{4 \cancel{\text{qt}}}$$

$$= \frac{9 \text{ gal}}{4} = 2\frac{1}{4} \text{gal}$$

Example 4

$$\begin{array}{r@{\;}r r}
 & 1 \text{ gal} & 2 \text{ qt} \\ \hline
3) & 4 \text{ gal} & 2 \text{ qt} \\
 & -3 \text{ gal} & \\ \hline
 & 1 \text{ gal} = & 4 \text{ qt} \\ \hline
 & & 6 \text{ qt} \\
 & & -6 \text{ qt} \\ \hline
 & & 0
\end{array}$$

Example 6

Strategy To find the number of gallons of water needed:

- Find the number of quarts required by multiplying the number of quarts one student needs (1) by the number of students (5) by the number of days (3).
- Convert the number of quarts to gallons.

Solution

$$\begin{array}{r} 5 \\ \times\ 1 \text{ qt} \\ \hline 5 \text{ qt} \end{array} \qquad \begin{array}{r} 5 \text{ qt} \\ \times 3 \\ \hline 15 \text{ qt} \end{array}$$

$$15 \cancel{\text{qt}} \cdot \frac{1 \text{ gal}}{4 \cancel{\text{qt}}} = 3\frac{3}{4} \text{ gal}$$

The students should take $3\frac{3}{4}$ gal of water.

SECTION 8.4 *pages 315–316*

Example 2

$$4.5 \text{ BTU} = 4.5 \cancel{\text{BTU}} \times \frac{778 \text{ ft} \cdot \text{lb}}{1 \cancel{\text{BTU}}}$$

$$= 3501 \text{ ft} \cdot \text{lb}$$

Example 4 $800 \text{ lb} \times 16 \text{ ft} = 12{,}800 \text{ ft} \cdot \text{lb}$

Example 6 $56{,}000 \text{ BTU} =$

$$56{,}000 \cancel{\text{BTU}} \times \frac{778 \text{ ft} \cdot \text{lb}}{1 \cancel{\text{BTU}}} =$$

$43{,}568{,}000 \text{ ft} \cdot \text{lb}$

Example 8 $\text{Power} = \dfrac{90 \text{ ft} \times 1200 \text{ lb}}{24 \text{ s}} = 4500 \dfrac{\text{ft} \cdot \text{lb}}{\text{s}}$

Example 10 $\dfrac{3300}{550} = 6 \text{ hp}$

SOLUTIONS to Chapter 9 Examples

SECTION 9.1 *pages 329–330*

Example 2 3.07 m = 307 cm

Example 4 7 cm = 0.07 m

3 m 7 cm = 3 m + 0.07 m = 3.07 m

Example 6

```
   3  m = 3.00 m
 −42 cm = 0.42 m
 ---------------
          2.58 m
```

Example 8

Strategy To find the cost of the shelves:
- Multiply the length of the bookcase (1 m 75 cm) by the number of shelves (4).
- Multiply the product by the cost per meter.

Solution

```
  1 m 75 cm = 1.75 m
×     4     =    4
 ------------------
              7.00 m
```

```
  $11.75
×      7
 -------
  $82.25
```

The cost is $82.25.

SECTION 9.2 *pages 333–334*

Example 2 42.3 mg = 0.0423 g

Example 4 3 g 54 mg = 3054 mg

Example 6

```
  4 g 620 mg = 4.620 g
×      8     =     8
 --------------------
              36.960 g
```

Example 8

Strategy To find how much fertilizer is required:
- Convert 300 g to kilograms.
- Multiply the number of kilograms by the number of trees (400).

Solution 300 g = 0.3 kg

```
   400
 ×0.3 kg
 --------
 120.0 kg
```

To fertilize the trees, 120 kg of fertilizer are required.

SECTION 9.3 *pages 337–338*

Example 2 2 kl 167 L = 2167 L

Example 4 $325\text{ cm}^3 = 325\text{ ml} = 0.325\text{ L}$

Example 6

$$12\overline{)22.992\text{ kl}} \quad 1.916\text{ kl}$$

Example 8

Strategy To find the profit:
- Find the number of containers by dividing the total amount of lotion (5 L) by the amount per container (125 ml).
- Multiply the number of containers by the price per container ($2.79).
- Subtract the cost ($42.50) from the total income.

Solution $\frac{5\text{ L}}{125\text{ ml}} = \frac{5\text{ L}}{0.125\text{ L}} = 40$

40 × $2.79 = $111.60

$111.60 − $42.50 = $69.10

The profit is $69.10.

SECTION 9.4 *pages 341–342*

Example 2

Strategy To find the number of calories:
- Multiply to find the number of hours worked in 5 days.
- Multiply the product by the number of calories burned off per hour.

Solution $1\frac{1}{2} \times 5 = \frac{3}{2} \times 5 = \frac{15}{2} = 7\frac{1}{2}$

$7\frac{1}{2} \times 240 = 1800\text{ cal}$

1800 cal are used.

Example 4

Strategy To find the number of kilowatthours:
- Multiply to find the number of watthours used.
- Convert to kilowatthours.

Solution 150 W × 200 h = 30,000 Wh

30,000 Wh = 30 kWh

30 kWh of energy are used.

Example 6

Strategy To find the cost:
- Convert 20 min to hours.
- Multiply to find the total number of hours the oven is used.
- Multiply the number of hours used by the number of watts to find the watthours.
- Convert to kilowatthours.
- Multiply the number of kilowatthours by the cost per kilowatthour.

Solution $20 \text{ min} = 20 \cancel{\text{min}} \times \frac{1 \text{ h}}{60 \cancel{\text{min}}} = \frac{20}{60}\text{h} = \frac{1}{3}\text{h}$

$\frac{1}{3}\text{h} \times 30 = 10 \text{ h}$

$10 \text{ h} \times 500 \text{ W} = 5000 \text{ Wh}$

$5000 \text{ Wh} = 5 \text{ kWh}$

$5 \times 8.7¢ = 43.5¢$

The cost is 43.5¢.

SECTION 9.5 *pages 345–346*

Example 2 $10 \text{ c} \approx 10 \cancel{\text{c}} \times \frac{1 \cancel{\text{qt}}}{4 \cancel{\text{c}}} \times \frac{1 \text{ L}}{1.06 \cancel{\text{qt}}} =$

$\frac{10 \text{ L}}{4.24} = 2.36 \text{ L}$

$10 \text{ c} \approx 2.36 \text{ L}$

Example 4 $\frac{60 \text{ ft}}{\text{s}} \approx \frac{60 \cancel{\text{ft}}}{\text{s}} \times \frac{1 \text{ m}}{3.28 \cancel{\text{ft}}} =$

$\frac{60 \text{ m}}{3.28 \text{ s}} = 18.29 \text{ m/s}$

$60 \text{ ft/s} \approx 18.29 \text{ m/s}$

Example 6 $\frac{\$1.86}{\text{gal}} \approx \frac{\$1.86}{\cancel{\text{gal}}} \times \frac{\cancel{\text{gal}}}{4 \cancel{\text{qt}}} \times \frac{1.06 \cancel{\text{qt}}}{1 \text{ L}} =$

$\frac{\$1.9716}{4 \text{ L}} \approx \frac{\$0.49}{\text{L}}$

$\$1.86/\text{gal} \approx \$0.49/\text{L}$

Example 8 $45 \text{ cm} \approx 45 \cancel{\text{cm}} \times \frac{0.39 \text{ in.}}{1 \cancel{\text{cm}}} = 17.55 \text{ in.}$

$45 \text{ cm} \approx 17.55 \text{ in.}$

Example 10 $\frac{75 \text{ km}}{\text{h}} \approx \frac{75 \cancel{\text{km}}}{\text{h}} \times \frac{1 \text{ mi}}{1.61 \cancel{\text{km}}}$

$= 46.58 \text{ mi/h}$

$75 \text{ km/h} \approx 46.58 \text{ mi/h}$

Example 12 $\frac{\$1.50}{\text{L}} \approx \frac{\$1.50}{\cancel{\text{L}}} \times \frac{1 \cancel{\text{L}}}{1.06 \cancel{\text{qt}}} \times \frac{4 \cancel{\text{qt}}}{1 \text{ gal}} =$

$\frac{\$6}{1.06 \text{ gal}} = \$5.66/\text{gal}$

$\$1.50/\text{L} \approx \$5.66/\text{gal}$

SOLUTIONS to Chapter 10 Examples

SECTION 10.1 *pages 359–366*

Example 2
$RS = QT - QR - ST$
$RS = 62 - 24 - 17$
$RS = 62 - 41$
$RS = 21$

Example 4 $90° - 67° = 23°$

Example 6 $180° - 32° = 148°$

Example 8
$\angle a = 118° - 68°$
$\angle a = 50°$

Example 10 $90° - 7° = 83°$
The other angles are 90° and 83°.

Example 12 $62° + 45° = 107°$
$180° - 107° = 73°$
The third angle is 73°.

Example 14 $\text{Radius} = \frac{1}{2} \times \text{diameter}$

$= \frac{1}{2} \times 8 \text{ in.}$

$= 4 \text{ in.}$

Example 16 Angles a and b are supplementary angles.

Angle $b = 180° - 125° = 55°$

Example 18 Angles a and c are corresponding angles. Therefore angle $c = 120°$.

Angles b and c are supplementary angles. Therefore $\angle b = 180° - 120° = 60°$.

SECTION 10.2 *pages 371–374*

Example 2

$$\begin{aligned} \text{Perimeter} &= (2 \cdot \text{length}) + (2 \cdot \text{width}) \\ &= (2 \cdot 2 \text{ m}) + (2 \cdot 0.85 \text{ m}) \\ &= 4 \text{ m} + 1.7 \text{ m} \\ &= 5.7 \text{ m} \end{aligned}$$

Example 4

$$\begin{aligned} \text{Circumference} &= \pi \cdot \text{diameter} \\ &\approx 3.14 \cdot 6 \text{ in.} \\ \text{Circumference} &\approx 18.84 \text{ in.} \end{aligned}$$

Example 6

$$\begin{aligned} \text{Perimeter} &= (2 \cdot \text{length}) + (\pi \cdot \text{diameter}) \\ &\approx (2 \cdot 8 \text{ in.}) + (3.14 \cdot 3 \text{ in.}) \\ &= 16 \text{ in.} + 9.42 \text{ in.} \\ \text{Perimeter} &\approx 25.42 \text{ in.} \end{aligned}$$

Example 8

Strategy To find the cost:

- Find the perimeter of the table.
- Multiply the perimeter by the per-meter cost of the stripping.

Solution

$$\begin{aligned} \text{Perimeter} &= (2 \cdot \text{length}) + (2 \cdot \text{width}) \\ &= (2 \cdot 3 \text{ m}) + (2 \cdot 0.74 \text{ m}) \\ &= 6 \text{ m} + 1.48 \text{ m} \\ &= 7.48 \text{ m} \end{aligned}$$

$\$1.76 \times 7.48 = \$13.1648$

The cost is $13.16.

SECTION 10.3 *pages 379–382*

Example 2

$$\begin{aligned} \text{Area} &= \frac{1}{2} \cdot \text{base} \cdot \text{height} \\ &= \frac{1}{2} \cdot 24 \text{ in.} \cdot 14 \text{ in.} = 168 \text{ in.}^2 \end{aligned}$$

Example 4

$$\begin{aligned} \text{Area} &= \text{area of rectangle} - \text{area of triangle} \\ &= (\text{length} \cdot \text{width}) - \left(\frac{1}{2} \cdot \text{base} \cdot \text{height}\right) \\ &= (10 \text{ in.} \times 6 \text{ in.}) - \left(\frac{1}{2} \cdot 6 \text{ in.} \times 4 \text{ in.}\right) \\ &= 60 \text{ in.}^2 - 12 \text{ in.}^2 \\ &= 48 \text{ in.}^2 \end{aligned}$$

Example 6

Strategy To find the area of the rug:

- Find the area in square feet.
- Convert to square yards.

Solution

$$\begin{aligned} \text{Area} &= \text{length} \cdot \text{width} \\ &= 12 \text{ ft} \cdot 9 \text{ ft} \\ &= 108 \text{ ft}^2 \end{aligned}$$

$108 \cancel{\text{ft}^2} \times \frac{1 \text{ yd}^2}{9 \cancel{\text{ft}^2}} = \frac{108}{9} \text{ yd}^2 = 12 \text{ yd}^2$

The area of the room is 12 yd².

SECTION 10.4 *pages 387–392*

Example 2

$$\text{Volume} = (\text{side})^3 = (5\text{ cm})^3 = 125\text{ cm}^3$$

Example 4

$$\text{Volume} = \pi(\text{radius})^2 \cdot \text{height} \approx \frac{22}{7}(7\text{ in.})^2 \cdot (15\text{ in.})$$

$$\text{Volume} \approx 2310\text{ in}^3.$$

Example 6

$$\text{Volume} = \frac{4}{3}\pi(\text{radius})^3 \approx \frac{4}{3} \cdot 3.14(3\text{ m})^3$$

$$\text{Volume} \approx 113.04\text{ m}^3$$

Example 8

Volume = volume of rectangular solid + volume of cylinder

$$= (\text{length} \cdot \text{width} \cdot \text{height}) + \pi(\text{radius})^2 \cdot \text{height}$$

$$\approx 1.5\text{ m} \cdot 0.4\text{ m} \cdot 0.4\text{ m} + 3.14 \cdot (0.8\text{ m})^2 \cdot 0.2\text{ m}$$

$$= 0.24\text{ m}^3 + 0.40192\text{ m}^3$$

$$\text{Volume} \approx 0.64192\text{ m}^3$$

Example 10

Volume = volume of rectangular solid + $\frac{1}{2}$ of the volume of cylinder

$$= (\text{length} \cdot \text{width} \cdot \text{height}) + \frac{1}{2}\pi(\text{radius})^2 \cdot \text{height}$$

$$\approx (24\text{ in.} \cdot 6\text{ in.} \cdot 4\text{ in.}) + \frac{1}{2} \cdot 3.14 \cdot (3\text{ in.})^2 \cdot 24\text{ in.}$$

$$= 576\text{ in.}^3 + 339.12\text{ in.}^3$$

$$\text{Volume} \approx 915.12\text{ in.}^3.$$

Example 12

Strategy To find the volume of the freezer, use the formula for the volume of a rectangular solid.

Solution

$$\text{Volume} = \text{length} \cdot \text{width} \cdot \text{height} = 7\text{ ft} \cdot 2.5\text{ ft} \cdot 3\text{ ft} = 52.5\text{ ft}^3$$

The volume of the freezer is 52.5 ft^3.

Example 14

Strategy To find the volume of the channel iron, add the volumes of the three rectangular solids.

Solution

$$\text{Volume} = 2(\text{length} \cdot \text{width} \cdot \text{height}) + (\text{length} \cdot \text{width} \cdot \text{height})$$

$$= 2(10\text{ ft} \cdot 0.5\text{ ft} \cdot 0.3\text{ ft}) + (10\text{ ft} \cdot 0.2\text{ ft} \cdot 0.2\text{ ft})$$

$$= 3\text{ ft}^3 + 0.4\text{ ft}^3$$

$$= 3.4\text{ ft}^3$$

The volume of the channel iron is 3.4 ft^3.

SECTION 10.5 *pages 397–400*

Example 2 $\sqrt{16} = 4$ $\sqrt{169} = 13$

Example 4 $\sqrt{32} \approx 5.657$

$\sqrt{162} \approx 12.728$

Example 6

$$\begin{aligned}\text{Hypotenuse} &= \sqrt{(\text{leg})^2 + (\text{leg})^2}\\ &= \sqrt{(8\text{ in.})^2 + (11\text{ in.})^2}\\ &= \sqrt{64\text{ in.}^2 + 121\text{ in.}^2}\\ &= \sqrt{185\text{ in.}^2}\end{aligned}$$

Hypotenuse $\approx$ 13.601 in.

Example 8

$$\begin{aligned}\text{Leg} &= \sqrt{(\text{hypotenuse})^2 - (\text{leg})^2}\\ &= \sqrt{(12\text{ ft})^2 - (5\text{ ft})^2}\\ &= \sqrt{144\text{ ft}^2 - 25\text{ ft}^2}\\ &= \sqrt{119\text{ ft}^2}\end{aligned}$$

Leg $\approx$ 10.909 ft

Example 10

Strategy To find the distance between the holes, use the Pythagorean Theorem. The hypotenuse is the distance between the holes. The length of each leg is given (3 cm and 8 cm).

Solution

$$\begin{aligned}\text{Hypotenuse} &= \sqrt{(\text{leg})^2 + (\text{leg})^2}\\ &= \sqrt{(3\text{ cm})^2 + (8\text{ cm})^2}\\ &= \sqrt{9\text{ cm}^2 + 64\text{ cm}^2}\\ &= \sqrt{73\text{ cm}^2}\end{aligned}$$

Hypotenuse $\approx$ 8.544 cm

The distance is 8.544 cm.

SECTION 10.6 *pages 403–406*

Example 2 $\dfrac{4\text{ \cancel{cm}}}{7\text{ \cancel{cm}}} = \dfrac{4}{7}$

Example 4

$$\frac{AB}{DE} = \frac{AC}{DF}$$

$$\frac{7\text{ \cancel{cm}}}{14\text{ \cancel{cm}}} = \frac{3\text{ cm}}{DF}$$

$7 \times DF = 14 \times 3$ cm

$7 \times DF = 42$ cm

$DF = 42\text{ cm} \div 7$

$DF = 6$ cm

Example 6 $AC = DF$, Angle BAC = Angle EDF, but $AB \neq DE$, therefore the triangles are not congruent.

Example 8

Strategy To find the height FG:

- Solve a proportion to find the height.

Example 10

Strategy To find the perimeter of triangle ABC:

- Solve a proportion to find the lengths of sides BC and AC.
- Use the formula Perimeter = side AB + side BC + side AC.

Solution $\frac{AC}{DF} = \frac{\text{height } CH}{\text{height } FG}$

$$\frac{10\ \cancel{\text{cm}}}{15\ \cancel{\text{cm}}} = \frac{7\text{ cm}}{\text{height}}$$

$10 \times \text{height} = 15 \times 7\text{ cm}$

$10 \times \text{height} = 105\text{ cm}$

$\text{height} = 105\text{ cm} \div 10$

$\text{height} = 10.5\text{ cm}$

The height FG is 10.5 cm.

Solution $\frac{BC}{EF} = \frac{AB}{DE}$

$$\frac{BC}{10\text{ in.}} = \frac{4\text{ in.}}{8\text{ in.}}$$

$BC \times 8 = 10\text{ in.} \times 4$

$BC \times 8 = 40\text{ in.}$

$BC = 40\text{ in.} \div 8$

$BC = 5\text{ in.}$

$$\frac{AC}{DF} = \frac{AB}{DE}$$

$$\frac{AC}{6\text{ in.}} = \frac{4\ \cancel{\text{in.}}}{8\ \cancel{\text{in.}}}$$

$AC \times 8 = 6\text{ in.} \times 4$

$AC \times 8 = 24\text{ in.}$

$AC = 24\text{ in.} \div 8$

$AC = 3\text{ in.}$

$$\text{Perimeter} = 4\text{ in.} + 5\text{ in.} + 3\text{ in.}$$
$$= 12\text{ in.}$$

The perimeter of triangle ABC is 12 in.

SOLUTIONS to Chapter 11 Examples

SECTION 11.1 *pages 419–420*

Example 2 -232

Example 4

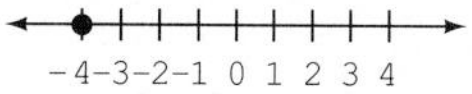

Example 6
a. $-12 < -8$
b. $-5 < 0$

Example 8
$|-7| = 7$
$|21| = 21$

Example 10
$|2| = 2$
$|-9| = 9$

Example 12 $-|-12| = -12$

SECTION 11.2 *pages 423–426*

Example 2
$-154 + (-37)$
-191

Example 4
$-36 + 17 + (-21)$
$-19 + (-21)$
-40

Example 6
$-5 + (-2) + 9 + (-3)$
$-7 + 9 + (-3)$
$2 + (-3)$
-1

Example 8
$-8 - 14$
$-8 + (-14)$
-22

Example 10
$3 - (-4) - 15$
$3 + 4 + (-15)$
$7 + (-15)$
-8

Example 12
$4 - (-3) - 12 - (-7) - 20$
$4 + 3 + (-12) + 7 + (-20)$
$7 + (-12) + 7 + (-20)$
$-5 + 7 + (-20)$
$2 + (-20)$
-18

Example 14 $17 - 10 - 2 - (-6) - 9$
$17 + (-10) + (-2) + 6 + (-9)$
$7 + (-2) + 6 + (-9)$
$5 + 6 + (-9)$
$11 + (-9)$
2

SECTION 11.3 *pages 429–432*

Example 2 $(-3) \cdot 4 \cdot (-5)$
$(-12) \cdot (-5)$
60

Example 4 $-38 \cdot 51$
-1938

Example 6 $-6 \cdot 8 \cdot (-11) \cdot 3$
$-48 \cdot (-11) \cdot 3$
$528 \cdot 3$
1584

Example 8 $-7(-8)(9)(-2)$
$56(9)(-2)$
$504(-2)$
-1008

Example 10 $(-135) \div (-9)$
15

Example 12 $-72 \div 4$
-18

Example 14 $84 \div (-6)$
-14

Example 16

Strategy To find the difference, subtract the average temperature throughout the earth's stratosphere (−70°) from the average temperature on earth's surface (57°).

Solution $57 - (-70)$
$57 + 70$
127

The difference is 127°F.

Example 18

Strategy To find the average daily low temperature:

- Add the seven temperature readings.
- Divide by 7.

Solution $-6 + (-7) + 1 + 0 + (-5) + (-10) + (-1)$
$-13 + 1 + 0 + (-5) + (-10) + (-1)$
$-12 + 0 + (-5) + (-10) + (-1)$
$-12 + (-5) + (-10) + (-1)$
$-17 + (-10) + (-1)$
$-27 + (-1)$
-28

$-28 \div 7 = -4$

The average daily low temperature was −4°.

SECTION 11.4 *pages 437–440*

Example 2 The LCM of 9 and 12 is 36.

$$\frac{5}{9} - \frac{11}{12} = \frac{20}{36} - \frac{33}{36} = \frac{20}{36} + \frac{-33}{36} =$$

$$\frac{20 - 33}{36} = \frac{-13}{36} = -\frac{13}{36}$$

Example 4 The LCM of 8, 6 and 3 is 24.

$$-\frac{7}{8} - \frac{5}{6} + \frac{2}{3} = \frac{-21}{24} - \frac{20}{24} + \frac{16}{24} =$$

$$\frac{-21}{24} + \frac{-20}{24} + \frac{16}{24} = \frac{-21 - 20 + 16}{24} =$$

$$-\frac{25}{24} = -1\frac{1}{24}$$

Example 6

$$\begin{array}{r} 3.907 \\ 4.9 \\ +\ 6.63 \\ \hline 15.437 \end{array}$$

Example 8

$$\begin{array}{r} 67.910 \\ -16.127 \\ \hline 51.783 \end{array}$$

$16.127 - 67.91 = -51.783$

Example 10

$2.7 + (-9.44) + 6.2$
$-6.74 + 6.2$
-0.54

Example 12

The product is positive.

$\left[-\frac{2}{3}\right]\left[-\frac{9}{10}\right] = \frac{2 \cdot 9}{3 \cdot 10} = \frac{18}{30} = \frac{3}{5}$

Example 14

The quotient is negative.

$-\frac{5}{8} \div \frac{5}{40} = -\frac{5}{8} \cdot \frac{40}{5} =$

$-\frac{5 \cdot 40}{8 \cdot 5} = -\frac{200}{40} = -5$

Example 16

$$\begin{array}{r} 5.44 \\ \times\ 3.8 \\ \hline 4352 \\ 1632 \\ \hline 20.672 \end{array}$$

$-5.44 \times 3.8 = -20.672$

Example 18

$3.44 \times (-1.7) \times 0.6$
$(-5.848) \times 0.6$
-3.5088

Example 20

$$\begin{array}{r} 0.231 \\ 1.7.\overline{)0.3.940} \\ -3\,4 \\ \hline 54 \\ -51 \\ \hline 30 \\ -17 \\ \hline 13 \end{array}$$

$-0.394 \div 1.7 \approx -0.23$

SECTION 11.5 *pages 445–446*

Example 2

$9 - 9 \div (-3)$
$9 - (-3)$
$9 + 3$
12

Example 4

$8 \div 4 \cdot 4 - (-2)^2$
$8 \div 4 \cdot 4 - 4$
$2 \cdot 4 - 4$
$8 - 4$
$8 + (-4)$
4

Example 6

$8 - (-15) \div (2 - 7)$
$8 - (-15) \div (-5)$
$8 - 3$
$8 + (-3)$
5

Example 8

$(-2)^2 \times (3 - 7)^2 - (-16) \div (-4)$
$(-2)^2 \times (-4)^2 - (-16) \div (-4)$
$4 \times 16 - (-16) \div (-4)$
$64 - 4$
$64 + (-4)$
60

Example 10

$7 \div \left(\frac{1}{7} - \frac{3}{14}\right) - 9$

$7 \div \left(-\frac{1}{14}\right) - 9$

$-98 - 9$
$-98 + (-9)$
-107

SOLUTIONS to Chapter 12 Examples

SECTION 12.1 *pages 459–464*

Example 2

$6a - 5b$
$6(-3) - 5(4)$
$-18 - 20$
$-18 + (-20)$
-38

Example 4

$-3s^2 - 12 \div t$
$-3(-2)^2 - 12 \div 4$
$-3(4) - 12 \div 4$
$-12 - 12 \div 4$
$-12 - 3$
$-12 + (-3)$
-15

Example 6

$-\frac{2}{3}m + \frac{3}{4}n^3$
$-\frac{2}{3}(6) + \frac{3}{4}(2)^3$
$-\frac{2}{3}(6) + \frac{3}{4}(8)$
$-4 + 6$
2

Example 8

$-3yz - z^2 + y^2$
$-3\left(-\frac{2}{3}\right)\left(\frac{1}{3}\right) - \left(\frac{1}{3}\right)^2 + \left(-\frac{2}{3}\right)^2$
$-3\left(-\frac{2}{3}\right)\left(\frac{1}{3}\right) - \frac{1}{9} + \frac{4}{9}$
$\frac{2}{3} - \frac{1}{9} + \frac{4}{9}$
1

Example 10

$5a^2 - 6b^2 + 7a^2 - 9b^2$
$5a^2 + (-6)b^2 + 7a^2 + (-9)b^2$
$5a^2 + 7a^2 + (-6)b^2 + (-9)b^2$
$12a^2 + (-15)b^2$
$12a^2 - 15b^2$

Example 12

$-6x + 7 + 9x - 10$
$(-6)x + 7 + 9x + (-10)$
$(-6)x + 9x + 7 + (-10)$
$3x + (-3)$
$3x - 3$

Example 14

$\frac{3}{8}w + \frac{1}{2} - \frac{1}{4}w - \frac{2}{3}$
$\frac{3}{8}w - \frac{1}{4}w + \frac{1}{2} - \frac{2}{3}$
$\frac{3}{8}w - \frac{2}{8}w + \frac{3}{6} - \frac{4}{6}$
$\frac{1}{8}w - \frac{1}{6}$

Example 16

$5(a - 2)$
$5a - (2)$
$5a - 10$

Example 18

$8s - 2(3s - 5)$
$8s + (-2)(3s) - (-2)(5)$
$8s + (-6s) + 10$
$2s + 10$

Example 20

$4(x - 3) - 2(x + 1)$
$4x - 4(3) - 2x - 2(1)$
$4x - 12 - 2x - 2$
$4x - 2x - 12 - 2$
$2x - 14$

SECTION 12.2 *pages 469–474*

Example 2

$$x(x+3) = 4x+6$$

$(-2)(-2+3)$	$4(-2)+6$
$(-2)(1)$	$(-8)+6$

$$-2 = -2$$

Yes, -2 is a solution.

Example 4

$$x^2 - x = 3x+7$$

$(-3)^2-(-3)$	$3(-3)+7$
$9+3$	$-9+7$

$$12 \neq -2$$

No, -3 is not a solution.

Example 6

$$\begin{aligned} -2+y &= -5 \\ -2+2+y &= -5+2 \\ 0+y &= -3 \\ y &= -3 \end{aligned}$$

The solution is -3.

Example 8

$$\begin{aligned} 7 &= y+8 \\ 7-8 &= y+8-8 \\ -1 &= y+0 \\ -1 &= y \end{aligned}$$

The solution is -1.

Example 10

$$\begin{aligned} \frac{1}{5} &= z+\frac{4}{5} \\ \frac{1}{5}-\frac{4}{5} &= z+\frac{4}{5}-\frac{4}{5} \\ -\frac{3}{5} &= z+0 \\ -\frac{3}{5} &= z \end{aligned}$$

The solution is $-\frac{3}{5}$.

Example 12

$$\begin{aligned} 4z &= -20 \\ \frac{4z}{4} &= \frac{-20}{4} \\ 1z &= -5 \\ z &= -5 \end{aligned}$$

The solution is -5.

Example 14

$$\begin{aligned} 8 &= \frac{2}{5}n \\ \left(\frac{5}{2}\right)(8) &= \left(\frac{5}{2}\right)\frac{2}{5}n \\ 20 &= 1n \\ 20 &= n \end{aligned}$$

The solution is 20.

Example 16

$$\begin{aligned} \frac{2}{3}t-\frac{1}{3}t &= -2 \\ \frac{1}{3}t &= -2 \\ \left(\frac{3}{1}\right)\frac{1}{3}t &= \left(\frac{3}{1}\right)(-2) \\ 1t &= -6 \\ t &= -6 \end{aligned}$$

The solution is -6.

Example 18

Strategy To find the discount, replace the variables S and R in the formula by the given values and solve for D.

Solution

$$\begin{aligned} S &= R-D \\ 22 &= 30-D \\ 22-30 &= 30-30-D \\ -8 &= -D \\ -8(-1) &= -D(-1) \\ 8 &= D \end{aligned}$$

The discount is \$8.

Example 20

Strategy To find the interest, replace the variables P and A in the formula by the given values and solve for I.

Solution

$$\begin{aligned} A &= P+I\cdot P \\ 1120 &= 1000+1000I \\ 1120-1000 &= 1000-1000+1000I \\ 120 &= 1000I \\ 120\left(\frac{1}{1000}\right) &= \left(\frac{1}{1000}\right)1000I \\ 0.12 &= I \end{aligned}$$

The interest rate is 12%.

SECTION 12.3 *pages 479–480*

Example 2

$$5x + 8 = 6$$
$$5x + 8 - 8 = 6 - 8$$
$$5x = -2$$
$$\frac{5x}{5} = \frac{-2}{5}$$
$$x = -\frac{2}{5}$$

The solution is $-\frac{2}{5}$.

Example 4

$$7 - x = 3$$
$$7 - 7 - x = 3 - 7$$
$$-x = -4$$
$$(-1)(-x) = (-1)(-4)$$
$$x = 4$$

The solution is 4.

Example 6

Strategy To find the Celsius temperature, replace the variable F in the formula by the given value and solve for C.

Solution

$$F = \frac{9}{5}C + 32$$
$$-22 = \frac{9}{5}C + 32$$
$$-22 - 32 = \frac{9}{5}C + 32 - 32$$
$$-54 = \frac{9}{5}C$$
$$\left(\frac{5}{9}\right)(-54) = \left(\frac{5}{9}\right)\frac{9}{5}C$$
$$-30 = C$$

The Celsius temperature is $-30°$.

Example 8

Strategy To find the cost per unit, replace the variables T, N, and F in the formula by the given values and solve for U.

Solution

$$T = U \cdot N + F$$
$$4500 = 250U + 1500$$
$$4500 - 1500 = 250U + 1500 - 1500$$
$$3000 = 250U$$
$$\frac{3000}{250} = \frac{250U}{250}$$
$$12 = U$$

The cost per unit is \$12.

SECTION 12.4 *pages 485–486*

Example 2

$$\frac{1}{5}x - 2 = \frac{2}{5}x + 4$$
$$-\frac{2}{5}x + \frac{1}{5}x - 2 = -\frac{2}{5}x + \frac{2}{5}x + 4$$
$$-\frac{1}{5}x - 2 = 4$$
$$-\frac{1}{5}x - 2 + 2 = 4 + 2$$
$$-\frac{1}{5}x = 6$$
$$(-5)\left(-\frac{1}{5}x\right) = (-5)6$$
$$x = -30$$

The solution is -30.

Example 4

$$4(x - 1) - x = 5$$
$$4x - 4 - x = 5$$
$$3x - 4 = 5$$
$$3x - 4 + 4 = 5 + 4$$
$$3x = 9$$
$$\frac{3x}{3} = \frac{9}{3}$$
$$x = 3$$

The solution is 3.

SECTION 12.5 *pages 491–492*

Example 2 $t - 2t$

Example 4 $\frac{5}{7x}$

Example 6 Twelve decreased by some number
The unknown number: x
$12 - x$

Example 8 The product of a number and one-half of the number
The unknown number: n
One-half the number: $\frac{n}{2}$
$(n)\left(\frac{n}{2}\right)$

SECTION 12.6 *pages 495–498*

Example 2 The unknown number: x

$$\begin{aligned} x + 4 &= 12 \\ x + 4 - 4 &= 12 - 4 \\ x &= 8 \end{aligned}$$

The number is 8.

Example 4 The unknown number: x

$$\begin{aligned} 2x &= 10 \\ \frac{2x}{2} &= \frac{10}{2} \\ x &= 5 \end{aligned}$$

The number is 5.

Example 6 The unknown number: x

$$\begin{aligned} 3x + 6 &= 4 \\ 3x + 6 - 6 &= 4 - 6 \\ 3x &= -2 \\ \frac{3x}{3} &= \frac{-2}{3} \\ x &= -\frac{2}{3} \end{aligned}$$

The number is $-\frac{2}{3}$.

Example 8 The unknown number: x

$$\begin{aligned} 4x + 3 &= 10 \\ 4x + 3 - 3 &= 10 - 3 \\ 4x &= 7 \\ \frac{4x}{4} &= \frac{7}{4} \\ x &= \frac{7}{4} \end{aligned}$$

The number is $1\frac{3}{4}$.

Example 10

Strategy To find the regular price, write and solve an equation using R to replace the regular price of the slacks.

Solution

$$\begin{aligned} 18.95 &= R - 6 \\ 18.95 + 6 &= R - 6 + 6 \\ 24.95 &= R \end{aligned}$$

The regular price of the slacks is \$24.95.

Example 12

Strategy To find the rpm, write and solve an equation using R to represent the rpm of the engine when in third gear.

Solution

$$\begin{aligned} 2500 &= \frac{2}{3}R \\ \frac{3}{2}(2500) &= \left(\frac{3}{2}\right)\frac{2}{3}R \\ 3750 &= R \end{aligned}$$

The rpm of the engine when in third gear is 3750.

Example 14

Strategy To find the total sales, write and solve an equation using S to represent the total sales.

Solution

$$2500 = 800 + 0.08S$$
$$2500 - 800 = 800 - 800 + 0.08S$$
$$1700 = 0.08S$$
$$\frac{1700}{0.08} = \frac{0.08S}{0.08}$$
$$21{,}250 = S$$

The total sales are $21,250.

Example 16

Strategy To find the number of hours, write and solve an equation using H to represent the number of hours of labor required.

Solution

$$300 = 100 + 12.50H$$
$$300 - 100 = 100 - 100 + 12.50H$$
$$200 = 12.50H$$
$$\frac{200}{12.50} = \frac{12.50H}{12.50}$$
$$16 = H$$

The number of hours of labor required is 16.

ANSWERS to Chapter 1 Odd-Numbered Exercises

SECTION 1.1 *pages 7–8*

1. 0 1 2 3 4 5 6 7 8 9 10 11 12 **3.** 0 1 2 3 4 5 6 7 8 9 10 11 12 **5.** $37 < 49$ **7.** $101 > 87$

9. $245 > 158$ **11.** $0 < 45$ **13.** $815 < 928$ **15.** $7003 < 7020$ **17.** Eight hundred five **19.** Four hundred eighty-five **21.** Two thousand six hundred seventy-five **23.** Forty-two thousand nine hundred twenty-eight **25.** Three hundred fifty-six thousand nine hundred forty-three **27.** Three million six hundred ninety-seven thousand four hundred eighty-three **29.** 85 **31.** 3456 **33.** 609,948 **35.** 4,003,002 **37.** 5000 + 200 + 80 + 7 **39.** 50,000 + 8000 + 900 + 40 + 3 **41.** 200,000 + 10,000 + 7000 + 500 + 80 + 6 **43.** 50,000 + 900 + 40 + 3 **45.** 200,000 + 500 + 80 + 3 **47.** 400,000 + 3000 + 700 + 5 **49.** 930 **51.** 1400 **53.** 7000 **55.** 44,000 **57.** 650,000

SECTION 1.2 *pages 13–18*

1. 11 **3.** 8 **5.** 19 **7.** 49 **9.** 28 **11.** 125 **13.** 693 **15.** 1099 **17.** 729 **19.** 1489 **21.** 14,399 **23.** 15,865 **25.** 156,499 **27.** 14,395 **29.** 1,786,688 **31.** 1,394,799 **33.** 18 **35.** 16,776 **37.** 149,999 **39.** 168,574 **41.** 129,843 **43.** 1584 **45.** 1219 **47.** 11,066 **49.** 5143 **51.** 102,317 **53.** 79,326 **55.** 1804 **57.** 1579 **59.** 19,740 **61.** 7420 **63.** 120,570 **65.** 207,453 **67.** 309,172 **69.** 343,263 **71.** 31 **73.** 534 **75.** 634 **77.** 14,302 **79.** 24,218 **81.** 11,974 **83.** 9323 **85.** 77,139 **87.** 10,417 **89.** 8217 **91.** Est.: 17,700 Cal.: 17,754 **93.** Est.: 2900 Cal.: 2872 **95.** Est.: 101,000 Cal.: 101,712 **97.** Est.: 158,000 Cal.: 158,763 **99.** Est.: 260,000 Cal.: 261,595 **101.** Est.: 940,000 Cal.: 946,718 **103.** Est.: 33,000,000 Cal.: 32,691,621 **105.** Est.: 34,000,000 Cal.: 34,420,922 **107.** The total attendance was 5792 people. **109.** The total number of yards gained by passing was 307. **111.** There were 8181 computers manufactured during the 3 months. **113a.** The total paid attendance at the sixth and seventh games was 131,348 people. **113b.** The total paid attendance for the entire series was 421,006 people. **115a.** The total amount deposited was $1659. **115b.** The new checking account balance is $3007. **117.** The total costs for the month were $5232.

SECTION 1.3 *pages 23–26*

1. 4 **3.** 4 **5.** 10 **7.** 7 **9.** 11 **11.** 9 **13.** 22 **15.** 60 **17.** 66 **19.** 33 **21.** 501 **23.** 962 **25.** 5002 **27.** 1513 **29.** 9 **31.** 7 **33.** 31 **35.** 47 **37.** 925 **39.** 4561 **41.** 3205 **43.** 1222 **45.** 1311 **47.** 3401 **49.** 1501 **51.** 65 **53.** 17 **55.** 8 **57.** 37 **59.** 353 **61.** 57 **63.** 160 **65.** 155 **67.** 24 **69.** 547 **71.** 45 **73.** 754 **75.** 378 **77.** 430 **79.** 3139 **81.** 3621 **83.** 738 **85.** 3545 **87.** 5258 **89.** 4887 **91.** 749 **93.** 5343 **95.** 66,463 **97.** 16,590 **99.** 52,404 **101.** 38,777 **103.** 4367 **105.** 4889 **107.** 2639 **109.** 1215 **111.** 5048 **113.** 628 **115.** 6532 **117.** 4286 **119.** 4042 **121.** 5209 **123.** 10,378 **125.** Est.: 60,000 Cal.: 60,427 **127.** Est.: 20,000 Cal.: 21,613 **129.** Est.: 360,000 Cal.: 358,979 **131.** Est.: 100,000 Cal.: 127,757 **133.** Est.: 500,000 Cal.: 456,767 **135.** Est.: 200,000 Cal.: 157,864 **137.** Est.: 1,000,000 Cal.: 1,082,767 **139.** Est.: 6,000,000 Cal.: 5,577,842 **141.** Est.: 2,000,000 Cal.: 1,625,369 **143.** There is $165 left in your checking account. **145.** The amount that remains to be paid is $899. **147.** The odometer reading at the start of the trip was 1645 miles. **149a.** The total amount deducted was $276. **149b.** The steelworker's take home pay is $596. **151.** There is $582 left in the budget. **153.** The team's profit for the season was $1,084,140.

SECTION 1.4 *pages 31–34*

1. 12 **3.** 35 **5.** 25 **7.** 0 **9.** 72 **11.** 21 **13.** 30 **15.** 54 **17.** 328 **19.** 220 **21.** 623 **23.** 711 **25.** 405 **27.** 280 **29.** 1143 **31.** 4010 **33.** 1500 **35.** 7248 **37.** 635 **39.** 2236 **41.** 5460 **43.** 1860 **45.** 1101 **47.** 1685 **49.** 46,963 **51.** 59,976 **53.** 19,120 **55.** 19,790 **57.** 144,759 **59.** 560,399 **61.** 651,000 **63.** 559,542 **65.** 336 **67.** 910 **69.** 640 **71.** 800 **73.** 1794 **75.** 1541 **77.** 63,063 **79.** 33,520 **81.** 380,834 **83.** 541,164 **85.** 400,995 **87.** 105,315 **89.** 66,000 **91.** 115,623 **93.** 428,770 **95.** 260,000 **97.** 344,463 **99.** 41,808 **101.** 189,500 **103.** 401,880 **105.** 1,052,763 **107.** 4,198,388 **109.** 1,232,000 **111.** 1,302,725 **113.** Est.: 540,000 Cal.: 550,935 **115.** Est.: 1,200,000 Cal.: 1,138,134 **117.** Est.: 600,000 Cal.: 605,022 **119.** Est.: 3,000,000 Cal.: 3,174,425 **121.** Est.: 18,000,000 Cal.: 19,212,228 **123.** Est.: 72,000,000 Cal.: 73,427,310 **125.** The plane used 4325 gallons of jet fuel. **127.** The machine can fill and cap 168,000 bottles of soft drink. **129.** The total cost of the car is $11,232. **131a.** The company can manufacture 192 video cameras. **131b.** There will be 4539 video cameras in stock. **133.** The total income was $824. **135.** There are 307,200 pixels on the screen. **137.** The oil company was purchased for $8,664,480,000.

SECTION 1.5 *pages 43–46*

1. 2 **3.** 6 **5.** 5 **7.** 16 **9.** 80 **11.** 210 **13.** 44 **15.** 206 **17.** 530 **19.** 902 **21.** 390 **23.** 1056 **25.** 2607 **27.** 9800 **29.** 21,560 **31.** 3580 **33.** 2 r1 **35.** 5 r2 **37.** 13 r1 **39.** 10 r3 **41.** 90 r2 **43.** 230 r1 **45.** 204 r3 **47.** 1347 r3 **49.** 778 r2 **51.** 391 r4 **53.** 2417 r1 **55.** 461 r8 **57.** 1160 r4 **59.** 708 r2 **61.** 3825 **63.** 9044 r2 **65.** 3 r15 **67.** 2 r3 **69.** 21 r36 **71.** 34 r2 **73.** 8 r8 **75.** 4 r49 **77.** 200 r25 **79.** 203 r2 **81.** 185 r2 **83.** 67 r70 **85.** 601 r8 **87.** 200 r9 **89.** 35 r47 **91.** 271 **93.** 4484 r6 **95.** 608 **97.** 15 r7 **99.** 1 r563 **101.** 50 r92 **103.** 40 r7 **105.** Est.: 5000 Cal.: 5129 **107.** Est.: 20,000 Cal.: 21,968 **109.** Est.: 22,500 Cal.: 24,596 **111.** Est.: 3000 Cal.: 2836 **113.** Est.: 3000 Cal.: 3024 **115.** Est.: 30,000 Cal.: 32,036 **117.** Est.: 80 Cal.: 83 **119.** Est.: 75 Cal.: 68 **121.** Each organization received $137,000. **123.** Each disc can store 368,640 bytes of information. **125.** The consultant charged $32 per hour. **127a.** The remaining balance to be paid is $8976. **127b.** The monthly payment is $187. **129.** Each charity received $63,553. **131.** There are 4216 employees participating in the plan.

SECTION 1.6 *pages 49–50*

1. 2^3 **3.** $6^3 \cdot 7^4$ **5.** $6^3 \cdot 11^5$ **7.** $3 \cdot 10^4$ **9.** $2^2 \cdot 3^3 \cdot 5^4$ **11.** $2 \cdot 3^2 \cdot 7^2 \cdot 11$ **13.** $2^2 \cdot 7^4 \cdot 11^4$ **15.** 8 **17.** 400 **19.** 900 **21.** 972 **23.** 120 **25.** 360 **27.** 0 **29.** 90,000 **31.** 540 **33.** 4050 **35.** 11,025 **37.** 25,920 **39.** 4,320,000 **41.** 92,160 **43.** 5 **45.** 10 **47.** 47 **49.** 8 **51.** 5 **53.** 8 **55.** 6 **57.** 53 **59.** 44 **61.** 19 **63.** 37 **65.** 168 **67.** 27 **69.** 14 **71.** 10 **73.** 9 **75.** 8 **77.** 7 **79.** 18

SECTION 1.7 *pages 53–54*

1. 1, 2, 4 **3.** 1, 2, 5, 10 **5.** 1, 5 **7.** 1, 2, 3, 4, 6, 12 **9.** 1, 2, 4, 8 **11.** 1, 13 **13.** 1, 2, 3, 6, 9, 18 **15.** 1, 5, 25 **17.** 1, 2, 3, 4, 6, 9, 12, 18, 36 **19.** 1, 2, 4, 7, 14, 28 **21.** 1, 29 **23.** 1, 2, 11, 22 **25.** 1, 2, 4, 11, 22, 44 **27.** 1, 7, 49 **29.** 1, 37 **31.** 1, 3, 19, 57 **33.** 1, 2, 3, 4, 6, 8, 12, 16, 24, 48 **35.** 1, 3, 29, 87 **37.** 1, 2, 23, 46 **39.** 1, 2, 5, 10, 25, 50 **41.** 1, 2, 3, 6, 11, 22, 33, 66 **43.** 1, 2, 4, 5, 8, 10, 16, 20, 40, 80 **45.** 1, 2, 3, 4, 6, 7, 12, 14, 21, 28, 42, 84 **47.** 1, 5, 17, 85 **49.** 1, 101 **51.** 1, 2, 3, 4, 5, 6, 8, 10, 12, 15, 20, 24, 30, 40, 60, 120 **53.** 1, 5, 25, 125 **55.** 1, 2, 3, 4, 6, 11, 12, 22, 33, 44, 66, 132 **57.** 1, 3, 5, 9, 15, 27, 45, 135 **59.** 1, 3, 7, 9, 21, 27, 63, 189 **61.** $2 \cdot 3$ **63.** prime **65.** $2 \cdot 2 \cdot 2 \cdot 2$ **67.** $2 \cdot 2 \cdot 3$ **69.** $3 \cdot 3$ **71.** $2 \cdot 2 \cdot 3 \cdot 3$ **73.** prime **75.** $2 \cdot 5 \cdot 5$ **77.** $5 \cdot 13$ **79.** $2 \cdot 2 \cdot 2 \cdot 2 \cdot 5$ **81.** $2 \cdot 3 \cdot 3$ **83.** $2 \cdot 2 \cdot 7$ **85.** prime **87.** $2 \cdot 3 \cdot 7$ **89.** $3 \cdot 3 \cdot 3 \cdot 3$ **91.** $2 \cdot 11$ **93.** prime **95.** $2 \cdot 5 \cdot 7$ **97.** $2 \cdot 43$ **99.** $5 \cdot 19$ **101.** prime **103.** $5 \cdot 11$ **105.** $2 \cdot 61$ **107.** $2 \cdot 59$ **109.** $2 \cdot 3 \cdot 5 \cdot 5$ **111.** $2 \cdot 2 \cdot 2 \cdot 3 \cdot 5$ **113.** $2 \cdot 2 \cdot 2 \cdot 2 \cdot 2 \cdot 5$ **115.** $7 \cdot 7 \cdot 7$ **117.** $2 \cdot 2 \cdot 2 \cdot 2 \cdot 5 \cdot 5$ **119.** $3 \cdot 3 \cdot 5 \cdot 5$

CHAPTER REVIEW *pages 57–58*

1. $101 > 87$ **2.** Two hundred seventy-six thousand fifty-seven **3.** 2,011,044 **4.** 10,000 + 300 + 20 + 7 **5.** 1081 **6.** 12,493 **7.** The total income from commissions was $2567. **8.** The total amount deposited was $301. The new checking account balance is $817. **9.** 1749 **10.** 5409 **11.** The odometer reading was 51,059 miles. **12.** There is $396 left in your budget. **13.** 22,761 **14.** 619,833 **15.** The total of the car payments is $1476. **16.** The assistant's total pay for last week was $384. **17.** 2135 **18.** 1306 r59 **19.** The car drove 27 miles per gallon of gasoline. **20.** The monthly car payment is $155. **21.** $2^4 \cdot 5^3$ **22.** $5^2 \cdot 7^5$ **23.** 600 **24.** 2 **25.** 17 **26.** 8 **27.** 1, 2, 3, 6, 9, 18 **28.** 1, 2, 3, 5, 6, 10, 15, 30 **29.** $2 \cdot 3 \cdot 7$ **30.** $2 \cdot 2 \cdot 2 \cdot 3 \cdot 3$

CHAPTER TEST *pages 59–60*

1. $21 > 19$ (1.1A) **2.** Two hundred seven thousand sixty-eight (1.1B) **3.** 1,204,006 (1.1B) **4.** 900,000 + 6000 + 300 + 70 + 8 (1.1C) **5.** 75,000 (1.1D) **6.** 96,798 (1.2A) **7.** 135,915 (1.2B) **8a.** They drove 855 miles. **8b.** The odometer reading was 48,481 miles. (1.2C) **9.** 9333 (1.3A)

10. 10,882 (1.3B) **11.** The remaining balance is \$13,955. (1.3C) **12.** 726,104 (1.4A) **13.** 6,854,144 (1.4B) **14.** The investor will receive \$2844 over 12 months. (1.4C) **15.** 703 (1.5A) **16.** 8710 r2 (1.5B) **17.** 1121 r27 (1.5C) **18.** The farmer needed 3000 boxes. (1.5D) **19.** $3^3 \cdot 7^2$ (1.6A) **20.** 432 (1.6A) **21.** 9 (1.6B) **22.** 7 (1.6B) **23.** 1, 2, 4, 5, 10, 20 (1.7A) **24.** $2 \cdot 2 \cdot 3 \cdot 7$ (1.7B) **25.** $2 \cdot 3 \cdot 17$ (1.7B)

ANSWERS to Chapter 2 Odd-Numbered Exercises

SECTION 2.1 *pages 65–66*

1. 40 **3.** 24 **5.** 30 **7.** 12 **9.** 24 **11.** 60 **13.** 56 **15.** 9 **17.** 32 **19.** 36 **21.** 660 **23.** 9384 **25.** 24 **27.** 30 **29.** 24 **31.** 576 **33.** 1680 **35.** 72 **37.** 90 **39.** 300 **41.** 120 **43.** 1 **45.** 3 **47.** 5 **49.** 25 **51.** 1 **53.** 4 **55.** 4 **57.** 2 **59.** 48 **61.** 2 **63.** 8 **65.** 2 **67.** 1 **69.** 2 **71.** 3 **73.** 5 **75.** 8 **77.** 1 **79.** 25 **81.** 7 **83.** 8

SECTION 2.2 *pages 69–70*

1. $\frac{3}{4}$ **3.** $\frac{4}{7}$ **5.** $1\frac{1}{2}$ **7.** $2\frac{5}{8}$ **9.** $3\frac{3}{5}$ **11.** $\frac{5}{4}$ **13.** $\frac{8}{3}$ **15.** $\frac{28}{8}$ **17.** $2\frac{3}{4}$ **19.** 5 **21.** $1\frac{1}{8}$ **23.** $2\frac{3}{10}$ **25.** 3 **27.** $1\frac{1}{7}$ **29.** $2\frac{1}{3}$ **31.** 16 **33.** $2\frac{1}{8}$ **35.** $2\frac{2}{5}$ **37.** 1 **39.** 9 **41.** $\frac{7}{3}$ **43.** $\frac{13}{2}$ **45.** $\frac{41}{6}$ **47.** $\frac{37}{4}$ **49.** $\frac{21}{2}$ **51.** $\frac{73}{9}$ **53.** $\frac{58}{11}$ **55.** $\frac{21}{8}$ **57.** $\frac{13}{8}$ **59.** $\frac{100}{9}$ **61.** $\frac{27}{8}$ **63.** $\frac{85}{13}$

SECTION 2.3 *pages 73–74*

1. $\frac{5}{10}$ **3.** $\frac{9}{48}$ **5.** $\frac{12}{32}$ **7.** $\frac{9}{51}$ **9.** $\frac{12}{16}$ **11.** $\frac{27}{9}$ **13.** $\frac{20}{60}$ **15.** $\frac{44}{60}$ **17.** $\frac{12}{18}$ **19.** $\frac{35}{49}$ **21.** $\frac{10}{18}$ **23.** $\frac{21}{3}$ **25.** $\frac{35}{45}$ **27.** $\frac{60}{64}$ **29.** $\frac{21}{98}$ **31.** $\frac{30}{48}$ **33.** $\frac{15}{42}$ **35.** $\frac{102}{144}$ **37.** $\frac{153}{408}$ **39.** $\frac{340}{800}$ **41.** $\frac{30}{72}$ **43.** $\frac{119}{210}$ **45.** $\frac{1}{3}$ **47.** $\frac{1}{2}$ **49.** $\frac{1}{6}$ **51.** $\frac{2}{3}$ **53.** $1\frac{1}{2}$ **55.** 0 **57.** $\frac{9}{22}$ **59.** 3 **61.** $\frac{4}{21}$ **63.** $\frac{5}{11}$ **65.** $\frac{2}{9}$ **67.** $\frac{3}{4}$ **69.** $1\frac{1}{3}$ **71.** $\frac{3}{5}$ **73.** $\frac{1}{11}$ **75.** 4 **77.** $\frac{1}{3}$ **79.** $\frac{3}{5}$ **81.** $2\frac{1}{4}$ **83.** $\frac{1}{5}$ **85.** $\frac{19}{51}$ **87.** $\frac{27}{56}$

SECTION 2.4 *pages 79–82*

1. $\frac{3}{7}$ **3.** 1 **5.** $1\frac{4}{11}$ **7.** $3\frac{2}{5}$ **9.** $2\frac{4}{5}$ **11.** $2\frac{1}{4}$ **13.** $1\frac{3}{8}$ **15.** $1\frac{7}{15}$ **17.** $\frac{15}{16}$ **19.** $1\frac{4}{11}$ **21.** $1\frac{1}{6}$ **23.** $\frac{13}{14}$ **25.** $1\frac{1}{45}$ **27.** $\frac{17}{18}$ **29.** $\frac{35}{48}$ **31.** $\frac{23}{60}$ **33.** $\frac{56}{57}$ **35.** $\frac{13}{21}$ **37.** $\frac{33}{98}$ **39.** $\frac{37}{56}$ **41.** $1\frac{17}{18}$ **43.** $1\frac{11}{48}$ **45.** $1\frac{9}{20}$ **47.** $1\frac{109}{180}$ **49.** $1\frac{91}{144}$ **51.** $2\frac{5}{72}$ **53.** $1\frac{20}{33}$ **55.** $2\frac{11}{24}$ **57.** $5\frac{7}{10}$ **59.** $5\frac{11}{16}$ **61.** $8\frac{23}{60}$ **63.** $18\frac{8}{9}$ **65.** $11\frac{1}{36}$ **67.** $24\frac{11}{12}$ **69.** $6\frac{33}{40}$ **71.** $11\frac{19}{24}$ **73.** $12\frac{17}{22}$ **75.** $15\frac{29}{160}$ **77.** $20\frac{37}{48}$ **79.** $44\frac{17}{84}$ **81.** $11\frac{7}{18}$ **83.** $8\frac{1}{12}$ **85.** $12\frac{71}{105}$ **87.** $15\frac{1}{4}$ **89.** $12\frac{7}{24}$ **91.** $29\frac{227}{240}$ **93.** $14\frac{19}{24}$ **95.** The fractional amount of their income spent is $\frac{17}{24}$. **97.** The price per share of the stock at the end of six months is \$$40\frac{3}{8}$. **99.** The length of the shaft is $1\frac{5}{16}$ inches. **101a.** The total hours worked is 20 hours. **101b.** The total wages for the five days is \$140. **103.** The total snowfall for the year is $30\frac{7}{8}$ in.

SECTION 2.5 *pages 87–90*

1. $\frac{2}{17}$ **3.** $\frac{1}{3}$ **5.** $\frac{1}{10}$ **7.** $\frac{7}{11}$ **9.** $\frac{1}{4}$ **11.** $\frac{2}{7}$ **13.** $\frac{1}{2}$ **15.** $\frac{19}{56}$ **17.** $\frac{1}{2}$ **19.** $\frac{4}{45}$ **21.** $\frac{11}{18}$

23. $\frac{5}{48}$ **25.** $\frac{1}{12}$ **27.** $\frac{20}{57}$ **29.** $\frac{17}{36}$ **31.** $\frac{37}{51}$ **33.** $\frac{17}{70}$ **35.** $\frac{49}{120}$ **37.** $\frac{25}{126}$ **39.** $\frac{11}{48}$ **41.** $\frac{23}{72}$

43. $\frac{81}{232}$ **45.** $3\frac{1}{6}$ **47.** $56\frac{4}{23}$ **49.** $4\frac{1}{3}$ **51.** $3\frac{2}{3}$ **53.** $1\frac{3}{5}$ **55.** $8\frac{2}{3}$ **57.** $7\frac{43}{45}$ **59.** $1\frac{2}{5}$ **61.** $14\frac{4}{5}$

63. $20\frac{17}{24}$ **65.** $3\frac{1}{8}$ **67.** $25\frac{8}{15}$ **69.** $6\frac{32}{45}$ **71.** $2\frac{1}{160}$ **73.** $1\frac{169}{200}$ **75.** $48\frac{31}{70}$ **77.** $85\frac{2}{9}$ **79.** $9\frac{5}{13}$

81. $10\frac{169}{180}$ **83.** $32\frac{43}{63}$ **85.** $68\frac{1}{60}$ **87.** $6\frac{83}{100}$ **89.** The remaining piece of moulding is $5\frac{1}{3}$ feet long.

91. There are $2\frac{3}{4}$ hours remaining before landing. **93.** The bridge shrank $\frac{1}{8}$ foot. **95.** The missing dimension is $8\frac{19}{24}$ feet. **97a.** The hikers will travel $17\frac{17}{24}$ miles the first two days. **97b.** There are $9\frac{19}{24}$ miles left to hike. **99.** The patient must lose $9\frac{1}{8}$ pounds.

SECTION 2.6 *pages 95–98*

1. $\frac{7}{12}$ **3.** $\frac{7}{48}$ **5.** $\frac{5}{12}$ **7.** $\frac{1}{3}$ **9.** $\frac{11}{20}$ **11.** $\frac{1}{7}$ **13.** $\frac{1}{8}$ **15.** $\frac{5}{12}$ **17.** $\frac{9}{50}$ **19.** $\frac{5}{32}$ **21.** 6

23. $\frac{2}{3}$ **25.** $\frac{15}{44}$ **27.** $\frac{2}{45}$ **29.** $\frac{5}{14}$ **31.** 10 **33.** $\frac{1}{15}$ **35.** $\frac{10}{51}$ **37.** $\frac{45}{328}$ **39.** $\frac{10}{27}$ **41.** 4

43. $\frac{100}{357}$ **45.** $1\frac{1}{2}$ **47.** 4 **49.** $\frac{4}{9}$ **51.** $\frac{1}{2}$ **53.** $16\frac{1}{2}$ **55.** 10 **57.** $6\frac{3}{7}$ **59.** $18\frac{1}{3}$ **61.** $1\frac{5}{7}$

63. $3\frac{1}{2}$ **65.** $25\frac{5}{8}$ **67.** $1\frac{11}{16}$ **69.** 16 **71.** 3 **73.** $8\frac{1}{4}$ **75.** $27\frac{1}{2}$ **77.** 7 **79.** $24\frac{1}{2}$ **81.** 22

83. 33 **85.** $8\frac{37}{40}$ **87.** $6\frac{19}{28}$ **89.** $75\frac{5}{8}$ **91.** $8\frac{1}{4}$ **93.** $3\frac{1}{3}$ **95.** 8 **97.** $8\frac{2}{3}$ **99.** $19\frac{1}{2}$ **101.** $3\frac{1}{2}$

103. 2 **105.** $7\frac{2}{5}$ **107.** 16 **109.** $32\frac{4}{5}$ **111.** $28\frac{1}{2}$ **113.** 18 **115.** $8\frac{8}{15}$ **117.** $34\frac{2}{5}$ **119.** 519

121. 270 **123.** $80\frac{1}{3}$ **125.** The car can travel $11\frac{3}{4}$ miles. **127.** The car can travel 361 miles.

129. The cook should use 7 cups of flour. **131.** The cost of 40 shares is $2735. **133a.** The amount budgeted for housing and utilities is $1280. **133b.** The amount remaining for other purposes is $1920. **135.** The total cost of 24 choir robes is $504.

SECTION 2.7 *pages 103–106*

1. $\frac{5}{6}$ **3.** $1\frac{5}{88}$ **5.** 0 **7.** $1\frac{1}{3}$ **9.** $8\frac{4}{5}$ **11.** $2\frac{73}{256}$ **13.** $1\frac{1}{9}$ **15.** $2\frac{1}{13}$ **17.** $\frac{2}{3}$ **19.** $\frac{9}{10}$ **21.** 2

23. 2 **25.** $\frac{1}{6}$ **27.** 6 **29.** $\frac{1}{15}$ **31.** 2 **33.** $2\frac{1}{2}$ **35.** 3 **37.** 6 **39.** $\frac{1}{2}$ **41.** 6 **43.** $20\frac{2}{3}$

45. $\frac{1}{30}$ **47.** $1\frac{4}{5}$ **49.** $8\frac{8}{9}$ **51.** $\frac{1}{6}$ **53.** $1\frac{1}{11}$ **55.** $\frac{16}{21}$ **57.** $\frac{7}{11}$ **59.** $2\frac{13}{16}$ **61.** $\frac{1}{9}$ **63.** 8

65. $\frac{11}{28}$ **67.** $\frac{5}{86}$ **69.** 72 **71.** $\frac{3}{22}$ **73.** $\frac{7}{44}$ **75.** $\frac{33}{40}$ **77.** $4\frac{4}{9}$ **79.** $\frac{35}{47}$ **81.** $53\frac{1}{2}$ **83.** $10\frac{2}{7}$

85. $9\frac{39}{40}$ **87.** $\frac{12}{53}$ **89.** $2\frac{6}{43}$ **91.** 0 **93.** $1\frac{31}{38}$ **95.** $1\frac{4}{15}$ **97.** $\frac{1}{144}$ **99.** $\frac{10}{11}$ **101.** 39

103. $2\frac{8}{39}$ **105.** $13\frac{25}{49}$ **107.** $4\frac{28}{61}$ **109.** $4\frac{62}{191}$ **111.** 68 **113.** 0 **115.** 6 **117.** $3\frac{13}{49}$

119. 4 **121.** $1\frac{61}{62}$ **123.** $\frac{1}{2}$ **125.** 800 boxes can be filled with cereal. **127.** The sales executive can travel 25 miles on 1 gallon of gas. **129.** The cost of the diamond would be $1920. **131.** You can buy 60 shares of stock. **133a.** There are $8\frac{1}{4}$ acres available for building. **133b.** There will be 33 parcels of land available for sale. **135.** 33 packages can be made from this shipment.

SECTION 2.8 *pages 109–110*

1. $\frac{11}{40} < \frac{19}{40}$ **3.** $\frac{2}{3} < \frac{5}{7}$ **5.** $\frac{5}{8} > \frac{7}{12}$ **7.** $\frac{7}{9} < \frac{11}{12}$ **9.** $\frac{13}{14} > \frac{19}{21}$ **11.** $\frac{7}{24} < \frac{11}{30}$ **13.** $\frac{9}{64}$ **15.** $\frac{8}{729}$ **17.** $\frac{2}{9}$ **19.** $\frac{3}{125}$ **21.** $\frac{8}{245}$ **23.** $\frac{1}{121}$ **25.** $\frac{16}{1225}$ **27.** $\frac{4}{49}$ **29.** $\frac{9}{125}$ **31.** $\frac{27}{88}$ **33.** $\frac{4}{121}$ **35.** $\frac{1}{30}$ **37.** $1\frac{1}{2}$ **39.** $\frac{20}{21}$ **41.** $\frac{12}{125}$ **43.** $\frac{11}{32}$ **45.** $\frac{17}{24}$ **47.** $\frac{14}{15}$ **49.** $2\frac{7}{10}$ **51.** $\frac{21}{44}$ **53.** $\frac{25}{39}$ **55.** $\frac{5}{24}$ **57.** $\frac{8}{9}$

CHAPTER REVIEW *pages 113–114*

1. 36 **2.** 54 **3.** 4 **4.** 5 **5.** $\frac{13}{4}$ **6.** $1\frac{7}{8}$ **7.** $3\frac{2}{5}$ **8.** $\frac{19}{7}$ **9.** 24 **10.** 32 **11.** $\frac{2}{3}$ **12.** $\frac{4}{11}$ **13.** $1\frac{1}{8}$ **14.** $1\frac{13}{18}$ **15.** $5\frac{7}{8}$ **16.** $18\frac{13}{54}$ **17.** The total rainfall is $21\frac{7}{24}$ inches. **18.** $\frac{1}{3}$ **19.** $\frac{25}{48}$ **20.** $14\frac{19}{42}$ **21.** $10\frac{1}{8}$ **22.** The second checkpoint is $4\frac{3}{4}$ miles from the finish line. **23.** $\frac{1}{15}$ **24.** $\frac{1}{8}$ **25.** $9\frac{1}{24}$ **26.** $16\frac{1}{2}$ **27.** The car can travel 243 miles. **28.** 2 **29.** $\frac{3}{4}$ **30.** 2 **31.** $3\frac{1}{3}$ **32.** The cost of each acre was $36,000. **33.** $\frac{11}{18} < \frac{17}{24}$ **34.** $\frac{5}{16}$ **35.** $\frac{5}{36}$ **36.** $\frac{1}{15}$

CHAPTER TEST *pages 115–116*

1. 120 (2.1A) **2.** 8 (2.1B) **3.** $\frac{11}{4}$ (2.2A) **4.** $3\frac{3}{5}$ (2.2B) **5.** $\frac{49}{5}$ (2.2B) **6.** $\frac{45}{72}$ (2.3A) **7.** $\frac{5}{8}$ (2.3B) **8.** $1\frac{11}{12}$ (2.4A) **9.** $1\frac{61}{90}$ (2.4B) **10.** $22\frac{4}{15}$ (2.4C) **11.** The total rainfall for the 3 months was $21\frac{11}{24}$ inches. (2.4D) **12.** $\frac{1}{4}$ (2.5A) **13.** $\frac{7}{48}$ (2.5B) **14.** $13\frac{81}{88}$ (2.5C) **15.** The value of one share is $\$27\frac{7}{8}$. (2.5D) **16.** $\frac{4}{9}$ (2.6A) **17.** 8 (2.6B) **18.** The electrician's total earnings are $420. (2.6C) **19.** $1\frac{3}{7}$ (2.7A) **20.** $2\frac{2}{19}$ (2.7B) **21.** There were 11 lots available for sale. (2.7C) **22.** $\frac{3}{8} < \frac{5}{12}$ (2.8A) **23.** $\frac{1}{6}$ (2.8B) **24.** $\frac{7}{24}$ (2.8C) **25.** $\frac{5}{6}$ (2.8C)

CUMULATIVE REVIEW *pages 117–118*

1. 290,000 (1.1D) **2.** 291,278 (1.3B) **3.** 73,154 (1.4B) **4.** 540 r12 (1.5C) **5.** 1 (1.6B) **6.** $2 \cdot 2 \cdot 11$ (1.7B) **7.** 210 (2.1A) **8.** 20 (2.1B) **9.** $\frac{23}{3}$ (2.2B) **10.** $6\frac{1}{4}$ (2.2B) **11.** $\frac{15}{48}$ (2.3A) **12.** $\frac{2}{5}$ (2.3B) **13.** $1\frac{7}{48}$ (2.4B) **14.** $14\frac{11}{48}$ (2.4C) **15.** $\frac{13}{24}$ (2.5B) **16.** $1\frac{7}{9}$ (2.5C) **17.** $\frac{7}{20}$ (2.6A) **18.** $7\frac{1}{2}$ (2.6B) **19.** $1\frac{1}{20}$ (2.7A) **20.** $2\frac{5}{8}$ (2.7B) **21.** $\frac{1}{9}$ (2.8B) **22.** $5\frac{5}{24}$ (2.8C) **23.** There is $862 in the checking account at the end of the week. (1.3C) **24.** The total income is $705. (1.4C) **25.** The total weight is $12\frac{1}{24}$ pounds. (2.4D) **26.** The length of the remaining piece is $4\frac{17}{24}$ feet. (2.5D) **27.** The car can travel 225 miles. (2.6C) **28.** 25 parcels can be sold from the remaining land. (2.7C)

ANSWERS to Chapter 3 Odd-Numbered Exercises

SECTION 3.1 *pages 123–124*

1. Twenty-seven hundredths **3.** One and five thousandths **5.** Thirty-six and four tenths **7.** Thirty-five hundred-thousandths **9.** Thirty-two and one hundred seven thousandths **11.** Twelve and one thousand two hundred ninety-four ten-thousandths **13.** Ten and seven thousandths **15.** Fifty-two and ninety-five hundred-thousandths **17.** Two hundred ninety-three ten-thousandths **19.** Six and three hundred twenty-four thousandths **21.** Two hundred seventy-six and three thousand two hundred ninety-seven ten-thousandths **23.** Two hundred sixteen and seven hundred twenty-nine ten-thousandths

25. Four thousand six hundred twenty-five and three hundred seventy-nine ten-thousandths **27.** One and one hundred-thousandth **29.** 0.762 **31.** 0.000062 **33.** 8.0304 **35.** 304.07 **37.** 362.048 **39.** 3048.2002 **41.** 28.04375 **43.** 1308.0045 **45.** 7.4 **47.** 23.0 **49.** 22.68 **51.** 480.33 **53.** 1.039 **55.** 1946.375 **57.** 0.0299 **59.** 2 **61.** 0.010290

SECTION 3.2 *pages 127–128*

1. 150.1065 **3.** 95.8446 **5.** 69.644 **7.** 92.883 **9.** 113.205 **11.** 317.69398 **13.** 0.37 **15.** 3.107 **17.** 40.9820 **19.** 764.667 **21.** Est.: 608 Cal.: 608.245 **23.** Est.: 954 Cal.: 953.473 **25.** Est.: 1.5 Cal.: 1.47 **27.** Est.: 2.7 Cal.: 2.5906 **29.** The sales executive's monthly income is $2709.50. **31.** The length of the shaft is 5.65 inches. **33.** The total rainfall was 30.41 inches. **35a.** The car was driven 202.3 miles. **35b.** The odometer reading was 2019.7 miles. **37.** The total income was $310.77. **39.** The average number of points scored per game was 108 points.

SECTION 3.3 *pages 131–132*

1. 5.627 **3.** 113.6427 **5.** 6.7098 **7.** 215.697 **9.** 53.8776 **11.** 72.7091 **13.** 4.685 **15.** 10.0365 **17.** 0.014 **19.** 4.088 **21.** 0.7727 **23.** 3.273 **25.** 791.247 **27.** 547.951 **29.** 403.8557 **31.** 22.479 **33.** Est.: 600 Cal.: 590.25 **35.** Est.: 30 Cal.: 35.194 **37.** Est.: 3 Cal.: 2.74506 **39.** Est.: 7 Cal.: 7.14925 **41.** The amount of sales was $469.61. **43.** The assistant's raise was $0.49 per hour. **45.** The missing dimension is 2.59 feet. **47a.** The total amount of the checks written was $607.36. **47b.** The new checking account balance is $422.38. **49.** Rainfall was 0.86 inches below normal. **51.** The after-tax income was $3,699,644.14.

SECTION 3.4 *pages 137–140*

1. 0.36 **3.** 0.30 **5.** 0.25 **7.** 0.45 **9.** 6.93 **11.** 1.84 **13.** 4.32 **15.** 0.74 **17.** 39.5 **19.** 2.72 **21.** 0.603 **23.** 0.096 **25.** 13.50 **27.** 79.80 **29.** 4.316 **31.** 1.794 **33.** 0.06 **35.** 0.072 **37.** 0.1323 **39.** 0.03568 **41.** 0.0784 **43.** 0.076 **45.** 34.48 **47.** 580.5 **49.** 20.148 **51.** 0.04255 **53.** 0.17686 **55.** 0.19803 **57.** 14.8657 **59.** 0.0006608 **61.** 53.9961 **63.** 0.536335 **65.** 0.429 **67.** 2.116 **69.** 0.476 **71.** 1.022 **73.** 37.96 **75.** 2.318 **77.** 3.2 **79.** 6.5 **81.** 6285.6 **83.** 3200 **85.** 35,700 **87.** 6.3 **89.** 3.9 **91.** 49,000 **93.** 6.7 **95.** Est.: 90 Cal.: 91.2 **97.** Est.: 0.8 Cal.: 1.0472 **99.** Est.: 4.5 Cal.: 3.897 **101.** Est.: 12 Cal.: 11.2406 **103.** Est.: 0.32 Cal.: 0.371096 **105.** Est.: 30 Cal.: 31.8528 **107.** Est.: 2000 Cal.: 1941.0694 **109.** Est.: 0.00005 Cal.: 0.0000435 **111.** The cost to rent the car is $20.52. **113.** The amount received from recycling is $14.06. **115.** The broker's fee is $173.25. **117a.** The amount of the payments is $4590. **117b.** The total cost of the car is $6590. **119.** The total cost to rent the car is $64.20. **121.** The rocket will travel 639,360 miles in one day.

SECTION 3.5 *pages 145–148*

1. 0.82 **3.** 4.8 **5.** 89 **7.** 60 **9.** 0.06 **11.** 69.7 **13.** 84.3 **15.** 32.3 **17.** 5.06 **19.** 1.3 **21.** 0.11 **23.** 3.8 **25.** 10.5 **27.** 6.1 **29.** 680 **31.** 290 **33.** 3.6 **35.** 4.7 **37.** 1.3 **39.** 6.5 **41.** 0.8 **43.** 0.30 **45.** 0.08 **47.** 22.70 **49.** 0.55 **51.** 0.07 **53.** 0.249 **55.** 0.970 **57.** 0.101 **59.** 0.007 **61.** 4 **63.** 6 **65.** 4 **67.** 41 **69.** 116 **71.** 0.0039 **73.** 3.897 **75.** 2.37835 **77.** 82.547 **79.** 0.835 **81.** 8.765 **83.** 0.02954 **85.** 0.28932 **87.** 0.000139 **89.** Est.: 10 Cal.: 9.6944 **91.** Est.: 750 Cal.: 773.4940 **93.** Est.: 2 Cal.: 2.0069 **95.** Est.: 100 Cal.: 103.5556 **97.** Est.: 100 Cal.: 103.3774 **99.** Est.: 80 Cal.: 67.2037 **101.** There were 6.23 yards gained per carry. **103.** There were 102 gallons used. **105.** There can be 3 complete shelves cut from the board. **107a.** The amount to be paid in monthly payments is $842.58. **107b.** The amount of each monthly payment is $46.81. **109.** The store's total income from electric shavers is $7126.20. **111.** Each person would have to pay $3333.

SECTION 3.6 *pages 151–152*

1. 0.625 **3.** 0.667 **5.** 0.167 **7.** 0.417 **9.** 1.750 **11.** 1.500 **13.** 4.000 **15.** 0.003 **17.** 7.080 **19.** 37.500 **21.** 0.375 **23.** 0.208 **25.** 3.333 **27.** 5.444 **29.** 0.313 **31.** 2.214 **33.** $\frac{4}{5}$ **35.** $\frac{8}{25}$ **37.** $\frac{1}{8}$ **39.** $1\frac{1}{4}$ **41.** $16\frac{9}{10}$ **43.** $8\frac{2}{5}$ **45.** $8\frac{437}{1000}$ **47.** $2\frac{1}{4}$ **49.** $\frac{23}{150}$ **51.** $\frac{703}{800}$

53. $1\frac{17}{25}$ **55.** $\frac{9}{200}$ **57.** $16\frac{18}{25}$ **59.** $\frac{33}{100}$ **61.** $\frac{1}{3}$ **63.** $\frac{48}{175}$ **65.** $0.15 < 0.5$ **67.** $6.65 > 6.56$
69. $2.504 > 2.054$ **71.** $\frac{3}{8} > 0.365$ **73.** $\frac{2}{3} > 0.65$ **75.** $\frac{5}{9} > 0.55$ **77.** $0.62 > \frac{7}{15}$ **79.** $0.161 > \frac{1}{7}$
81. $0.86 > 0.855$ **83.** $1.005 > 0.5$ **85.** $\frac{5}{24} > 0.202$ **87.** $0.429 < \frac{4}{9}$

CHAPTER REVIEW *pages 155–156*

1. Twenty-two and ninety-two ten thousandths **2.** Three hundred forty-two and thirty-seven hundredths **3.** 22.0092 **4.** 3.06753 **5.** 7.94 **6.** 0.05678 **7.** 36.714 **8.** 834.474 **9.** $478.02 **10.** 22.8635 **11.** 4.8785 **12.** The new checking account balance is $661.51. **13.** 8.932 **14.** 25.7446 **15.** The amount of income tax paid was $5600. **16.** 54.5 **17.** 6.594 **18.** The amount of each monthly payment is $123.45. **19.** 0.778 **20.** 2.33 **21.** $\frac{3}{8}$ **22.** $\frac{2}{3}$ **23.** $0.055 < 0.1$ **24.** $\frac{5}{8} > 0.62$

CHAPTER TEST *pages 157–158*

1. Forty-five and three hundred two ten-thousandths (3.1A) **2.** 209.07086 (3.1A) **3.** 7.095 (3.1B) **4.** 0.0740 (3.1B) **5.** 255.957 (3.2A) **6.** 458.581 (3.2A) **7.** The total income was $1543.57. (3.2B) **8.** 4.087 (3.3A) **9.** 27.76626 (3.3A) **10.** 1.37 inches (3.3B) **11.** 0.00548 (3.4A) **12.** 2.4723 (3.4A) **13.** The cost of the long distance call is $4.63. (3.4B) **14.** 232 (3.5A) **15.** 1.538 (3.5A) **16.** The amount of each monthly payment is $142.85. (3.5B) **17.** 0.692 (3.6A) **18.** $\frac{33}{40}$ (3.6B) **19.** $0.66 < 0.666$ (3.6C) **20.** $\frac{3}{8} > 0.35$ (3.6C)

CUMULATIVE REVIEW *pages 159–160*

1. 235 r17 (1.5C) **2.** 128 (1.6A) **3.** 3 (1.6B) **4.** 72 (2.1A) **5.** $4\frac{2}{5}$ (2.2B) **6.** $\frac{37}{8}$ (2.2B) **7.** $\frac{25}{60}$ (2.3A) **8.** $1\frac{17}{48}$ (2.4B) **9.** $8\frac{35}{36}$ (2.4C) **10.** $5\frac{23}{36}$ (2.5C) **11.** $\frac{1}{12}$ (2.6A) **12.** $9\frac{1}{8}$ (2.6B) **13.** $1\frac{2}{9}$ (2.7A) **14.** $\frac{19}{20}$ (2.7B) **15.** $\frac{3}{16}$ (2.8B) **16.** $2\frac{5}{18}$ (2.8C) **17.** Sixty-five and three hundred nine ten-thousandths. (3.1A) **18.** 504.6991 (3.2A) **19.** 21.0764 (3.3A) **20.** 55.26066 (3.4A) **21.** 2.154 (3.5A) **22.** 0.733 (3.6A) **23.** $\frac{1}{6}$ (3.6B) **24.** $\frac{8}{9} < 0.98$ (3.6C) **25.** There were 234 passengers on the continuing flight. (1.3C) **26.** The value of each share was $\$32\frac{7}{8}$. (2.5D) **27.** The checking account balance was $617.38. (3.3B) **28.** The resulting thickness of the bushing was 1.395 inches. (3.3B) **29.** The amount of income tax paid last year was $6008.80 (3.4B) **30.** The monthly payment is $23.87. (3.5B)

ANSWERS to Chapter 4 Odd-Numbered Exercises

SECTION 4.1 *pages 165–166*

1. $\frac{1}{5}$, 1:5, 1 TO 5 **3.** $\frac{2}{1}$, 2:1, 2 TO 1 **5.** $\frac{3}{8}$, 3:8, 3 TO 8 **7.** $\frac{37}{24}$, 37:24, 37 TO 24 **9.** $\frac{1}{1}$, 1:1, 1 TO 1 **11.** $\frac{7}{10}$, 7:10, 7 TO 10 **13.** $\frac{1}{2}$, 1:2, 1 TO 2 **15.** $\frac{2}{1}$, 2:1, 2 TO 1 **17.** $\frac{3}{4}$, 3:4, 3 TO 4 **19.** $\frac{5}{3}$, 5:3, 5 TO 3 **21.** $\frac{2}{3}$, 2:3, 2 TO 3 **23.** $\frac{2}{1}$, 2:1, 2 TO 1 **25.** $\frac{1}{1}$, 1:1, 1 TO 1 **27.** $\frac{1}{2}$, 1:2, 1 TO 2 **29.** $\frac{3}{2}$, 3:2, 3 TO 2 **31.** The ratio of housing cost to total income is $\frac{1}{3}$. **33.** The ratio of utilities cost to food cost is $\frac{3}{8}$. **35.** The ratio of people attending on Friday to people attending on Saturday was $\frac{16}{13}$. **37.** The ratio of turns in the primary coil to turns in the secondary coil is $\frac{1}{12}$. **39a.** The amount of increase is $20,000. **39b.** The ratio of the increase to the original value is $\frac{2}{9}$. **41.** The ratio of increase in price to original price is $\frac{2}{7}$.

SECTION 4.2 *pages 169–170*

1. $\frac{3 \text{ pounds}}{4 \text{ people}}$ **3.** $\frac{\$20}{3 \text{ boards}}$ **5.** $\frac{20 \text{ miles}}{1 \text{ gallon}}$ **7.** $\frac{5 \text{ children}}{2 \text{ families}}$ **9.** $\frac{8 \text{ gallons}}{1 \text{ hour}}$ **11.** 2.5 feet/second **13.** $325/week **15.** 110 trees/acre **17.** $4.71/hour **19.** 52.4 miles/hour **21.** 28 miles/gallon **23.** $1.65/pound **25.** $595.00/ton **27.** The student drives 28.4 miles per gallon of gas. **29.** The rocket uses 213,600 gallons of fuel per minute. **31.** The dividend is $2.90 per share. **33a.** There were 4878 compact discs meeting company standards. **33b.** The cost was $5.44 per disc. **35.** The profit was $0.60 per box of strawberries. **37.** Light travels 186,000 miles per second.

SECTION 4.3 *pages 175–178*

1. True **3.** Not true **5.** Not true **7.** True **9.** True **11.** Not true **13.** True **15.** True **17.** True **19.** Not true **21.** True **23.** Not true **25.** Not true **27.** True **29.** $n = 3$ **31.** $n = 6$ **33.** $n = 20$ **35.** $n = 2$ **37.** $n = 8$ **39.** $n = 4$ **41.** $n \approx 4.38$ **43.** $n = 9.6$ **45.** $n \approx 10.67$ **47.** $n \approx 6.67$ **49.** $n = 26.25$ **51.** $n = 96$ **53.** $n \approx 9.78$ **55.** $n \approx 3.43$ **57.** $n \approx 1.34$ **59.** $n = 50.4$ **61.** $n \approx 112.24$ **63.** $n \approx 56.95$ **65.** There are 168 calories in the serving of cereal. **67.** There are 50 pounds of fertilizer used. **69.** The monthly payment for the loan is $162.60. **71.** The property tax is $2500. **73.** It would take 2496 bricks. **75.** The distance between the two cities is 114 miles. **77.** 1.5 ounces of medication are required. **79.** There would be 160,000 people voting in the election. **81.** There would be 960 braking defects. **83.** You own 400 shares of stock. **85.** Students used the computer 300 hours. **87.** 1750 pounds of salt can be recovered.

CHAPTER REVIEW *pages 181–182*

1. $\frac{2}{7}$, 2:7, 2 TO 7 **2.** $\frac{2}{5}$, 2:5, 2 TO 5 **3.** $\frac{1}{1}$, 1:1, 1 TO 1 **4.** $\frac{2}{5}$, 2:5, 2 TO 5 **5.** The ratio of the high temperature to the low temperature is 2:1. **6.** The ratio of increase to original value is $\frac{1}{2}$. **7.** The ratio of decrease in price to original price is $\frac{2}{5}$. **8.** The ratio of T.V. advertising to newspaper advertising is $\frac{5}{2}$. **9.** $\frac{100 \text{ miles}}{3 \text{ hours}}$ **10.** $\frac{\$15}{4 \text{ hours}}$ **11.** 62.5 miles/hour **12.** $7.50/hour **13.** 27.2 miles/gallon **14.** $1.75/pound **15.** The cost is $44.75 per share. **16.** You drove an average of 56.8 miles/hour. **17.** The cost is $.68/pound. **18.** The cost per radio is $37.50. **19.** True **20.** True **21.** True **22.** Not true **23.** $n = 36$ **24.** $n = 68$ **25.** $n = 65.45$ **26.** $n = 19.44$ **27.** There are 22.5 pounds of fertilizer used. **28.** The cost of the insurance is $193.50. **29.** The property tax is $2400. **30.** It would take 1344 blocks to build the wall.

CHAPTER TEST *pages 183–184*

1. $\frac{3}{2}$, 3:2, 3 TO 2 (4.1A) **2.** $\frac{3}{5}$, 3:5, 3 TO 5 (4.1A) **3.** $\frac{1}{3}$, 1:3, 1 TO 3 (4.1A) **4.** $\frac{1}{6}$, 1:6, 1 TO 6 (4.1A) **5.** The ratio of the city temperature to the desert temperature is $\frac{43}{56}$. (4.1B) **6.** The ratio of the cost of radio advertising to the total cost of advertising is $\frac{8}{13}$. (4.1B) **7.** $\frac{9 \text{ supports}}{4 \text{ feet}}$ (4.2A) **8.** $\frac{\$27}{4 \text{ boards}}$ (4.2A) **9.** $1836.40/month (4.2B) **10.** 30.5 miles/gallon (4.2B) **11.** The cost of the lumber is $1.73/foot. (4.2C) **12.** The plane's speed is 538 miles/hour. (4.2C) **13.** True (4.3A) **14.** Not true (4.3A) **15.** $n = 40.5$ (4.3B) **16.** $n = 144$ (4.3B) **17.** The student's body contains 132 pounds of water. (4.3C) **18.** The person requires 0.875 ounce of medication. (4.3C) **19.** The property tax is $2800. (4.3C) **20.** The dividend would be $625. (4.3C)

CUMULATIVE REVIEW *pages 185–186*

1. 9158 (1.3B) **2.** $2^4 \cdot 3^3$ (1.6A) **3.** 3 (1.6B) **4.** $2 \cdot 2 \cdot 2 \cdot 2 \cdot 2 \cdot 5$ (1.7B) **5.** 36 (2.1A) **6.** 14 (2.1B) **7.** $\frac{5}{8}$ (2.3B) **8.** $8\frac{3}{10}$ (2.4C) **9.** $5\frac{11}{18}$ (2.5C) **10.** $2\frac{5}{6}$ (2.6B) **11.** $4\frac{2}{3}$ (2.7B) **12.** $\frac{23}{30}$ (2.8C) **13.** Four and seven hundred nine ten-thousandths (3.1A) **14.** 2.10 (3.1B)

15. 1.990 (3.5A) **16.** $\frac{1}{15}$ (3.6B) **17.** $\frac{1}{8}$ (4.1A) **18.** $\frac{29¢}{2 \text{ bars}}$ (4.2A) **19.** 33.4 miles/gallon (4.2B) **20.** $n = 4.25$ (4.3B) **21.** 57.2 miles/hour (4.2C) **22.** $n = 36$ (4.3B) **23.** The new balance is $744. (1.3C) **24.** The monthly payment is $370. (1.5D) **25.** There are 105 pages remaining to be read. (2.6C) **26.** The cost for each acre is $36,000. (2.7C) **27.** The amount of change is $19.62. (3.3B) **28.** The player's batting average is 0.271. (3.5B) **29.** There will be 25 inches eroded. (4.3C) **30.** The person needs 1.6 ounces of medication. (4.3C)

ANSWERS to Chapter 5 Odd-Numbered Exercises

SECTION 5.1 *pages 191–192*

1. $\frac{1}{4}$, 0.25 **3.** $1\frac{3}{10}$, 1.30 **5.** 1, 1.00 **7.** $\frac{73}{100}$, 0.73 **9.** $3\frac{83}{100}$, 3.83 **11.** $\frac{7}{10}$, 0.70 **13.** $\frac{22}{25}$, 0.88 **15.** $\frac{8}{25}$, 0.32 **17.** $\frac{2}{3}$ **19.** $\frac{5}{6}$ **21.** $\frac{1}{9}$ **23.** $\frac{5}{11}$ **25.** $\frac{3}{70}$ **27.** $\frac{1}{15}$ **29.** 0.065 **31.** 0.0055 **33.** 0.0825 **35.** 0.0675 **37.** 0.0045 **39.** 0.804 **41.** 16% **43.** 5% **45.** 1% **47.** 70% **49.** 124% **51.** 0.4% **53.** 0.6% **55.** 310.6% **57.** 54% **59.** 33.3% **61.** 62.5% **63.** 16.7% **65.** 17.5% **67.** 177.8% **69.** 30% **71.** $23\frac{1}{3}\%$ **73.** $57\frac{1}{7}\%$ **75.** $27\frac{3}{11}\%$ **77.** $237\frac{1}{2}\%$ **79.** $216\frac{2}{3}\%$

SECTION 5.2 *pages 195–196*

1. $n = 8$ **3.** $n = 10.8$ **5.** $n = 0.075$ **7.** $n = 80$ **9.** $n = 51.895$ **11.** $n = 7.5$ **13.** $n = 13$ **15.** $n = 3.75$ **17.** $n = 20$ **19.** $n = 210$ **21.** 5% of 95 **23.** 84% of 32 **25.** 15% of 80 **27.** $n \approx 25{,}805.032$ **29.** The amount deducted for income tax is $403.20. **31.** The profit is $4000. **33.** The rebate is $1680. **35a.** The sales tax is $570. **35b.** The total cost including sales tax is $10,070. **37.** The total number of employees needed is 671. **39.** The real estate taxes are $50,092.

SECTION 5.3 *pages 199–200*

1. $n = 32\%$ **3.** $n = 16\frac{2}{3}\%$ **5.** $n = 200\%$ **7.** $n = 37.5\%$ **9.** $n = 18\%$ **11.** $n = 0.25\%$ **13.** $n = 20\%$ **15.** $n = 400\%$ **17.** $n = 2.5\%$ **19.** $n = 37.5\%$ **21.** $n = 0.25\%$ **23.** $n = 2.4\%$ **25.** $n = 9.6\%$ **27.** $n \approx 43.01367\%$ **29.** The company spends 5% of its budget for advertising. **31.** The down payment is 20% of the selling price. **33.** The 1990 value is 180% of the 1984 value. **35a.** The price increase was $31,000. **35b.** The increase is 25% of the original price. **37.** Yes, the account executive answered enough questions to pass. **39.** The land area is 29% of the total area of the earth.

SECTION 5.4 *pages 203–204*

1. $n = 75$ **3.** $n = 50$ **5.** $n = 100$ **7.** $n = 85$ **9.** $n = 1200$ **11.** $n = 19.2$ **13.** $n = 7.5$ **15.** $n = 32$ **17.** $n = 200$ **19.** $n = 80$ **21.** $n = 9$ **23.** $n = 504$ **25.** $n = 108$ **27.** $n \approx 7122.1547$ **29.** The estimated life of the brakes is 50,000 miles. **31.** The city's population was 56,000 people. **33.** The price at the competitor's store is $46.50. **35a.** The manufacturer tested 3000 computer boards. **35b.** There were 2976 computer boards that were not defective. **37.** The increase in the number of boxes was 1680. **39.** The increase in earnings was $1,085,400.

SECTION 5.5 *pages 207–208*

1. $n = 65$ **3.** 25% **5.** $n = 75$ **7.** 12.5% **9.** $n = 400$ **11.** $n = 19.5$ **13.** 14.8% **15.** $n = 62.62$ **17.** $n = 5$ **19.** $n = 45$ **21.** $n = 15$ **23.** 300% **25.** $n = 3200$ **27.** 2.4% **29.** $n \approx 12{,}323.661$ **31.** The total amount collected is $24,500. **33.** The cost of the calculator was $71.25. **35.** 12% of the alarms were false alarms. **37a.** The labor cost was $95.70. **37b.** The total cost of parts was $78.30. **39.** The percent of volunteers showing improvement was 92.5%. **41.** The new contract offered $1,464,375.

CHAPTER REVIEW *pages 211–212*

1. $\frac{3}{25}$ **2.** $\frac{1}{6}$ **3.** 0.42 **4.** 0.076 **5.** 38% **6.** 150% **7.** $n = 60$ **8.** $n = 5.4$ **9.** $n = 19.36$ **10.** $n = 77.5$ **11.** The company spent $4500 for T.V. advertising. **12.** The total cost is $1041.25. **13.** $n = 150\%$ **14.** $n = 20\%$ **15.** $n = 7.3\%$ **16.** $n = 613.3\%$ **17.** The dividend is 4.6% of the stock price. **18.** The percent increase in value was 64%. **19.** $n = 75$ **20.** $n = 10.9$ **21.** $n = 157.5$ **22.** $n = 504$ **23.** The city's population was 70,000 people. **24.** The previous year's batting average was 0.252. **25.** $n = 198.4$ **26.** $n = 160\%$ **27.** The computer cost $3000. **28.** The student answered 85% of the questions correctly.

CHAPTER TEST *pages 213–214*

1. 0.973 (5.1A) **2.** $\frac{1}{6}$ (5.1A) **3.** 30% (5.1B) **4.** 163% (5.1B) **5.** 150% (5.1B) **6.** $66\frac{2}{3}\%$ (5.1B) **7.** $n = 50.05$ (5.2A) **8.** $n = 61.36$ (5.2A) **9.** 76% of 13 (5.2A) **10.** 212% of 12 (5.2A) **11.** The company spends $4500 for advertising. (5.2B) **12.** There were 1170 pounds of vegetables unspoiled. (5.2B) **13.** 25% (5.3A) **14.** 216.67% (5.3A) **15.** The store's temporary employees were 16% of the total number of permanent employees. (5.3B) **16.** The student answered 91.3% of the questions correctly. (5.3B) **17.** $n = 80$ (5.4A) **18.** $n \approx 28.3$ (5.4A) **19.** There were 32,000 transistors tested. (5.4B) **20.** The increase is 60% of the original price. (5.4B) **21.** $n \approx 143.0$ (5.5A) **22.** 1000% (5.5A) **23.** The increase in the hourly wage is $1.02. (5.5B) **24.** The population now is 220% of the population 10 years ago. (5.5B) **25.** The value of the car is $6500. (5.5B)

CUMULATIVE REVIEW *pages 215–216*

1. 4 (1.6B) **2.** 240 (2.1A) **3.** $10\frac{11}{24}$ (2.4C) **4.** $12\frac{41}{48}$ (2.5C) **5.** $12\frac{4}{7}$ (2.6B) **6.** $\frac{7}{24}$ (2.7B) **7.** $\frac{1}{3}$ (2.8B) **8.** $\frac{13}{36}$ (2.8C) **9.** 3.08 (3.1B) **10.** 1.1196 (3.3A) **11.** 34.2813 (3.5A) **12.** 3.625 (3.6A) **13.** $1\frac{3}{4}$ (3.6B) **14.** $\frac{3}{8} < 0.87$ (3.6C) **15.** $n = 53.3$ (4.3B) **16.** $9.60/hour (4.2B) **17.** $\frac{11}{60}$ (5.1A) **18.** $83\frac{1}{3}\%$ (5.1B) **19.** 19.56 (5.2A) **20.** $133\frac{1}{3}\%$ (5.3A) **21.** 9.92 (5.4A) **22.** 342.9% (5.5A) **23.** The mechanic's take-home pay is $592. (2.6C) **24.** The monthly payment is $92.25. (3.5B) **25.** There were 420 gallons of gasoline used during the month. (3.5B) **26.** The real estate tax is $3000. (4.3C) **27.** The sales tax is 6% of the purchase price. (5.3B) **28.** 45% of the people surveyed did not favor the candidate. (5.3B) **29.** The 1990 value is 250% of the 1985 value. (5.5B) **30.** 18% of the children tested had levels of lead that exceeded federal standards. (5.5B)

ANSWERS to Chapter 6 Odd-Numbered Exercises

SECTION 6.1 *pages 221–222*

1. The unit cost is $0.056 per ounce. **3.** The unit cost is $0.149 per ounce. **5.** The unit cost is $0.033 per screw. **7.** The unit cost is $0.195 per foot. **9.** The unit cost is $0.037 per ounce. **11.** The unit cost is $2.995 per clamp. **13.** The more economical purchase is 12 ounces for $1.79. **15.** The more economical purchase is 12 ounces for $2.64. **17.** The more economical purchase is 46 ounces for $0.69. **19.** The more economical purchase is 2 quarts for $4.79. **21.** The more economical purchase is 200 tablets for $9.98. **23.** The more economical purchase is 16 ounces for $1.32. **25.** The total cost is $8.37. **27.** The total cost is $2.94. **29.** The total cost is $3.51. **31.** The total cost is $0.40. **33.** The total cost is $2.01. **35.** The change is $5.25.

SECTION 6.2 *pages 229–232*

1. The percent increase is 5.5%. **3.** The percent increase is 8%. **5.** The percent increase is $33\frac{1}{3}\%$. **7a.** The amount of increase is $2925. **7b.** The new salary is $35,425. **9a.** The amount of increase is $0.24. **9b.** The total dividend per share of stock is $1.84. **11.** The home is 240 square feet larger.

13. The markup is \$71.25. **15.** The markup is \$12.60. **17.** The markup rate is 30%. **19a.** The markup is \$77.76. **19b.** The selling price is \$239.76. **21a.** The markup is \$21.70. **21b.** The selling price is \$83.70. **23.** The selling price is \$227.20. **25.** The percent decrease is 40%. **27.** The percent decrease is 40% **29.** The car loses \$3360 in value. **31a.** The amount of the decrease is 12 cameras. **31b.** The percent decrease is 60%. **33a.** The amount of decrease was \$15.20. **33b.** The average monthly bill is \$60.80. **35.** The percent decrease is 20%. **37.** The discount rate is $33\frac{1}{3}\%$. **39.** The discount is \$68. **41.** The discount rate is 30%. **43a.** The discount is \$7.20. **43b.** The sale price is \$37.80. **45a.** The discount is \$40. **45b.** The discount rate is 25%. **47.** The sale price is \$35.70.

SECTION 6.3 *pages 235–236*

1. The simple interest due is \$9000. **3.** The simple interest due is \$13,500. **5.** The interest owed is \$16. **7a.** The interest due is \$72,000. **7b.** The monthly payment is \$6187.50. **9a.** The interest charged is \$1080. **9b.** The monthly payment is \$545. **11.** The monthly payment is \$1332.50. **13.** The amount in the account after one year is \$820.62. **15.** The value of the investment is \$134,253. **17.** The value of the investment is \$11,235.93. **19a.** The value of the investment will be \$111,446.25. **19b.** The interest earned will be \$36,446.25. **21a.** The value of the investment will be \$52,008.25. **21b.** Yes, the restaurant will have enough money in its account. **23.** The amount of interest earned is \$15,967.58.

SECTION 6.4 *pages 241–242*

1. The mortgage is \$82,450. **3.** The down payment is \$7500. **5.** The down payment is \$62,500. **7.** The loan origination fee is \$3750. **9a.** The down payment is \$4750. **9b.** The mortgage is \$90,250. **11.** The mortgage is \$86,400. **13.** The monthly mortgage payment is \$1363.05. **15.** No, the couple cannot afford to buy the home. **17.** The monthly property tax is \$53.75. **19a.** The monthly mortgage payment is \$1960.22. **19b.** The interest is \$1018.50 per month. **21.** The total monthly payment (mortgage and property tax) is \$531.73. **23.** The monthly mortgage payment is \$676.15.

SECTION 6.5 *pages 245–246*

1. The receptionist does not have enough money for the down payment. **3.** The sales tax is \$742.50. **5.** The license fee is \$250. **7a.** The sales tax is \$420. **7b.** The total cost of the sales tax and the license fee is \$595. **9a.** The down payment is \$550. **9b.** The amount financed is \$1650. **11.** The amount financed is \$28,000. **13.** The monthly truck payment is \$361.84. **15.** The cost to operate the car is \$5120. **17.** The cost per mile is \$0.11. **19.** The interest is \$74.75. **21a.** The amount financed is \$30,600. **21b.** The monthly truck payment is \$650.16. **23.** The monthly car payment is \$503.37.

SECTION 6.6 *pages 249–250*

1. The clerk earns \$220. **3.** The commission earned is \$2955. **5.** The commission received was \$84. **7.** The teacher receives \$3244 per month. **9.** The commission earned is \$416. **11.** The commission earned was \$540. **13.** The installer receives \$280. **15.** The consultant's hourly wage is \$125 per hour. **17a.** The therapist's hourly overtime wage is \$21.56. **17b.** The therapist earns \$344.96. **19a.** The increase in pay is \$0.93 per hour. **19b.** The clerk's hourly wage is \$7.13. **21a.** The amount of sales over \$100,000 is \$250,000. **21b.** The commission earned is \$35,000. **23.** The attendant's hourly wage is \$7.36.

SECTION 6.7 *pages 257–260*

1. The new balance is \$486.32. **3.** The balance in the account is \$1222.47. **5.** The current balance is \$825.27. **7.** The current balance is \$3000.82. **9.** Yes, there is enough money to purchase a refrigerator. **11.** Yes, there is enough money to make both purchases. **13.** The bank statement and checkbook balance. **15.** The bank statement and checkbook balance.

CHAPTER REVIEW *pages 263–264*

1. 14.5¢/ounce **2.** 60 ounces for $8.40 **3.** The percent increase was 30.4%. **4.** The yearly percent increase was 200%. **5.** The selling price is $2079. **6.** The percent increase in earnings is 15%. **7.** The markup is $72. **8.** The sale price is $141. **9.** The simple interest due is $8250. **10.** The simple interest due is $1650. **11.** The value of the investment is $61,483.25. **12.** The value of the investment will be $55,731. **13.** The down payment is $18,750. **14.** The loan origination fee is $1875. **15.** The total monthly payment is $578.53. **16.** The monthly mortgage payment is $1232.11. **17.** The total cost is $1158.75. **18.** The cost for the 4 items is 12.1¢/mile. **19.** The amount of the interest is $97.33. **20.** The monthly payment is $343.59. **21.** The commission received was $3240. **22.** The nurse's total income was $655.20. **23.** The current balance is $943.68. **24.** The current balance is $8866.58.

CHAPTER TEST *pages 265–266*

1. The cost per foot is $6.92. (6.1A) **2.** The more economical purchase is 5 pounds for $1.65. (6.1B) **3.** The cost is $14.53. (6.1C) **4.** The percent increase is 20%. (6.2A) **5.** The percent increase was 150%. (6.2A) **6.** The selling price is $301. (6.2B) **7.** The selling price is $6.25. (6.2B) **8.** The percent decrease was 57.6%. (6.2C) **9.** The percent decrease was 20%. (6.2C) **10.** The sale price is $209.30. (6.2D) **11.** The discount rate is 40%. (6.2D) **12.** The simple interest due is $2750. (6.3A) **13.** The interest earned was $43,055.70. (6.3B) **14.** The loan origination fee is $3350. (6.4A) **15.** The monthly mortgage payment is $1108.61. (6.4B) **16.** The amount financed is $14,350. (6.5A) **17.** The monthly car payment is $320.02. (6.5B) **18.** The nurse earns $703.50. (6.6A) **19.** The current balance is $6612.25. (6.7A) **20.** The bank statement and checkbook balance. (6.7B)

CUMULATIVE REVIEW *pages 267–268*

1. 13 (1.6B) **2.** $8\frac{13}{24}$ (2.4C) **3.** $2\frac{37}{48}$ (2.5C) **4.** 9 (2.6B) **5.** 2 (2.7B) **6.** 5 (2.8C) **7.** 52.2 (3.5A) **8.** 1.417 (3.6A) **9.** $51.25/hour (4.2B) **10.** $n = 10.94$ (4.3B) **11.** 62.5% (5.1B) **12.** 27.3 (5.2A) **13.** 0.182 (5.1A) **14.** 42% (5.3A) **15.** 250 (5.4A) **16.** 154.76 (5.5A) **17.** The total rainfall was $13\frac{11}{12}$ inches. (2.4D) **18.** The amount the family paid in taxes was $570. (2.6C) **19.** The ratio of decrease in price to original price is $\frac{3}{5}$. (4.1B) **20.** The car drove 33.4 miles per gallon. (4.2C) **21.** The unit cost is $0.93. (4.2C) **22.** The dividend is $280. (4.3C) **23.** The sale price is $720. (5.2B) **24.** The selling price is $119. (6.2B) **25.** The percent increase is 8%. (6.2A) **26.** The simple interest due is $6000. (6.3A) **27.** The monthly payment is $381.60. (6.5B) **28.** The new balance is $2243.77. (6.7A) **29.** The cost per mile is $0.20. (6.5B) **30.** The monthly mortgage payment is $743.18. (6.4A)

ANSWERS to Chapter 7 Odd-Numbered Exercises

SECTION 7.1 *pages 275–278*

1. There were 19,000 gallons sold. **3.** The percent of gasoline sold in week 1 was 21.1%. **5.** The percent of the budget that comes from the federal government is 15%. **7.** The total number of fish caught was 4000 fish. **9.** The percent of fish caught in 1990 was 27.5%. **11.** The ratio is $\frac{15}{83}$. **13.** The ratio is $\frac{86}{45}$. **15.** The ratio is $\frac{60}{13}$. **17.** The ratio is $\frac{11}{18}$. **19.** The ratio is $\frac{314}{229}$. **21.** The ratio is $\frac{11}{212}$. **23.** The amount spent on food was $5112. **25.** The amount spent on clothing and entertainment was $3408. **27.** There are 81,600 people between the ages of 31 and 45. **29.** There are 124,800 people between the ages of 31 and 65. **31.** The amount budgeted for administrative costs is $1,077,600. **33.** The amount budgeted for administrative costs and teacher salaries is $12,347,500.

SECTION 7.2 *pages 281–282*

1. There were 40 cars sold in May. **3.** The ratio is $\frac{3}{10}$. **5.** The amount of premiums earned in 1987 was $70,000. **7.** The amount of benefits exceeded the premiums earned in 1986. **9.** There were 200 burglaries committed per 100,000 people in 1985. **11.** The ratio is $\frac{10}{7}$. **13.** The amount of January snowfall was 60 inches. **15.** The total snowfall during November and December was 85 inches. **17.** The average high temperature during January was 40°. **19.** The difference between the average high temperature in Honolulu and New York during July was 2°. **21.** The difference between the number of business and residential calls made is 25,000 calls. **23.** There were 195,000 residential calls made.

SECTION 7.3 *pages 285–286*

1. There were 13 students who scored between 60 and 80. **3.** There were 11 students who scored above 80. **5.** There were 36 customers who made purchases between $20 and $30. **7.** There were 66 customers who made purchases of more than $30. **9.** There were 20 cars sold. **11.** The ratio was $\frac{4}{7}$. **13.** There were 4 runners completing the 100-yard dash in less than 11 seconds. **15.** There were 19 runners who ran the race between 11 and 12 seconds. **17.** There were 30 families who spent between 4% and 8% on vacations. **19.** There were 15 families who spent less than 4% of their income on vacations. **21.** There were 100 students scoring between 600 and 700. **23.** There were 475 students scoring below 600.

SECTION 7.4 *pages 289–290*

1. Your mean grade is 83. **3.** The mean height of the basketball team is 77.4 inches. **5.** The mean low temperature is 17.4°. **7.** The mean monthly paycheck is $1240. **9.** The mean braking distance is 218 feet. **11.** The mean annual rainfall is 18 inches. **13.** The median temperature for the week is 95°. **15.** The median time for running the most popular programs is 56 seconds. **17.** The median age is 26 years. **19.** The median number of tickets given out is 11.5. **21.** The median amount of coffee obtained is 3.965 ounces. **23.** The median population is 13,550,000.

CHAPTER REVIEW *pages 293–294*

1. There were 44 students receiving grades. **2.** The ratio is $\frac{7}{3}$. **3.** The percent of students receiving C grades is 29.5%. **4.** The teams' total income is $965,000,000. **5.** The ratio is $\frac{20}{9}$. **6.** The percent of income from national broadcasting is 22.8%. **7.** The difference in temperatures was 30°. **8.** The ratio was $\frac{7}{4}$. **9.** Wednesday had the lowest temperature. The temperature was 15°. **10.** The difference in profit was $20 million. **11.** The total profit made was $85 million. **12.** The ratio was $\frac{5}{8}$. **13.** There were 9 trees over 72 inches tall. **14.** The ratio was $\frac{9}{10}$. **15.** The percent of trees between 66 and 69 inches tall was 31.7%. **16.** Fewer than 100 points were scored in 28 games. **17.** The ratio was $\frac{5}{8}$. **18.** The percent of games in which over 120 points were scored was 7.3%. **19.** Your mean score is 82.4. **20.** The median number of students is 26.5.

CHAPTER TEST *pages 295–296*

1. The total number of students is 30. (7.1A) **2.** The ratio is $\frac{1}{2}$. (7.1A) **3.** The percent of students receiving a B grade is 30%. (7.1A) **4.** The ratio is $\frac{1}{16}$. (7.1B) **5.** The percent of the budget from state funds is 53.75% (7.1B) **6.** The ratio is $\frac{43}{5}$. (7.1B) **7.** In November 1990 27,000 cars were sold. (7.2A)

8. The difference in the number of cars sold was 5000. (7.2A) **9.** The difference in the first 3 months' sales was 19,000 cars. (7.2A) **10.** The third quarter income for 1991 is \$5,000,000. (7.2B) **11.** The ratio is $\frac{3}{4}$. (7.2B) **12.** The difference in fourth quarter incomes is \$1,000,000. (7.2B) **13.** The number of employees receiving a salary over \$25,000 is 21. (7.3A) **14.** The ratio is $\frac{10}{51}$. (7.3A) **15.** The percent of employees whose salaries are between \$20,000 and \$30,000 is 52.9%. (7.3A) **16.** 55 families watched between 15 and 25 hours of television a week. (7.3B) **17.** The ratio is $\frac{3}{5}$. (7.3B) **18.** The percent of families watching over 15 hours of television each week was 60%. (7.3B) **19.** The mean number of miles driven is 189 miles. (7.4A) **20.** The median score was 77. (7.4B)

CUMULATIVE REVIEW *pages 297–298*

1. 540 (1.6A) **2.** 14 (1.6B) **3.** 120 (2.1A) **4.** $\frac{5}{12}$ (2.3B) **5.** $12\frac{3}{40}$ (2.4C) **6.** $4\frac{17}{24}$ (2.5C) **7.** 2 (2.6B) **8.** $\frac{64}{85}$ (2.7B) **9.** $8\frac{1}{4}$ (2.8C) **10.** 209.305 (3.1A) **11.** 2.82348 (3.4A) **12.** 16.67 (3.6A) **13.** 26.4 miles/gallon (4.2B) **14.** $n = 3.2$ (4.3B) **15.** 80% (5.1B) **16.** 80 (5.4A) **17.** 16.34 (5.2A) **18.** 40% (5.3A) **19.** The income for the week was \$650. (6.6A) **20.** The cost for \$50,000 of life insurance is \$207.50. (4.3C) **21.** The interest due on the loan is \$6875. (6.3A) **22.** The markup rate is 55%. (6.2B) **23.** The family budgeted \$570 for food. (7.1B) **24.** The difference between the number of problems the two students answered correctly was 12 problems. (7.2B) **25.** The mean high temperature for the week was 69.6°. (7.4A) **26.** The median salary is \$21,650. (7.4B)

ANSWERS to Chapter 8 Odd-Numbered Exercises

SECTION 8.1 *pages 305–306*

1. 72 in. **3.** $2\frac{1}{2}$ ft **5.** 39 ft **7.** $5\frac{1}{3}$ yd **9.** 84 in. **11.** $3\frac{1}{3}$ yd **13.** 10,560 ft **15.** $\frac{5}{8}$ ft **17.** 57 in. **19.** 8 ft 4 in. **21.** 1 ft 3 in. **23.** 13 ft 5 in. **25.** 5 yd 2 ft **27.** $14\frac{2}{3}$ ft **29.** $4\frac{1}{6}$ in. **31.** 10 yd 1 ft **33.** 2 yd 2 ft **35.** $1\frac{7}{8}$ yd **37.** There can be 14 tiles placed along one row of the counter top. **39.** The missing dimension is $1\frac{5}{6}$ ft. **41.** The length of material needed is $7\frac{1}{2}$ in. **43.** Each piece of board is $1\frac{2}{3}$ ft long. **45.** The length of framing needed is 6 ft 6 in. **47.** The length of the wall is $33\frac{3}{4}$ ft.

SECTION 8.2 *pages 309–310*

1. 4 lb **3.** 64 oz **5.** $1\frac{3}{5}$ tons **7.** 12,000 lb **9.** $4\frac{1}{8}$ lb **11.** 24 oz **13.** 2600 lb **15.** $\frac{1}{4}$ ton **17.** $11\frac{1}{4}$ lb **19.** 4 tons 1000 lb **21.** 2 lb 8 oz **23.** 5 tons 400 lb **25.** 1 ton 1700 lb **27.** 33 lb **29.** 14 lb **31.** 9 oz **33.** 1 lb 7 oz **35.** The weight of the bricks is 2000 lb. **37.** The weight of the tiles is 63 lb. **39.** The weight of the case of soft drinks is 9 lb. **41.** There is 1 lb 5 oz of shampoo in each container. **43.** The cost of the ham roast is \$13.50. **45.** The cost of mailing the manuscript is \$8.75.

SECTION 8.3 *pages 313–314*

1. $7\frac{1}{2}$ c **3.** 24 fl oz **5.** 4 pt **7.** 7 c **9.** $5\frac{1}{2}$ gal **11.** 9 qt **13.** $3\frac{3}{4}$ qt **15.** $1\frac{1}{4}$ pt **17.** $4\frac{1}{4}$ qt **19.** 3 gal 2 qt **21.** 2 qt 1 pt **23.** 7 qt **25.** 6 fl oz **27.** $17\frac{1}{2}$ pt **29.** $\frac{7}{8}$ gal **31.** 5 gal 1 qt **33.** 1 gal 2 qt **35.** $2\frac{3}{4}$ gal **37.** There should be $7\frac{1}{2}$ gal of coffee prepared. **39.** There are $4\frac{1}{4}$ qt of the final solution. **41.** The farmer used $8\frac{3}{4}$ gal of oil. **43.** There are 36 c of grape juice in a case. **45.** The more economical purchase is 1 qt for \$1.20. **47.** The store makes \$43.50 profit on each package of hand lotion.

SECTION 8.4 *pages 317–318*

1. 19,450 ft · lb **3.** 19,450,000 ft · lb **5.** 1500 ft · lb **7.** 29,700 ft · lb **9.** 30,000 ft · lb **11.** 25,500 ft · lb **13.** 35,010,000 ft · lb **15.** 9,336,000 ft · lb **17.** 2 hp **19.** 8 hp **21.** 2750 ft · lb/s **23.** 3850 ft · lb/s **25.** 500 ft · lb/s **27.** 4800 ft · lb/s **29.** 1440 ft · lb/s **31.** 3 hp **33.** 12 hp

CHAPTER REVIEW *pages 321–322*

1. 48 in. **2.** $4\frac{2}{3}$ yd **3.** 9 ft 3 in. **4.** 1 yd 2 ft **5.** 13 ft 4 in. **6.** 2 ft 6 in. **7.** 3 ft 6 in. **8.** 54 oz **9.** $1\frac{1}{5}$ tons **10.** 9 lb 3 oz **11.** 1 ton 1000 lb **12.** 43 lb **13.** 2 lb 7 oz **14.** It costs $6.30 to mail the book. **15.** 3 qt **16.** 40 fl oz **17.** There are $13\frac{1}{2}$ qt in a case. **18.** There were 16 gal of milk sold. **19.** 38,900 ft · lb **20.** 1600 ft · lb **21.** The furnace releases 27,230,000 ft · lb of energy in one hour. **22.** 7 hp **23.** 1375 ft · lb/s **24.** The engine has 480 ft · lb/s of power.

CHAPTER TEST *pages 323–324*

1. 30 in. (8.1A) **2.** 2 ft 5 in. (8.1B) **3.** Each piece is $1\frac{1}{3}$ ft long. (8.1C) **4.** The length of the wall is 48 ft. (8.1C) **5.** 46 oz (8.2A) **6.** 2 lb 8 oz (8.2A) **7.** 17 lb 1 oz (8.2B) **8.** 1 lb 11 oz (8.2B) **9.** The total weight of the workbooks is 750 lb. (8.2C) **10.** The class received $28.13 for the cans. (8.2C) **11.** $3\frac{1}{4}$ gal (8.3A) **12.** 28 pt (8.3A) **13.** $12\frac{1}{4}$ gal (8.3B) **14.** 8 gal 1 qt (8.3B) **15.** The number of cups in a case is 60 c. (8.3C) **16.** The mechanic makes $126 profit. (8.3C) **17.** The energy required is 3,750 ft · lb. (8.4A) **18.** The furnace releases 31,120,000 ft · lb of energy in one hour. (8.4A) **19.** The lifting power needed is 160 ft · lb/s. (8.4B) **20.** The horsepower of the motor is 4 hp. (8.4B)

CUMULATIVE REVIEW *pages 325–326*

1. 180 (2.1A) **2.** $5\frac{3}{8}$ (2.2B) **3.** $3\frac{7}{24}$ (2.5C) **4.** 2 (2.7B) **5.** $4\frac{3}{8}$ (2.8C) **6.** 2.10 (3.1B) **7.** 0.038808 (3.4A) **8.** $n = 8.8$ (4.3B) **9.** 1.25 (5.2A) **10.** 42.86 (5.4A) **11.** $2.15/pound (6.1A) **12.** $8\frac{11}{15}$ inches. (8.1B) **13.** 1 lb 8 oz (8.2B) **14.** 31 lb 8 oz (8.2B) **15.** $2\frac{1}{2}$ qt (8.3B) **16.** 1 lb 12 oz (8.2B) **17.** The dividend to be received is $280. (4.3C) **18.** Your new checking balance is $642.79 (6.7A) **19.** The executive's total monthly income is $3100. (6.6A) **20.** The amount of carrots that could be sold is 2425 lb. (5.2B) **21.** The percent receiving a score between 80% and 90% is 18%. (7.3A) **22.** The selling price is $308. (6.2B) **23.** The interest paid on the loan is $14,666.67. (6.3A) **24.** Each student received $1267. (8.2C) **25.** The cost of mailing the books was $10.80. (8.2C) **26.** The better buy is 36 oz for $2.70. (6.1B) **27.** The contractor saves $1080. (8.3C) **28.** The energy required is 3200 ft · lb. (8.4A) **29.** The power required is 400 ft · lb/s. (8.4B)

ANSWERS to Chapter 9 Odd-Numbered Exercises

SECTION 9.1 *pages 331–332*

1. 420 mm **3.** 8.1 cm **5.** 6.804 km **7.** 2109 m **9.** 4.32 m **11.** 88 cm **13.** 6.42 m; 642 cm **15.** 42.6 cm; 426 mm **17.** 8.42 m **19.** 78.85 m **21.** 29.6 cm **23.** 0.414 km **25.** 18.126 km **27.** 901,161.28 km **29.** The total length of ceiling joists needed is 69 m. **31.** The missing dimension is 7.8 cm. **33.** The race was 30.5 km long. **35.** The distance between rivets is 17.9 cm. **37.** There are 15.6 m of fencing left on the roll. **39.** Light travels 9.4608×10^{15} m in 1 year.

SECTION 9.2 *pages 335–336*

1. 0.420 kg **3.** 0.127 g **5.** 4200 g **7.** 450 mg **9.** 1.856 kg **11.** 4.057 g **13.** 3.922 kg; 3922 g **15.** 7.891 g; 7891 mg **17.** 3.308 g **19.** 4.692 kg **21.** 736 g **23.** 32.571 g **25.** 1.720 kg

27. 805,025.566 g **29.** The increase in weight was 4.85 kg. **31.** The weight of the blocks is 2820 kg. **33.** Each scout will carry 20.92 kg. **35.** The cost of the ham is $19.74. **37.** The cost for the extra luggage will be $21.60. **39.** The total weight of the automobiles is 21,075 kg.

SECTION 9.3 *pages 339–340*

1. 4.2 L **3.** 3420 ml **5.** 423 cm^3 **7.** 642 ml **9.** 0.042 L **11.** 435 cm^3 **13.** 1267 cm^3 **15.** 3023 cm^3 **17.** 35 L **19.** 3.042 L; 3042 ml **21.** 3.004 kl; 3004 L **23.** 12.438 L **25.** 37.944 L **27.** 6.254 L **29.** 1131.372 L **31.** 0.5545072 L **33.** The auxiliary needs 6.6 L of punch. **35.** 4000 patients can be immunized. **37.** The students will use 1.08 L of acid. **39.** The amount of oxygen in 50 L of air is 9.5 L. **41.** There are 7 bottles of cough syrup still in stock. **43.** The profit made was $8715.

SECTION 9.4 *pages 343–344*

1. You can eliminate 3300 cal. **3.** The person would need 3000 cal per day. **5.** The percent of the total daily intake of calories left is 44.4%. **7.** You burn up 10,125 cal in 30 days. **9.** You lost 410 cal. **11.** You would have to hike 2.6 h. **13.** The oven used 1250 Wh of energy. **15.** The TV set uses 2.205 kWh of energy. **17.** The cost of listening to the stereo is $.32. **19.** The hair dryer uses 6 kWh of energy each week. **21.** The percent decrease is 17.0%. **23.** The cost of using the welder is $109.98.

SECTION 9.5 *pages 347–348*

1. 91.74 m **3.** 1.93 m **5.** 5.45 kg **7.** 235.85 ml **9.** 104.65 km/h **11.** $3.72/kg **13.** $0.39/L **15.** The weight loss will be 1 lb. **17.** 40,068.07 km **19.** 328 ft **21.** 1.59 gal **23.** 4920 ft **25.** 12.72 gal **27.** 1.86 pt **29.** 49.69 mi/h **31.** $1.45/gal **33.** 0.43 lb can be lost. **35.** 901 gal

CHAPTER REVIEW *pages 351–352*

1. 3.7 mm **2.** 1250 m **3.** 9.55 m **4.** 20.5 cm **5.** 36.4 m **6.** 3.5 km **7.** There are 37.2 m of wire fence left on the roll. **8.** 450 mg **9.** 4050 g **10.** 14.243 g **11.** 12.46 kg **12.** 37.95 kg **13.** 2.519 g **14.** The total cost of the turkey is $20.37. **15.** 5.6 ml **16.** 1200 cm^3 **17.** 10.613 L **18.** 6.597 L **19.** 102.44 L **20.** 1.81 L **21.** There should be 50 L of coffee prepared. **22.** You can eliminate 2700 cal. **23.** It costs $3.42 to run the T.V. set. **24.** 1098.9 yd **25.** 4.19 lb **26.** The ham costs $7.48/kg. **27.** It is necessary to cycle 8.75 h.

CHAPTER TEST *pages 353–354*

1. 4.26 cm (9.1A) **2.** 5038 m (9.1A) **3.** 4.65 m (9.1B) **4.** 22.16 m (9.1B) **5.** The total length of rafters needed is 114 m. (9.1C) **6.** The distance between the rivets is 17.5 cm. (9.1C) **7.** 3290 g (9.2A) **8.** 3.089 g (9.2A) **9.** 24.36 kg (9.2B) **10.** 7.297 g (9.2B) **11.** The weight of the box of tiles is 36 kg. (9.2C) **12.** It costs $540 to fertilize the apple orchard. (9.2C) **13.** 3250 ml (9.3A) **14.** 1600 cm^3 (9.3A) **15.** 0.75 L (9.3B) **16.** 3.931 L (9.3B) **17.** The clinic needs 5.2 L of vaccine. (9.3C) **18.** It costs $16.32 to operate the TV set. (9.4A) **19.** The record ski jump for men is 193.90 m. (9.5A) **20.** The record ski jump for women is 360.80 feet. (9.5B)

CUMULATIVE REVIEW *pages 355–356*

1. 6 (1.6B) **2.** $12\frac{13}{36}$ (2.4C) **3.** $\frac{29}{36}$ (2.5C) **4.** $3\frac{1}{14}$ (2.7B) **5.** 1 (2.8B) **6.** 2.0702 (3.3A) **7.** $n = 31.3$ (4.3B) **8.** 175% (5.1B) **9.** 145 (5.4A) **10.** 2.25 gallons (8.3A) **11.** 18.75 m (9.1A) **12.** 2.528 km (9.1B) **13.** 5.05 kg (9.2A) **14.** 3.672 g (9.2B) **15.** 0.83 kg (9.2B) **16.** 5.548 L (9.3B) **17.** There is $1683 left after the rent is paid. (2.6C) **18.** The amount of income tax paid is $7207.20. (3.4B) **19.** The property tax is $1500. (4.3C) **20.** The new car buyer would receive a $1620 rebate. (5.2B) **21.** The dividend is 6.5% of the investment. (5.3B) **22.** Your average grade is 76. (7.4A) **23.** The salary next year will be $25,200. (5.2B) **24.** The discount rate is 22%. (6.2D) **25.** The length of the wall is 36 feet. (8.1C) **26.** The number of quarts of apple juice is 12. (8.3C) **27.** The profit is $59.20. (8.3C) **28.** There are 24 L of chlorine used. (9.3C) **29.** It costs $1.89 to operate the hair dryer. (9.4A) **30.** The conversion is 96.6 km/h. (9.5A)

ANSWERS to Chapter 10 Odd-Numbered Exercises

SECTION 10.1 *pages 367–370*

1. angle **3.** 180° **5.** 30 **7.** 21 **9.** 14 **11.** 59° **13.** 108° **15.** 77° **17.** 53° **19.** 77° **21.** 118° **23.** 133° **25.** 86° **27.** 180° **29.** rectangle or square **31.** cube **33.** parallelogram, rectangle, or square **35.** cylinder **37.** 102° **39.** 90° and 45° **41.** 14° **43.** 90° and 65° **45.** 8 in. **47.** $4\frac{2}{3}$ ft **49.** 7 cm **51.** 2 ft 4 in. **53.** $a = 106°, b = 74°$ **55.** $a = 112°, b = 68°$ **57.** $a = 38°, b = 142°$ **59.** $a = 58°, b = 58°$ **61.** $a = 152°, b = 152°$ **63.** $a = 130°, b = 50°$

SECTION 10.2 *pages 375–378*

1. 56 in. **3.** 20 ft **5.** 92 cm **7.** 47.1 cm **9.** 9 ft 10 in. **11.** 50.24 cm **13.** 240 m **15.** 157 cm **17.** 121 cm **19.** 50.56 m **21.** 3.57 ft **23.** 139.3 m **25.** They need $4\frac{1}{2}$ mi of fencing. **27.** They need 15.7 m of binding. **29.** The bicycle travels 10π ft. **31.** The length of weather stripping is 20.71 ft. **33.** The total cost to fence the lot is $8022.50. **35.** The distance for one revolution is 584,040,000 mi.

SECTION 10.3 *pages 383–386*

1. 144 ft^2 **3.** 81 in.^{2} **5.** 50.24 ft^2 **7.** 20 in.^{2} **9.** 2.13 cm^2 **11.** 16 ft^2 **13.** 817 in.^{2} **15.** 154 in.^{2} **17.** 26 cm^2 **19.** 2220 cm^2 **21.** 150.72 in.^{2} **23.** 8.851323 ft^2 **25.** The area is 1728 yd.^{2} **27.** The area is 10,000 π in.^{2}. **29.** It will cost $570.90. **31.** The area is 68 m^2. **33.** The increase in area is 113.04 in.^{2}. **35.** The total area is 125.1492 mi^2.

SECTION 10.4 *pages 393–396*

1. 144 cm^3 **3.** 512 in.^{3} **5.** 2143.57 in.^{3} **7.** 150.72 cm^3 **9.** 6.4 m^3 **11.** 5572.45 mm^3 **13.** 3391.2 ft^3 **15.** $42\frac{7}{8}$ ft^3 **17.** 82.26 in.^{3} **19.** 1.6688 m^3 **21.** 69.08 in.^{3} **23.** The volume is 40.5 m^3. **25.** The volume is 17,148.59 ft.^{3} **27.** The volume is 2304π ft^3. **29.** The aquarium will be filled by 15.0 gal of water. **31.** The volume is 809,516.25 ft^3. **33.** The total weight of the water is 4,056,000 lb.

SECTION 10.5 *pages 401–402*

1. 2.646 **3.** 6.481 **5.** 12.845 **7.** 13.748 **9.** 5 in. **11.** 8.602 cm **13.** 11.180 ft **15.** 4.472 cm **17.** 12.728 yd **19.** The lengths of the two legs are 6 ft and 10.392 ft. **21.** 21.21 cm **23.** 8 cm **25.** 11.314 yd **27.** The length of the cable is 35 m. **29.** The length of the steel plate is 5.5 in. **31.** The distance between the holes is 6.32 in. **33.** It cost $109.20 to fence the plot. **35.** The total length of the conduit is 35 ft.

SECTION 10.6 *pages 407–408*

1. $\frac{1}{2}$ **3.** $\frac{3}{4}$ **5.** The two triangles are congruent. **7.** The two triangles are congruent. **9.** 7.2 cm **11.** 3.3 m **13.** The height of the flagpole is 18 ft. **15.** The height of the building is 16 m. **17.** The perimeter is 12 m. **19.** The area is 56.25 cm^2.

CHAPTER REVIEW *pages 411–412*

1. 3 **2.** 75° **3.** 0.75 m **4.** 58° and 90° **5.** $a = 135°, b = 45°$ **6.** $a = 100°, b = 80°$ **7.** 26 ft **8.** 31.4 cm **9.** 47.7 in. **10.** The bicycle travels 73.3 ft. **11.** 55 m^2 **12.** 63.585 cm^2 **13.** 57.12 in.^{2} **14.** The area is 28 yd.^{2} **15.** 200 ft^3 **16.** 267.9 ft^3 **17.** 240 in.^{3} **18.** The volume is 1144.53 ft^3. **19.** 3.873 **20.** 26 cm **21.** The ladder will reach up to 15 ft. **22.** 16 cm **23.** 64.8 m^2

CHAPTER TEST *pages 413–414*

1. 58° (10.1A) **2.** 90° and 50° (10.1B) **3.** 150° (10.1C) **4.** 180° (10.1C) **5.** 6.8 m (10.2A) **6.** 15.85 ft (10.2B) **7.** It will cost $556.84 to carpet the area. (10.2C) **8.** $3\frac{1}{7}$ m² (10.3A) **9.** 10.125 ft² (10.3B) **10.** The cross-sectional area is 103.82 ft². (10.3C) **11.** There are 113.04 in.² more of the pizza. (10.3C) **12.** 169.56 m³ (10.4A) **13.** 1406.72 cm³ (10.4B) **14.** The volume is 860.625 in³. (10.4C) **15.** $\sqrt{189} \approx 13.748$ (10.5A) **16.** 9.747 ft (10.5B) **17.** The length of the rafter is 15 ft. (10.5C) **18.** $1\frac{1}{5}$ ft (10.6A) **19.** The width of the canal is 25 ft. (10.6B) **20.** 10 m (10.6A)

CUMULATIVE REVIEW *pages 415–416*

1. 48 (2.1B) **2.** $7\frac{41}{48}$ (2.4C) **3.** $\frac{39}{56}$ (2.7B) **4.** $\frac{2}{15}$ (2.8C) **5.** 38.5 (3.5A) **6.** $8.72/hr (4.2B) **7.** $n = 37.5$ (4.3B) **8.** $\frac{3}{8}$ (5.1A) **9.** 8% (5.3A) **10.** 85 (5.4A) **11.** 1.5 ft (8.1A) **12.** 800 lb (8.2A) **13.** 32,500 m (9.1A) **14.** 31.58 m (9.1B) **15.** 1.270 g (9.2A) **16.** 8.727 L (9.3B) **17.** The monthly payment is $181. (1.5D) **18.** The sales tax is $47.06. (4.3C) **19.** The operator's hourly wage was $14.60. (5.4B) **20.** The sale price is $54. (6.2D) **21.** The value after 20 years would be $184,675.75. (6.3B) **22.** The weight of the package is 54 lb. (8.2C) **23.** The distance between the rivets is 22.5 cm. (9.1C) **24.** 14° (10.1A) **25.** $a = 74°$, $b = 106°$ (10.1A) **26.** 29.42 cm (10.2B) **27.** 60 in.² (10.3B) **28.** 92.86 in.³ (10.4B) **29.** 10.63 ft (10.5B) **30.** 36 cm (10.6B)

ANSWERS to Chapter 11 Odd-Numbered Exercises

SECTION 11.1 *pages 421–422*

1. $3 < 5$ **3.** $-2 > -5$ **5.** $-16 < 1$ **7.** $3 > -7$ **9.** $-11 < -8$ **11.** $-1 > -6$ **13.** $0 > -3$ **15.** $6 > -8$ **17.** $-14 < 16$ **19.** $35 > 28$ **21.** $-42 < 27$ **23.** $21 > -34$ **25.** $-27 > -39$ **27.** $-87 < 63$ **29.** $68 > -79$ **31.** $-62 > -84$ **33.** $94 > 83$ **35.** $59 > -67$ **37.** $-93 < -55$ **39.** $-88 < 57$ **41.** $0 < 129$ **43.** $-131 < 101$ **45.** $-194 < -180$ **47.** −16 **49.** 3 **51.** −45 **53.** 88 **55.** 2 **57.** 6 **59.** 5 **61.** 1 **63.** −8 **65.** 0 **67.** 19 **69.** 22 **71.** −20 **73.** −18 **75.** 23 **77.** −27 **79.** −41 **81.** 25 **83.** 30 **85.** −34 **87.** −45 **89.** 36 **91.** 61 **93.** −52 **95.** −93 **97.** 119

SECTION 11.2 *pages 427–428*

1. −2 **3.** 20 **5.** −11 **7.** −9 **9.** −3 **11.** 1 **13.** −5 **15.** −30 **17.** 9 **19.** 1 **21.** −10 **23.** −28 **25.** −41 **27.** −392 **29.** 8 **31.** −7 **33.** −9 **35.** 9 **37.** −3 **39.** 18 **41.** −9 **43.** 11 **45.** 0 **47.** 11 **49.** 2 **51.** −138 **53.** −8 **55.** −337

SECTION 11.3 *pages 433–436*

1. 42 **3.** −24 **5.** 6 **7.** 18 **9.** −20 **11.** −16 **13.** 25 **15.** 0 **17.** −72 **19.** −102 **21.** 140 **23.** −228 **25.** −320 **27.** −156 **29.** −70 **31.** 162 **33.** 120 **35.** 36 **37.** 192 **39.** −108 **41.** −2100 **43.** 0 **45.** −251,636 **47.** −2 **49.** 8 **51.** 0 **53.** −9 **55.** −9 **57.** 9 **59.** −24 **61.** −12 **63.** 31 **65.** 17 **67.** 15 **69.** −13 **71.** −18 **73.** 19 **75.** 13 **77.** −19 **79.** 17 **81.** 26 **83.** 23 **85.** 25 **87.** −34 **89.** 11 **91.** −13 **93.** 13 **95.** 12 **97.** −14 **99.** 290 **101.** The temperature rose to −1°C. **103.** Your score was −15 points. **105.** The difference in elevation is 20,602 ft. **107.** The difference in elevation is 30,314 ft. **109.** The mean daily low temperature was −3°.

SECTION 11.4 *pages 441–444*

1. $1\frac{1}{12}$ **3.** $-\frac{5}{24}$ **5.** $-\frac{19}{24}$ **7.** $\frac{5}{26}$ **9.** $\frac{7}{24}$ **11.** $\frac{7}{24}$ **13.** $-\frac{8}{15}$ **15.** $-\frac{1}{12}$ **17.** $\frac{3}{4}$ **19.** $\frac{17}{18}$ **21.** $-\frac{47}{48}$ **23.** $\frac{3}{8}$ **25.** $-\frac{7}{60}$ **27.** $-\frac{1}{16}$ **29.** $-\frac{7}{24}$ **31.** $\frac{13}{24}$ **33.** $-1\frac{1}{10}$ **35.** −1.63 **37.** 3.5 **39.** 1.91 **41.** −1.22 **43.** −181.51 **45.** −22.804 **47.** −0.6 **49.** −10.03 **51.** −37.19

53. −17.5 **55.** −1.867 **57.** −12.932 **59.** −2.0867 **61.** $-\frac{3}{8}$ **63.** $\frac{1}{10}$ **65.** $-\frac{4}{9}$ **67.** $-\frac{7}{30}$ **69.** $\frac{2}{27}$ **71.** $-\frac{45}{328}$ **73.** 10 **75.** $\frac{45}{328}$ **77.** $-\frac{3}{7}$ **79.** $1\frac{1}{2}$ **81.** $-\frac{8}{9}$ **83.** $1\frac{3}{4}$ **85.** $\frac{5}{6}$ **87.** $-2\frac{2}{7}$ **89.** $-13\frac{1}{2}$ **91.** 3.78 **93.** 31.15 **95.** 74.88 **97.** 0.117 **99.** 0.0363 **101.** 2.94 **103.** −79.8 **105.** 0.504 **107.** −0.11 **109.** 9.91 **111.** −84.3 **113.** −49.8 **115.** 7.3 **117.** −3.8 **119.** −22.70 **121.** 0.07 **123.** 0.55 **125.** −520.26 **127.** −0.962325 **129.** 0.0000277 **131.** −0.009814

SECTION 11.5 *pages 447–448*

1. 4 **3.** 0 **5.** 12 **7.** −6 **9.** −5 **11.** 2 **13.** 8 **15.** 18 **17.** −12 **19.** −3 **21.** 1 **23.** 2 **25.** −3 **27.** 14 **29.** −4 **31.** 33 **33.** −13 **35.** −12 **37.** 17 **39.** 0 **41.** 30 **43.** 94 **45.** −8 **47.** 39 **49.** 0.21 **51.** −0.96 **53.** −0.29 **55.** −1.76 **57.** 2.1 **59.** $\frac{3}{16}$ **61.** $\frac{7}{16}$ **63.** $-\frac{5}{16}$ **65.** $-\frac{5}{8}$

CHAPTER REVIEW *pages 451–452*

1. $-2 > -40$ **2.** $0 > -3$ **3.** 4 **4.** −22 **5.** 5 **6.** −6 **7.** −26 **8.** 1 **9.** −26 **10.** 6 **11.** The temperature rose to −4°. **12.** $-\frac{5}{24}$ **13.** $-\frac{1}{4}$ **14.** $\frac{17}{24}$ **15.** $\frac{1}{36}$ **16.** 0.21 **17.** −7.4 **18.** $-\frac{1}{4}$ **19.** $\frac{3}{40}$ **20.** $\frac{8}{25}$ **21.** $\frac{15}{32}$ **22.** $1\frac{5}{8}$ **23.** −0.042 **24.** −1.28 **25.** −0.08 **26.** 10 **27.** 8 **28.** 18 **29.** 2 **30.** 24 **31.** 1.33 **32.** $-\frac{7}{18}$ **33.** $\frac{1}{3}$

CHAPTER TEST *pages 453–454*

1. $-8 > -10$ (11.1A) **2.** $0 > -4$ (11.1A) **3.** −2 (11.1B) **4.** −7 (11.2A) **5.** −14 (11.2A) **6.** 3 (11.2B) **7.** 10 (11.2B) **8.** −48 (11.3A) **9.** 90 (11.3A) **10.** −9 (11.3B) **11.** The temperature rose to 7°C. (11.3C) **12.** The average low temperature was −2°. (11.3C) **13.** $\frac{1}{15}$ (11.4A) **14.** $\frac{1}{12}$ (11.4A) **15.** $\frac{7}{24}$ (11.4A) **16.** $\frac{3}{10}$ (11.4A) **17.** $\frac{1}{12}$ (11.4B) **18.** $-\frac{4}{5}$ (11.4B) **19.** −1.88 (11.4A) **20.** −4.014 (11.4A) **21.** 3.213 (11.4A) **22.** −0.0608 (11.4B) **23.** 3.4 (11.4B) **24.** −26 (11.5A) **25.** −4 (11.5A)

CUMULATIVE REVIEW *pages 455–456*

1. 0 (1.6B) **2.** $4\frac{13}{14}$ (2.5C) **3.** $2\frac{7}{12}$ (2.7B) **4.** $1\frac{2}{7}$ (2.8C) **5.** 1.80939 (3.3A) **6.** $n = 18.67$ (4.3B) **7.** 13.75 (5.4A) **8.** 1 gal 3 qt (8.3A) **9.** 6.692 L (9.3A) **10.** 1.28 m (9.5A) **11.** 132 cm (10.2A) **12.** 113.04 in^3. (10.4A) **13.** −3 (11.2A) **14.** $-3\frac{3}{8}$ (11.4A) **15.** $-10\frac{13}{24}$ (11.4A) **16.** 19 (11.5A) **17.** 3.488 (11.4B) **18.** $31\frac{1}{2}$ (11.4B) **19.** −7 (11.3B) **20.** $\frac{25}{42}$ (11.4B) **21.** −4 (11.5A) **22.** −4 (11.5A) **23.** The length of board remaining is $2\frac{1}{3}$ ft. (2.5D) **24.** Your new checkbook balance is $803.31. (6.7A) **25.** The percent decrease in price is 27.3% (6.2C) **26.** There should be 10 gal of coffee prepared. (8.3C) **27.** The perimeter is 42.4 m. (10.2C) **28.** The area of the figure is 18.28 cm^2. (10.3B) **29.** The height of the platform is 5 ft. (10.5C) **30.** The average high temperature was −4°. (11.3C)

ANSWERS to Chapter 12 Odd-Numbered Exercises

SECTION 12.1 *pages 465–468*

1. -33 **3.** -12 **5.** 15 **7.** 16 **9.** 3 **11.** 11 **13.** -4 **15.** -3 **17.** -22 **19.** -40
21. -1 **23.** 14 **25.** 32 **27.** 6 **29.** 7 **31.** 49 **33.** -24 **35.** 63 **37.** 9 **39.** 4
41. $-\frac{2}{3}$ **43.** 0 **45.** 0 **47.** $-3\frac{1}{4}$ **49.** 8.5023 **51.** -18.950658 **53.** 3.5961 **55.** $16z$
57. $9m$ **59.** $12at$ **61.** $3yt$ **63.** $-3x - 12y$ **65.** $-2t^2$ **67.** $13c - 5$ **69.** $-2t$ **71.** xy
73. $-6uv$ **75.** $4x^2 - xy$ **77.** $3y^2 - 2$ **79.** $14w - 8u$ **81.** $10 - 23xy$ **83.** $-11v^2$ **85.** $-8bc - 7c$
87. $-11ab$ **89.** $-3y^2 - x^2$ **91.** $-8ab - 3a$ **93.** $4y^2 - y$ **95.** $-3a - 2b^2$ **97.** $7x^2 - 8x$
99. $-3s + 6t$ **101.** $-3m + 10n$ **103.** $5ab + 4ac$ **105.** $\frac{2}{3}a^2 + \frac{3}{5}b^2$ **107.** $6.994x$ **109.** $1.56m - 3.77n$
111. $5x + 20$ **113.** $4y - 12$ **115.** $-2a - 8$ **117.** $9s - 9t$ **119.** $12a + 24$ **121.** $8m - 14$
123. $-4w + 12$ **125.** $7m + 4z$ **127.** $3z - 12$ **129.** $5w + 6$ **131.** $-2m + 12$ **133.** $8m + 6$
135. $6z + 14$ **137.** $-10 + 2a$ **139.** $14x + 4$ **141.** $-2y - 8$ **143.** $9y - 6$ **145.** $10x + 2$
147. $y - 6$ **149.** $8x - 9t - 21$

SECTION 12.2 *pages 475–478*

1. no **3.** yes **5.** no **7.** no **9.** yes **11.** yes **13.** yes **15.** yes **17.** yes **19.** yes
21. yes **23.** yes **25.** yes **27.** no **29.** no **31.** no **33.** yes **35.** $x = -2$ **37.** $z = 14$
39. $x = 2$ **41.** $w = -4$ **43.** $m = -5$ **45.** $t = 0$ **47.** $x = 2$ **49.** $y = -3$ **51.** $y = 10$
53. $t = -5$ **55.** $z = -12$ **57.** $x = -7$ **59.** $x = \frac{2}{3}$ **61.** $x = -1$ **63.** $x = \frac{1}{3}$ **65.** $y = -\frac{1}{8}$
67. $t = -\frac{3}{4}$ **69.** $y = -1\frac{1}{12}$ **71.** $w = -2.98407$ **73.** $y = -8.3482$ **75.** $x = -2.2663$ **77.** $x = 6$
79. $z = -9$ **81.** $t = -5$ **83.** $y = 14$ **85.** $y = 8$ **87.** $y = -3$ **89.** $y = 20$ **91.** $y = -21$
93. $y = -15$ **95.** $x = 16$ **97.** $x = -42$ **99.** $y = 15$ **101.** $y = -38\frac{2}{5}$ **103.** $x = 9\frac{3}{5}$ **105.** $t = -10\frac{4}{5}$
107. $y = 1\frac{17}{18}$ **109.** $a = -2\frac{2}{3}$ **111.** $b = 3$ **113.** $x = -8.5948926$ **115.** $y = -0.42159$
117. $x = 7.322665$ **119.** The increase in value is \$3420. **121.** The depreciation is \$6400. **123.** The car travels 38 mi/gal. **125.** The distance traveled was 648 mi. **127.** The markup on each shirt is \$6.30. **129.** The cost is \$10.30.

SECTION 12.3 *pages 481–484*

1. $x = 3$ **3.** $n = 5$ **5.** $w = -1$ **7.** $z = -4$ **9.** $x = 1$ **11.** $x = 3$ **13.** $x = -2$ **15.** $x = -3$
17. $x = -2$ **19.** $x = -1$ **21.** $x = 3$ **23.** $x = -1$ **25.** $x = 7$ **27.** $x = 2$ **29.** $x = 2$ **31.** $x = 0$
33. $x = 2$ **35.** $x = 0$ **37.** $x = \frac{2}{3}$ **39.** $x = 9$ **41.** $x = -\frac{5}{8}$ **43.** $x = 1$ **45.** $x = \frac{1}{2}$ **47.** $x = -2\frac{1}{2}$
49. $x = 2\frac{2}{7}$ **51.** $x = -\frac{1}{6}$ **53.** $x = 7$ **55.** $x = -2\frac{1}{2}$ **57.** $x = 1\frac{1}{2}$ **59.** $x = 2$ **61.** $x = -2$
63. $x = 2\frac{1}{3}$ **65.** $x = -1$ **67.** $x = \frac{2}{7}$ **69.** $x = 1\frac{1}{12}$ **71.** $x = 5$ **73.** $x = 10$ **75.** $w = 5$
77. $t = 36$ **79.** $y = -6$ **81.** $x = -9$ **83.** $v = -4\frac{4}{5}$ **85.** $z = -8\frac{3}{4}$ **87.** $x = 36$ **89.** $x = 1\frac{1}{3}$
91. $x = 2$ **93.** $t = 4$ **95.** $x = 9$ **97.** $y = 4$ **99.** $x = -3.125$ **101.** $t = -4$ **103.** $x = 3$
105. $a = -4$ **107.** $x = -2$ **109.** $x = 10.755154$ **111.** $x = 0.5482538$ **113.** $x = -0.0862068$
115. The Celsius temperature is -40°C. **117.** The time required is 15 s. **119.** The cost per unit was \$130. **121.** The mechanic's monthly salary is \$1800. **123.** Total sales for the month were \$32,000. **125.** The executive's commission rate was 4.5%.

SECTION 12.4 *pages 487–490*

1. $x = \frac{1}{2}$ **3.** $x = -1$ **5.** $x = -4$ **7.** $x = -1$ **9.** $x = -4$ **11.** $x = 3$ **13.** $x = -1$ **15.** $x = 2$
17. $x = 3$ **19.** $x = 0$ **21.** $x = -2$ **23.** $x = -1$ **25.** $x = -\frac{2}{3}$ **27.** $x = -4$ **29.** $x = \frac{1}{3}$

31. $x = -\frac{3}{4}$ **33.** $x = \frac{1}{4}$ **35.** $x = -7$ **37.** $x = -3$ **39.** $x = 3$ **41.** $x = 2\frac{1}{6}$ **43.** $x = 0$ **45.** $x = 5$ **47.** $x = 1\frac{2}{7}$ **49.** $x = 3\frac{1}{2}$ **51.** $x = 1$ **53.** $x = -16$ **55.** $x = 1\frac{1}{8}$ **57.** $x = 1\frac{1}{3}$ **59.** $x = -\frac{2}{5}$ **61.** $x = 10$ **63.** $x = -22$ **65.** $x = -0.2640613$ **67.** $x = 1$ **69.** $x = 3$ **71.** $x = -3$ **73.** $x = 2$ **75.** $x = -21$ **77.** $x = 3\frac{2}{3}$ **79.** $x = 0$ **81.** $x = -\frac{1}{3}$ **83.** $x = 5$ **85.** $x = 1$ **87.** $x = 3$ **89.** $x = 5$ **91.** $x = -1$ **93.** $x = 2\frac{1}{2}$ **95.** $x = 1\frac{1}{5}$ **97.** $x = -5$ **99.** $x = 3\frac{1}{4}$ **101.** $x = 30$ **103.** $x = -3\frac{1}{2}$ **105.** $x = \frac{1}{3}$ **107.** $x = 6\frac{1}{2}$ **109.** $x = 1\frac{11}{12}$ **111.** $x \approx -1.5251141$ **113.** $x = 6.57$

SECTION 12.5 *pages 493–494*

1. $y - 9$ **3.** $z + 3$ **5.** $\frac{2}{3}n + n$ **7.** $\frac{m}{m-3}$ **9.** $9(x + 4)$ **11.** $n - (-5)n$ **13.** $(c)\left(\frac{1}{4}c\right)$ **15.** $m^2 + 2m^2$ **17.** $2(t + 6)$ **19.** $\frac{x}{x+9}$ **21.** $3(b + 6)$ **23.** x^2 **25.** $\frac{x}{20}$ **27.** $4x$ **29.** $\frac{3}{4}x$ **31.** $4 + x$ **33.** $5x - x$ **35.** $x(x + 2)$ **37.** $7(x + 8)$ **39.** $x^2 + 3x$ **41.** $(x + 3) + \frac{1}{2}x$ **43.** $\frac{3x}{x}$

SECTION 12.6 *pages 499–502*

1. $x + 7 = 12; x = 5$ **3.** $3x = 18; x = 6$ **5.** $x + 5 = 3; x = -2$ **7.** $6x = 14; x = 2\frac{1}{3}$ **9.** $\frac{5}{6}x = 15; x = 18$ **11.** $3x + 4 = 8; x = 1\frac{1}{3}$ **13.** $4x - 7 = 9; x = 4$ **15.** $\frac{x}{9} = 14; x = 126$ **17.** $\frac{x}{4} - 6 = -2; x = 16$ **19.** $7 - 2x = 13; x = -3$ **21.** $9 - \frac{x}{2} = 5; x = 8$ **23.** $\frac{3}{5}x + 8 = 2; x = -10$ **25.** $\frac{x}{4.186} - 7.92 = 12.529;$ $x = 85.599514$ **27.** $37.35 - x = 3.5x; x = 8.3$ **29.** The competitor's price is \$76.75. **31.** The value of the camper last year was \$18,750. **33.** The selling price is \$1050. **35.** The family's monthly income is \$2720. **37.** The monthly output one year ago was 500 computers. **39.** The regular price is \$200. **41.** It took the plumber 3 h. **43.** The markup rate is 40%. **45.** The total wall space is 9000 ft.^{2} **47.** The original rate was 5 gal/min. **49.** Total monthly sales were \$42,540. **51.** The repair job requires 21 h of labor.

CHAPTER REVIEW *pages 505–506*

1. 13 **2.** 6 **3.** $-4bc$ **4.** $\frac{71}{30}x^2$ **5.** $-2a + 2b$ **6.** $-3x + 4$ **7.** yes **8.** no **9.** $x = -4$ **10.** $x = -5$ **11.** $x = -9$ **12.** $x = \frac{5}{4}$ **13.** The car traveled 23 mi/gal. **14.** $x = 7$ **15.** $x = 10$ **16.** $x = -18$ **17.** $x = \frac{54}{5}$ **18.** The Celsius temperature is 37.8°C. **19.** $x = -\frac{5}{2}$ **20.** $x = 5$ **21.** $x = 5$ **22.** $x = -\frac{1}{3}$ **23.** $n + \frac{n}{5}$ **24.** $(n + 5) + \frac{1}{3}n$ **25.** $9 - 2x = 5; x = 2$ **26.** $5x = 50, x = 10$ **27.** The sale price is \$228.

CHAPTER TEST *pages 507–508*

1. -38 (12.1A) **2.** $-5\frac{5}{6}$ (12.1A) **3.** $-4y - 11x$ (12.1B) **4.** $8y - 7$ (12.1C) **5.** no (12.2A) **6.** $x = 26$ (12.2B) **7.** $x = -2\frac{4}{5}$ (12.2C) **8.** $x = -16$ (12.2C) **9.** The monthly payment is \$137.50. (12.2D) **10.** $x = -2$ (12.3A) **11.** $x = 95$ (12.3A) **12.** $x = -\frac{1}{2}$ (12.3A) **13.** There were 4000 clocks made. (12.3B) **14.** $x = 3\frac{1}{5}$ (12.4A) **15.** $x = 0$ (12.4B) **16.** $x + \frac{1}{3}x$ (12.5A) **17.** $5(x + 3)$ (12.5B) **18.** $2x - 3 = 7; x = 5$ (12.6A) **19.** $3x + 5 = x - 2; x = -3\frac{1}{2}$ (12.6A) **20.** The executive's total sales were \$40,000. (12.6B)

CUMULATIVE REVIEW *pages 509–510*

1. 41 (1.6B) **2.** $1\frac{7}{10}$ (2.5C) **3.** $\frac{11}{18}$ (2.8C) **4.** 0.047383 (3.4A) **5.** \$4.20/hr (4.2B) **6.** $n = 26.67$ (4.3B) **7.** $\frac{4}{75}$ (5.1A) **8.** 140% (5.3A) **9.** 6.4 (5.4A) **10.** 18 ft 9 in (8.1B) **11.** 22 oz (8.2A) **12.** 2.082 g (9.2A) **13.** −1 (11.2A) **14.** 19 (11.2B) **15.** 6 (11.5A) **16.** −48 (12.1A) **17.** $-10x + 8z$ (12.1B) **18.** $3y + 23$ (12.1C) **19.** $x = -1$ (12.3A) **20.** $x = -6\frac{1}{4}$ (12.4B) **21.** $x = -7\frac{1}{2}$ (12.2C) **22.** $x = -21$ (12.3A) **23.** The percent was 17.6%. (5.3B) **24.** The price of the pottery was \$39.90. (6.2B) **25.** The perimeter is 19.71 m. (10.2B) **26.** The simple interest due is \$2933.33. (6.3A) **27.** The mathematical expression is $3n + 4$. (12.5B) **28.** The average speed is 53 mi/h. (12.6B) **29.** The total monthly sales is \$32,500. (12.6B) **30.** The number is 2. (12.6A)

FINAL EXAMINATION *pages 511–514*

1. 3259 (1.3B) **2.** 519 (1.5C) **3.** 16 (1.6B) **4.** 144 (2.1A) **5.** $1\frac{49}{120}$ (2.4B) **6.** $3\frac{29}{48}$ (2.5C) **7.** $6\frac{3}{14}$ (2.6B) **8.** $\frac{4}{9}$ (2.7B) **9.** $\frac{1}{6}$ (2.8B) **10.** $\frac{1}{13}$ (2.8C) **11.** 164.177 (3.2A) **12.** 60.205 (3.3A) **13.** 0.027918 (3.4A) **14.** 0.69 (3.5A) **15.** $\frac{9}{20}$ (3.6B) **16.** 24.5 mi/gal (4.2B) **17.** $n = 54.9$ (4.3B) **18.** $\frac{9}{40}$ (5.1A) **19.** 135% (5.1B) **20.** 125% (5.1B) **21.** 36 (5.2A) **22.** $133\frac{1}{3}\%$ (5.3A) **23.** 70 (5.4A) **24.** 20 in. (8.1A) **25.** 1 ft 4 in. (8.1B) **26.** 2.5 lb (8.2A) **27.** 6 lb 6 oz (8.2B) **28.** 2.25 gal (8.3A) **29.** 1 gal 3 qt (8.3B) **30.** 248 cm (9.1A) **31.** 23.10 m (9.1B) **32.** 1.614 kg (9.2A) **33.** 1.258 kg (9.2B) **34.** 2067 ml (9.3A) **35.** 0.9105 L (9.3B) **36.** 88.55 km (9.5A) **37.** The perimeter is 3.9 m. (10.2A) **38.** The area is 45 in^2. (10.3A) **39.** The volume is 1200 cm^3. (10.4A) **40.** −4 (11.2A) **41.** −15 (11.2B) **42.** $-\frac{1}{2}$ (11.4B) **43.** $-\frac{1}{4}$ (11.4B) **44.** 6 (11.5A) **45.** $-x + 17$ (12.1C) **46.** $x = -18$ (12.2C) **47.** $x = 5$ (12.3A) **48.** $x = 1$ (12.4A) **49.** Your new balance is \$959.93. (6.7A) **50.** There will be 63,750 people voting in the election. (4.3C) **51.** The price was \$4500. (5.4B) **52.** The average income was \$3794. (7.4A) **53.** The simple interest due is \$9000. (6.3A) **54.** The percent is 41.4%. (7.1B) **55.** The discount rate is 28%. (6.2D) **56.** The weight of the tiles is 81 lb. (8.2C) **57.** The perimeter is 28.56 in. (10.2B) **58.** The area is 16.86 cm^2. (10.3B) **59.** The number is 16. (12.6A)

Index

Absolute value, 420
Acute angle, 361
Addend, 9
Addition
  Associative Property of, 9, 461
  carrying, 10-11
  Commutative Property of, 9, 461
  of decimals, 125-128
  defined, 9
  of fractions, 75-82
  of integers, 423-425
  and large numbers, 10
  of mixed numbers, 76-77
  on the number line, 9, 423
  and Order of Operations Agreement, 48
  Property of Equations, 470, 479
  Property of Zero, 9, 470
  rational numbers, 437-438
  related to subtraction, 19
  table of basic facts, A2
  and variable expressions, 460
  verbal phrases for, 491
  of whole numbers, 9-18
Adjacent angle, 365
Algebra, introduction to, 457-514
  introduction to equations, 469-478
  translating sentences into equations and solving, 495-498
  translating verbal expressions into mathematical expressions, 491-494
  variable expressions, 459-468
Alternate exterior angles, 366
Alternate interior angles, 366
Amortization schedule, 261
Angles, 360-362
  acute, 361
  adjacent, 365
  alternate exterior, 366
  alternate interior, 366
  complementary, 361
  corresponding, 366
  degrees, 360
  formed by intersecting lines, 365-366
  obtuse, 361
  right, 360
  straight, 361
  supplementary, 361
  vertex of, 360
  vertical, 365
Application problems
  area, 382, 385-386
  bank statements, 257-260
  capacity, 312, 314, 338, 340
  car expenses, 245-246
  circle graphs, 277-278
  decimals, 126, 128, 130, 132, 136, 140, 144, 148
  division, 41-42, 46
  energy, 343-344
  equations, 474, 478, 480, 483-484, 497-498, 500-502
  fractions, 78, 81-82, 86, 90, 94, 98, 101-102, 106
  integers, 432, 436
  interest, 235-236
  length, 304, 306, 330, 332
  mass, 334, 336
  multiplication, 30, 34
  percent decrease, 231-232
  percent increase, 229-230
  percents, 194, 196, 198, 200, 201-202, 204, 206, 208
  perimeter, 374, 377-378
  proportions, 174, 177-178
  purchasing, 221-222
  Pythagorean Theorem, 400, 402
  rates, 168, 170
  ratios, 164, 166
  real estate expenses, 241-242
  solids, 392, 395-396
  strategy for, 12
  subtraction, 22, 26
  triangles, 408
  wages, 249-250
  weight, 308, 310
  whole numbers, 12, 17-18
Approximately equal to (≈), 11
Area, 379-381
Associative Property
  of Addition, 9, 461
  of Multiplication, 27, 463
Astronomical units, 300
Average, 287
Avoirdupois Weight System, 328

Balancing a checkbook, 252-256
Bank statements, 251-260
Bar graph, 279, 283
Base
  of a cylinder, 364
  in percent equations, 201-204
  of a triangle, 362
Basic percent equation, 193
Borrowing in subtraction
  mixed numbers, 84
  whole numbers, 20-21
British Thermal Unit (BTU), 315
Broken-line graph, 280, 284
BUFFON NEEDLE PROBLEM, 409
Business and consumer applications, 217-268
  bank statements, 251-260
  car expenses, 243-246
  interest, 233-236
  percent decrease, 226-228
  percent increase, 223-225
  purchasing, 219-222
  real estate expenses, 237-242
  wages, 247-250

Calculators
  change sign key, 449
  and estimation, 11
  and fractions, 153
  memory key, 179
  and Order of Operations Agreement, 55
  percent key, 209
  and powers, 503
  and rates, 179
  and unit rates, 179
  yx key, 503
Calorie (cal), 341
Capacity
  Metric System of Measurement, 337-340
  U.S. Customary System of Measurement, 311-314
Car
  owning, 244
  purchasing, 243
Carrying in addition, 10-11
Center
  of a circle, 363
  of a sphere, 363
Change sign key, and calculators, 449
Chapter reviews
  business and consumer applications, 263-264
  decimals, 155-156
  fractions, 113-114
  geometry, 411-412
  introduction to algebra, 505-506

Metric System of Measurement, 351-352
percents, 211-212
ratio and proportion, 181-182
rational numbers, 451-452
statistics, 293-294
U.S. Customary System of Measurement, 321-322
whole numbers, 57-58
Chapter summaries, 55-56, 111-112, 153-154, 180, 210, 261-262, 292, 320, 350, 409-410, 450, 504
Chapter tests, 59-60, 115-116, 157-158, 183-184, 213 214, 265-266, 295-296, 323-324, 353-354, 413-414, 453-454, 507-508
Checkbook, balancing, 251-256
Circle
area of, 380
center of, 363
circumference, 371-372
diameter, 363
radius, 363
Circle graph, 273-274
Circumference of a circle, 371-372
Class frequency, 283
Class interval, 283
Class midpoint, 284
Coefficient, 417
Commission, 247
Common denominator, 75
Common factor, 64
Common multiple, 63
Commutative Property
of Addition, 9, 461
of Multiplication, 27, 463
Comparison
bar graphs, 279
broken-line graphs, 280
pictographs, 271
rates, 167-170
ratios, 163-166
unit costs, 219-220
Complementary angles, 361
Composite geometric figures
area of, 381
perimeter of, 373
Composite geometric solids, 390-391
Composite number, 52
Compound interest, 234
formula, 503
table of, A4-A5
Computers
and amortization, 261
BUFFON NEEDLE PROBLEM, 409
and fractions, 111
and mean, 291
and Metric System of Measurement, 349
and U.S. Customary System of Measurement, 319
Computer simulation, 409
Congruent objects, 403
Congruent triangles, 403-404
Constant term, 461
Consumer applications, *see* Business and consumer applications
Conversion
decimals to fractions, 149-150
fractions to decimals, 149
metric to/from U.S. Customary, 345-346
of percents, 189-190
Corresponding angles, 366
Cost, 224
total, 220
unit, 219-220
Cross products, 171
Cube, 363
volume of, 388
Cube of a number, 47
Cubic centimeter, 337
Cubic units, 387
Cumulative reviews, 117-118, 159-160, 185-186, 215-216, 267-268, 297-298, 325-326, 355-356, 415-416, 455-456, 509-510
Cup, 311
Cylinder
height of, 364
volume of, 388-389

Data, 271
Decimal fractions, 120
Decimal notation, 121
Decimal point, 121
Decimals, 119-160
addition of, 125-128
converting to/from fractions, 149-150
division of, 141-148
estimation of, 125, 130, 135, 143
multiplication of, 133-140
and number line, 150
order relation, 150
and percents, 189-190
and power of 10, 133-135
rounding, 122, 149
standard form, 121
subtraction of, 129-132
terminating, 437
word form, 121
writing, 121-122
Decrease, percent, 226-228
Degree, 360
Denominator, 67
common, 75
least common, 75, 83
Diameter
of a circle, 363
of a sphere, 363
Difference, 19, 491
Digits, rounding, 6
Discount, 227-228
Discount rate, 227-228
Distance
Metric System of Measurement, 329-332
U.S. Customary System of Measurement, 301
Distributive Property, 463, 486
Dividend, 35
Division
of decimals, 141-148
defined, 35
estimation of, 41
and factors, 51
of fractions, 99-107
of integers, 430-431
and large numbers, 39
of mixed numbers, 99-100
and Order of Operations Agreement, 48
by powers of ten, 133-134
Properties of Zero in, 35
and rational numbers, 439-440
as related to multiplication, 35
rounding, 141
verbal phrases for, 491
of whole numbers, 35-46
writing as fractions, 133
Divisor, 35
Down payment
car purchase, 243
house purchase, 237

Egyptian fractions, 62
Electrical energy, measuring, 341
Energy
Metric System of Measurement, 341-344
U.S. Customary System of Measurement, 315-316
Equal, verbal phrases for, 495
Equal signs, history of, 458
Equations
Addition Property of, 470, 479
algebraic, 469-478
$ax = b$ form, 472-473
$ax + b = c$ form, 479-484
$ax + b = cx + d$ form, 485
false, 469
Multiplication Property of, 472, 479, 485
parentheses in, 486
solution of, 469

solving, 470-490, 495-498
translating sentences into, 495-498
true, 469
variable = constant, 470-471
$x + a = b$ form, 470-471
Equivalent fractions, 71
Estimation
adding decimals, 125
calculators, 11
dividing decimals, 143
division, 41
multiplication, 29
multiplying decimals, 135
subtracting decimals, 130
subtraction, 21
whole numbers, 11
Expanded form of a number, 4-5
Exponential notation, 47
Exponents, 47
and fractions, 107-108
and Order of Operations Agreement, 48
and signed numbers, 445
simplifying expressions containing, 107-108
Expressions
equality of, 469
Order of Operations Agreement, 48
translating verbal expressions into mathematical expressions, 491-494
variable, 459-468

Factor(s), 51
common, 64
greatest common, 64
in multiplication, 27
and square root, 397
Factorization, prime, 52, 63, 64
Fluid ounce, 311
Foot, 301
Foot pound, 315
per second, 316
Fraction bar, 67, 491
Fractions, 61-118
addition of, 75-82
and calculators, 153
and computers, 111
converting to/from decimals, 133, 149
decimal, 120
division of, 99-107
Egyptian, 62
equivalent, 71
and exponents, 107-108
and greatest common factor, 64
improper, 67-68
inverting, 99
and least common multiple, 63
mixed numbers, *see* Mixed numbers
multiplication of, 91-98
and Multiplication Property of One, 71
on the number line, 107
and Order of Operations Agreement, 108
order relation, 107
and percents, 189
proper, 67
reciprocal of, 99
simplest form, 72
subtraction of, 83-90
writing, 67
Frequency polygon, 284

Gallon, 311
Geometric figures, 362-364
area of, 379-380
composite, 373, 381, 390-391
perimeter of, 371-372
Geometric solids, 363, 387-389
Geometry, 357-416
angles, 359-362, 365-366
area, 379-386
geometric figures, 362-364
lines, 359-360, 365, 366
perimeter, 371-378
Pythagorean Theorem, 398-402
similar triangles, 403-408
volume, 387-396
Grade point average, calculation of, 291
Gram, 333, 349
Graph(s)
bar, 279
broken-line, 280
circle, 273-274
frequency polygon, 284
histogram, 283
on the number line, 3, 107, 423
whole number, 3
Greater than, 3, 419
Greatest common factor (GCF), 64

Heat energy, measuring, 341
Height
of a cylinder, 364
of a parallelogram, 362
of a triangle, 362
Hindu-Arabic system, 2
Histogram, 283
Home
owning, 238-240
purchasing, 237-238
Horsepower, 316
Hourly wage, 247
Hypotenuse, 362, 398

Improper fractions, 67-68
and mixed numbers, 68
and whole numbers, 68
Inch, 301
Increase, percent, 223-225
Integers, 419-436
addition of, 423-425
division of, 430-431
multiplication of, 429-430
negative, 419
order relation, 419
positive, 419
subtraction of, 425-426
writing, 423
Interest
compound, 234
simple, 233
Interest rate, 233
Intersecting lines, 359
Inverting fractions, 99

Klein bottle, 358

Least common denominator (LCD), 75, 83
Least common multiple (LCM), 63, 437
Leg of a triangle, 362
and Pythagorean Theorem, 398
Length
Metric System of Measurement, 329-332
U.S. Customary System of Measurement, 301-302
Less than, 3, 419
Letters, frequency of occurrence, 270
Light years, 300
Like terms, 461
Line(s)
angles formed by intersection of, 365-366
intersecting, 359
parallel, 359-360
perpendicular, 360
transversal, 365
Line segment, 359
Liquid measure, 311
Liter, 337, 349
Loan
amortization of, 261
origination fee, 237-238

Markup, 224-225
Markup rate, 224-225
Mass, 333-336
Mathematical expressions, translating verbal expressions into, 491-494
Mean, 287
and computers, 291

weighted, 291
Measurement
angles, 360-361
area, 379
capacity, 311-314, 337-340
circumference, 371
energy, 315-316, 341-344
length, 301-302, 329-332
mass, 333-336
perimeter, 371-372
power, 316
systems of, *see* Metric system of measurement; U.S. Customary System of Measurement
volume, 387-391
weight, 307-310, 333-334
Median, 288
Memory key, on calculators, 179
Metals, measuring weight of, 328
Meter, 329, 349
Metric system of measurement, 327-356
capacity, 337-340
computers and, 349
converting to/from U.S. Customary System of Measurement, 345-346
energy, 341-344
length, 329-332
mass, 333-336
prefixes, 329
Mile, 301
Minuend, 19
Mixed numbers, 67
addition of, 76-77
division of, 99-100
and improper fractions, 68
multiplication of, 92-93
subtraction of, 84-85
Mobius strip, 358
Monthly payment table, A6
Mortgage, 237-240
Multiple
common, 63
least common, 63, 437
Multiplication
Associative Property of, 27, 463
Commutative Property of, 27, 463
of decimals, 133-140
defined, 27
estimation of, 29
of fractions, 91-98
of integers, 429-430
large whole numbers, 28-29
of mixed numbers, 92-93
number line and, 27
Order of Operations Agreement and, 48
powers of ten, 47, 133
Property of Equations, 472, 479, 485
Property of One, 27, 71, 72, 472
Property of Reciprocals, 472
Property of Zero, 27
rational numbers, 439-440
table of basic facts, A2
verbal phrases for, 491
of whole numbers, 27-34
zeros in, 28-29
Musical scales, 162

Natural numbers, 419
Negative integers, 419
writing, 423
Negative numbers, 419
absolute value of, 420
history of, 418
Negative sign, 425
Number line, 3
and absolute value, 420
addition, 9, 423
decimals, 150
fractions, 107
graphs on, 3, 107, 423
integers, 419, 423
multiplication, 27
subtraction, 19
whole numbers, 3
Number(s)
absolute value of, 420
composite, 52
cubed, 47
decimal notation, 121
expanded form of, 4-5
factoring, 51-54
family tree for, 2
fractions, *see* Fractions
large, 10
mixed, *see* Mixed numbers
multiples of, 63
natural, 419
negative, 419
opposite of, 420
positive, 419
prime, 52-54
rational, *see* Rational numbers
squared, 47
square root of, 397-398
whole, *see* Whole numbers
Numerator, 67
Numerical coefficient, 461

Obtuse angle, 361
One, Multiplication Property of, 27, 71, 72, 472
Operation symbol, 491
Opposite, 420
Order of Operations Agreement, 48
and calculators, 55
fractions, 108
rational numbers, 445-449
variable expressions, 459
Order relation
decimals, 150
fractions, 107
integers, 419
whole numbers, 3
Ounce, 307
fluid, 311

Parallel lines, 359-360
Parallelogram, 362-363
Parentheses
in equations, 486
in variable expressions, 463-464
Percent, 187-216
in circle graphs, 273
commissions, 247
and decimals, 189-190
of decrease, 226-228
defined, 189
discount, 227-228
and fractions, 189
home loans, 237-238
of increase, 223-225
interest, 233
ratios, 205
symbol for, 188
writing, 189-190
Percent equations
basic, 193
proportion method, 205-208
solving for amount, 193-196, 273
solving for base, 201-204
solving for percent, 197-200
Percent key, on calculator, 209
Perfect square, 397
Perimeter, 371-378
Period, 4
Perpendicular lines, 360
Pi (π), 380, 409
Pint, 311
Place value
decimals, 121, 122, 125, 129
expressed as a power of ten, 47
rounding to a given, 5-6, 122
and subtraction, 20
whole numbers, 4
Place-value chart, 4
decimals, 125
division, 36
expanded form, 4
multiplication, 29
whole numbers, 4-5
Plane, 359
Plane figure, 359, 371
Points, 237-238
Positive integers, 419
Positive numbers, 419
absolute value of, 420

Pound, 307
Power, 316
Powers of ten, 47
  on calculators, 503
  and decimals, 133-135
  and division, 133-134
  and multiplication, 47, 133
  and place value, 47
Prime factorization, 52
  greatest common factors and, 64
  least common multiples and, 63
Prime number, 52-54
Principal, 233
Product, 27, 491
  cross, 171
Proper fraction, 67
Properties
  of addition, 9, 461, 470, 479
  Associative
    of Addition, 9, 461
    of Multiplication, 27, 463
  Commutative
    of Addition, 9, 461
    of Multiplication, 27, 463
  Distributive, 463, 486
  of Equations
    Addition, 470, 479
    Multiplication, 472, 479, 485
  of One, Multiplication, 27, 71, 72, 472
  of Reciprocals, Multiplication, 472
  of Zero
    in addition, 9, 470
    in division, 35
    in multiplication, 27
Property tax, 239
Proportions, 171-178
  cross products, 171
  defined, 171
  not true, 171
  and percent problems, 205-208
  solving, 172-173
  true, 171
Purchasing, 219-222
Pythagorean Theorem, 398-402

Quadrilateral, 362
Quantity
  comparison of, 163-170
  unknown, 459
Quart, 311
Quotient, 35, 37-38, 491

Radius
  of a circle, 363
  of a sphere, 363-364
Rate(s), 167-170
  and calculators, 179
  and proportion, 171
  simplest form, 167
  unit, 167, 179
  writing, 167
Rational numbers
  adding or subtracting, 437-438
  least common multiple, 437
  multiplying and dividing, 439-440
  operations with, 437-444
  Order of Operations Agreement, 445-449
Ratio(s), 163-166, 491
  circle graphs and, 273
  percents, 205
  pictographs and, 271
  proportion and, 171
  simplest form, 163
  writing, 163
Ray, 360
Real estate expenses, 237-242
Reciprocals, 99
  Multiplication Property of, 472
Rectangle, 363
  area of, 379
  perimeter of, 371
Rectangular solid, 363
  volume of, 387
Remainder, 37-38
Right angle, 360
Right triangle, 362
  finding unknown side of, using Pythagorean Theorem, 398-399
  square of the hypotenuse of, 398
  30°-60°-90°, 399
  45°-45°-90°, 398-399
Rounding
  decimals, 122, 149
  in division, 141
  and place value, 5-6, 122
  whole numbers, 5-6, 29

Salary, 247
Sale price, 227-228
SAS Rule, 404
Sector, 273
Selling price, 224-225
Sentences, translating into equations, 495-498
Side-Angle-Side Rule (SAS), and triangles, 404
Side-Side-Side Rule (SSS), and triangles, 404
Signed numbers, *see* Integers
Similar objects, 403
Similar triangles, 403-408
Simple interest, 233
Simplest form rate, 167
Simplest form ratio, 163
Simplifying
  fractions, 72
  variable expressions, 461
Solid(s), 359
  composite geometric, 390-391
  geometric, 363
  measurement of, 307, 387
  rectangular, 363
  U.S. Customary System of Measurement, 307
Solution of an equation, 469
Solving equations, 470-490
Solving proportions, 172-173
Space, 359
Space figures, 363-364
Sphere, 363-364
  volume of, 388
Square, 363
  area of, 379
  perfect, 397
  perimeter of, 371
Square of a number, 47
Square roots, 397-398
  table of, A3
Square units, 379
SSS Rule, 404
Standard form
  decimal, 121
  whole number, 4
Statistics, 269-298
  average, 287
  bar graph, 279
  broken-line graph, 280
  circle graph, 273-274
  defined, 271
  frequency polygon, 284
  histogram, 283
  mean, 287
  median, 288
  pictograph, 271-272
Straight angle, 361
Subtraction
  borrowing, 20-21
  checking, 19
  of decimals, 129-132
  defined, 19
  estimation of, 21
  of fractions, 83-90
  of integers, 425-426
  of mixed numbers, 84-85
  on the number line, 19
  operation sign, 425
  Order of Operations Agreement and, 48
  rational numbers, 437-438
  related to addition, 19
  verbal phrases for, 491
  of whole numbers, 19-26
  zero and, 20
Subtrahend, 19
Sum, 9, 491
Supplementary angles, 361
Symbols
  absolute value (||), 420

approximately equal to (≈), 11
degree (°), 360
division (÷, −), 36
greater than (>), 3
less than (<), 3
multiplication (×, ·), 27
percent (%), 188
Pi (π), 380
right angle (∟), 347
square root ($\sqrt{\ }$), 397

Tables of basic facts
addition, A2
compound interest, A4-A5
monthly payment, A6
multiplication, A2
square roots, A3
T-diagram, 52
Ten, powers of, *see* Powers of ten
Term, 460-461
Terminating decimals, 437
Time, circle graphs and, 273
Ton, 307
Total cost, 220
Transversal, 365
Triangles, 362
area of, 380
base of, 362
congruent, 403-404
height of, 362
perimeter of, 371
right, *see* Right triangles
Side-Angle-Side Rule and, 404
Side-Side-Side Rule and, 404
similar, 403-408
Troy Weight System, 328

Unit, 163
Unit cost, 219-220
Unit rate, 167
and calculators, 179
U.S. Customary System of Measurement, 299-326, 328
capacity, 311-314
and computers, 319
conversion to/from metric units, 345-346
energy, 315-316
length, 301-302
power, 316
weight, 307-310

Variable, 459
Variable = constant equation, 470-471
Variable expressions, 459-468
containing no parentheses, 460-461
containing parentheses, 463-464
evaluating, 459-460
simplifying, 460-464
terms of, 460
Variable part, 461
Variable term, 461
Verbal expressions, translating into mathematical expressions, 491-494
Vertex, of angle, 360
Vertical angle, 365
Volume, 387-396
of composite geometric solids, 390-391
of geometric solids, 387-389

Wage, 247-250
Watthour, 341
Weight
Metric System of Measurement, 333-334
U.S. Customary System of Measurement of, 307-310
Weighted mean, 291
Whole numbers, 1-60
addition of, 9-18
borrowing, 20-21
carrying, 10-11
definition of, 3
division of, 35-46
estimating quotient of, 41
estimation of, 11
expanded form, 4-5
graph of, 3
and improper fractions, 68
multiplication of, 27-34
on the number line, 3
Order of Operations Agreement, 48
order relation, 3
reading, 4
reciprocal of, 99
rounding, 5-6, 29
standard form, 4
subtraction of, 19-26
word form, 4
writing, 4-5
Word form
decimals, 121
whole numbers, 4

Yard, 301

Zero
Addition Property of, 9, 470
Multiplication Property of, 27
properties of, in division, 35
and subtraction with borrowing, 20